REWAS 2013

Enabling Materials

Resource

Sustainability

REWAS 2013

Enabling Materials Resource Sustainability

Proceedings of the symposium, REWAS 2013: Enabling Materials Resource Sustainability, sponsored by the EPD-LMD Recycling and Environmental Technology Committee and the TMS Materials and Society Committee; and the symposium, Battery Recycling, sponsored by the EMPMD Energy Conversion and Storage Committee and the EPD-LMD Recycling and Environmental Technology Committee

Held during the
TMS 2013 Annual Meeting & Exhibition
San Antonio, Texas, USA
March 3-7, 2013

Editors

Anne Kvithyld • Christina Meskers

Co-editors

Randolph Kirchain • Gregory Krumdick • Brajendra Mishra
Markus Reuter • Cong Wang • Mark Schlesinger
Gabrielle Gaustad • Diana Lados • Jeffrey Spangenberger

Editors
Anne Kvithyld
Christina Meskers

ISBN 978-3-319-48580-5 ISBN 978-3-319-48763-2 (eBook)
DOI 10.1007/978-3-319-48763-2

Chemistry and Materials Science: Professional

Published by Springer International Publishers, Switzerland, 2016
Reprint of the original edition published by John Wiley & Sons, Inc., 2013, 978-1-11860-587-5

Printed on acid-free paper

This Springer imprint is published by Springer Nature
The registered company is Springer International Publishing AG
The registered company address is: Gewerbestrasse 11, 6330 Cham, Switzerland

TABLE OF CONTENTS
REWAS 2013: Enabling Materials Resource Sustainability

Enabling Sustainability through Metal Production

Enabling Sustainability through Recycling & End-of-Pipe Solutions I

Enabling Sustainability through Process Design, Modeling & Simulation

Enabling Sustainability through Life Cycle Management, LCA and Industrial Ecology

Enabling Sustainability through Systems Modelling and Design, Life Cycle Management, LCA and Industrial Ecology

Battery Recycling

Enabling Sustainability through the Physics of Metals & Materials Processing

Enabling Sustainability through Education and Consumer Awareness

Enabling Sustainability through Recycling & End-of-Pipe Solutions II

Enabling Sustainability through Systems Modelling and Design

PREFACE

"The challenges facing engineering today are not those of isolated locales, but of the planet as a whole and all the planet's people. Meeting all those challenges must make the world not only a more technologically advanced and connected place, but also ... more sustainable, safe, healthy, and joyous...."

- National Academy of Engineering Grand Challenges Committee (2009)

This book collects selected papers presented at REWAS 2013: Enabling Materials Resource Sustainability, co-located with the 2013 TMS Annual Meeting.

REWAS 2013 follows in the footsteps of previous REWAS conferences organized by TMS; the first one took place in Spain in 1999. In terms of content, REWAS 2013 builds on the successful Sustainable Materials Processing and Production symposium held at the 2010 TMS Annual Meeting and the Materials and Society plenary session, "Linking Science and Technology for Global Energy Solutions" at the 2011 TMS Annual Meeting.

In the 2010 symposium we for the first time connected the metallurgical and materials science community to the broader, interdisciplinary topic of sustainability by making the link between technology, metrics, modelling, education, and the role materials play in the transition to a sustainable society.

REWAS 2013 again brings together the metallurgical and materials science community on the one hand and other stakeholders from across the materials and product life cycle on the other hand. By pulling in practitioners and educators from outside the TMS community we hope to further foster interdisciplinary approaches and system thinking to contribute to materials sustainability in relation to other societal challenges.

In the plenary session the theme is Realizing sustainable... where key sectors such as energy, water, buildings, materials, electronics, transport, and the trends, drivers, challenges and opportunities to achieve sustainability will be explored. One parallel session track will address Enabling sustainability through ... with talks on metals and materials processing, recycling technology, product design, and (thermodynamic) modeling and simulation. A special session is dedicated to Battery Recycling. Another parallel session track focuses on Understanding sustainability through... tools to understand the progress toward sustainability, large systems modeling and design. An additional session in this track focuses on the integration of sustainability and cross-disciplinary research in and between universities and companies, and on the communication of sustainability to the general public.

We hope that these proceedings capture the breadth and depth of REWAS 2013: Enabling Materials Resource Sustainability, and reflect the collaborative energy and spirit of the symposium.

All this would not have been possible without the support, contributions, ideas, and encouragement of the following

- The TMS family: TMS Programming staff and leadership, Extraction and Processing Division, Light Metals Division, Materials and Society Committee, and the Recycling and Environmental Technologies Committee.
- The international scientific committee: Gerardo Alvear (Xstrata Technology), Helmut Antrekowitsch (Montanuniversität Leoben), Diran Apelian (Worcester Polytechnic Institute), Bart Blanpain (KU Leuven), Liyuan Chai (Central South University), David DeYoung (Alcoa), Matthew Eckelman (Northeastern University), Daniel Mueller (Norwegian University of Science and Technology), Kazuki Morita (University of Tokyo), Shinichiro Nakamura (Waseda University), M. Christina Negri (Argonne National Laboratory), John Rankin (CSIRO), Matthias Schluep (EMPA), Maurits Van Camp (Umicore Group R&D), Antoinette Van Schaik (MARAS), Kari Heiskanen (Aalto University), and Yongxiang Yang (Delft University of Technology).
- Our plenary speakers, invited speakers, and all REWAS 2013 speakers and authors.
- The Center for Resource Recovery and Recycling (CR3).
- The conference sponsors.

A big thanks to you all.

Lead Editors

Anne Kvithyld
Christina Meskers

Co-editors

Randolph Kirchain
Gregory Krumdick
Brajendra Mishra
Markus Reuter
Cong Wang
Mark Schlesinger
Gabrielle Gaustad
Diana Lados
Jeffrey Spangenberger

---

[1] R. Kirchain, C. Meskers, M. Reuter, D. Lados, "TMS 2010 Materials and Society Symposium: Making a "sustainable" world", *JOM*, 61 (11) 2009, pp. 17-18.
M.A. Reuter, C. Meskers, R. Kirchain, D. Lados, "Sustainable Materials Processing and Production – TMS 2010 Materials and Society Symposium", *JOM*, 62 (8) 2010, pp. 20-23
R. Kirchain and C. Meskers, "Sustainability panel discussion: Voices of industry, academia and national labs", *JOM*, 62 (8) 2010, pp. 25-32.
Sustainable Materials Processing and Production proceedings included in *EPD Congress 2010* proceedings

TMS

WILEY

REWAS 2013

Enabling Materials Resource Sustainability

Lead Editors

Anne Kvithyld

Christina Meskers

Co-Editors

Randolph Kirchain

Gregory Krumdick

Brajendra Mishra

Markus Reuter

Cong Wang

Mark Schlesinger

Gabrielle Gaustad

Diana Lados

Jeffrey Spangenberger

BROWSE Papers

SUBJECT Index

SEARCH CD

TMS Website

ABOUT Editors

CD Help

COPYRIGHT ©2013

TMS

WILEY

REWAS 2013

Enabling Materials Resource Sustainability

Anne Kvithyld

Anne Kvithyld is a senior research scientist at SINTEF in Trondheim, Norway, where she has been since 2007. Her research interests are centered around production of metals, in particular refining and recycling. She earned her Ph.D. from the Norwegian University of Science and Technology in 2003. She has been a Visiting Scholar at the Colorado School of Mines, USA and was the recipient of the 2011 Vittorio de Nora Award for Environmental Improvements in Metallurgical Industries.

REWAS 2013

Enabling Materials Resource Sustainability

Christina Meskers

Christina Meskers leads the business development and marketing team at Umicore Precious Metals Refining in Hoboken, Belgium after completing her Ph.D. in metallurgy at Delft University of Technology in 2008. She is responsible for the market research for the Business unit and part of her activities is bringing Umicore's ideas and expertise on closed loop material cycles, recycling and sustainability forward in numerous interactions with stakeholders, such as universities and research institutes, industry associations, governmental working groups and other actors in the supply chains.

Christina currently serves as member of the StEP (Solving the E-waste Problem) Initiative Steering Committee, the TMS (The Minerals, Metals & Materials Society) Recycling Committee and is past vice-chair of the TMS Materials & Society Committee. In 2008 she was a recipient of the TMS Young Leader Award of the Extraction and Processing Division of TMS.

REWAS 2013
Enabling Materials Resource Sustainability

Randolph Kirchain

Randolph Kirchain's research and teaching explores the impact of materials-technology decisions on the economic and environmental performance of the products into which those materials are transformed and the systems in which they are produced, used, and eventually discarded. The choice of material has sweeping implications on the realization of a product. To address this, Dr. Kirchain's research deals with the development of methods to model two critical aspects of technological performance: (1) life cycle economics and (2) materials system sustainability - particularly resource efficiency and recyclability. Dr. Kirchain has authored over 80 publications in refereed journals and conferences. He has been awarded the American Iron and Steel Institute's Top Technical Achievement Award, the General Motors Technical Achievement Award, and the TMS Recycling Technology Award.

REWAS 2013

Enabling Materials Resource Sustainability

Gregory Krumdick

Gregory Krumdick is a principal systems engineer in the Energy Systems Division at Argonne National Laboratory. He earned his M.S. degree in Bioengineering from the University of Illinois at Chicago, focusing on process control systems. Mr. Krumdick has spent the past 22 years with Argonne, where he has been the principal investigator on numerous industrial process scale-up projects and lead engineer on several pilot plant systems for the Process Technology Research section. Currently, Mr. Krumdick is leading Argonne's battery materials scale-up program and is overseeing Argonne's new Materials Engineering Research Facility.

REWAS 2013

Enabling Materials Resource Sustainability

Brajendra Mishra

Brajendra Mishra is a professor and the Assoc. Director of Kroll Institute for Extractive Metallurgy in Metallurgical & Materials Engineering at the Colorado School of Mines. Dr. Mishra is the Co-Director of the National Science Foundation's Industry/University Collaborative Research Center on Resource Recovery & Recycling – the first National Center of its kind. Brajendra received his Bachelor of Technology [1981] degree in Metallurgical Engineering from the Indian Institute of Technology in Kharagpur, India and his M.S. [1983] and Ph.D. [1986] in Materials Science from the University of Minnesota in Minneapolis.

Dr. Mishra has over 25 years of research experience in molten salt pyrometallurgy and electrochemistry and has many contributions to the application of these technologies to nuclear materials and critical materials waste processing. Dr. Mishra has authored over 400 technical publications in refereed journals and conference proceedings. He holds 10 patents and has authored/edited 20 books. Dr. Mishra is a Fellow of ASM (2001). Mishra received the Distinguished Service Award from The Minerals, Metals & Materials Society (2010) and the highest award of Honorary Membership from the Indian Institute of Metals (2008). Brajendra served as the 2006 President of the Minerals, Metals & Materials Society (TMS) and the 2011 President of American Institute of Mining, Metallurgical & Petroleum Engineers (AIME). Dr. Mishra was instituted as an Honorary Professor of Kazakh Natl. Technical University, Almaty, and has received the EPD Distinguished Lecturer Award from TMS (2012).

REWAS 2013

Enabling Materials Resource Sustainability

Markus Reuter

Markus Reuter is a Director of Technology Management at Outotec Finland. He is Adjunct Professor at Aalto University (Finland), Professorial Fellow at the University of Melbourne (Australia) and Guest Professor at Central South University Changsha (China). He was a professor at TU Delft (Netherlands) and Adjunct Professor at University of Stellenbosch. He holds a D. Eng. degree from Stellenbosch University, a Dr. habil. degree from RWTH Aachen (Germany), and a Ph.D. degree from the University of Stellenbosch. He has also worked for Mintek and Anglo American Corporation in South Africa. Reuter has over 380 publications and is a co-author of the book *Metrics of Material and Metal Ecology.* He is the lead author of a second report on recycling for UNEP. His research interests include design for recycling, process metallurgy, recycling, systems engineering and process control.

REWAS 2013
Enabling Materials Resource Sustainability

Cong Wang

Cong Wang is Senior Research Engineer of Saint-Gobain Innovative Materials R&D. Prior to joining Saint-Gobain, he worked at the Alcoa Technical Center. He obtained his Ph.D. from Carnegie Mellon University, M.S. from the Institute of Metal Research, Chinese Academy of Sciences, and B.S. from Northeastern University with distinctions, respectively. His specialties are in materials processing, micro-structure characterization, mechanical testing, and electrochemistry.

REWAS 2013

Enabling Materials
Resource
Sustainability

Mark Schlesinger

Mark Schlesinger is Professor of Metallurgical Engineering at the Missouri University of Science and Technology. He is a former Fulbright Scholar (KTH, Sweden) and Leif Eiriksson Fellow (NTNU, Norway). He is the author of five books, including *Extractive Metallurgy of Copper* (5th ed.), and *Aluminum Recycling.*

BROWSE Papers
SUBJECT Index
SEARCH CD
TMS Website
ABOUT Editors
CD Help
COPYRIGHT ©2013
TMS
WILEY

REWAS 2013
Enabling Materials Resource Sustainability

Gabrielle Gaustad

Gabrielle Gaustad received her Ph.D. in Material Science from the Massachusetts Institute of Technology and is currently an Assistant Professor in the Golisano Institute for Sustainability at the Rochester Institute of Technology. Her research focuses on evaluating the economic trade-offs for materials at their end-of-life and enabling resource recovery via blending plans, upgrading technologies, and design for recycling. Dr. Gaustad is active in promoting participation of women and minorities in the STEM disciplines including involvement in the McNair Scholars program, ACS SEED, and the TMS Women in Science committee.

REWAS 2013
Enabling Materials Resource Sustainability

Diana Lados

Diana Lados is Associate Professor at Worcester Polytechnic Institute (WPI) and Founding Director of the Integrative Materials Design Center (iMdc), an industry-government-university alliance dedicated to advancing the frontiers of sustainable materials-process design and manufacturing for high-performance, reliability, and recyclability. She received her B.S./M.S. in Mechanical Engineering from Polytechnic University of Bucharest in 1997, her second M.S. in Mechanical Engineering from Southern Illinois University in 1999, and her Doctorate in Materials Science and Engineering from WPI in 2004. Dr. Lados is credited with significant research contributions in the areas of fatigue, fatigue crack growth, high-temperature behavior, and fracture of engineering materials, as well as for her original work in materials processing.

Dr. Lados is the recipient of numerous awards and honors, including the 2012 Silver Medal from ASM; a 2012 "Woman to Watch" in New England for innovation and leadership; the 2011 Early Career Faculty Fellow Award from TMS; and the 2010 Robert Lansing Hardy Award from TMS. She was chosen in 2010 by the National Academy of Engineering to participate in the prestigious Frontiers of Engineering symposia for exceptional research and innovative education. She has 60 publications and over 120 technical presentations, seminars, and keynote lectures, and recently contributed a chapter on "Design for Fatigue Crack Growth Resistance in Aluminum Alloys" to the book titled, *Fundamentals of Aluminum Metallurgy*.

REWAS 2013
Enabling Materials Resource Sustainability

Jeffrey Spangenberger

Jeff Spangenberger is an Engineering Specialist, Senior at Argonne National Laboratory with a B.S. in chemical engineering from Iowa State University. Working in the Energy Systems Division's Process Technology Group Jeff has lead research activities ranging in scale from bench top testing through full scale plant installations and process efficiency studies. While he works to find ways of reducing energy demands in many energy intensive processes, much of his work relates to the separation and recovery of materials in waste streams destined for landfill. This research has resulted in the conceptualization, design, and construction of equipment, pilot scale processes and industrial scale plants resulting in numerous patents and awards.

REWAS 2013
Enabling Materials Resource Sustainability

Enabling Sustainability through Metal Production

Session Chairs

Kari Heiskanen
Bart Blanpain

REWAS 2013: Enabling Materials Resource Sustainabilit
*Edited by: Anne Kvithyld, Christina Meskers, Randolph Kirchain, Gregory Krumdick, Brajendra Mishra
Markus Reuter, Cong Wang, Mark Schlesinger, Gabrielle Gaustad, Diana Lados, and Jeffrey Spangenberge*
TMS (The Minerals, Metals & Materials Society), 201

Highly efficient slag cleaning – latest results from pilot-scale tests

Roland König[1], Axel Weyer[1], Dr. Rolf Degel[1], Jürgen Schmidl[2], Dr. Harald Kadereit[2], Dr. Andreas Specht[2]

[1] SMS Siemag AG
Eduard-Schloemann-Strasse 4, Düsseldorf, 40237, Germany
[2] Aurubis AG
Hovestrasse 50, Hamburg, 20539, Germany

Keywords: copper separation, slag cleaning, SAF, iron silicate

Abstract

Modern life would not be possible without copper. The high worldwide demand for copper raises several questions for the copper industry in terms of economic and ecological considerations regarding the main product copper as well as the by-product iron silicate. The future challenge for copper smelters is to increase the yield by reducing the copper losses and to obtain the iron-silicate as a marketable by-product. This can be achieved by further treatment of the slag. A newly developed slag cleaning technology has been evaluated in a specially designed pilot furnace integrated into an industrial process. Applying a magnetic field across a DC field improves stirring, thereby fostering the settling of entrained copper droplets. The results showed that a 30 % to 50 % reduction of the Cu content in the iron silicate product is feasible, depending on the composition of the incoming copper slag. This makes the process economically attractive.

Introduction – Current Stage of Technology

Submerged Arc Furnaces (SAF) have been utilized in copper production for several decades now. They are used for cleaning the slag produced, downstream of flash smelters or converters. The residual copper content in the slag is reduced and the final product iron silicate is used for road construction and river embankments. The main areas of application correspond to the locations of the copper producers in South America and Europe, but also increasingly in the Southern Africa region and parts of the Asian region. The furnaces customarily used are those developed from submerged arc furnaces for ferroalloy manufacture, i.e., round alternating current furnaces with three electrodes and equipped with proven and reliable electrode columns. Although the furnaces are adapted individually to the requirements, it can be said that on average they have a shell diameter of approx.. 10,000 mm and a power of approx. 10 MVA. At the prompting of SMS Siemag, a reappraisal of the furnace shape has taken place during the last few years. An ever greater number of slag cleaning furnaces are being designed as rectangular furnaces, with three or six electrodes in line depending on the power, since in many cases a better ratio of furnace hearth volume to rated power is thus achieved, particularly in combination with ISA smelter plants. Even though these latest copper slag cleaning furnaces often attain a final copper content in the slag of only 0.70 wt%, an overall appraisal shows that in many furnaces the average final copper content is around 1.00 wt%, not least because of the fluctuating quality of the raw materials charged in practical application.

Motivation

Over the past few years the prices for copper have risen sharply, in fact by an average of 100 % compared with the price level at the time the many round copper slag cleaning furnaces were built some 20 years ago. But it is not only the copper that is in the limelight; the increasingly significant question is being raised as to what should happen with the slag produced. Is disposal at dumping sites required or is further utilization possible? In parallel with the rising prices for copper as a metal, the requirements for residual materials have also increased, with a view to them being re-used as recyclable materials. On principle, a pure iron silicate rock is expected to be obtained, and not residual slag. Environmental requirements are also becoming stricter. Now that raw materials are becoming scarcer, it is becoming increasingly important in society to ensure that resources are conserved, which means that metal losses in slag need to be minimized even further. This, however, must remain economically feasible. The question therefore arises as to how the slag cleaning process can be further improved, with good payback conditions and with enhanced usability and operational reliability achieved by reducing the fluctuations in product quality described above.

Idea – Physical background

The settling behavior of the slag limits the reduction of the metal losses in the slag. Small copper droplets remain "suspended" in the slag, while larger ones are deposited as a result of gravity separation. Figure 1 shows the REM image of slag from a conventional copper slag cleaning furnace.

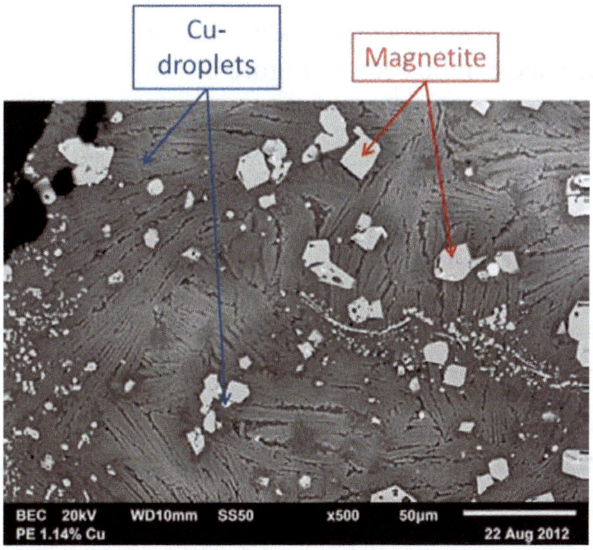

Figure 1: REM image of conventional copper slag (c_{Cu} = 1.14 wt%; X: 500)

Even though a certain amount of retention time of the slag in the furnace is necessary, the increase in the dwell time over and above a certain magnitude will only result in minimal improvements in settling behavior. Enlarging the size of the furnaces accordingly can even cause a negative effect, i.e., skull build-ups and cold zones in the furnace, leading in turn to operational problems. The correct approach therefore has to be one of producing as many large copper particles as possible in the slag, in other words to reduce the quantity of small particles. A new route became apparent for the first time in 2006 – the combination of a conventional slag cleaning furnace with a downstream electromagnetic stirring reactor. Figure 2 illustrates the suggested process route published in the patent of SMS Siemag WO2006/131372 A1.

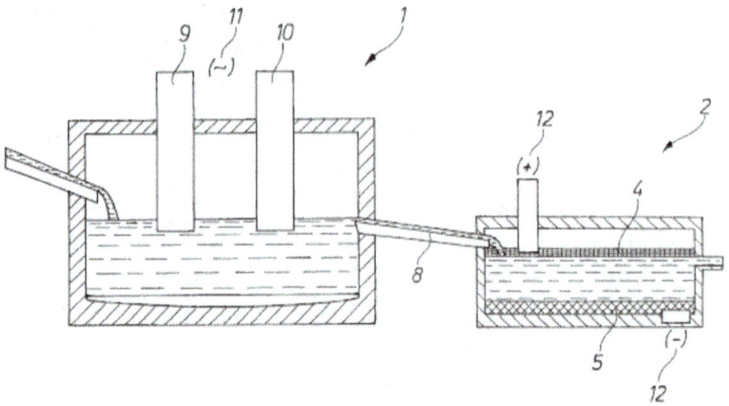

Figure 2: Suggested combination of conventional slag cleaning furnace and electromagnetic stirring reactor [1]

In physical terms, the Lorenz force is used, which arises and affects the particles when these are present in a direct-current field that is intersected at right angles by a magnetic field. This effect can be understood more easily by applying the well-known three-finger rule, or Fleming's rule. The effect in question is independent of whether or not the particle is magnetic, and copper of course is non-magnetic. The copper particles are therefore made to rotate and, owing to differing velocities of the particles, this rotational movement causes the particles to collide and conglomerate, thus enabling them to be deposited once they have agglomerated to a bigger grain size. This briefly described physical process makes an innovative furnace technology necessary and the layout of such a furnace can be seen in Figure 3.

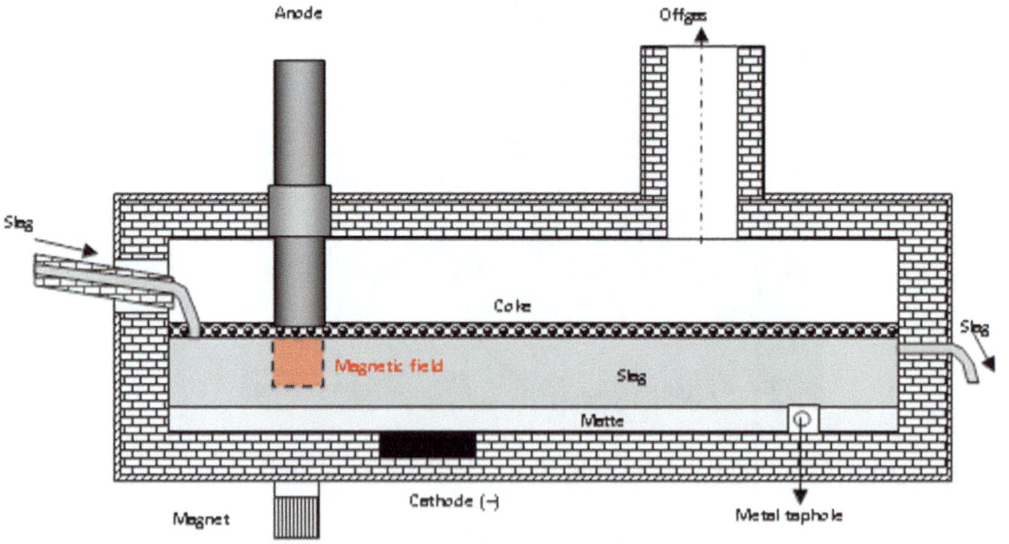

Figure 3: Layout of the innovative electromagnetic stirring reactor [2]

The new concept requires the move away from an AC - SAF towards a DC - SAF which is divided into two zones. In the first zone of the rectangular furnace the electromagnets are surrounded by the yoke of the shell, so that a force is applied which also penetrates the cell width and the refractory lining. Therefore, the furnace shell must also be constructed partly from nonmagnetizable special steel. In the second stage, no electromagnetic stirring is applied and the copper droplets are sedimented.

First steps

Since the new process was physically coherent and appeared to be feasible, it very quickly became the subject of intensive further development. The route towards a pilot furnace first progressed via simulations and laboratory trials. Some of the work took place simultaneously. The assumptions made were confirmed and the initial small trials showed positive results (Figure 4).

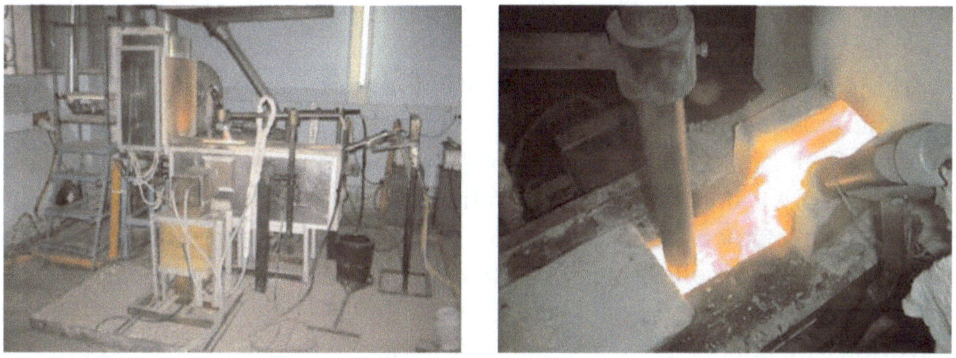

Figure 4: Experimental set-up for the laboratory scale trials carried out by Prof. Warczok at Universidad de Chile

The laboratory furnace had a holding capacity of 30 kg and the tests confirmed the initial assumptions that copper droplets coagulated due to electromagnetic stirring.
Nevertheless, a new process also necessitates trials on a larger unit. Such a unit was erected in Chile after a short construction period at the Anglo American copper plant in Charges (Figure 5). The first larger-sized trials were conducted in 2007 in a furnace with a capacity of 1 t/h.

Figure 5: Image of the first large-scale pilot furnace at the Anglo American copper plant

The material charged was converter slag. Although a tendency towards effective slag cleaning became apparent, this unit proved to be operationally reliable and stable only to a limited extent, and the electrical system in particular became a bottleneck and prevented lengthy furnace campaigns. The target was to integrate the furnace realistically into an industrial plant. Already at that time, a possible commercial-sized installation was envisaged and therefore the scale-up factor was not intended to be as large as in the Chilean plant. One of the main problems of the plant in Chagres was the distance between Europe and Chile. It was difficult and expensive to operate trials by a larger team and the possible reaction time for personal contact and exchange of experience and advice was too long.

Ideal partner

Based on the experience with the mini-plant in Chile and the laboratory trials, a newly designed pilot facility should be installed. In light of the problems observed, a larger-scale pilot plant in Europe was the preferred option. By consistently maintaining earlier contacts, initial discussions with Aurubis, previously Norddeutsche Affinerie, soon led to cooperation in this field, as there had been a good level of exchange and a close relationship between the two partners for several years. This working environment was a result of continuous modernization work and ongoing maintenance of the existing facilities. Aurubis operates two slag cleaning furnaces from SMS Siemag, a 9 MVA - SAF built at the end of the 1970s operating in the primary copper production line, and an 8 MVA - SAF built in 1989, operating in the recycling line. Many factors favored trials on the planned new process under the most realistic possible conditions. To a large extent, Aurubis works as a custom smelter, meaning that the delivered raw materials are processed into copper or copper products on the basis of TC/RCs. In this way, trials can be carried out on various compositions of copper ores and slag. Another reason was the experienced operating personnel and metallurgists, a well-established development department with its own laboratories, a sophisticated logistics system and the necessary space for a new, small furnace installation acting as a pilot plant close to the existing SAF. Intensive planning on a mutual basis was made possible by geographical proximity, since a distance of 400 km is like a Sunday stroll for two global companies like SMS Siemag AG and Aurubis. The cooperation proved to be an ideal partnership, not least in view of the short distance between Düsseldorf, where SMS Siemag AG is located, and Hamburg, where Aurubis AG is located. Both companies also value the high environmental acceptability of their plants and thus, in spite of strict environmental standards at the Hamburg location, there were no problems with operating licenses. The aim was to fully integrate a pilot plant into the industrial process.

Hamburg pilot plant design

The planning thus began in the years 2008/2009. It quickly became clear that only a permanently installed pilot plant in a fixed bay would meet the needs regarding safety, slag handling and environmental protection. Although this necessitated a somewhat higher expenditure on planning than had initially been assumed, it finally created the possibility to test the process under very real practical conditions as well as to examine individual components in a purposeful manner. The main challenge, however, was the electromagnet. The difficulty did not lie in providing a magnetic force of the approx. 200 Gauss required by means of an electromagnet. Instead, the great technical feat was to ensure that this could be done in the middle of an air gap of 1400 mm. A competent partner in this venture was SMS Elotherm GmbH in Remscheid, close to Cologne, along with SMS Meer, another member of the large SMS group. The main specialization of Elotherm is hardening and heating by means of magnetic forces. Although this is not SMS Elotherm's standard field, intensive consultations, simulations and calculations finally resulted in a large, 3.5 t heavy electromagnet in a U-shape with dimensions of approx. 2.5 m x 2.0 m. It comprises compressed air-cooled main and auxiliary coils and variable power control. The height of the magnet can be precision-adjusted at the vessel, so in practice various height positions can be examined and an optimum alignment between the electrode and the copper

matte bath can be achieved. In order to avoid any loss of magnetic power or undesired magnetization of individual components, the entire furnace shell was ultimately constructed from stainless steel [3], [4].
The aim was to achieve an average throughput of 2t/h of slag up to a maximum of 3 t/h. For this, a DC rectangular furnace was designed, equipped with two electrodes on electrode arms with hydraulic regulation, a bottom electrode and an inside hearth diameter of approx. 2700 mm x 600 mm. The power supply was provided by the works network via an 825 kVA transformer and a rectifier for each electrode, thus enabling a highly variable power input. The complete plant was erected in a currently unused, partially enclosed bay. Charging is performed with specially converted ladles which are conveyed by the existing special-purpose lift trucks from the slag tapping area of the SAF to the stirring reactor and then charged by crane. The crane driver performs the charging. The aim here is to achieve a constant mass inlet flow. As a semi-enclosed slag cleaning furnace, the furnace is designed with an overflow, i.e., causing the inflowing slag to displace the cleaned slag present in the furnace. The system is completed by an emergency overflow and a matte taphole, which is also used at the end of the trial to empty the furnace.
For reasons of simplification, in this pilot furnace the bottom electrode was constructed as a carbon block in the bottom lining, with a graphite electrode screwed into it so as to enable the current to be directed upwards again laterally through the brick lining. The plant control system was designed as a semi-automatic system, with elaborate monitoring technology and data recording. A screenshot of the operating surface is shown in Figure 6.

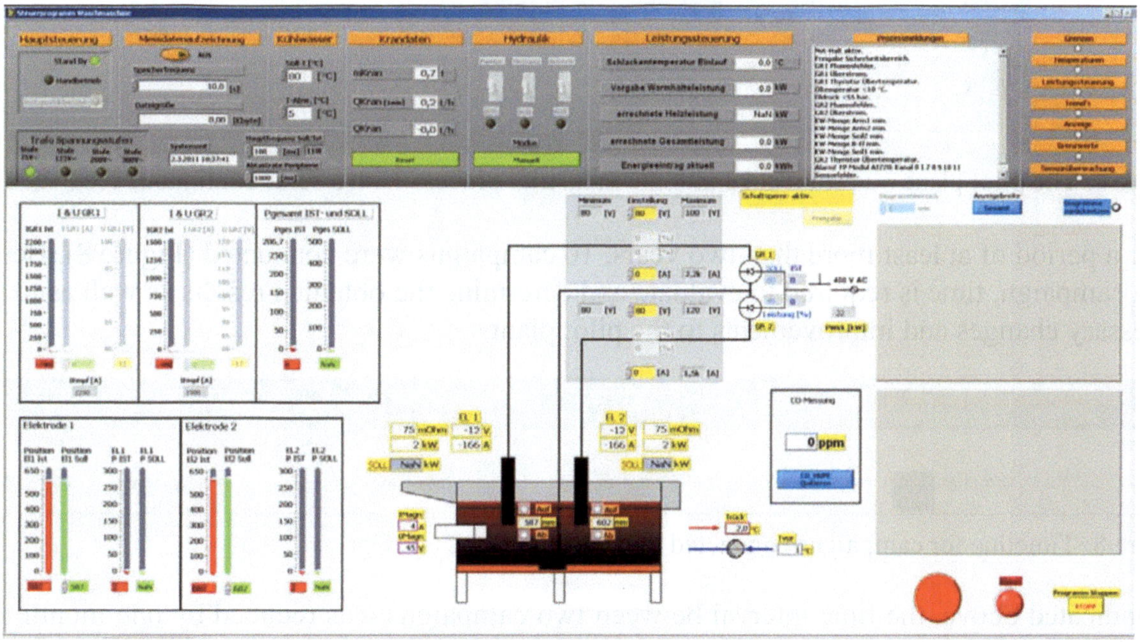

Figure 6: Image of the furnace control system developed jointly by Aurubis and SMS Siemag

The HMI was developed jointly by Aurubis and SMS Siemag and thus adapted optimally to the plant. The control system was further extended on several occasions in subsequent steps. The analysis results from introduced and outflowing slag are utilized for the control and planning of further operation. In spite of only partial automation, the operation of the pilot plant requires only four or five people per shift. The offgas is extracted from the charging side and the slag launder and conducted via pipeline to a central waste-gas cleaning system belonging to Aurubis. The refractory lining was produced by using several grades of bricks in cooperation with RHI, Vienna.

Hamburg campaigns

The pilot plant in Hamburg was operated in campaigns. The duration of each campaign was 8 to 10 days, including the heat-up phase for 2-3 days and the operating period. During the operating period the furnace was fed with slag from the industrial slag cleaning furnace semi-continuously via crane.

Figure 7: The pilot plant was embedded into the industrial process with a semi-continuous feed via crane

Over a period of at least more than two years, 10 campaigns were conducted (Figure 8). Between each campaign, time is required to evaluate and determine the obtained results as well as to make necessary changes and improvements to the pilot plant.

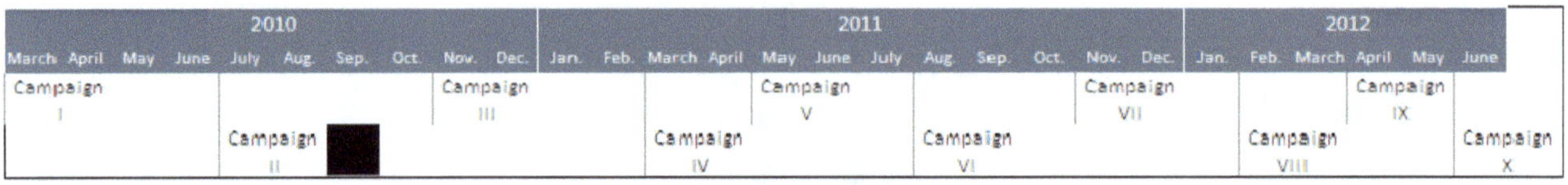

Figure 8: Timeline for campaigns conducted in the pilot plant

As indicated above, the time interval between two campaigns was reduced by one month to 1-2 months after campaign IV. Significant changes/improvements in terms of operating the pilot plant are

- Optimizing energy input by developing a power control software tool
- Optimizing the material flow and decreasing the downtime period of rectifiers by relocating the slag overflow launder
- Increasing the idle time of the slag launder by optimizing the geometry
- Installation of a coke charging device
- Optimized temperature control via a thermo-camera system

These factors gradually increased the overall Power-ON time from 70 % during the first 2 campaigns to > 90 % after campaign V. During 1566 hours of operating time 647 ladles of slag were treated. Consequently, this amounted to 1294 chemical assays of slag feed material and overflow iron silicate product.

Parallel to the test trials in the pilot plant in Hamburg, the obtained results were evaluated by CFD simulation. The optimized interaction between simulation and physical test work provided the opportunity to select relevant parameters which were subsequently tested in the trial campaign (Figure 9).

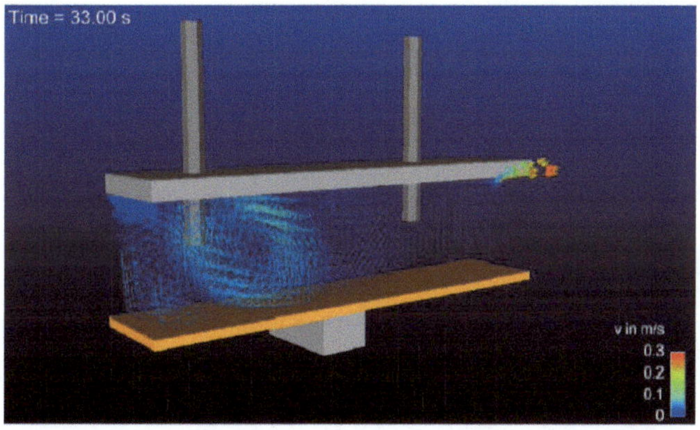

Figure 9: CFD simulations made it possible to select parameters, which were then tested in the trial campaigns.

The investigated parameters can be divided into three groups, with several sub-items for each group:

- Magnetic field application
  - Position
  - Operating mode
  - Magnetic force
- Slag feed parameters
  - Copper concentration
  - Magnetite concentration
  - Iron silica ratio
  - Slag basicity
- Process parameters
  - Retention time
  - Temperature
  - Power supply between the 2 electrodes

Latest results

All investigated parameters focus on one determining indicator – the remaining copper content in the overflow product iron silicate. To evaluate the relative efficiency of copper separation, a comparison to a reference status without any enhanced magnetic stirring is required. Figure 10 shows the histogram for the obtained results without enhanced stirring and with applied enhanced stirring. Since the copper separation depends on the copper concentration of the feed material, it is necessary to compare both measurement series with respect to average and standard deviation before determining the differences of the achieved results. The comparison of these two parameters showed no significant deviation between the two investigated states.

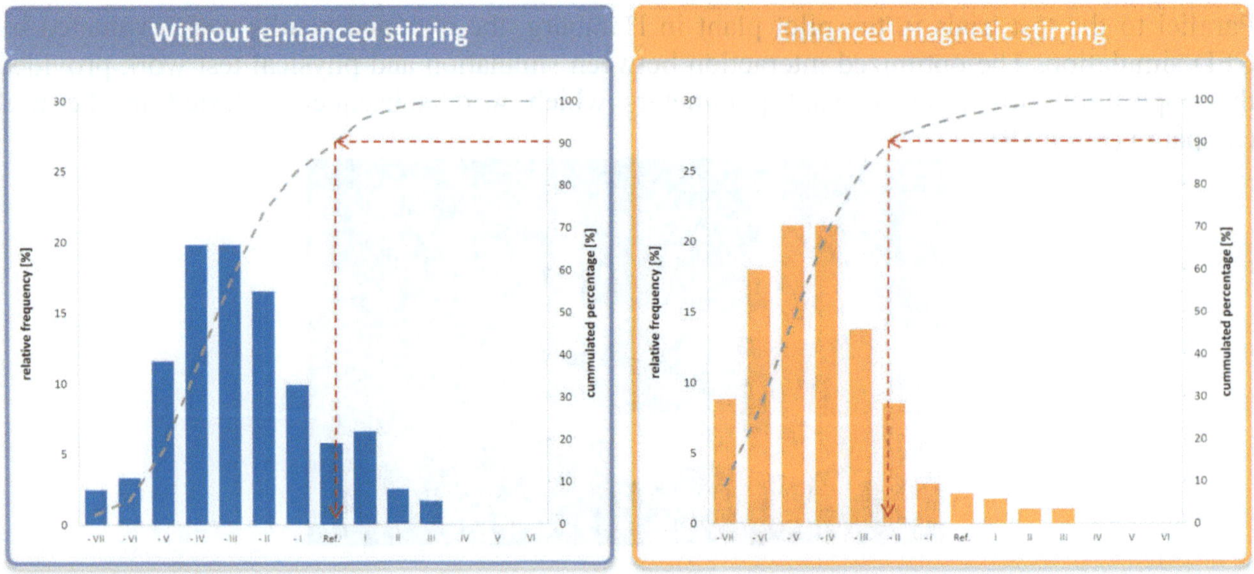

Figure 10: Comparison of results obtained without enhanced stirring and with applied enhanced stirring [5]

On the x-axis the copper concentration of the overflow is divided into 14 histogram classes. The axis is scaled to the average copper content (Ref.) of the produced iron silicate product from the industrial process. Each class represents a change to the reference level of 6.25 %. The primary y-axis shows the relative frequency distribution for the investigated series of measured data, while the cumulated percentage is plotted on the secondary y-axis.

The reference status without enhanced stirring shows a broad distribution of relative frequencies over the observed classes. The highest frequency, 20 % each, is obtained in the classes - III and - IV according to a decreased copper level in the final product of 18.75 % and 25.00 %. A reduced copper concentration more than 30 % can be observed for 17.4 % of the analyzed charges.

The graph for the enhanced magnetic stirring shows an accumulation of values in the classes - V and below. The relative frequency of copper concentration in the iron silicate product is nearly tripled compared to the status without enhanced stirring to 47.9 % of the investigated charges. In general, the results obtained during the application of the magnetic field show a smaller variety of the achieved Cu concentration in the overflow.

The red dotted line illustrates the difference between the cumulative percentages for both measurement series. Assuming that a reliability of the gained copper concentration of 90 % would be sufficient for industrial operation, it is evident that without enhanced stirring the results are comparable to the reference state and it would need the enhanced stirring to get to even better results. Summarizing these results, it can be stated that the application of a magnetic field led to a significantly lower copper concentration in the remaining iron silicate product.

Consequently, the achieved result raises other questions in terms of dependencies of operating parameters. Several parameters like temperature, concentration of magnetite, slag basicity and others were taken into account, but one main influencing factor remained – the copper concentration of the feed material.

Figure 11 illustrates the average copper separation between the overflow iron silicate product and the feed material. A description in the form of a histogram was chosen in this case as well. The difference is that on the x-axis the classification of the copper concentration for the feed material is used. On the y-axis the average copper separation was determined for each process status on the one hand and for each histogram class on the other hand.

$$\Delta Cu_i = \frac{\overline{c}_{Cu_{inlet}} - \overline{c}_{Cu_{overflow}}}{\overline{c}_{Cu_{Ref.}}} \cdot 100\%$$

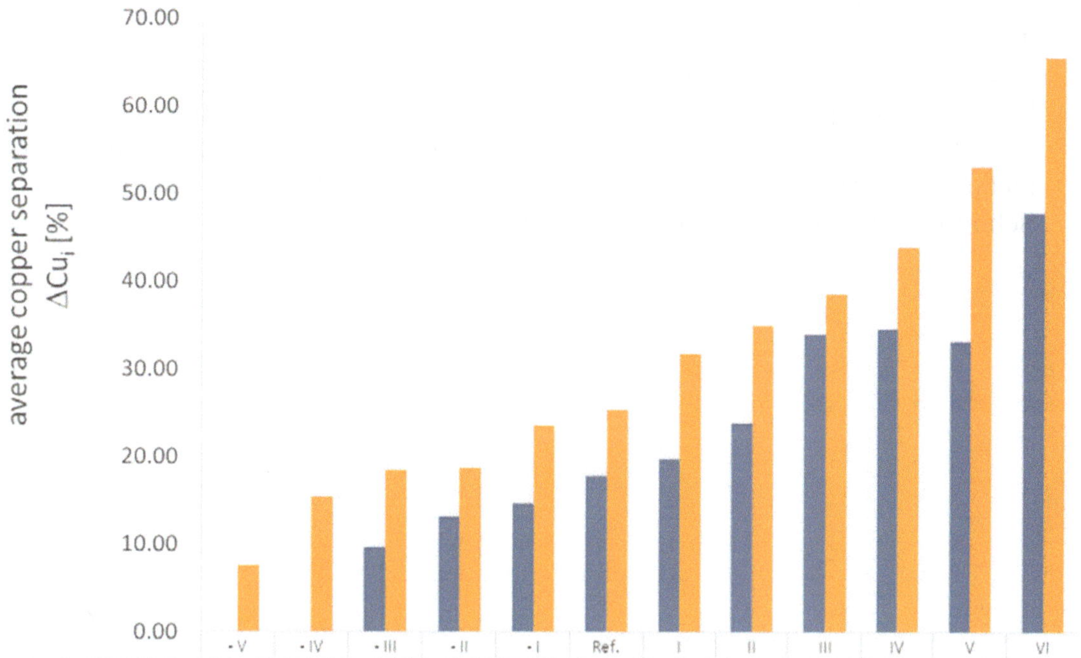

	- V	- IV	- III	- II	- I	Ref.	I	II	III	IV	V	VI
without enhanced stirring [%]	0	0	10	13	15	18	20	24	34	35	33	48
enhanced magnetic stirring [%]	7	15	18	19	24	25	32	35	39	44	53	66

Figure 11: Comparison of copper separation without and with enhanced magnetic stirring depending on the copper concentration of the feed material [5]

The driving force of separating copper from the feed material increases with a rising copper concentration. This means that even at high copper concentrations in the feed, sufficient separation to low copper concentrations in the overflow iron silicate product is feasible. Determining the copper separation ratio between the enhanced stirring status and the status without stirring, the value is > 1 for every histogram class. This means that independently of the copper concentration of the feed material, the application of the magnetic field leads to a better copper separation than the reference without enhanced stirring. At low copper concentrations (- IV) in the feed, copper can be separated by applying the magnetic field, while this was not observed without the additional stirring.

Summary & outlook

The principle of increased copper separation from slag by applying a magnetic field across a DC field was evaluated during 10 campaigns in a pilot-scale operation. Even at a high copper concentration V, VI) in the feed material, sufficient copper separation was observed, just as it was for low concentrations - IV, - V). Summarizing the determined results, it can be said that the application of the magnetic field amounts to 1.5 times better separation than without enhanced stirring.

The results at hand show the technical feasibility of the investigated process. Consequently, the next steps are a scale-up to industrial operation and the determination of the economic feasibility. Furthermore, this technology could also be interesting for other metallurgical industries. Other applications have to be tested (e.g., PGM, nickel, ferroalloys, etc.).

References

[1] Kunze J., Degel R., Borgwardt D., Warczok A., Riveros G.; Method and device for extracting a metal from slag containing the metal; WO2006/131372 A1

[2] Degel R., et al.; Latest results of the slag cleaning reactor for copper recovery and its potential for the PGM industry

[3] Degel R., et al., Latest results of the intensive slag cleaning reactor for metal recovery on the basis of copper, June 2010, Cu2010 in Hamburg, Proceedings

[4] Warczok, A., Riveros, T., Degel, R., Kunze, J., Oterdoom: Slag cleaning in circular and rectangular furnaces, September 2007, Cu2007 in Toronto, Proceedings

[5] Schmidl J., Zschiesche C., Zervos J.; internal report, August 2012

REWAS 2013: Enabling Materials Resource Sustainability
Edited by: Anne Kvithyld, Christina Meskers, Randolph Kirchain, Gregory Krumdick, Brajendra Mishra, Markus Reuter, Cong Wang, Mark Schlesinger, Gabrielle Gaustad, Diana Lados, and Jeffrey Spangenberger
TMS (The Minerals, Metals & Materials Society), 2013

THE REVIVAL OF ONAHAMA SMELTER & REFINERY FROM THE DISASTER BY THE GREAT EAST JAPAN EARTHQUAKE

Naoki Horihata, Shoji Kawashima, Tetsuro Sakai

Onahama Smelter & Refining Co., LTD
Onahama Smelter & Refinery
1-1 Nagisa Onahama Iwaki-City Fukushima JAPAN

Keyword: Great East Japan Earthquake, Copper Smelter & Refinery, Recycling,

Abstract

The earthquake with seismic intensity six trembled Iwaki City, Fukushima Prefecture, Japan on March 11 2011 and a subsequent tsunami attacked the coast on which Onahama Smelter & Refinery was located. The earthquake damaged many facilities and liquefied the ground. The tsunami flooded one third of the plant area. Furthermore, the repair work had to be suspended by the accident happened at Fukushima Daiichi Nuclear Power Station. The repair work was re-started in April and progressed smoothly. The operation was restarted on June 30 and the revival of Onahama Smelter & Refinery from the disaster by the Great East Japan Earthquake was completed. Radioactivity measurements have been programmed in detail and applied all over the materials covering from raw materials to products in order to supply safe and relief products to customers. The radioactive contamination to the products has not been detected at all.

Introduction

Onahama Smelter & Refinery was founded in 1963 as the first Cooperation Copper Smelter & Refinery in Japan. Until now, each of the facilities has been strengthened year by year and 60,000Mt/M copper concentrate was treated and 25,000 Mt/M electrolytic copper was produced. Furthermore 15,000 Mt/M combustible waste materials such as shedder residue from used automobile, electrical home appliances and vending machines were treated by substitution of fossil fuel used at two reverberatory furnaces. Table I shows capacity of each of plants.
The Great East Japan Earthquake with seismic intensity nine as seismic centers located at the northeastern offing was occurred at 14:46 on March 11 2011. The earthquake with seismic intensity six and a subsequent tsunami attacked Onahama Smelter & Refinery. The revival was full of difficulty by addition to the big damage by rumors caused by an accident in Fukushima Daiichi Nuclear Power Station afterwards. In spite of these difficulties, the operation was restarted on June 30 after every single person engaged in the revival had devoted efforts for 3.5 months.

The following items will be described in the present paper:
1) Damage received by the Great East Japan Earthquake and revival plan,
2) Current operation after the revival.

Table I. Capacity of Each of Plants

Smelter	60,000Mt/month Copper Concentrate Mitsubishi S Furnace, 2-Reverberatory Furnaces, 5-P.S. Converters, 3-Anode Furnaces
Refinery	25,000Mt/month Electrolytic Copper Starting Sheet Tank House No.1&No.2 Tank Houses (12,000mT/Month) No.3 Tank House (13,000mT/month)
Acid Plant	55,400Mt/month (98%H_2SO_4) No1 & No3 Acid Plants (Single contact converters)
Gypsum Plant	28,000Mt/month
Fine Casting Plant	8,000Mt/month, Billets&Cakes

Damage received by the Great East Japan Earthquake and revival plan

Electricity, drinking water, plant water, seawater stopped promptly after the earthquake occurrence. After the urgent stop measures were executed in the trembling caused by an intense earthquake, all workers evacuated to neighboring ground. Fortunately there was not environmental and work related accident at all with the help of appropriate urgent correspondence and the evacuation procedure drilled regularly twice a year. The damage of each of plants is listed in Table II and photos of miserable damaged facilities are demonstrated in Figure 1.

The repair work was limited without the restoration of the lifelines which were supplied by Tohoku Electric Power, Fukushima prefecture and Iwaki city. It took 74 days though all lifelines recovered because the lifelines were damaged all over the Tohoku district. On the other hand, almost all facilities in Onahama Smelter & Refinery were also damaged. Therefore, it was judged that each of facilities was restarted one by one.

The revival plan was summarized as follows.

1. Repairs of the utility such as pits and plumbing of plant water and sea water were taken priority. Especially, an outlet port of discharge was completely collapsed by the tsunami and had to be remodeled to earthquake-resistant structure before sea water was supplied. Figure 2 shows photos of outlet port of discharge drainage.
2. The fine casting plant was restarted prior to other plants. Because the operation could be possible without supply of sea water and by treated purchased electrolytic copper which was delivered by Mitsubishi Materials Corporation. First of all, it was necessary to satisfy customer needs to the maximum as much as possible.
3. Mitsubishi S furnace, one reverberatory furnace, three PS converters and two anode furnaces which were necessary for partial operation, shall be restored first at the smelter section.
4. Starting sheet tank house and No.2 commercial tank house which were with a little damage comparing with No.1 & No.3 commercial tank houses shall be restored first at the refinery section. Figure 3 shows photos of No.3 tank house.
5. No.3 acid plant which was less damaged than No.1 acid plant shall be restored.

Table II. Damage of Each of Plants

Smelter S furnace, reverberatory furnaces, PS converters and anode furnaces stopped with full melt. The ceiling and the sidewall bricks of the reverberatory furnaces dropped more than 20,000 pieces.
Refinery More than 140,000 sheets of anodes and cathods dropped into cells and inside lining of more than 80% cells was injured.
Acid Plant & Gypsum Plant Converters, drying towers and absorption towers stopped with SO_2 gas. Many cracks and breakings were observed in towers, plumbing and gas duct .
Utilites Sea water pit was breaking. Many pipes of drinking water, plant water and sea water cracked. Outlet port of discharge drainage was collapsed. Numberless cave-ins and cracks were observed in all roads.

Figure 1 Photos of damaged facilities

Figure 2 Photos of outlet port of discharge drainage

Figure 3 Photos of No.3 tank house

Table III. Process of Revival

Mar. 11	Hit by the Great East Japan Earthquake
Mar. 23	Received electricity from Tohoku Electric Power
Mar. 28	Restrated supply of drinking water by Iwaki city
Apr. 11,12	Hit by the aftershocks of intensity 6 twice
Apr.22	Restrated supply of plant water by Iwaki city
May. 16	Restarted fine casting plant
May. 24	Restarted supply of sea water by Fukushima prefecture
Jun. 29	Restarted refining operation at No2 tank house
Jun. 30	Restarted smelting operation at 65% of capacity
Jul. 2	Increased smelting operation to 90% of capacity
Aug. 23	Returned to smelting operation at full capacity
Aug. 31	Returned to refining operation at full capacity

410 employees, 350 contractors in plants and 12 staffs who were dispatched from Mitsubishi Materials Corporation as a parent company, were engaged in revival. Furthermore, the total number of contractors which were temporarily called for repair work from the all over Japan reached to more than 36,000 man-days.
The process of revival was shown in Table III. Partial operation at 65% of the capacity was restarted on June 30 and full operation was realized on 30 August when the original revival plane was moved up more than one month before. As a result, the revival of Onahama Smelter & Refinery from the disaster by the Great East Japan Earthquake was completed.

Current operation after the revival

Onahama Smelter & Refinery was located at the position of 53km south-southwest from Fukushima Daiichi Nuclear Power Station and the location has been out of a refuge area. However there was concern of the radioactive contamination to products after the operation was restarted. New system for radiation control has been introduced to cope with inquiry about safety from customers and secured safety of products.
Main correspondence to radioactive contamination carrying out was shown as follows.

1. As of raw materials, all recyclable materials such as shredder residue have been checked over loading trucks utilizing radiation detector. Only recyclable materials less than the standard value had been received. As radiation of fly ash generated from incinerators was higher than the standard value, treatment of fly ash has been suspended since accident of Fukushima Daiichi Nuclear Power Station.
2. As of products, radiation dose of products was measured by the third party. Fortunately, radiation dose of products such as electrolytic copper, sulfuric acid, gypsum, copper sulfate, crude nickel sulfate was lower than detection limit.
3. Radiation dose of discharge water from waste water treatment plant and off gas from stack was measured once a month. The value always showed lower than detection limit. Space dose of radioactivity was measured once a month and the result was posted in order to let the employees feel relieved.

By such severe management, the radioactive contamination to products has not been found at all and the measured value has been reported to customers if requested.
Figure 4 shows amount of treated copper concentrate and produced electrolytic copper after revival. It was forced that the treatment of copper concentrate sometimes decreased or stopped by blackout or water supply troubles caused by the frequent aftershock, however damage of facilities in plants was not occurred at all by an effect of the earthquake-resistant remodeling. The supply of lifelines was recovered month by month, and improved further. Consequently the production also increased and the treatment of copper concentrates has achieved a new record of 68,425 metric-tons in August 2012. On the other hand, the collection of combustible waste materials such as shedder residue had decreased sharply after the operation was restarted as shown in Figure 5. It consisted of two reasons.

1) The earthquake disaster passed throughout East Japan and many scrap factories which generated shredder residue were driven into the shutdown.
2) Many people had to buy used cars immediately because many cars were carried away by the tsunami. As a result, a disposal car decreased sharply and the raw materials of the shredder dust also decreased sharply.

However, the collected amount of shredder residue gradually recovered and returned to the amount before the earthquake disaster since April 2012.

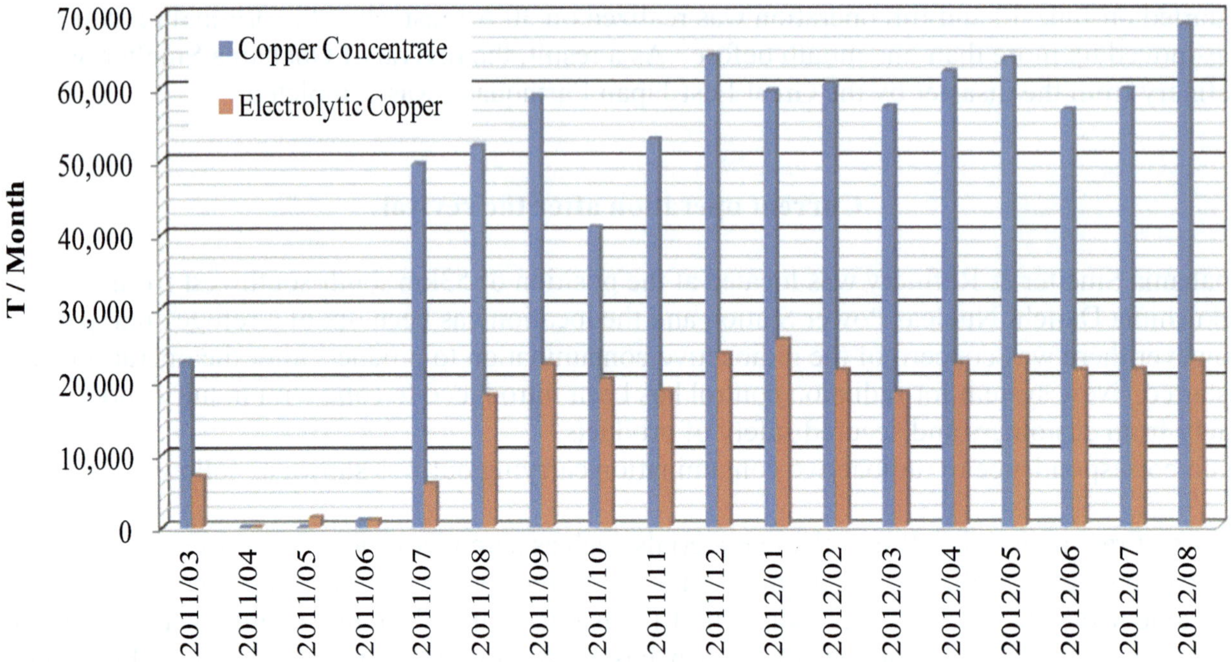

Figure 4 Amount of treated copper concentrate and produced electrolytic copper after revival.

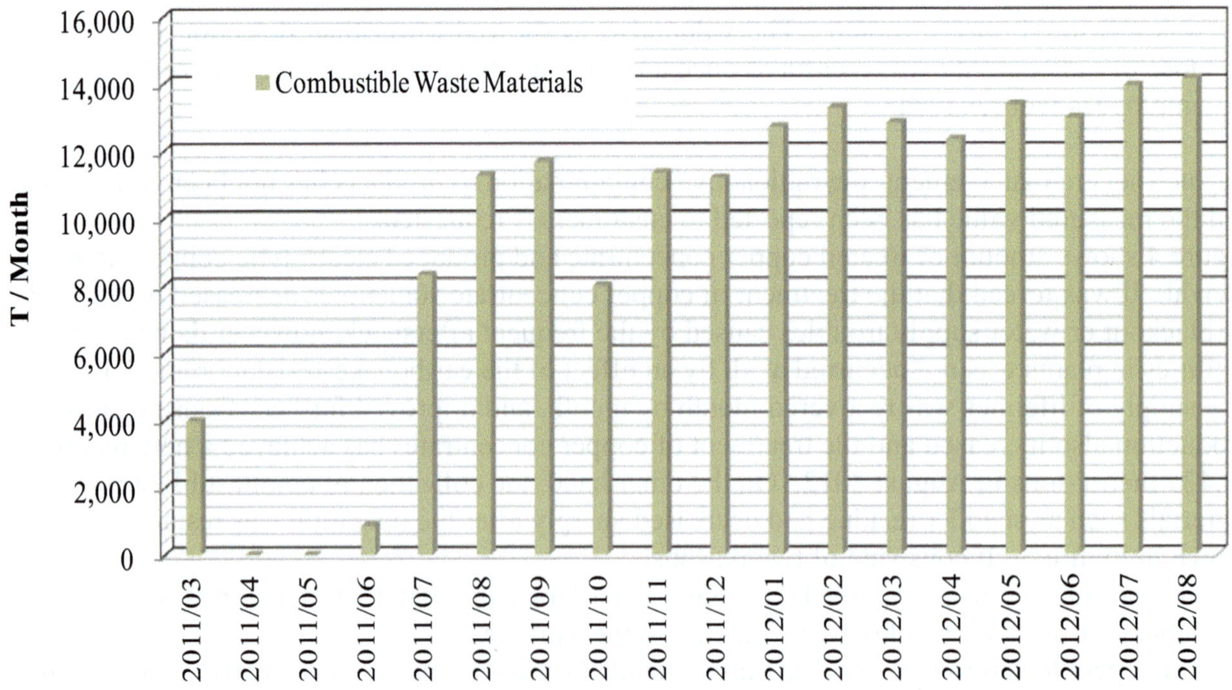

Figure 5 Amount of treated combustible waste materials after revival.

Conclusions

1. The earthquake disaster was an unhappy event for Onahama Smelter & Refinery. However, the employees grew up vigorously and gained a lot of irreplaceable experience through revival process. It was possible that restart of operation was made largely early as a result of desperate effort of all people.
2. There has never been radioactive contamination to products after the nuclear accident because new system for severe radiation control has been conducted rigidly.
3. The operation after the revival is favorable as a result of improvement of the facilities including earthquake-resistant remodeling together. The new record of amount of treated copper concentrates was established in August 2012 and collection of combustible waste materials also turned upward.

References

1. H.Asao et al., "New Era of Onahama Smelter with New O-SR Process-Onahama Type Direct Laundered S-Furnace and Reverb Furnaces-", Processing Materials for Properties, TMS, 2009, 625-630

2. O.Iida et al., "The initial years of the O-SR process", Proceeding of Cu 2010 MMIJ Fall Meeting (Fukuoka Japan)

3. T.Sakai et al., "Establishment of New Copper Smelting and Recycling Process with O-SR Process", Journal of MMIJ Vol.128 2012, 495-499

4. M.Nishiwaki and S.Hayashi , "Stabilization of Recycling Technology for Shredder Residue in Copper Smelting Process", Journal of MMIJ Vol.121 2005, 357-362

5. T.kawai et al., "Copper Concentrate Smelting in Peirce-Smith Converters at Onahama Smelter, Converting and Fire Refining Practices, TMS, 2005, 119-123

LEACHING OF URANIUM AND VANADIUM FROM KOREAN DOMESTIC ORE

Joon Soo Kim, Kyeong Woo Chung, Hoo In Lee, Jin-Young Lee, J. Rajesh Kumar*

Korea Institute of Geoscience & Mineral Resources (KIGAM), Daejeon 305-350, Korea
*e-mail: rkumarphd@kigam.re.kr, rajeshkumarphd@rediffmail.com

Keywords: Leaching, Uranium, Vanadium, Korean ore

Abstract

Countries like Korea having very limited uranium resources and founded deposits having low grade metal values. Uranium is the main source to generate the nuclear power as cheap and more quantity of the electricity will generate. For this reasons the upcoming researchers in developed/developing countries are establishing more research and development on extraction and separation technologies for uranium. The present scientific study focused on leaching process of Korean domestic ore. The following experiments are carryout for optimization of the leaching process. Acid influence on leaching process was tested and noted that 2.0 M sulfuric acid concentration is the optimized conditions for present study. The time influence on leaching process was observed and its optimized 2 h for complete leaching process. The temperature influence tested and optimized the 80^0C for complete leaching process and pulp density is 50% (wt %).

Introduction

Uranium need going very fast in developed countries and Korea having low grade uranium ores associated with valuable metal vanadium and other impurities like iron, aluminium, silicon predominantly. The present study focused on both metals such as uranium and vanadium recovery and separation for country needs. For the recovering process leaching is the first step in hydrometallurgical methodology. The first stage targeted both metals leaching in next stage separate the each other.

Before began the any process the literature review will give the past reported information as well as new ideas for present study. The literature review noted the following reported methodologies on uranium leaching process. Uranium content in phosphate rock was leached with alkaline solutions like ammonium carbonate and bicarbonate and produced uranyl carbonate by Guzman el at [1]. The uranium leaching process established by using $Fe_2(SO_4)_3$ with H_2SO_4 as one set then sulfuric acid having low nitric acid combination as another set by Shakir et al [2], further the leach liquor proceeded by liquid-gel extraction. Buck and coworkers from Argonne National Laboratory developed a procedure to remove the uranium(IV) by carbonate leaching process in presence of oxygen [3]. Uranium in 4th oxidation stage leached by above said process was follows:

$$UO_2 + 1/2\ O_2 + 3CO_3^{2-} = [UO_2(CO_3)_3]^{4-} + 2OH^- \quad (1)$$

The new leaching system hydrogen peroxide and sodium sulfate with sulfuric acid was applied for uranium. Advantage of this methodology is to reduce the sulfuric acid concentration 700 to 400 g acid kg^{-1} [4]. Uranium content in various oxidation states such as four and six was selectively leached by HCl [5]. Nanomaterials used for in situ chemical leaching of uranium ore and developed mechanisms [6]. From the various waste sources uranium was leached long term by Patra et al [7]. The present leaching procedure developed for both metal ions uranium and vanadium leached and further recovery and each other separation procedures going to establish by using extraction process.

Materials and Methods

Apparatus and reagents

Uranium and vanadium metals analysis was carryout using inductively coupled plasma optimal emission spectrometer (ICP-OES) Perkin Elmer Model Optima 2000 Dr. All reagents used were analytical reagent grade. Korean domestic ore sample present as Fig. 1.

Figure 1. Korean domestic ore sample

The chemical composition of the Korean domestic ore sample was presented in Table I.

Table I Chemical composition of Korean domestic ore sample

Component	Content (%)	Component	Content (%)
U_3O_8	0.058	P_2O_5	0.25
V_2O_5	0.16	ZnO	0.03
SiO_2	55.1	NiO	0.08
Al_2O_3	7.45	S	0.19
Fe_2O_3	3.47	Fixed carbon	26.9
CaO	0.35	H_2O	~5.1
MgO	0.85		

Leaching Procedure

The experimental set up for the leaching process made by pyrex material to face the sudden temperature changes in sulfuric acid leaching process. During a leaching process density of the water evaporation will occurred to prevent this reflux condenser was set to on upper portion (Fig. 2). Loaded thermocouple will measures the temperature of the reaction water and set the temperature ±2°C inside and outside of the room. After reaching the optimum temperature condition the uranium ore sample was added to sulfuric acid solution by systematically. Mechanical agitator used for proper mixing the sample with acid solution. The agitation speed keep 500 rpm for all experiments in the present study. Analysis of the metals s was recorded by ICP-OES and leach rate was calculated. The general agreement between the percent leaching obtained was within ± 2%. Collected

Figure 2. Experimental set up of the leaching process

Results and Discussions

Effect of time

The first experiment tested the time influence on leaching process. The obtained results presented as Fig. 3. The preliminary studies on time optimized 2 h time requires for both metals leaching process and all other experiments keep leaching process up to 2 h.

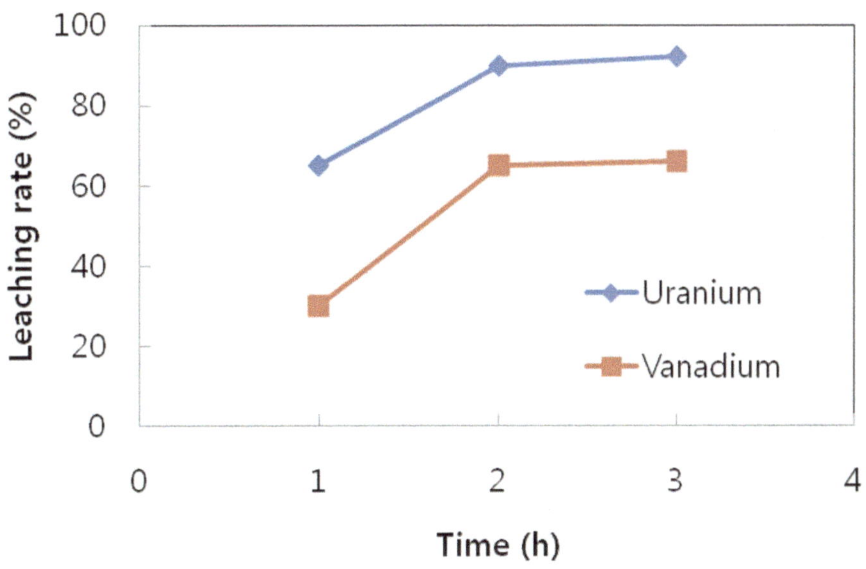

Figure 3. Effect of time on Korean domestic ore leaching process

Effect of particle size

The particle size of the Korean domestic ore sample was tested and obtained results are presented as Fig. 4. The results clearly demonstrate that, -48 mesh size is ideal for both metals.

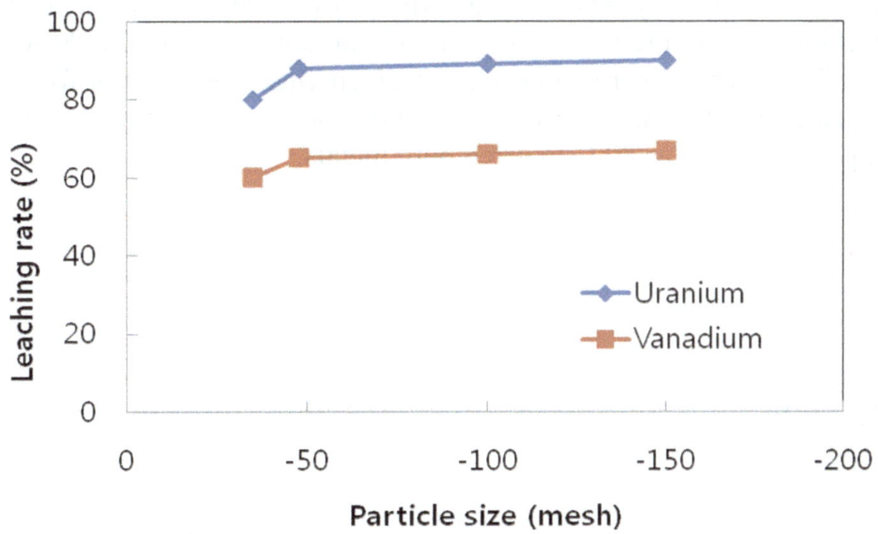

Figure 4. Particle size effect on Korean domestic ore leaching process

Effect of acid concentration

The effect of sulfuric acid concentration on uranium and vanadium leaching process from uranium ore was carried out in between 1.0 to 7.0 M of acid concentration. The metal leaching rate increased with increases acid concentration and sharp increase in leaching rate with variation from 1.0 to 2.0 M of sulfuric acid concentration. Very slow increase in leaching rate with variation from 2.0 to 6.0 M sulfuric acid. The present study concludes that, 2.0 M acidity is good to precede further experiments. The results were presented in Figure 5.

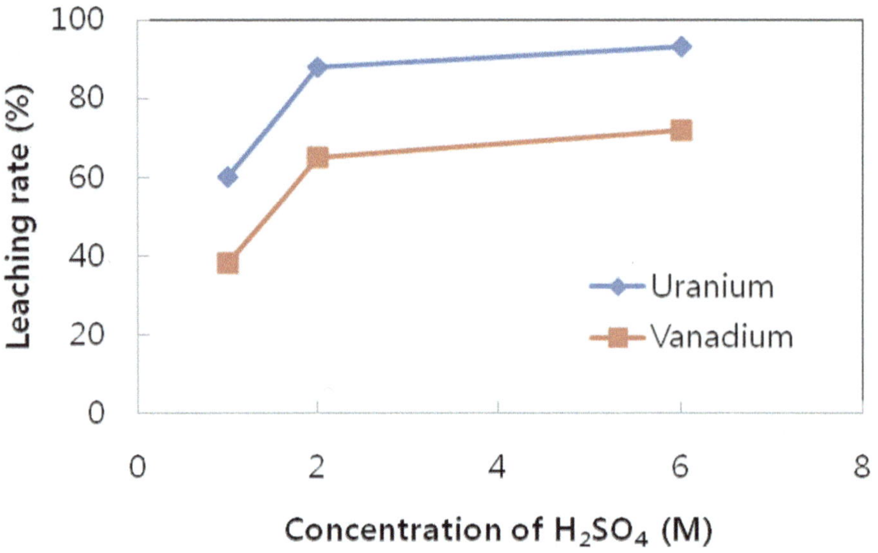

Figure 5. Effect of sulfuric acid concentration on uranium vanadium leaching process from Korean domestic ore

Effect of temperature

Temperature influence play key role in leaching process and the present study carry out temperature effect on uranium ore processing. The temperature varied 60 to 80°C and the results are presented as Fig. 6. The experimental results given final conclusions for uranium highest leaching rate observed at 80°C where as for vanadium 60°C temperature was suitable. Further experiments temperature fixed as 80°C to leach both metals.

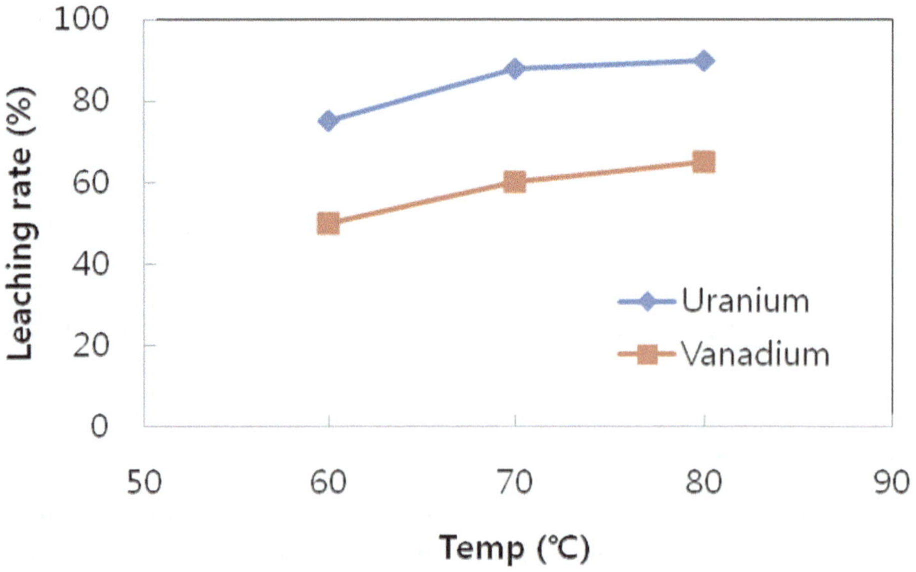

Figure. 6 Effect of temperature on uranium vanadium leaching process from Korean domestic ore

Effect of pulp density

Productivity of the leaching process was interlinked with density. For this reason it's necessary to test the density effect on the leaching operations and optimize the process. Very little variations in leaching rate was observed at various pulp densities (PD) from 20 to 60 wt. %, for further experiments PD was fixed as ~50 wt. % (Fig. 7) .

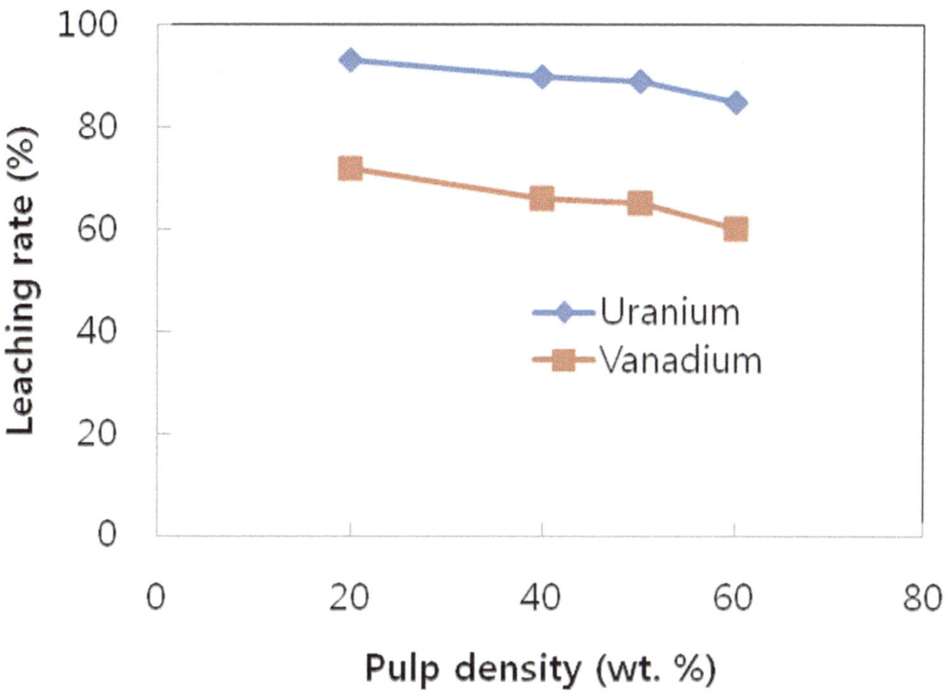

Figure 7. Effect of pulp density on uranium vanadium leaching process from Korean domestic ore.

Conclusions

The following conclusions drawn from the present investigations on leaching process for uranium and vanadium metals from Korean domestic ore:

1. The leaching process experimental conditions optimized for uranium and vanadium metals from Korean domestic ore and developed the basic experimental procedures such as time, particle size, acid influence, temperature effect and pulp density behavior
2. The basic time effect experiment concludes that 2 h time is sufficient for uranium and vanadium leaching process.
3. Even the moderate sulfuric acid concentration i.e 2 M acid leached ~ 90 to 94 % of uranium leached whereas same experimental condition ~65 to 72% of vanadium was leached from ore sample
4. The target metals uranium and vanadium was leached above 90% of uranium at 80°C where as ~65% leached vanadium at same temperature
5. The ideal pulp density optimized from present study was 50% for both title metal ions leaching process
6. The ideal conditions optimized for Korean domestic ore sample by present study:

Parameter	Optimized Condition
Time	2 h
Particle size	-48 mesh
Acidity	2 M H_2SO_4
Temperature	80°C
Pulp density (PD)	50%
Agitation speed	500 rpm

7. Finally, present research paper concludes that, the proper leaching conditions (above said conditions) will full fill the uranium and vanadium leaching process from Korean domestic ore

References

1. E. T. R. Guzman, E. O. Regil, and G. P. Malagon, "Uranium leaching from phosphate rock", *Journal of the Radioanalytical and Nuclear Chemistry Letters*, 201 (1995) 313-320.

2. K. Shakir, M. Aziz, and Sh. G. Beheir, "Studies on uranium recovery from a uranium bearing phosphatic sandstone by a combined heap leaching-liquid-gel extraction process.1 Heap leaching, Hydrometallurgy, 31 (1992) 29-40.

3. E. C. Buck, N. R. Brown, and N. L. Dietz, "Contaminant uranium phases and leaching at the Femald site in Ohio", Environmental *Science and Technology,* 30 (1996) 81-88.

4. A. Yuksel, M. Eral, and S. Olmez, "Leaching of uranium with the H_20_2-Na_2SO_4-H_2SO_4 (HSS) system and the efficiency of acigol lake water", *Journal of the Radioanalytical and Nuclear Chemistry Letters*, 200 (1995) 169-179.

5.H. K. Fouad, "Determination of hexavalent and tetravalent uranium in phosphate ores through hydrochloric acid selective leaching", *Journal of the Radioanalytical and Nuclear Chemistry* 285 (2010) 193-197.

6. S. Klimkova, et al., "Zero-valent iron nanoparticles in treatment of acid mine water from in situ uranium leaching", *Chemosphere,* 82 (2011) 1178–1184.

7. A. C. Patra, et al., "Long-term leaching of uranium from different waste matrices", *Journal of Environmental Management,* 92 (2011) 919-925.

Acknowledgments

This work was supported by the Energy Efficiency & Resources Program (Physical and Chemical mineral dressing/refining of low grade uranium ore) of the Korea Institute of Energy Technology Evaluation and Planning (KETEP) grant funded by the Korea Government Ministry of Knowledge Economy.

STUDY OF ADSORPTION PROPERTY OF Ga(III) ONTO STRONGLY BASIC RESIN FOR Ga EXTRACTION FROM BAYER LIQUOR

Zhuo Zhao[1], Yongxiang Yang[1], Hao Lu[1], Zhongsheng Hua[1], Xiaoling Ma[2]

[1]Anhui University of Technology; No. 59 Hudong Road; Maanshan, Anhui, 243002, China
[2]Shimadzu (China) Co. Ltd; No. 570 West Huaihai Road; Shanghai, 200052, China

Keywords: gallium recovery, Bayer liquor, adsorption, ion exchange, resin

Abstract

Ion-exchange is the main technology used in industry for gallium recovery from Bayer liquor, the largest gallium production resource. However, the co-extraction of vanadium and the degradation of resins are the major issues. Further investigations related to fundamental theory are needed. This paper reports the study of the adsorption properties of a strongly basic resin having a combination of one =NOH group and another active group -NH_2 for Ga(III) extraction. The influence of operational conditions such as contact time, initial Ga(III) concentration and temperature on Ga(III) adsorption were extensively investigated. The results revealed that the resin has high adsorption capacity and Ga(III) selectivity. The optimal adsorption condition was obtained at temperatures of 40-50°C and contact time of 40-60 min. The Ga(III) adsorption data on the resin fit well with the pseudo second-order kinetics. Langmuir and Freundlich models were used to describe Ga(III) adsorption isotherms on the resin.

1 Introduction

Gallium is a metallic element in Group IIIA of the periodic table. It was discovered in 1875 by a French chemist Paul-Émile Lecoq de Boisbaudran [1]. This soft and silvery-blue metal has an atomic weight of 69.723, a hardness of 1.5 mohs, and a specific gravity (i.e., density) of 5.904 and 6.905 g/cm^3 for the solid and liquid respectively. A solid piece of gallium will liquefy when being placed in one's hand. It exhibits an unusually large liquid range due to its low melting point of 302.98 K and high boiling point of 2676 K. Solid gallium has an orthorhombic crystal structure and displays a conchoidal fracture similar to glass. Gallium can exist in the form of six isotopes of which only two are stable. The two stable isotopes are ^{69}Ga (60.4%) and ^{71}Ga (39.6%). Gallium can form many substances such as bromides, chlorides, hydrides, iodides, nitrides, oxides, selenides, sulphides and tellurides. Gallium has the oxidation states of +1 and +3 valence. Although the element is stable with water it reacts vigorously with halogens even at low temperatures. The element rapidly dissolves in either aqua regia or concentrated sodium hydroxide in an aqueous medium. It also reacts with strong bases and oxidants. The element is suited to readily form alloys (e.g., eutectic alloys) with most metals in conjunction with being a component in low-melting alloys.

Since the 1970s gallium metal has attracted a lot of interests. Gallium combined with elements of group 15 displayed semiconducting properties [2, 3], such as gallium arsenide (GaAs) and gallium nitride (GaN). They are valuable compounds employed in advanced semiconductors for microwave transceivers, DVD's, laser diodes in compact discs and other electronic applications [4]. The supply and demand of gallium-bearing products have gradually increased during the past decade. Hence, an effective extraction of gallium is an important investigation topic.

Gallium is a very widespread trace element. Its content in the lithosphere is about 10 ppm. Gallium is a cryptomorphic element whose average clarkes (geochemical density indicators) in the lithosphere are relatively high. Due to the practical inability to form its own minerals in natural conditions, it is considered quite rare [5]. Given the geochemical affinity between Al and Ga, the latter occurs mainly concentrated in bauxites and hosted by diaspore, various aluminosilicates (such as clays), apatite, nepheline and frequently alunite. Ga also has a chalcophile affinity, and thus, may occur as gallite ($CuGaS_2$), and is frequently substituted for Zn and Cu in sulphides, mainly in sphalerite (ZnS), germanite ($Cu_{26}Fe_4Ge_4S_{32}$), and chalcopyrite ($CuFeS_2$) [6]. It is therefore not economical to mine any minerals for merely recovering gallium. Generally, Gallium is recovered as by-product mainly from the production of alumina, and to a lesser extent from zinc production [7, 8, 9]. Other alternative potential sources of gallium are coal fly ash, which can contain as much as 100 ppm of the metal [10, 11], and the recycling of industrial electronic scrap [12].

Bayer liquor obtained during alumina production from bauxite is the most important source of gallium. It was estimated that about 90% of world primary gallium is produced from Bayer liquor [13]. Gallium concentration in the bauxite ores is in a range from 20 to 80 ppm. In the Bayer process about 70 % of the gallium is leached from the bauxite and follows the aluminum into the caustic soda solution, and the remaining 30 % is disposed with the red mud. Gallium accumulates in the Bayer liquor due to several successive circulations , reaching a concentration of 100-300 mg/l [14].

Four kinds of methods have been developed to recover gallium from Bayer solutions [15], including fractional precipitation, electrochemical, solvent extraction, and ion exchange. Fractional precipitation method is based on Al-Ga precipitation with CO_2 and further separation of Al and Ga with lime milk or sodium aluminate solutions. This approach is more environmental friendly and with low cost, but its flowsheet is usually complicated. Electrochemical method includes both mercury cathode electrolysis, and cementation. The electrolysis with mercury cathode has been prohibited in most countries because of high toxicity of mercury. Cementation is an

electrochemical process realized by the displacement reaction between gallium and the reductants, such as sodium amalgam, aluminum and aluminum-gallium alloy. Solvent extraction is an efficient method and by using Kelex 100 system about 80 % of the gallium in Bayer liquor can be extracted. However, the kinetics of extraction has been proved to be very slow, which usually requires several hours. Ion exchange is the main method applied in industry for gallium recovery from Bayer liquor.
Duolite ES-346 and DHG586 exhibit good extracting properties for gallium, and are used as industrial resins. Unfortunately, the co-extraction of vanadium and the degradation of the amidoxime groups still remain as the main problems during industrial application. Further investigations related to fundamental theory are needed.

2 Experimental

2.1 Materials and Methods

A strongly basic resin (LSC-600), containing a combination of one =NOH group and another active group $-NH_2$, was used as raw material. The Bayer liquor was taken from the pilot plant of Zhengzhou Research Institute of Chalco. The composition is shown in Table I, in which α_k means the molar ratio between Na_2O_k and Al_2O_3 in the liquor. Other reagents and solvents were of analytical grade and used without further purification. The concentrations of metal ions were measured on a ICPE-9000.

Table I. Chemical Composition of Bayer Liquor (g/L)

Na_2O_k	Al_2O_3	α_k	Na_2O_T	SiO_2	V_2O_5	Ga
149.02	80.71	3.04	175.73	0.78	0.21	0.19

2.2 Adsorption Procedures

2.2.1 Adsorption Capacity

Adsorption capacities of the resin for gallium in Bayer liquor were determined by batch tests according to the following procedure. 1g of the resin was added to 200mL of Bayer liquor in a 250-mL glass bottle with a stopper. After being shaken at 298K for 24 h, the solution was separated from the resin. The concentration of gallium in the solution was determined by ICP. The adsorption capacity was calculated accordingto Eq. (1).

$$Q=\frac{(C_0-C)}{W}V \tag{1}$$

Where Q is the adsorption capacity (mg/g); C_0 and C are the initial concentration and the concentration at any time t, respectively, of gallium in solution (g/L); V is the solution volume (mL); and W is the dry weight of resins (g). Each determination in the adsorption procedures was repeated three times and the results were given as average values. The error bars were also indicated wherever necessary.

2.2.2 Adsorption Kinetics

To obtain the data of adsorption kinetics, 1g of the resin was added to 200mL of Bayer liquor. The mixture was shaken continuously in a thermostat-cum-shaking assembly at a pre-determined temperature. Aliquots of 1mL solution were withdrawn at desirable intervals and the concentration of gallium in the solution was determined by ICP. The adsorption capacity of the resin was calculated according to Eq. (1).

2.2.3 Isothermal Adsorption

The isothermal adsorption property of the resin was investigated also by batch tests. In a typical procedure, a series of 250-mL tubes were used. Each tube was filled with 1g of the resins and 200mL of metal ion solution of varying concentrations and desired temperature. The mixture was shaken for 24 h. The adsorption capacities were calculated also by using Eq. (1).

3 Results and Discussion

3.1 Adsorption Capacity

To investigate the affinity of resin for Ga(III) and V(V), an experiment was conducted by batch tests. The results were tabulated in Table II. From Table II, it could be note that, the resin had high adsorption capacity for Ga(III) but low adsorption capacity for V(V). The adsorption capacity of resin for Ga(III) and V(V) were 31.9 mg/g and 6.4 mg/g, respectively.

Table II. Adsorption Capacities of Resin for Metal Ions

Ions	Adsorption Capacity(mg/g)
Ga(III)	31.1
V(V)	6.4

3.2 Adsorption Kinetics

Fig. 1 showed the plots of adsorption capacity of LSC-600 toward Ga(III) versus the contact time at different temperatures, respectively. The data in Fig. 1 were modeled by the Lagergren's pseudo first-order and pseudo second-order rate equation given below as Eqs. (2) and (3), respectively.

$$\log(Q_0-Q)-\log Q_0 \quad \frac{k_1}{2.303}t \qquad (2)$$

$$\frac{t}{Q}=\frac{1}{k_2Q_0^2}+\frac{1}{Q_0}t \qquad (3)$$

where k_1 is the rate constant of pseudo first-order adsorption (h^{-1}); k_2 is the rate constant of pseudo second-order adsorption (g/mg); Q_0 and Q are the adsorption uptake at equilibrium and at time t, respectively (mg/g).

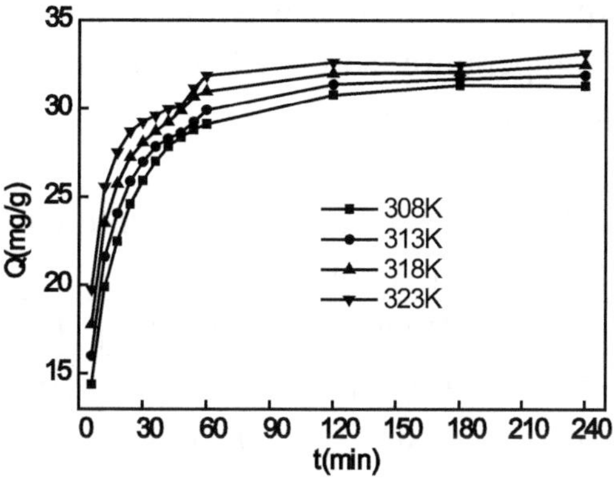

Figure 1. Adsorption capacities of Ga(III)versus time on LSC-600 at different temperatures.

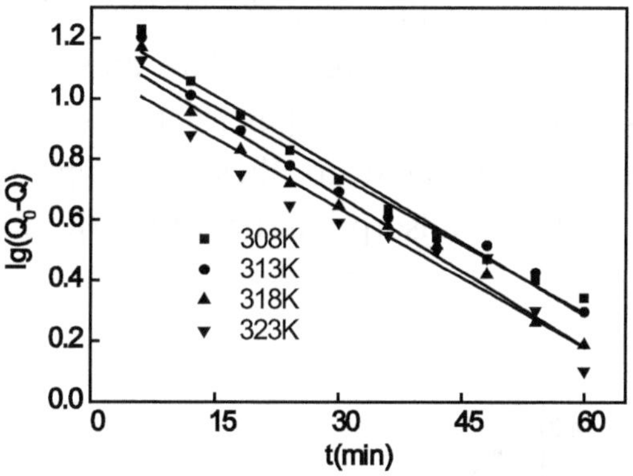

Figure 2. The pseudo first-order kinetic model of LSC-600 for Ga(III).

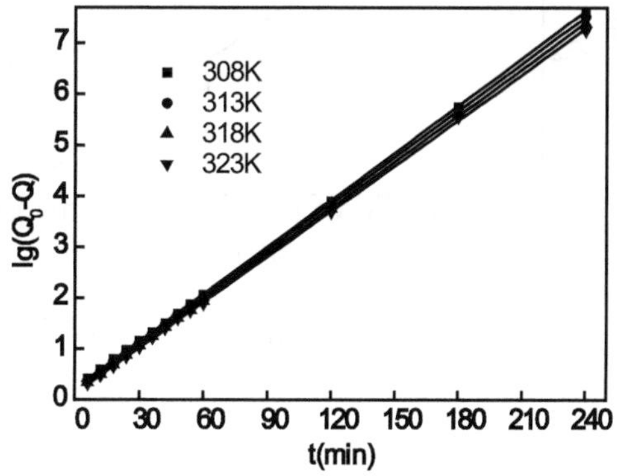

Figure 3. The pseudo second-order kinetic model of LSC-600 for Ga(III).

Both the models were used to fit the kinetics curves and the results showed that pseudo second-order model was more suitable since the values of R^2 could be regarded as a measure of the goodness-of-fit of experimental data on the kinetic models. The straight-lines of pseudo first-order and second-order kinetic model were shown in Figs. 2 and 3. The corresponding parameters calculated according to the models were tabulated in Table III. It could be seen that: The adsorption rates increased with the increasing temperature (see the values of k_2 in Table III), and the saturation adsorption capacities also increase with the increasing temperature as we expected. A possible explanation for this was that the diffusion rate of Ga(III) was enhanced by increasing temperature, so Ga(III) could approach and contact with the resin much faster at higher temperature.

Table III. The Kinetic Parameters of Ga(III) Adsorption on LSC-600 at Different temperatures

T(K)	Q_0(mg/g)	Pseudo First-order		Pseudo Second-order	
		$k_1(h^{-1})$	R^2	k_2(g/mg/h)	R^2
308	31.32	0.037	0.9793	0.0046	0.9999
313	31.90	0.035	0.9650	0.0054	0.9999
318	32.51	0.038	0.9748	0.0066	0.9999
323	33.13	0.035	0.9225	0.0079	0.9998

According to Arrhenius equation, $\ln k_2 = -E_a/RT + \ln A$, plotting $\ln k_2$ against $1/T$, a straight line could be obtained (see Fig. 4). The apparent activation energies of adsorption Ea calculated from the linear slopes for LSC-600 was 30.75kJ/mol. These low activation energies as compared to these of typical chemical reaction of 65–250kJ/mol implied that the adsorption of LSC-600 for Ga(III) was a facile procedure.

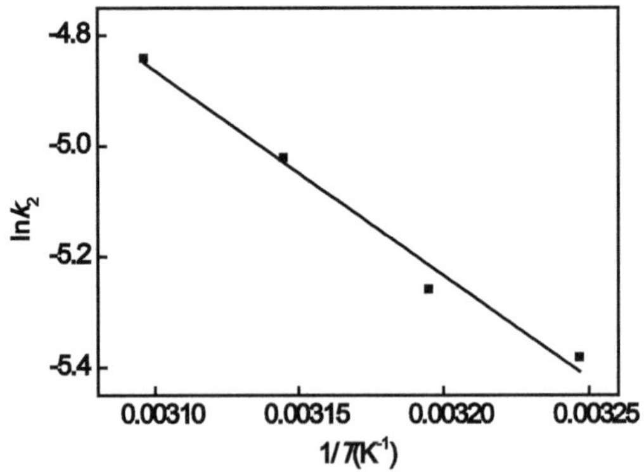

Figure 4. ln k_2 versus $1/T$ plot for the adsorption of LSC-600 for Ga(III).

3.3 Adsorption Isotherm

The adsorption isotherms of LSC-600 for Ga(III) were investigated at four different temperatures and the data were analyzed with Langmuir (4) and Freundlich (5) equations, respectively. Then, Figs. 5 and 6 were obtained.

$$\frac{C}{Q}=\frac{1}{bQ_0}+\frac{C}{Q_0} \tag{4}$$

$$\ln Q=\ln K_F+\frac{1}{n}\ln C \tag{5}$$

where Q is the adsorption capacity (mg/g), C the equilibrium concentration of Ga(III) (mg/L), Q_0 the saturated adsorption capacity (mg/g), b an empirical parameter, n the Freundlich constant, and K_F the binding energy constant reflecting the affinity of the resin for metal ion. As we known, the Langmuir equation is applicable to homogeneous adsorption, where the adsorption of each sorbate molecule on to the surface had equal adsorption activation energy. While, the Freundlich equation is employed to described heterogenous systems and reversible adsorption and is not restricted to the formation of monolayer.

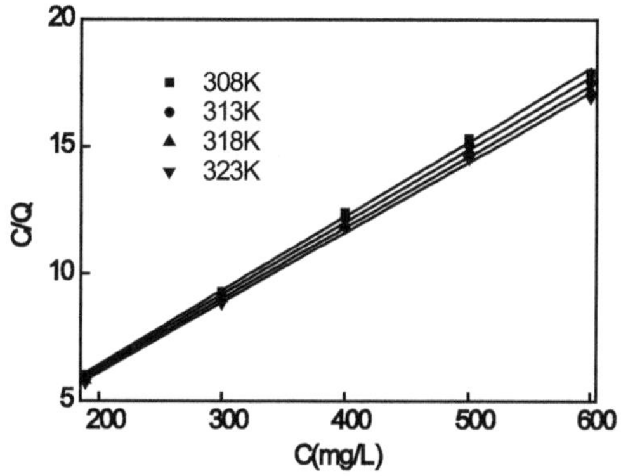

Figure 5. The Langmuir isotherms of LSC-600 for Ga(III).

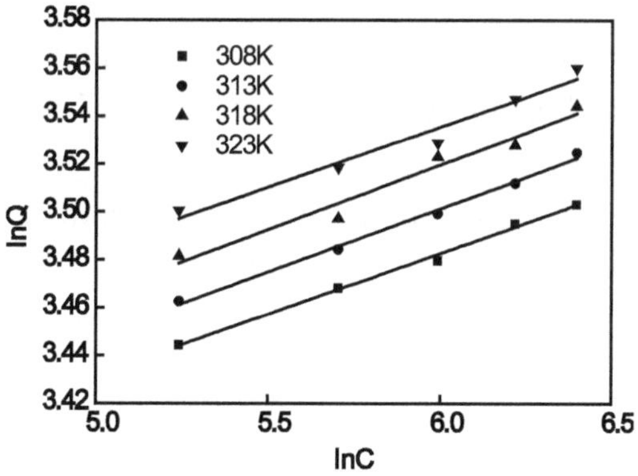

Figure 6. The Freundlich isotherms of LSC-600 for Ga(III).

The parameters for the two isotherms models obtained from Figs. 5 and 6 were summarized. It was found that the regression coefficients R^2 obtained from Langmuir or Freundlich models were greater than 0.9544, which implied that both Langmuir and Freundlich models could be used to describe the adsorption isotherms of Ga(III) on LSC-600 at the temperature range of 308-323 K. In addition, the values of n varied in the range between 1 and 10. This demonstrated that Ga(III) was easily adsorbed on the resin.

4 Conclusions

This paper reports the study of the adsorption properties of a strongly basic resin having a combination of one =NOH group and another active group $-NH_2$ for Ga(III) extraction. The influence of operational conditions such as contact time, initial Ga(III) concentration and temperature on Ga(III) adsorption were extensively investigated. The results revealed that the resin has high adsorption capacity and Ga(III) selectivity. The optimal adsorption condition was obtained at temperatures of 40-50°C and contact time of 40-60 min. The Ga(III) adsorption data on the resin fit well with the pseudo second-order kinetics. The apparent activation energies of adsorption *E*a calculated was 30.75kJ/mol. Langmuir and Freundlich models were used to describe Ga(III) adsorption isotherms on the resin.

Acknowledgements

This work was supported by the National Natural Science Foundation of China (51104002).

Reference

1. R.R. Moskalyk, "Gallium: the backbone of the electronics industry," *Minerals Engineering*, 16 (2003), 921-929.
2. W.L. Chou et al., "Removal of gallium (III) ions from acidic aqueous solution by supercritical carbon dioxide extraction in the green separation process," *Journal of hazardous materials*, 160 (2008), 6-12.
3. B. Gupta et al., "Separations and recovery of indium and gallium using bis(2,4,4-trimethylpentyl)phosphinic acid (Cyanex 272)," *Separation and Purification Technology*, 57 (2007), 294-303.
4. O. Font et al., "Recovery of gallium and vanadium from gasification fly ash," *Journal of hazardous materials*, 139 (2007), 413-423.
5. J. Poledniok, "Speciation of scandium and gallium in soil," *Chemosphere*, 73 (2008), 572-579.
6. J.E. Dutrizac, T.T. Chen, "The behaviour of gallium during jarosite precipitation," *Canadian Metallurgical Quarterly*, 39 (2000), 1-14.
7. D.O. Flamini et al., "Electrodeposition of gallium onto vitreous carbon," *Journal of Applied Electrochemistry*, 37 (2007), 467-471.
8. K. Xu et al., "Study on the recovery of gallium from phosphorus flue dust by leaching with spent sulfuric acid solution and precipitation," *Hydrometallurgy*, 86 (2007), 172-177.

9. R. Dumortier et al., "Removal and recovery of gallium from aqueous solutions by complexation with sodium di-(n-octyl) phosphinate," *Hydrometallurgy*, 76 (2005), 207-215.
10. Z. Fang, H.D. Gesser, "Recovery of gallium from coal fly ash," *Hydrometallurgy*, 41 (1996), 187-200.
11. J. Sturgill et al., "Pollution prevention in the semiconductor industry through recovery and recycling of gallium and arsenic from GaAs polishing wastes," *Clean Technologies and Environmental Policy*, 2 (2000), 18-27.
12. H.S. Lee, C.W. Nam, "A study on the extraction of gallium from gallium arsenide scrap," *Hydrometallurgy*, 49 (1998), 125-133.
13. X. Lu et al., "Research progress in gallium recovery technology," *Nonferrous Metals*, 60 (2008), 105-108.
14. A. Figueiredo et al., "Determination of lanthanides (La, Ce, Nd, Sm) and other elements in metallic gallium by instrumental neutron activation analysis," *Journal of Alloys and Compounds*, 344 (2002), 36-39.
15. Z. Zhao et al., "Recovery of gallium from Bayer liquor: A review," *Hydrometallurgy*, 125–126 (2012), 115-124.

PRE-DRYING *EUCALYPTUS SALIGNA* FOR CARBONIZATION

Marcelo Breda Mourao, Lina Maria Varon, Cyro Takano
Department of Metallurgical and Materials Engineering
University of São Paulo
05508-090 São Paulo, SP, Brazil

Keywords: Carbonization; Eucalyptus; wood charcoal ironmaking

Abstract

The growing search for forms of clean renewable energy has placed wood charcoal as a possible alternative to coal in the Brazilian steel industry. The carbonization of Eucalyptus Saligna is an important step in the production of charcoal. The aim of this study is to analyze the behavior of the endothermic and exothermic reactions involved in the carbonization process. The carbonization of logs of Eucalyptus Saligna has been performed in a stainless steel reactor at different temperatures (300 ° C, 400 ° C and 500 ° C) and the results were compared with the ones obtained through thermal analysis (TG / DSC) of small samples (36 mg and 54 mg) of the same wood. The TG/DSC results indicated that the process presents both endothermic and exothermic features, at different ranges of temperature and time, in connection with the degradation of the three major macromolecules contained in the wood: cellulose, hemicellulose and lignin.

1. Introduction

The steel industry intensively uses energy from fossil coals, causing significant environmental impact especially in the generation of CO_2, the main gas linked to anthropogenic greenhouse effect. The production of one ton of hot metal in a coke blast furnace releases between 1.7 and 1.9 tons of CO2, depending on the country's mode of energy production. Maintained current conditions and the mode of production, together with projected future increase in demand, it is estimated that in 2050 both steel production and the emission of CO_2 will double in relation to current levels [1].
Ongoing developments aim to drastically reduce this consumption, especially in the reduction step of iron ore, using innovative technologies in both the blast furnace and emerging processes. Another approach is the use of renewable energy, more specifically charcoal. In Brazil, there is a significant production of pig iron from charcoal blast furnace, mainly in small to medium scale plants. In order to increase the scale of wood charcoal production, the whole cycle of charcoal production need to be optimized. In this regard, the evolution of technologies in areas related to the production of charcoal, such as forests, wood drying and carbonization, can lead to a better economic and environmental sustainability of the steel from charcoal.
Those assigned to the planting of eucalyptus, are being expanded widely in Brazil, showing an increase of 7.1% between 2004 and 2009. This strong expansion in the amount of planted area is due to a number of factors, among which are: the rapid growth in a short rotation cycle, the high rates of photosynthesis by the plant. The increase in the number of research related to cultivation of eucalyptus forests with the goal of achieving a genetic improvement, has generated an increase in the growth rates, a greater adaptability of the plant to the climatic conditions of the soil, the increase in the quality of wood, as well as a greater tolerance to pests and diseases [2].

In this sense, one could argue that the genus Eucalyptus, shown quite favorable to the industry of carbonization, because it allows the selection of adequate and homogeneous material for the production of charcoal, in addition to being highly productive. To Bacha and Barros [4], the preference for the eucalyptus for the production of charcoal to be used for steel is due to the fact that it has a higher density, compared for example with a pine wood. The average values of the densities of the woods of eucalyptus related in the literature vary between 0.61 g / cm ^{3} (Eucalyptus grandis) and 0.81 g / cm ^{3} (eucalyptus tereticornis) as pine (Pinus taeda da) has a lower average density the order of 0.46 g / cm ^{3} [5].

Due to the importance of using wood charcoal in the steel industry and the prospects in Brazil for the production of charcoal from eucalyptus forests, and the critical role played by the variation in wood properties on the quality of charcoal, there is a great expectation to study each stage of production of charcoal.

This paper focuses on the carbonization of *Eucalyptus Saligna* at variable temperature with different heating rates looking to contribute with additional information about the heat effect of the reactions that occur in this process. In order to do this, samples were carbonized in a stainless steel reactor and compared with the results of TG/DSC analysis at two different heating rates.

The carbonization process was analyzed by Naso [2010], who proposed a theoretical DSC curve, based on the data supplied independently by the DSC reaults for the three macromolecules. It was considered dry eucalyptus with a typical composition of 50% cellulose, 27% lignin and 23% hemicelluloses. This study did not take into account the interactions between the macromolecules, which must be broken and require energy. Table 1 summarizes the determined type of heat effect as a function of the temperature.

Table 1 –Heat effect of the carbonization process as a function of the temperature, as determined by Naso [2010].

Heat effect	[T°C]
Exothermic	200-307
Endothermic	307-392
Endothermic	536-735
Exothermic	735-800

2. Experimental

Samples of *EUCALYPTUS saligna* were provided by the experimental station of the Agriculture School Luiz de Queiroz (Esalq). The samples, originated from seven years old trees, measured one meter long with different diameters. After the arrival to the Department of Metallurgy and Materials USP, the wood was placed in a covered place where it remained at ambient conditions. The carbonization experiments were conducted in a stainless steel reactor which is heated by electrical resistance. Samples were selected and cut in 0.20 m length and diameters between 0.14 and 0.15 m. Before carbonization, the samples were dried in an oven at 110 ± 5 ° C for 24 hours. A tube was set in the reactor lid to allow the exit of the gases released during the carbonization process. A thermocouple was placed inside the reactor to measure the temperature. The reactor with the wood was placed in the furnace turned off and at room temperature. The rate of heating varied between 3.5 ° C / min and 6.2 ° C / min. The furnace was turned off two hours after the reactor temperature reached the specified maximum value (300, 400 or 500 ° C). The reactor was then cooled to room temperature for collection of the samples.

The thermal analyses were carried out on a NETZSCH STA 409C equipment. The weight of the samples ranged between 36mg and 54 mg; the samples were heated to 950 ° C at rates of 10 ° C/min and 5 ° C/min under an argon flow of 50 ml/min.

3. Results and discussion

Fig. 1 In the carbonization experiments using a stainless steel retort heated by electrical resistance, time and temperature data were collected from measurements taken inside the retort. Samples heated in the furnace at temperatures of 300, 400 or 500 °C contained 19% moisture on a dry basis, i.e. at the equilibrium moisture content.

Furthermore, we conclude that apart from the temperatures measured during such experiments (300, 400 or 500 °C), a peak temperature of 500 °C – defined as the maximum temperature observed in the carbonization process - was eventually reached, as shown in Fig.1. Table 2 shows the results of yields obtained for different temperatures. It can be seen that the values of the charcoal yield are similar for the three different nominal temperatures employed. Close results were achieved on account of exothermic reactions that occurred during the carbonization process. Fig. 2 shows that despite great differences in furnace temperatures, real temperatures from inside the retort were very close to each other, and almost reached the same peak of 500 °C.

Table 2 – Technical data from the carbonization of eucalyptus logs at different constant temperatures in a stainless steel retort heated in an electric pit furnace

Operational temperature, °C	300	400	500
Dry wood (g)	1686,6	1720,6	1673,4
Charcoal (g)	518,4	571,3	517,3
Charcoal yield	30,7%	33,2%	30,9%

Consequently, the actual carbonization temperature of the samples seemed to bear no relation to that of the furnace. Carbonization maximum temperatures got really close to each other and were similarly high, about 500 °C, resulting in slight variations of both charcoal yields and their respective fixed carbon contents at 400 and 500 °C.

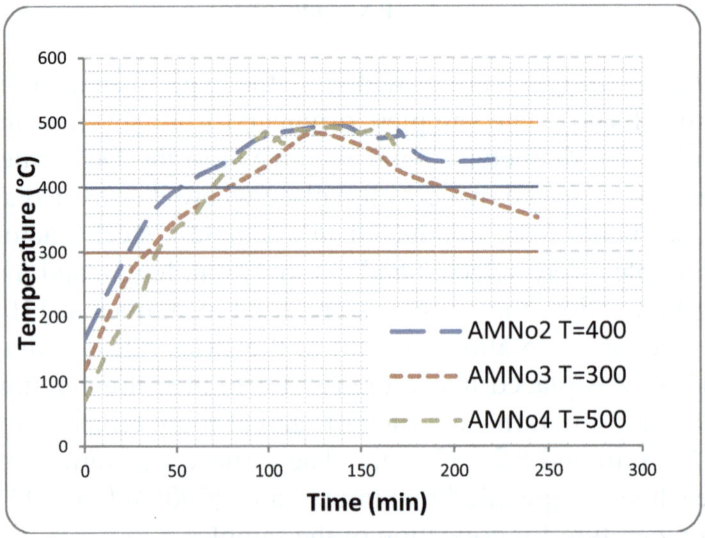

Fig. 1 - Carbonization of Eucalyptus logs ranging from 14 cm to 15 cm in diameter, 20 cm long, at different constant temperatures in a stainless steel retort heated in an electric pit furnace

As the temperature inside the retort can rise above those set by the furnace temperature controller, it can be concluded that the overall carbonization process is exothermic. However, due to the large size of the samples and the possibility of overlap of the several reactions occurring during the overall process, further investigation is necessary.

TG/DSC thermal analyses carried out at a heating rate of 5 ºC/min and 10 ºC/min has allowed to observe mass loss during carbonization, as shown in Fig. 2, as well as to identify the variation in mass loss rate as a function of time.

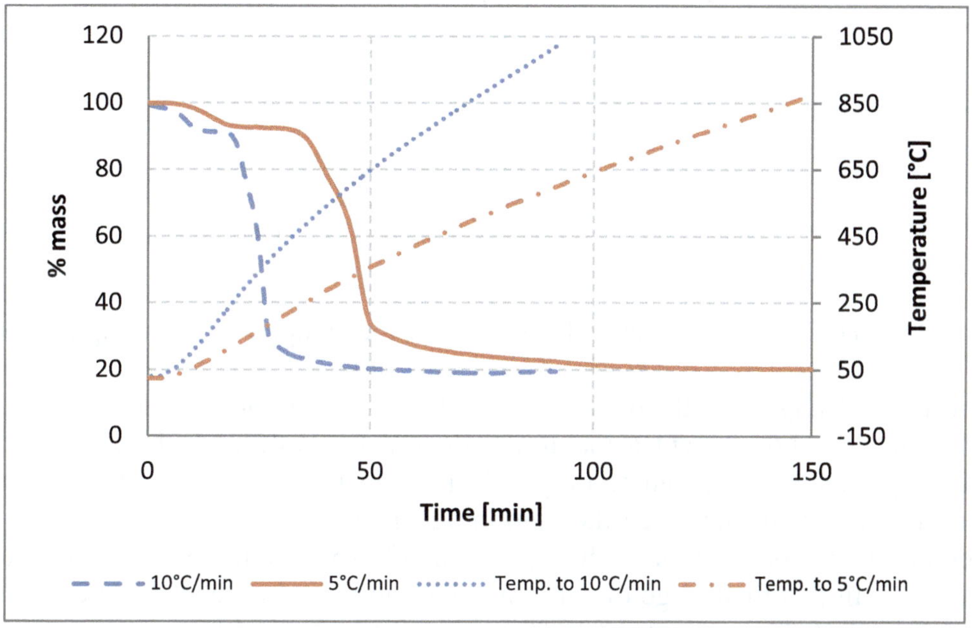

Fig. 2 – TG curves for Eucalyptus saligna wood samples at different heating rates

The mass loss as a function of time for both heating rates present similar behavior, although they are shifted along the horizontal axis. Both curves indicate a mass loss between the initial temperature and that of 103 ºC. This fact corresponds to the loss of hygroscopic water in the sample, accounting for 7,18% of its initial weight.

After loss of water, and until 250°C, there is no mass loss, for both heating rates. Around 250°C starts the hemicellulose degradation and production of oxygenated gases. This fact is shown in Fig. 3 by the first common peak for both heating rates of the DTG curves. The second peak, at around 355 ºC, was associated mainly to cellulose degradation and the production of hydrocarbons. This stage, which presented the maximum degradation rate, ended at 500 ºC with a mass loss of 22% for both samples, irrespective of the rate. Moreover, there was hydrocarbon dissociation and H2 production after 500 ºC. We could also observe a constant loss rate above this temperature.

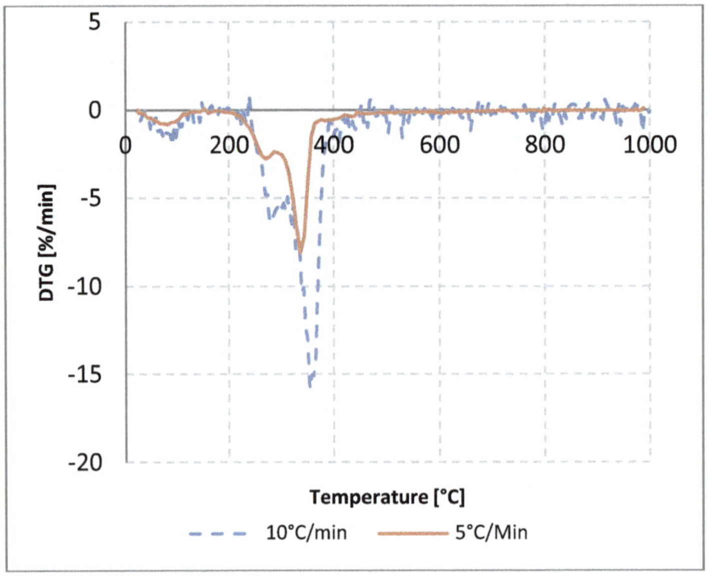

Fig. 3 - DTG curves for Eucalyptus saligna wood samples at a heating rate of 10°C/min and 5°C/min

According to Yang et al., the degradation of the third macromolecule that constitutes eucalyptus, i.e., lignin, does not present a maximum degradation rate and, therefore, cannot be identified within a particular temperature range. That happens because the lignin slow decomposition occurs at a constant rate since the beginning of the process. The maximum content of carbon presented in charcoal and the highest point in H2 production were both reached between 700 °C and 900 °C. In this final stage of carbonization, no mass loss was detected. At 950 °C, there was a mass loss of 19,36% at a rate of 10 °C/min, and 19,42% at 5 °C/min.

When comparing the decomposition of the three different macromolecules separately by DSC analysis, as shown in Fig. 4, Yang et al., it was possible to verify their endothermic and exothermic behaviors for different temperature ranges. However, when these three molecules were combined into wood, their endothermic and exothermic behaviors were more hardly perceptible. Thus, in order to clarify what actually happens during the carbonization process, it was performed a comparison between the behavior of each individual molecule, as in Fig. 4, with DSC results for the wood at two different carbonization rates, as shown in Fig. 5.

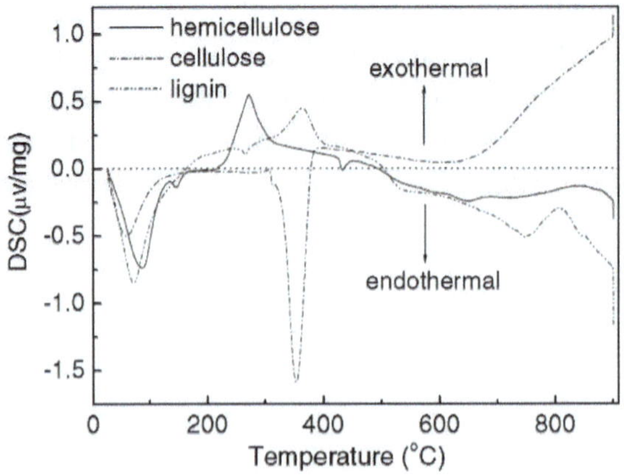

Fig. 4 - DSC curves for hemicellulose, cellulose, and lignin obtained from Yang et al.

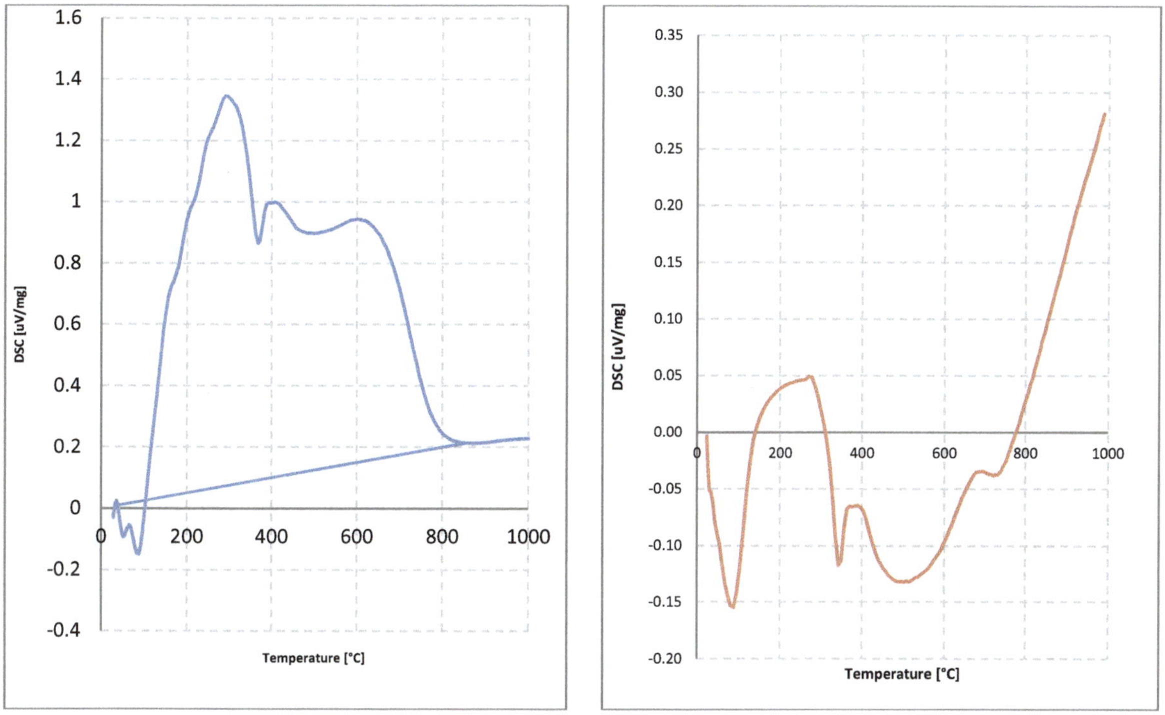

a) b)

Fig. 5 – DSC curves for Eucalyptus saligna wood samples at a heating rate of 10°C/min and 5°C/min

Fig. 4 shows endothermic behavior for all three components up to 200 ºC, as a consequence of moisture extraction from the samples. This behavior was displayed by *eucalyptus saligna* up to 103 ºC, as summarized in Fig. 5a and 5b. After such extraction, the sample carbonized at 10 ºC/min showed an exothermic behavior up to about 835 ºC (Fig. 5a), temperature at which it shifted to an endothermic range.

Fig. 5b shows that it is possible to perceive different endothermic and exothermic reactions by lowering the heating rate. The endothermic phase of moisture extraction was observed from the initial temperature up to 137 ºC, followed by an exothermic phase up to 312 ºC. Soon afterwards, a new endothermic phase took place up to 783 ºC. Under such carbonization conditions, the number of peaks of DSC curves is the same as for a heating rate of 10 ºC/min. Consequently, it is possible to conclude that it is necessary to lower the heating rate to determine the temperature ranges in which exothermic and endothermic reactions occur. It is worth pointing out that the theoretical behavior described in this study can be truly achieved only by lowering the heating rate, since increases in it might lead to inaccurate thermal control. Empirical evidences regarding endothermic and exothermic peak temperatures may differ according to the heating rate employed in the experiment.

4. Conclusions

A peak temperature of 500 ºC was achieved inside the retort during the carbonization process of samples measuring 20 cm long, and ranging from 14 cm to 15 cm in diameter, regardless of the furnace temperature previously set (300 ºC, 400 ºC or 500 ºC). In such temperature range, we clearly identified the occurrence of an exothermic carbonization reaction. Nevertheless, for TG/DSC analyses of wood samples measuring from 36 mg to 54 mg long, it was observed different exothermic and endothermic reactions for different temperature and time ranges, enabling some relation with the degradation of three main macromolecules that constitutes the wood: cellulose, hemicellulose, and lignin.

Acknowledgements
The authors would like to thank CNPq (National Center of Technological Development)

References

1. Harada, T.; Tanaka, H.: *ISIJ Int.,* vol.51 (2011), No. 8, pp. 1301-1307
2. SBS– Sociedade Brasileira de Silvicultura, 2008. São Paulo, 2008. Disponível em: <http://www.sbs.org.br/FatoseNumerosdoBrasilFlorestal.pdf>.
3. Naso P.G, Carvão vegetal: análise do processo de carvoejamento. Disertation in portugues..
4. Yang, H., Yan, R., Chen, H., Lee, D.H., Zheng, C. Fuel. 2007, Vol. 86, pp. 1781-1788.
5. Brito, J.O; Barrichelo G.L.E. IPEF: Instituto de Pesquisas e Estudos Florestais n.20, p.101-113, jun.1980.
6. Bacha, C. J. C.; Barros, A. L. M. de, 2004. Scientia Forestalis, n. 66, p. 191-203
7. Tremblay, C.: Cloutier, A.; Fortin, Y. Wood Science and Technology.New York, v. 34, 2000[a]
8. Vasconcelos, S.G; Jankowsky, R; Pontes, I; Ariel, A. Scientia Forestalis n. 63, p. 214-220, jun. 2003Basson, E.: *The steel industry in a sustainable society*, World Steel Association, April 2012

REWAS 2013

Enabling Materials Resource Sustainability

Enabling Sustainability through Recycling & End-of-Pipe Solutions I

Session Chairs

Markus Reuter
Mark Schlesinger

REWAS 2013: Enabling Materials Resource Sustainabili
Edited by: Anne Kvithyld, Christina Meskers, Randolph Kirchain, Gregory Krumdick, Brajendra Mishr
Markus Reuter, Cong Wang, Mark Schlesinger, Gabrielle Gaustad, Diana Lados, and Jeffrey Spangenberg
TMS (The Minerals, Metals & Materials Society), 201

THERMAL PROCESSING OF INDUSTRIAL ASHES FOR FERROVANADIUM PRODUCTION

Yanping Xiao[1]*, Yongxiang Yang[2], Alan Lai[2] and Rob Boom[2]

[1]School of Metallurgy and Resource, Anhui University of Technology
Hudong Rd 59, 243002 Ma'anshan, China
[2]Department of Materials Science and Engineering,
Delft University of Technology, Mekelweg 2, 2628 CD Delft, the Netherlands

Keywords: Petroleum fly ash, BOF flue dust, Thermal processing, Ferrovanadium (FeV)

Abstract

Petroleum fly ash generated from heavy oil fired power plant contains significant amount of vanadium and carbon. In the present study a recipe-based concept is introduced for metal recovery by making use of two waste ash residues: basic oxygen furnace (BOF) steelmaking flue dust, and petroleum fly ash. The carbon contained in the fly ash is used as the reductant and BOF flue dust provides iron source and partially slag formers. A ferrovanadium alloy (FeV) with 15-20% vanadium was produced through reduction smelting of the two ash mixture at 1550-1600°C in the lab . Some impurities (Ni, S and C) remain in the metal. The carbon to metal ratio has the largest effect on the metal quality. The slag properties such as basicity are important for metal yield and metal quality. The results prove that a recipe-based multi-waste co-processing can be a sustainable solution for converting wastes to valuable raw materials.

Introduction

Each year approximately 280,000 tons of vanadium is extracted from the earth either as metal bearing ores or vanadium containing oil. The annual vanadium production is approximately 61,000 tons in 2010, according to USGS [1]. Vanadium metal, mainly in the form of ferrovanadium, is primarily used as an alloying element in steel, which consumes 85% of the world's vanadium. The addition of vanadium increases the strength, hardness and fatigue resistance of the steel. The vanadium refines the grain size which increases the strength of the steel and also forms carbides which increase the hardness [2, 3]. There are many varieties of high speed tool steels with vanadium contents in the range of 0-5wt%. Adding vanadium in the micro-alloyed range up to 0.15wt% will increase the tensile strength of the steel [4]. Vanadium is also used as an alloying element in titanium, especially for aircraft alloys accounting for about 10% of the usage. The remaining 5% of vanadium usage is split into various chemical uses. Vanadium pentoxide (V_2O_5) is an important catalyst in sulfuric acid production [2]. New developmental batteries known as vanadium redox batteries use vanadium containing electrolytes. They have potential in large power storage applications [1]. There are also vanadium applications as protective materials in nuclear reactors [5]. However these applications do not consume large quantities of vanadium.

Most ferrovanadium (containing 45, 50 and 80% V) is produced using V_2O_5 as a feedstock. Aluminothermic reduction is used to produce 80% V ferrovanadium [1,2]. A mixture of V_2O_5, aluminum, iron (or iron oxide), and lime are mixed together [5]. The reduction of vanadium and iron with aluminum is a highly exothermic reaction so that the reaction is self-sustaining once started [6]. The reaction can be started with a trigger or simply by heating the charge to a sufficiently high temperature in an electric arc furnace. Production of lower grade ferrovanadium (45 and 50% V) uses a silicothermic reduction technique [1, 2].

The primary resource of world's vanadium comes from mined ore directly as mineral concentrates derived from vanadiferous titanomagnetite (VTM). Secondary sources include vanadium rich steelmaking slags (where the steel is produced from VTM), petroleum fly ash, spent catalyst, and bauxite sludges [1, 7]. During the steel production process using vanadiferous titanomagnetite ore, a vanadium rich slag is formed in the basic oxygen furnace. This contains approximately 12-24% vanadium [2]. Some crude oils in the world can contain vanadium varying between 10 ppm to 1400 ppm [7]. Specifically Venezuelen oils can contain up to 500 ppm of vanadium [8]. Oil distillation is a process where many fractions can be produced such as gasoline, kerosene, lubricants etc. One of the heavier fractions is known as heavy fuel oil, and inside this fraction most of the vanadium and other metals are concentrated. It is a thick viscous liquid that is used for power generation due to its high energy value. The heavy fuel oil is only burned in special power plants and on some ships. When it is burned it produces two kinds of ash: bottom ash and fly ash. Bottom ash is the heavy ash that remains on the bottom of the furnace. Fly ash is the light ash that is collected in the furnace off-gas cleaning system. The metals, specifically vanadium and nickel, are concentrated in the fly ash. The final concentration of vanadium in the fly ash ranges between 2% and 40% as V_2O_5. This fly ash is also known to contain large amounts of unburned carbon from incomplete combustion. Vanadium can also be found in small quantities in bauxite ores. The bauxite is refined to produce aluminum oxide in the Bayer process, and the Bayer sludge as the production waste can contain about 20% V_2O_5 [5].

Nearly all current vanadium recovery from heavy oil fly ash is made through hydrometallurgical methods. There have been a few attempts with pyrometallurgical treatment [9]. One of the methods for recovering vanadium from heavy oil fly ash was to simply burn off the carbon to concentrate the vanadium oxides. This method utilized a pretreatment step to remove carbon and sulfur. The pretreatment temperature was 1000°C which resulted in a sulfur free fly ash. This carbon and sulfur free fly ash was then mixed with either aluminum or ferrosilicon or both to act as the reducing agent. The main fluxing agent was lime but other additions were also investigated. The resulting alloys contained up to 15% vanadium and up to 6% nickel. There were large amounts of silicon in the final product ranging between 14-48%. The high silicon content is from the ferrosilicon used to provide a source of iron and as a source of silicon as reductant. A recovery of 89% vanadium was achieved and the whole process was carried out with a 100 kVA DC arc furnace with a charge of about 150 kg of fly ash. Another method of recovering vanadium from petroleum fly ash combines the fly ash with basic oxygen furnace (BOF) steelmaking flue dust to produce a ferrovanadium alloy [10]. The BOF dust contains mostly iron oxides, metallic iron, calcium carbonate and zinc oxide. This method allows two waste products, heavy fuel oil fly ash and BOF dust, to be combined to produce a valuable product of ferrovanadium. This can be considered more efficient processing because it will utilize carbon, vanadium and iron contained in waste products.

In the present study, experimental investigation has been conducted to recover the metal values for FeV production from two industrial waste resources (fly ash and flue dust) with a pyro-metallurgical treatment. The major component in the petroleum fly ash is $VOSO_4 \cdot 3H_2O$, which is considered as an important source of vanadium, instead of the traditional V_2O_5 used for FeV production. The high FeO_x and CaO contents in the BOF flue dust provide the necessary secondary resource for alloying and slag formation. The significant amount of carbon in the petroleum fly ash could serve as the reductant at least in the pre-reduction stage. A comprehensive recipe-based processing route is illustrated in Figure 1 for processing the 2 waste streams from the power and metallurgical industry so as to close the metals cycle.

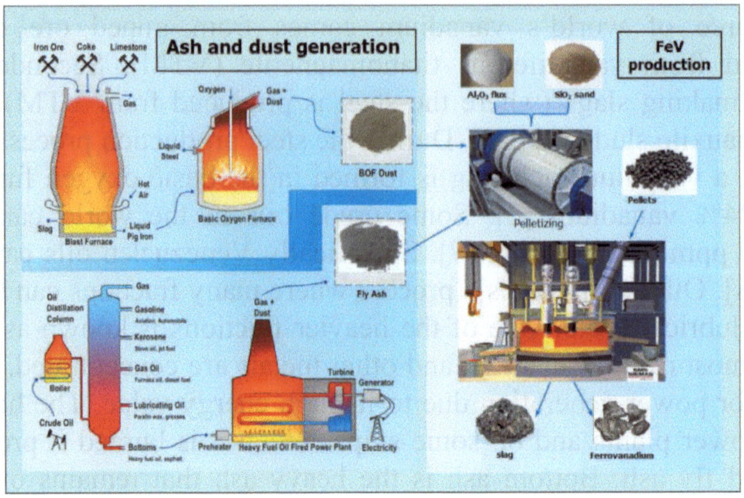

Figure 1. Concept flowsheet of direct pyrometallurgical production of ferrovanadium from petroleum fly ash and BOF flue dust

Experimental

Raw Materials and Characterization

Fly ash was provided by Greenshores N.V. from a heavy oil fired power plant located in Curaçao. The plant burns the heavy oil extracted from Venezuela, and the ash was collected from the dust collection system and was a fine black powder. The fly ash was dried at 105°C for 24 hours to remove moisture in the lab before any processing. Additionally some of the fly ash was put through two different pretreatments intended to remove carbon and sulfur. The first pretreatment was conducted for 6 hours in an oven in air at 400°C. The second was conducted for 6 hours in an oven in air at 600°C. The dried as-received ash will be referred to as FA-105-24hr, the 400°C pretreated ash as FA-400-6hr, and the 600°C pretreated ash as FA-600-6hr. Basic oxygen furnace dust was provided by Tata Steel Netherlands. It was the coarse fraction from the dust collection system and was light brown in color. The dust was dried at 105°C for 24 hours to remove moisture.

The moisture content of the FA-105-24hr and BOF dust was determined by heating the samples in a drying oven at 105°C for 24 hours and measuring the mass loss after drying. Additionally the mass loss during the fly ash pretreatments was also recorded. In order to determine the chemical and mineralogical composition of the fly ash and the BOF dust several characterization techniques were used. General chemical composition was determined with X-ray fluorescence spectroscopy (XRF) using a Phillips model PW2400 wavelength dispersive spectrometer. X-ray diffraction (XRD) was used to examine the mineralogical composition of the samples using a Siemens model D5005 diffractometer. LECO combustion was used to determine the carbon and sulfur content. To determine the physical structure of the ashes and dust they were mounted in epoxy and polished to be examined with an optical microscope and a scanning electron microscope (SEM). A small amount of ash or dust was mixed with epoxy and allowed to set so that the ash or dust was suspended in the epoxy. Then the hardened epoxy disc was polished by hand using ethanol and ethylene glycol as the lubricating fluid. This was done in order to reduce any leaching or decomposition from water during polishing. After polishing micrographs were taken with an optical microscope and a JEOL JSM 6500F scanning electron microscope. Prior to SEM analysis the samples were coated with carbon to make them conductive. Both microscopes revealed the microstructure of the samples.

Smelting Tests

Smelting tests were carried out in a vertical tube furnace using a protective argon atmosphere to prevent the samples from oxidizing. Different amounts of fly ash, BOF dust, and fluxing agents were mixed together and put into an Alsint crucible. This crucible was then placed into another Alsint holding crucible to protect the furnace in case the inner crucible failed. A lid was then put over both crucibles however it was not air tight and allowed gas to enter and escape. The crucibles with the charge inside were loaded into the furnace and the argon flow was started. The argon gas created an over pressure in the furnace and gradually removed all the air from the furnace. The furnace was heated to the desired temperature and allowed to stay for 2 hours before cooling commenced. After the furnace was cooled below 100°C the crucibles were removed. The metal was separated from the slag and weighed. The metal and a portion of the slag were mounted in epoxy and polished for microstructural analysis using SEM and compositional analysis with energy dispersive spectroscopy (EDS). Some of the slag was also milled into a powder for XRF analysis. After SEM analysis some of the metal samples were removed from the epoxy and analyzed with XRF, XRD, and LECO combustion. Three temperatures 1558°C, 1600°C, and 1635°C were used for the smelting tests (designated for 1550, 1600 and 1650°C). To investigate the effect of smelting conditions, the amounts and ratios of the charge materials were varied to examine the effect on the metal recovery and metal and slag quality, as given in Table 1. The carbon to metals (C/M) ratio in the charge is defined as the carbon in charge divided by the stoichiometric moles of required carbon, based on the reducible oxides compositions including Fe, Ni and V. The BL04 was designed to test the effect of C-free fly ash on the smelting performance.

Table 1: Smelting experimental conditions

Sample No.	**C/M ratio**	**Charge and flux Mass, g**						**Temp. °C**
		FA-105 – 24hr	**BOF Dust**	**SiO_2**	**Al_2O_3**	**CaO**	**FA-600 – 6hr**	
AL01	1.8	4	14	7				1635
AL02	1.8	4	14	7				1635
AL03	1.8	4	14	3	3			1600
AL04	1.8	4	14	3	3			1595
BL01	2.4	6.5	14	3	3			1595
BL02	2.4	6.5	14	3	3			1600
BL03	2.4	6.5	14	3	3	2		1600
BL04	2.2	8	14	3.6	3.6	1.6	2	1600
BL05	2.4	4.9	10.5	3.5	3.5	3.4		1600
CL01	3.1	16	20	12				1600
CL02	3.1	16	20	12				1558
CL03	2.8	9	14	7				1558
CL04	2.8	9	14	7				1635

Results and Discussion

Characterization

The moisture content and mass loss data from the pretreatment of the fly ash and the BOF dust are shown in Figure 2. During the carbon burning-off pretreatment, the carbon concentration decreased from the original 36.5% (dried sample) to 12.6% for the FA-400-6hr sample, and to zero for the FA-600-6hr sample. The sulfur content remained almost unchanged at 400 °C, but decreased slightly from 12.3% to about 10% at 600 °C. All the metals increased in concentration with higher temperature treatments, and vanadium increased from 27% to 37% at 400 °C, and to 43% at 600 °C. The XRD results indicate that the oxidation state of vanadium changed from 4^{+} to 5^{+} at higher temperature due to the oxidation in air. It is also important to notice that during the pretreatments significant amount of the valuable elements were also lost together with carbon burn-off based on the mass balance, for example with up to 50% loss of vanadium.

Table 2: Mass loss and contents of carbon and sulfur under different pretreatment conditions

Sample	Test	Mass Loss wt %	C, wt%	S, wt%	Color
FA-105-24hr	Drying	14.2	36.5	12.3	Black
FA-400-6hr	LOI	51.9	12.6	13.2	Green-Brown
FA-600-6hr	LOI	63.1	<0.05	9.8	Chocolate Brown
BOF Dust 105-24hr	Drying	16.4	0.6	<0.002	Light Brown

(a) FA-105-24hr (b) FA-400-6hr (c) FA-600-6hr (d) BOF Dust 105-24 hr

Figure 2: Processed fly ashes and BOF dust

Table 3 lists the compositions of the petroleum fly ash (FA) and the BOF dusts (after drying at 105°C for 48 hours). The fly ash contains mainly vanadium (27%), sulphur (12%), and nickel (6%). The BOF flue dust collected from a wet scrubber (a coarse fraction) after drying contains 42% calcium and 36% iron. According to the XRD analysis, vanadium in the petroleum-fly ash occurs mainly in the form of $VOSO_4 \cdot 3H_2O$, and the form of nickel is identified as $NiS_2O_6.6H_2O$. Significant amounts of carbon, and certain amounts of $MgSO_4 \cdot 6H_2O$ and $CaSO_4$ were detected. According to LECO analysis, the carbon content is 36.5 wt%, and sulphur content is 12.3 wt%. The main crystalline phases identified with XRD for the BOF dust (coarse fraction) is $Ca(OH)_2$, followed by $CaCO_3$, Fe_2O_3, FeO, and certain amounts of MgO.

Figure 3 shows the micrographs of the dry fly ash and the BOF flue dust, taken at 50x magnification. The fly ash showed a cenosphere structure composed mostly of carbon, which is also reported in literature [11]. The bright colored particles in the fly ash could be the metal compounds, which is close to the limit of EDS identification. The BOF dust was a much less homogenous material, compared to the fly ash. It contains metallic particles, oxides, and mixed metallic-oxide particles. The metallic particles show up as bright white in the micrographs and oxides appear as different shades of grey. Some of the out-of-focus areas are actually below the surface of the epoxy disk but are still visible through the epoxy.

Table 3. Compositions of the petroleum fly ash and the BOF flue dusts

Element, wt%	Fe	V	Ni	Si	Ca	Al	Mg	S	P	Na	K
FA-105-24hr	0.767	27.56	5.925	0.269	0.788	0.058	2.645	12.75	0.023	1.386	0.080
BOF Dust	36.03	0.033	0.008	0.705	42.55	0.222	7.400	0.049	0.008	0.053	0.045
Element, wt%	**Ti**	**Cr**	**Mn**	**Zn**	**Pb**	**Cu**	**Cd**	**Cl**	**C**	**S**	**SUM**
FA-105-24hr	0.133	0.000	0.022	0.052	0.059	0.049	0.000	0.000	36.5	12.3	89.1
BOF Dust	0.070	0.015	0.625	0.193	0.011	0.010	0.000	0.027	0.570	0.002	88.6

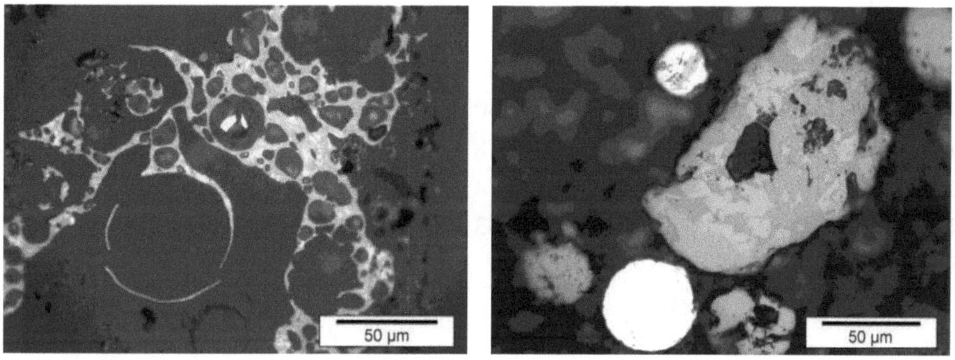

Figure 3: Microstructure of fly ash (left) and BOF dust (right)

Smelting

Various smelting conditions were investigated. The charge inputs and ratios were varied as well as the temperature. Table 4 gives the results from the smelting experiments under different conditions. It is clear that higher temperature (1600°C or above) is required in order to obtain a good metal-slag separation. The two tests conducted at 1558°C gave not well separated metal phase, and the metal beads were dispersed in the slag. The metal recovery (defined as total metal obtained respect to theoretical maximum reducible metals) varies from 52% to 92% at the smelting temperatures above 1600°C. The metal recovery was is defined as the mass of the metal recovered divided by the mass of total reducible metals (Fe, V, Ni) in the charge. The metal and slag separation depends also on the viscosity of both the metal and slag. Another way to improve the metal recovery is by adjusting the slag composition.

One of the main criteria for the product quality was how well the metal and slag were separated. As it is shown from the results in Table 4, the metal recovery is a rough indication because the metal product contains also significant amount of other elements, mainly carbon, silicon, and sulfur which would bring error in the recovery calculation.

For the metal sample from smelting test CL04, the SEM microstructure is shown in Figure 4. The carbide particles contain on average 88wt% V, 6wt% Fe, 1wt% Ni and 5.5wt% C. Comparing the carbon to metal ratio with the metal composition and the microstructure of the metal, it can be found that higher carbon to metal ratio leads to a higher vanadium content in the recovered metal. This is expected since more carbon will contribute to increase the vanadium reduction in the charge. Based on the slag analysis it was found that often sulfur was associated with calcium in almost equal mass proportions, which indicates the formation of compound CaS. The calcium to sulfur ratio (Ca:S) in the charge was calculated as a means of quantifying the amount of calcium in the charge. Figure 5 shows the near linear trend of relation between CaO in slag and sulphur concentration in the metal, with increasing the CaO in slag the amount of sulphur will be reduced in the metal (for the tests AL3, AL4 and BL1-4).

Table 4: Metal yield and compositions from smelting tests

Sample No.	C/M ratio	Metal composition wt% (SEM)						Metal composition wt% (XRF)					Recovery metal, %
		Fe	V	Ni	S	C	Si	Fe	V	Ni	S	C (LECO)	
AL01	1.8	76	6	0	12	6	0						89
AL02	1.8	82	6	2	11	0	0						92
AL03	1.8	87	3	2	7	1	0	89	2	1	7	0.02	81
AL04	1.8	83	0	3	11	2	0	81	0	1	9	0.3	52
BL01	2.4	77	13	3	4	3	1	80	8	2	2	1.6	88
BL02	2.4	71	15	5	7	5	0	78	11	1	7	1.6	85
BL04	2.2	68	18	4	5	4	0	74	15	3	4	1.2	81
BL05	2.4	74	0	3	15	8	0	83	0	2	12	0.5	54
CL04	2.8	63	23	4	1	4	6						86

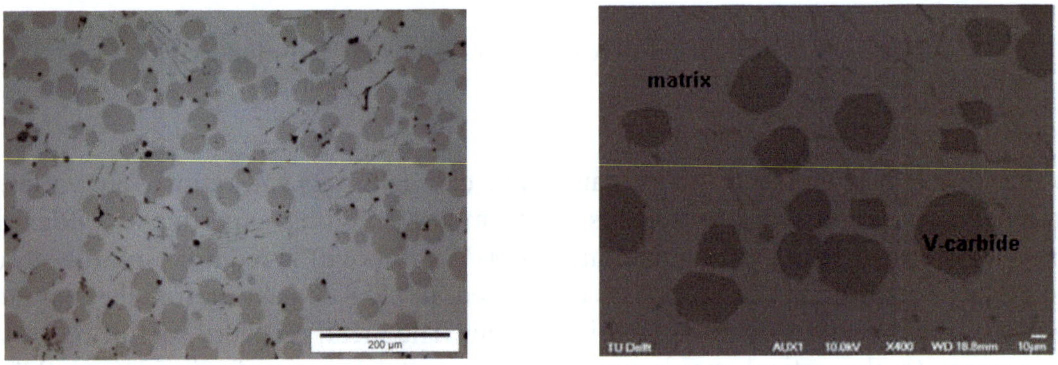

Figure 4: Microstructure of metal sample from test CL04.

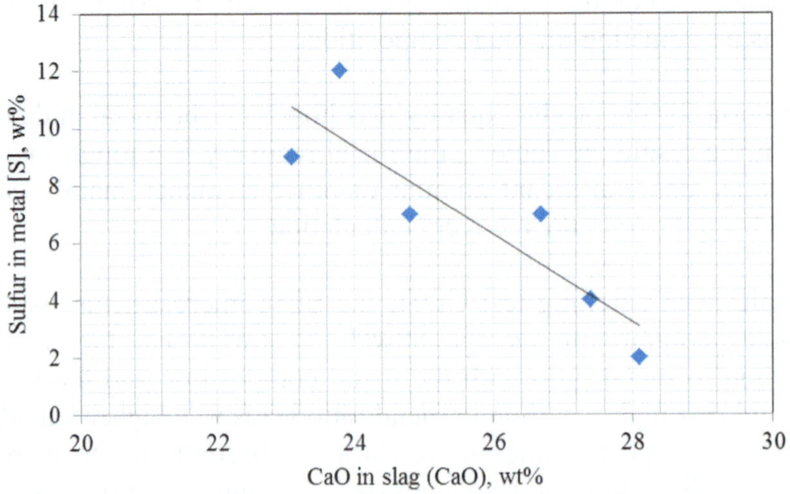

Figure 5: Relationship between CaO in slag and S in metal

The objective of test BL04 was to use C-free fly ash to increase the vanadium content in the metal product, in comparison to test BL02. The results show that the vanadium content in the metal increased from 11% to 15%, and the sulfur dropped from 7% to 4%.

In general, comparing to the previous results of 18 – 20wt% V in the recovered metal [10], the present vanadium loss during the smelting is high. This may be caused by the feed preparation during the smelting. The effect of charge compactness is significant, and pelletizing of the powder mixture used in previous study [10] can promote the vanadium reduction by carbon, thus increasing the metal recovery.

Evaluation of the Reaction System

Main Reactions and Thermodynamic Analysis

Various reaction stages were identified based on the thermodynamic predictions. Detailed thermodynamic analysis of the reaction systems have been published previously [10, 12] for direct FeV production from the mixture of petroleum fly ash BOF steelmaking flue dust. Below is a summary of 3 types of possible chemical reactions: (1) thermal decomposition of sulphates, (2) solid state carbothermic reduction, and (3) smelting reduction above melting temperatures of the ash and slag.

- Dehydration and sulphates decomposition of the petroleum fly ash: <1000°C

$$VOSO_4 \cdot 3H_2O = VO_2 + SO_3\,(g) + 3H_2O\,(g) \quad (1)$$
$$NiSO_4 \cdot 6H_2O = NiO + SO_3\,(g) + 6\,H_2O\,(g) \quad (2)$$
$$SO_3\,(g) = SO_2\,(g) + 0.5\,O_2\,(g) \quad (3)$$

- Solid-gas pre-reduction: 1000 – 1400°C

$$FeO + CO\,(g) = Fe + CO_2\,(g) \quad (4)$$
$$VO_2 + 2CO\,(g) = V + 2CO_2\,(g) \quad (4a)$$
$$NiO + CO\,(g) = Ni + CO_2\,(g) \quad (4b)$$
$$CO_2\,(g) + C\,(\text{residue in ash}) = 2CO\,(g) \quad (5)$$

- Smelting Reduction: >1400 °C (assuming ash melting and slag formation at 1400°C)

$$(FeO) + CO\,(g) = Fe_{(l)} + CO_2\,(g) \quad (6)$$
$$VO_2 + 2CO\,(g) = [V]_{Fe} + 2CO_2(g) \quad (6a)$$
$$NiO + CO\,(g) = [Ni]_{Fe} + CO_2(g) \quad (6b)$$
$$CO_2\,(g) + [C] = 2CO\,(g) \quad (7)$$
$$SiO_2 + CaO = 2CaO \cdot SiO_2 \quad (8)$$
$$[V] + C = VC \quad (9)$$

Figure 6 shows the equilibrium composition and complexity of the reactions of the fly ash – flue dust mixture in the absence of additional fluxing agent (SiO_2). It can be seen that reduction of iron and nickel oxides starts at quite early stage, but the reduction of VO_2, V_2O_3 and VO to vanadium carbides (VC and V_2C) start at higher temperatures. VC formation starts at slightly above 1100ºC, and at about 1400ºC VC is gradually converts to V_2C and metallic V. Conversion of VC to metallic V requires higher temperatures than 1600ºC. As can be seen here, vanadium carbide formation and decomposition is an issue based on thermodynamic calculations.

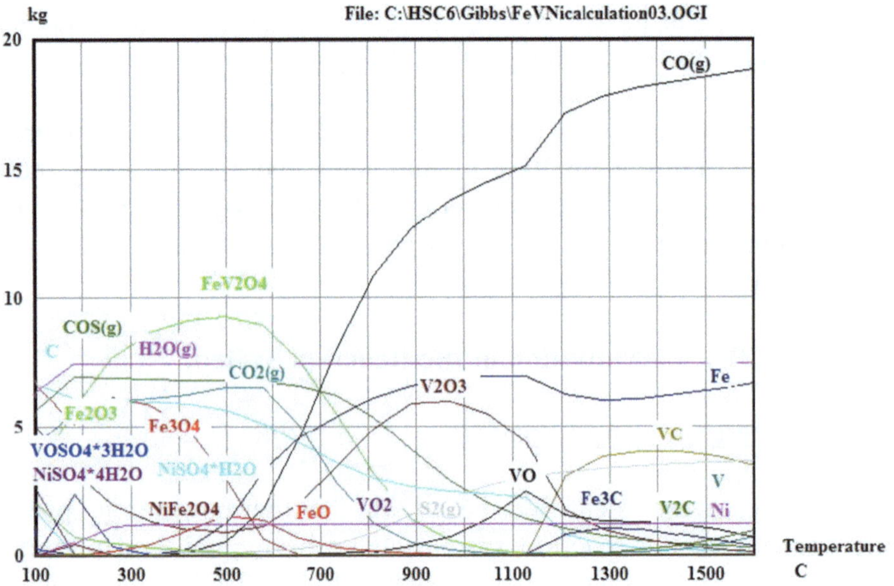

Figure 6. Equilibrium of carbothermic reduction of fly ash and BOF flue dust mixture.

Prior to high temperature smelting experiments of the fly ash and BOF flue dust, thermal behavior of the petroleum fly ash and BOF flue dusts were investigated: mixture without silica addition, mixture with extra silica, mixture with Al addition. A continuous weight loss (higher than 50 % in total) was observed for the mixture of petroleum fly ash and BOF flue dust without silica addition. The weight loss was reduced to about 35% with the addition of SiO_2 for slag formation. With the addition of aluminium in the system, there is further reduced weight loss with the Al_2O_3 formation (oxidation). At the low temperatures between approximately 25 and 800ºC, the weight loss was mostly caused by the volatilization, dehydration and decomposition of sulphates in the sample. When Al was added as extra reducing agent, the melting peak of metallic aluminium is shown in the DSC curve. The dramatic change in the DSC-curve starting at approximately 1350ºC, not only indicates a reduction reaction, responsible for the simultaneous weight loss, but probably also indicates the onset of melting of the sample mixtures. In addition, it is important to note that low melting point of V_2O_5 (690°C) may have some kinetic influence on the slow decomposition of $VOSO_4$ and further hinders the carbothermic reduction of vanadium oxides. The vanadium slag properties will be affected by the oxygen potentials in the system, and the formation of lower VO_x will have significant influence on the reducibility of the oxides and melting temperature of the slag.

Concluding Remarks

The laboratory smelting tests prove that it is possible to produce an iron-vanadium alloy using heavy fuel oil fly ash and BOF steelmaking flue dust without any additional reducing agents through a high temperature smelting operation. The best alloy produced in the present study was 74% Fe, 15%V, 3%Ni, 4%S and 1%C. The reducing agent comes directly from the fly ash as unburned carbon.

It is critical to control and optimize the ratio of carbon to metals in the charge. The slag is another important factor that needs to be further investigated. It is evidence that the CaO content of the slag plays an important role in controlling the amount of sulfur that reports to the metal and slag. The higher the CaO content in the slag, the lower the sulfur level in the metal product.

Acknowledgments

The authors would like to acknowledge Tata Steel Netherlands for providing the BOF flue dust samples and Greenshores N.V. for providing the petroleum fly ash for the experiments; J. van den Berg and R. Hendrikx at the Department of Materials Science and Engineering of Delft University of Technology are acknowledged for the LECO and X-ray analysis.

References

1. Désirée E. Polyak. 2011 Minerals Yearbook. U.S. Geological Survery, 2012.
2. R.R. Moskalyk and A.M. Alfantazi. Processing of vanadium: a review. Minerals Engineering, 16:793–805, 2003.
3. W.B. Morrison. Microalloy steels - the beginning. Materials Science and Technology, 25(9):1066–1073, 2009.
4. ASM Handbook, Volume 1, Properties and Selection: Irons, Steels, and Highperformance Alloys. ASM International, 1995.
5. T.K. Mekherjee S.P. Patinaik and C.K. Gupta. Ferrovanadium from a secondary source of vanadium. Metallurgical Transactions B, 14B:133–135, 1983.
6. F.A. Schmidt O.N. Carlson and W.E. Krupp. A process for preparing high-purity vanadium. Journal of Metals, pages 320–323, 1966.
7. Vanadium and Vanadium Compounds. Ullman's Encyclopedia of Industrial Chemistry. Wiley-VCH, Wienheim, Germany, 2005.
8. I. Saucedo J. Revilla R. Navarro, J. Guzman and E. Guibal. Vanadium recovery from oil fly ash by leaching, precipitation and solvent extraction processes. Waste Management, 27:425–438, 2007.
9. Masud A. Abdel-latif. Recovery of vanadium and nickel from petroleum fly ash. Minerals Engineering, 15:953–961, 2002.
10. Y. Xiao, H. Jalkanen, Y. Yang, C.R. Mambote and R. Boom, Ferrovanadium production from petroleum fly ash and BOF flue dust. Minerals Engineering, 23 (2010) 1155 – 1157.
11. Shang-Lin Tsai and Min-Shing Tsai. A study of the extraction of vanadium and nickel in oil-fired fly ash. Resources, Conservation and Recycling, 22:163–176, 1998.
12. Y. Xiao, H. Jalkanen, Y. Yang, C. R. Mambote (extern) and R. Boom. Thermal Processing of Petroleum Fly Ash for Vanadium Recovery. "XXV International Mineral Processing Congress (IMPC 2010)", The Australasian Institute of Mining and Metallurgy (AusIMM), ed. Publication Series No 7/2010. 6 – 10 September 2010, Brisbane, Australia. pp. 1643-1652.

REWAS 2013: Enabling Materials Resource Sustainabil
Edited by: Anne Kvithyld, Christina Meskers, Randolph Kirchain, Gregory Krumdick, Brajendra Mish
Markus Reuter, Cong Wang, Mark Schlesinger, Gabrielle Gaustad, Diana Lados, and Jeffrey Spangenberg
TMS (The Minerals, Metals & Materials Society), 20

CHARACTERIZATION OF COPPER SLAG

X. Wang[1,2*], D. Geysen[1], S.V. Padilla T.[1], N. D'Hoker[2], T. Van Gerven[2],B. Blanpain[1]

[1]Department of Metallurgy and Materials Engineering, KU Leuven, Belgium
[2]Department of Chemical Engineering, KU Leuven, Belgium

*Corresponding author:
Address: Kasteelpark Arenberg 44 - box 2450, 3001 Heverlee, Belgium.
Email: xuan.wang@cit.kuleuven.be
Tel: + 32 16 3 21780

Abstract

Copper slag can be treated as a secondary resource since it usually contains a substantial amount of copper and other valuable metals. Characterization of the slags to determine the expected metal recovery is essential for the design of separation flow sheets. In this study, the chemical and mineralogical composition of a copper slag from El Teniente, Chile were characterized. The copper content of the fayalite based slag is around 0.87 wt% which is higher than for certain copper ores. Copper exists as sulfides in the form of droplets in the slag. The content and particle size distribution of the major sulfide phases (bornite, chalcopyrite and chalcocite) were quantified using analytical scanning electron microscopy (QEMSCAN). The copper bearing particles have a wide particle size distribution from a few microns up to mm level. Large copper bearing particles (> 100 μm) are composed mainly of bornite and chalcocite and tend to accumulate in the lower part of the slag layer. As characterized with X-ray computed tomography (CT), around 70 volume% of copper value exist in these large copper bearing particles.

Key words: copper slag, recovery, copper sulfides, particles

1. Introduction

Each year, a huge amount of copper slag is produced worldwide. It has been estimated that about 2.2 tons of slag is generated for every ton of copper produced. This resulted in an annual worldwide slag production of approximately 24.6 million tons in the beginning of this century [1]. Most of this material is piled up in dumps and landfills and is therefore a burden for the environment and the metallurgical plants themselves. The interest in new and cost-effective waste management techniques have increased significantly in recent years due to strict environmental regulations, high waste treatment costs and limited availability of disposal sites. On the other hand, natural resources become rare and the value of metals is increasing. Copper slag usually contains a certain quantity of valuable metals, such as copper, nickel, cobalt and iron [1]. Therefore it can be treated as a secondary resource for metal extraction rather than as a waste. Numerous studies have been carried out on metal recovery from copper slag, using hydro- and/or pyrometallurgical methods [2-9].

Characterization of feed material to determine the expected concentrate grade and metal recovery is essential for the design of separation flow sheets and achieving sustainable development in the utilization of resources [10]. The composition, mineralogy and morphology of copper slags can be different due to various types of original ore, previous pyrometallurgical processes and cooling procedures. Extensive research has been performed to characterize the chemical, phase and granulometric composition of copper slag in view of metal recovery through various techniques, such as inductively coupled plasma (ICP) [11], scanning electron microscopy (SEM), X-ray diffraction (XRD) [1, 7], micro Raman spectroscopy [9], atomic absorption spectroscopy (AAS), electron probe micro-analyzer (EPMA) [12], and thermodynamic modelling [13], sieving and laser diffraction [5]. However, most of these studies investigated the composition and mineralogy of the slags without associating this information with the subsequent recovery process of valuable metal, such as leaching and flotation. I n addition, besides knowing the general chemical and mineralogical composition, a more detailed study on the distribution of valuable metals within the slag is important for the design of a tailored slag treating process.

Traditionally, the particle size distribution in minerals is studied in two dimensions by analysis of thin sections or polished sections using microscopy through optical or electronic microscopy [14]. The obtained 2D dimensional data then can be transformed into 3D results through stereological methods. However, it is difficult to provide accurate quantitative information about the internal 3D structures of minerals [15] since the calculation involves assumptions based on textural information or geometrical probability [16]. Moreover, verification with other valid techniques may be needed [17]. Finally, a highly polished surface is required when applying these techniques which will destruct the mineral particles [18]. X-ray computed tomography (CT) is a non-destructive characterizing technology which can provide direct volumetric data acquisition and quantitative analysis in 3D with short analysis time and limited sample preparation [19]. During the measurement, the sample is placed between a X-ray source and a detector. The sample is then scanned 360° around step by step. A projection of the linear attenuation coefficients is measured in each step. The reconstruction of the sample is carried out based on the projections, following the Radon transformation and its mathematical framework [20]. The different mineral phases in the sample can be differentiated because the linear attenuation coefficient μ (cm^{-1}) at each measuring point depends on the electron density, the effective atomic number Z_e of the material and the energy of the X-ray beam [21]. This can be described by the equation:

$$\mu = \rho\,(a + \frac{bZ_e^{3.8}}{E^{3.2}})$$

where ρ (g/cm^3) is the density of the material, a (cm^2/g) is an energy-dependent coefficients and b ($cm^2 \cdot eV/g$) is an constant for each mineral [14]. CT scanning has been applied to visualize and characterize the crack distribution, mineral dissemination, particle size distribution, porosity, particle shape and mineral exposure in many studies of different materials, such as sulfide ores and asphalt mixture [14, 18, 21, 22]. CT can also be applied to analyze the mineral composition of individual particles [16]. However, the accuracy can be limited due to the similarity in effective atomic number (Z_e) of different minerals, such as iron and copper sulfides [18].

Quantitative evaluation of minerals can be performed by scanning electron microscopy (QEMSCAN) which is an automated image analysis system using backscattered electron (BSE) and energy dispersive X-ray (EDX) signals from a scanning electron microscope (SEM) to produce coloured mineral maps, quantitative measurement of minerals, mineral grain size, mineral liberation and elemental distribution (in 2D). As a versatile rapid and economic analysis system, it has been applied in the study of ore flotation, coal utilisation and other geology and mineralogy studies [23-27]. However, the particle size obtained by QEMSCAN is a "section size" from a random section of the sample block. It is derived from two dimensional data and on average less than the maximum section size of the three dimensional particle [24]. And the population density of certain particles or pores could be rather unrealistic [28]. These intrinsic drawbacks of 2D analysis can cause errors in determination of particle size and volume distribution of valuable metals.

For the sake of metal recovery, the particle size and mineralogy of metal phases are the key metallurgical parameters. CT scanning and QEMSCAN are powerful tools to characterize these parameters and have been applied in many domains respectively, but not yet in the study of copper slag. In this work, the distribution of copper bearing grains was studied through CT scanning. In addition, the species and distribution of copper bearing phases were characterized through QEMSCAN. Once the mineralogy and distribution of copper bearing particles are understood, the suitable slag processing techniques and conditions can be designed in view of copper recovery based on these information.

2. Materials and methods

The copper slag studied in this project was produced by Caletones (El Teniente, Chile) copper refinery. The liquid slag was poured into a slag yard and then allowed to solidify. The solidified slag layer was then fragmented by a drill and transported to a landfilling spot. The slag was received in form of black solid material with size from around 100 μm up to 7 cm. In order to obtain representative results, slag samples were taken randomly and without selective sieving during comminution. The slag was first crushed with a jaw crusher with final size ranging from 5 to 10 mm. Further milling was performed with a tungsten carbide ring mill and McCrone micronizing mill down to below 10 μm.

The overall chemical composition of the slag was determined with X-Ray Fluorescence (XRF, Philips PW 2400) and Inductively Coupled Plasma Mass Spectroscopy (ICP-MS, Thermo Elemental type X Series). Around 3 g pulverised sample with a particle size below 10 μm was used for XRF, whereas 0.1 g ground slag with particle size below 125 μm was digested into solution with "three acid digestion method" (by order using HNO_3, $HClO_4$ and HF) [29] for ICP-MS measurement. The mineralogical composition of the slag was characterized with X-Ray Diffraction (XRD, Siemens D500). Samples of 2.7 g of milled slag with particle size below 10 μm mixed with 0.3 g (10 wt% of the total sample) zincite (ZnO) internal standard were used for XRD measurement. The minerals were identified using *Diffrac+ Eva* software and the quantification of phases was carried out with Rietveld refinement.

The images of the cross section of the slag layer were taken with a digital camera and the distribution of copper bearing particles inside the slag body was characterised with Computed Tomography (CT, GE Nanotom). The sample for macroscale observation was made by cutting through the vertical direction of the slag layer (around 5 cm thick). From this cross section, a cylinder with 5 mm diameter and length of 5 cm length was prepared for CT scanning. 9 sections were scanned from the top to the bottom of the cylinder. In each section, 2400 images with resolution of 2284 x 2304 pixels. And voxel size of 2x2x2 μm^3 were acquired. The scanning of each section takes 20 mins. The images were then reconstructed and analysed with CT-Analyser (SkyScan) and VGStudio Max 2.1 (Volume Graphics GMBH). The microstructure of the slag was characterized with Scanning Electron Microscopy (SEM, Philips XL30 FEG). The bulk slag was crushed into small pieces and embedded into a resin (Technovit 4004, Heraeus Kulzer GmbH). Then the sample was polished and coated with carbon for SEM characterization. Different phases in the slag can be identified through EDX analysis coupled with SEM. The distribution and portions of copper bearing phases was obtained through QEMSCAN. The areal fraction of a component within a random transverse section taken from the sample has been proved to be equivalent to the volume fraction of that component throughout the sample [30]. Hence, the surface percentage (2D) obtained from scanning is equal to the volume percentage (3D). A piece of slag, representing the structure of the whole slag layer, was cut through the vertical direction of the slag layer. The cross section was then polished and coated with carbon for QEMSCAN measurement. The scanning area was about 5 x 50 mm and the resolution was 1 μm^2 per pixel for a total measurement time of around 2 hours.

3. Results and discussion

3.1 Chemical composition

The chemical composition of the slag is presented in Table 1. Besides the elements listed in Table 1, the remaining weight percentage is attributed to oxygen present in metal oxides. The most abundant elements in the slag are iron (around 42 wt%) and silicon (around 14 wt%). This is in line with the fayalite based nature of the copper slag. The concentration of copper in slag is close to 0.9 wt%, which is higher than that of certain copper ores [31]. The other metal with potential economical value is molybdenum. The concentration of molybdenum in slag is higher than 0.1 wt% which lies in the range of the content of viable ore bodies (from 0.01 to 0.25 wt%) [32].

Table 1. Chemical composition of copper slag.

Major element	wt%	Standard deviation	Trace element	ppm	Standard deviation
Al	1.45	0.02	As	238	1
Cu	0.872	0.02	Ba	116	4
Fe	41.7	3	Ca	478	80
K	0.877	0.01	Cl*	586	300
Mg	0.563	0.004	Co	159	10
Na	0.280	0.02	Cr	71.0	2
Mo	0.121	0.003	Mn	162	5
S*	0.426	0.008	Sr	75.5	0.4
Si*	13.6	0.3	Ni	44.7	6
Ti	0.373	0.007	Pb	341	20
Zn	0.198	0.003	Sb	254	7
O**	39.3		Sr	75.5	0.4
			V	55.0	4
			W	92.5	6

*XRF data, ** calculated based on the listed data, all others by ICP-MS. The standard deviation was calculated with 3 replicates.

3.2 Mineralogical composition

The X-ray diffraction pattern of the copper slag is shown in Figure 1. The calculated quantitative XRD results after excluding internal standard material ZnO are listed in Table 2.

Table 2. Quantitative XRD results of the slag sample.

Mineral	Chemical formula	wt %
Fayalite	$2FeO \cdot SiO_2$	52
Magnetite	Fe_3O_4	13
Glassy phase		35

Fayalite is the major crystalline mineral present in the slag taking up to 52 wt% of the system. Fayalite has a hardness of 6.5 Mohs. This high hardness is responsible for difficult milling and high grinding energy input. The grindability index, which indicates the energy required for grinding, of copper slag was reported to be 26.8 kWh/ton which is about twice that of copper ores (13.13 kWh/ton) [9]. Besides fayalite, 13 wt% magnetite is present due to the overoxidation of iron oxide [33]. Due to the low contents of copper bearing phases, their characteristic peaks in the pattern is not strong and clear. Hence, they cannot be quantified through XRD. The quantitative results of copper bearing phases are arguable due to their low contents. The large amorphous phase content (35 wt%) suggests that the slag has undergone a relatively fast cooling process and/or has relatively high silica content.

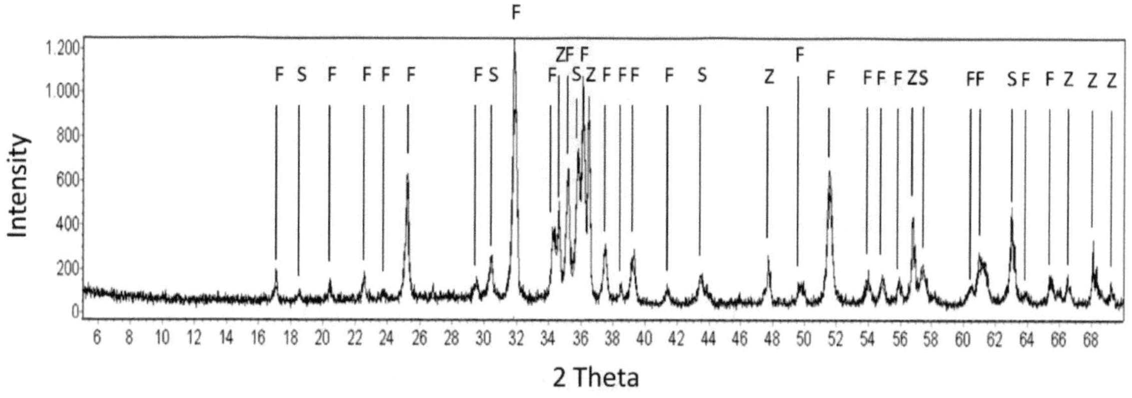

Figure 1. XRD pattern of the copper slag: (F) fayalite; (S) spinel (magnetite); (Z) zincite.

3.3 Morphology

The cross sections representing the internal structure are shown in Figure 2. Most of the slag exists as large bricks indicating that they originate from thick (around 5 cm) slag layers. As can be observed, large copper sulfide particles (shiny droplets) as well as pores and cracks can be observed. The size of the copper sulfide particles in the slag varies from a few microns (observed by SEM) up to 1 mm. The majority of the large copper sulfide particles is located in the lower part of the slag layer. This may be due to the long solidification time of the thick slag layer, which allows the sedimentation and accumulation of copper sulfide particles in the lower part of the slag layer, while the upper part of the slag layer is more quickly cooled by air. The effect of cooling rate on microstructure of slag and size distribution of copper bearing phases has been indicated by previous researches [11, 34]. However, the inhomogeneous distribution of copper sulfides particles in the vertical direction of the slag layer has not been mentioned previously. The distribution of copper sulfide particles in the horizontal direction of slag layer is rather homogenous. Interestingly, as shown in Figure 2, accumulation of large pores in the middle of slag layer (pore belt) can be observed. The formation of pores may be caused by the gas generated from the oxidation of sulphur in the sulfides or of remaining carbon from the cleaning stage. The rising gas bubbles may have been arrested in their ascent through the slag layer by the faster solidified slag at the upper part of the slag layer and therefore accumulated in the middle part of the slag layer.

(1) (2)

Figure 2. Cross section of slag fragmented slag.

Figure 3 shows the microstructure of the slag. Fayalite, magnetite (spinel), copper sulfides and glassy phase can be distinguished. This conclusion is in line with the results obtained with other techniques (XRD, XRF). The copper sulfide phases exist as spherical particles with diverse particle sizes (from a few microns to several hundred microns). All the copper bearing phases observed were sulfides, having an overall Cu-Fe-S composition. However, the Cu-Fe-S stoichiometry varies among the copper bearing particles. No metallic copper was observed in the slag.

As shown in Figure 3, different crystal shapes can be observed within the same slag fragment which may caused by the change in cooling rate at different positions. In the part close to the top of the slag layer, the liquid slag is in direct contact with the air. As a result, the cooling rate in this part is higher. Skeletal/dendritic fayalite crystals coexist with fine magnetite and copper sulfide particles could form during the solidification. The solidified slag may act as an insulator and slow down the cooling of the middle and bottom part of the slag layer [35]. Hence, slag in these parts can develop into larger grains.

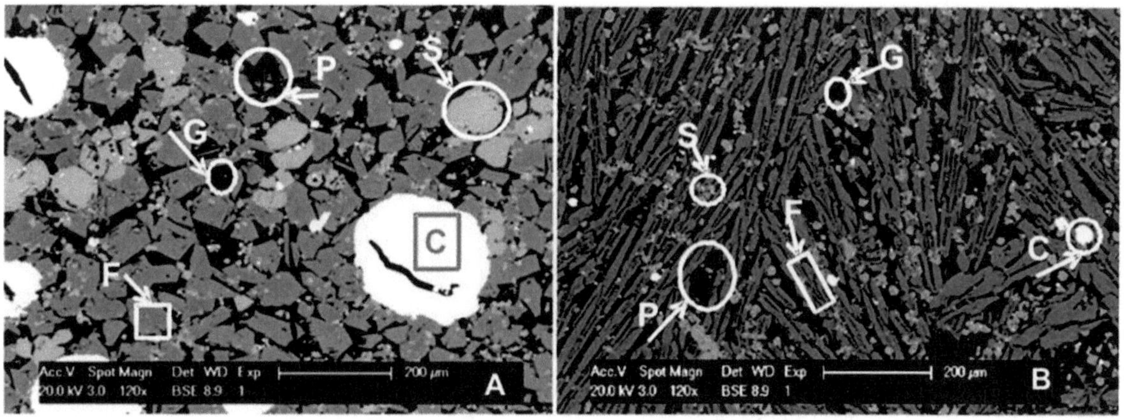

Figure 3. SEM-Backscattered Electron (BSE) image of copper slag: (F) fayalite; (S) magnetite (spinel); (G) glass; (C) copper sufides; (P) pores.

Detailed study of the distribution of copper bearing phases was carried out using a QEMSCAN system. As shown in Table 3, bornite was the most abundant copper sulfide phase, followed by chalcopyrite and chalcocite. Trace concentrations of idaite (Cu_5FeS_6) were also found. Hence, compared to XRD, QEMSCAN is more suitable for characterizing mineralogy of minerals in small amount.

Table 3. Volume abundance of copper bearing phases in slag

Mineral	Chemical formula	volume %
Bornite	Cu_5FeS_4	0.23
Chalcocite/Digenite	Cu_2S/ Cu_9S_5	0.14
Chalcopyrite	$CuFeS_2$	0.27
Idaite	Cu_5FeS_6	0.01
Copper bearing phases in total		0.65
Other phases in slag		99.35

Grain size analysis in two dimensions was performed to determine the size distribution of the various copper bearing grains. Figure 4 was compiled from data obtained across the entire scanned surface of the sample (5 mm x 45 mm). Only bornite and chalcocite/digenite occur in grain sizes of larger than 20 µm while chalcopyrite only appears in grain sizes below 20 µm. Chalcocite has a roughly bimodal size distribution, with a significant percentage of grains in size ranges larger than 40 µm.

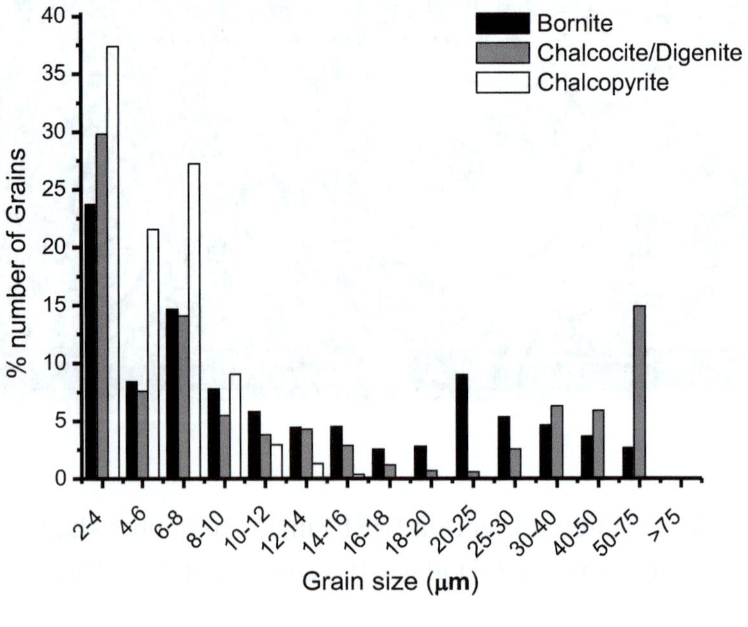

Figure 4. Copper sufide grain size distribution (2D).

However, the grain size here only refers to the size of the individual copper sulfide mineral. The copper bearing particles, especially large particles (> 100 μm), are aggregates of various copper sulfides in most cases. Hence, a copper bearing particle can be several times larger than a copper sulfide grain. Chalcocite and bornite are the major components of large copper bearing particles and are, therefore, potentially easier to be recovered. Chalcopyrite, on the other hand, is randomly distributed in the slag samples as fine grains. Figure 5 demonstrates a typical aggregate copper bearing particle with complex texture.

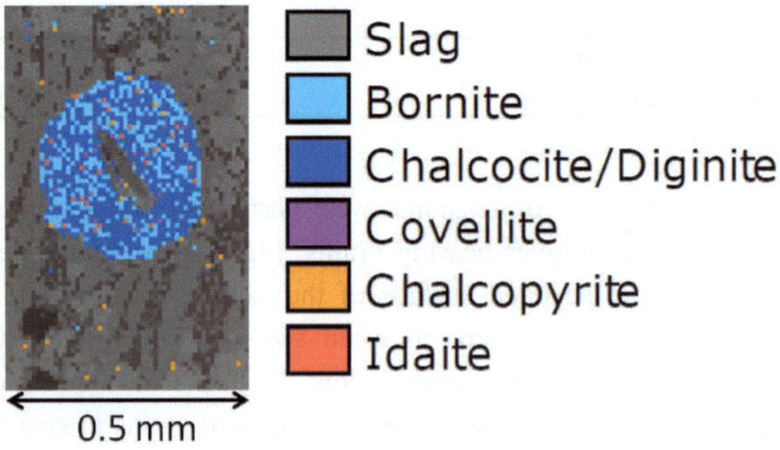

Figure 5. Large aggregate copper bearing particle composed of various copper sulfides.

Further study of the particle size distribution of copper bearing particles was performed through CT scanning. As can be observed in Figure 6, around 60% of copper bearing particles accumulate in the lower (depth > 50 %) part of the slag layer. The section near the bottom of slag layer (depth between 80 ~ 90 %) contains more than 25 % of the copper sulfides. As can be observed from Figure 6, large particles tend to accumulate in the lower part of the slag layer. There appears to be depletion around the middle of the depth. This may be caused by the accumulation of pores in the middle of slag layer (Figure 2b), which arguably hindered the settlement of copper bearing particles.

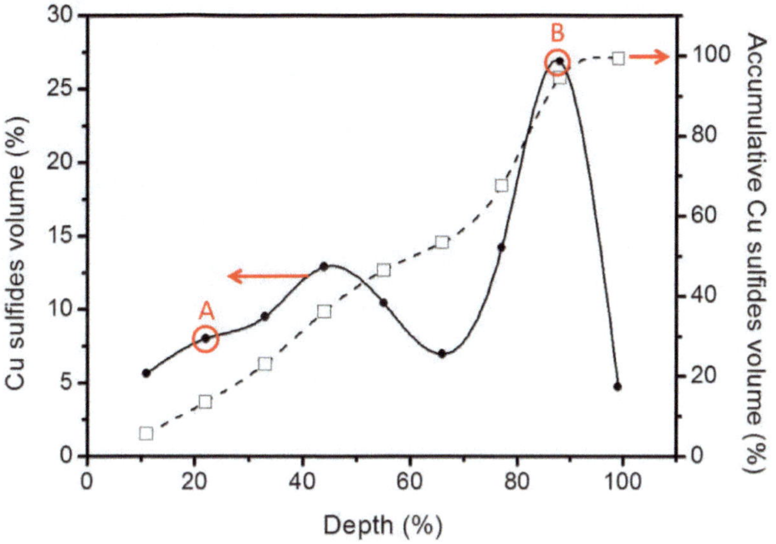

Figure 6. Copper sulfides volume distribution along depth of slag layer: (A) section near the top of slag layer; (B) section near bottom of the slag layer.

3D distributions of copper bearing particles in section A (depth between 11~22 %) and section B (depth between 78~89 %) are shown in Figure 7. Due to different linear attenuation coefficients for copper sulfides and the matrix material, copper bearing particles can be differentiated from the matrix material after segregation of reconstructed 3D structures. The matrix materials are set to be transparent and the copper bearing particles are coloured with gray. It can be observed that the copper bearing particles exist in the form of vast small particles in the top part of the slag layer and large spherical particles in the bottom part of the slag layer.

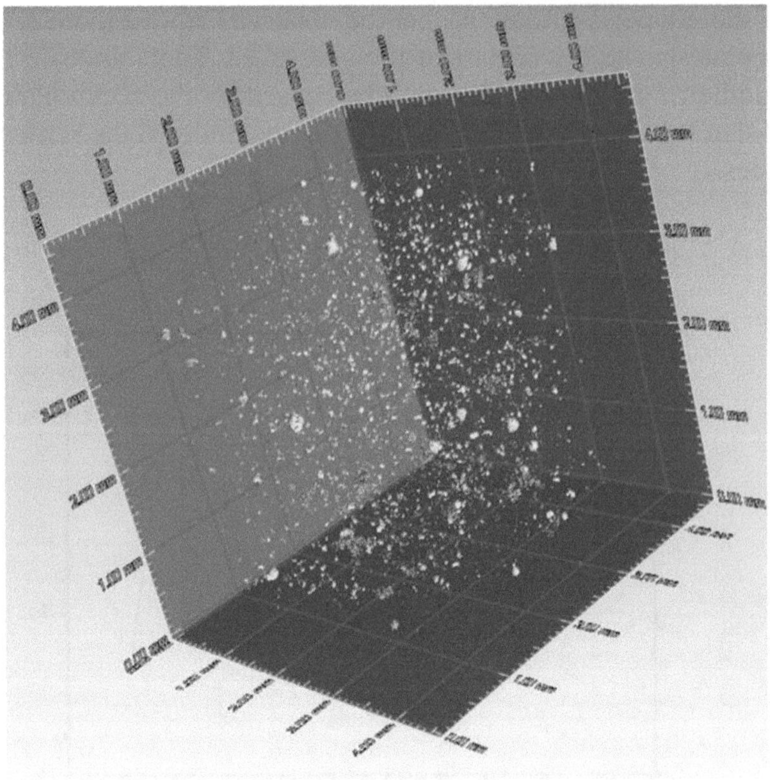

a

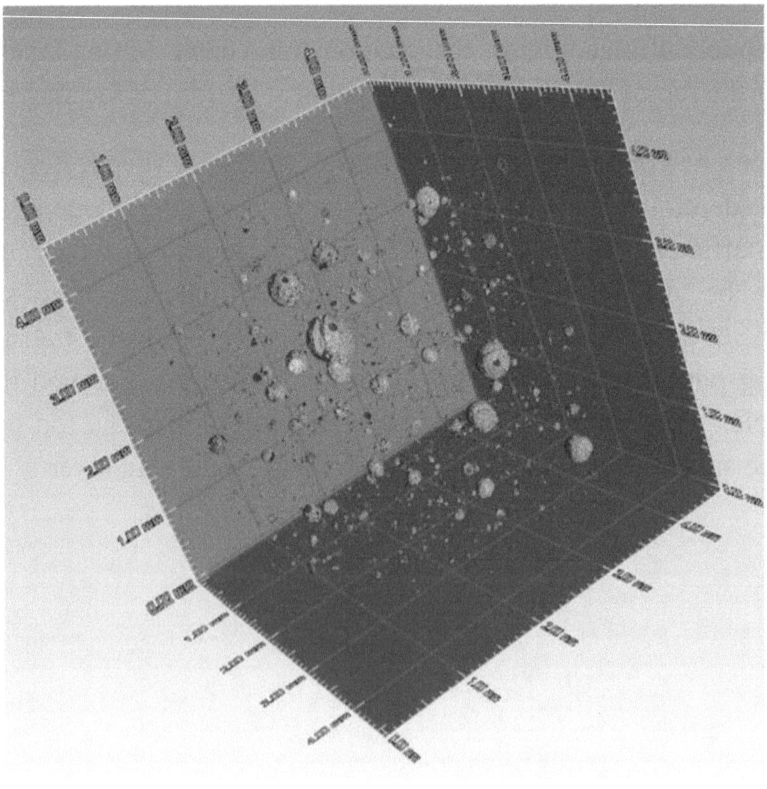

b

Figure 7. 3D distribution of copper bearing particles after phase segregation: (a) section A; (b) section B.

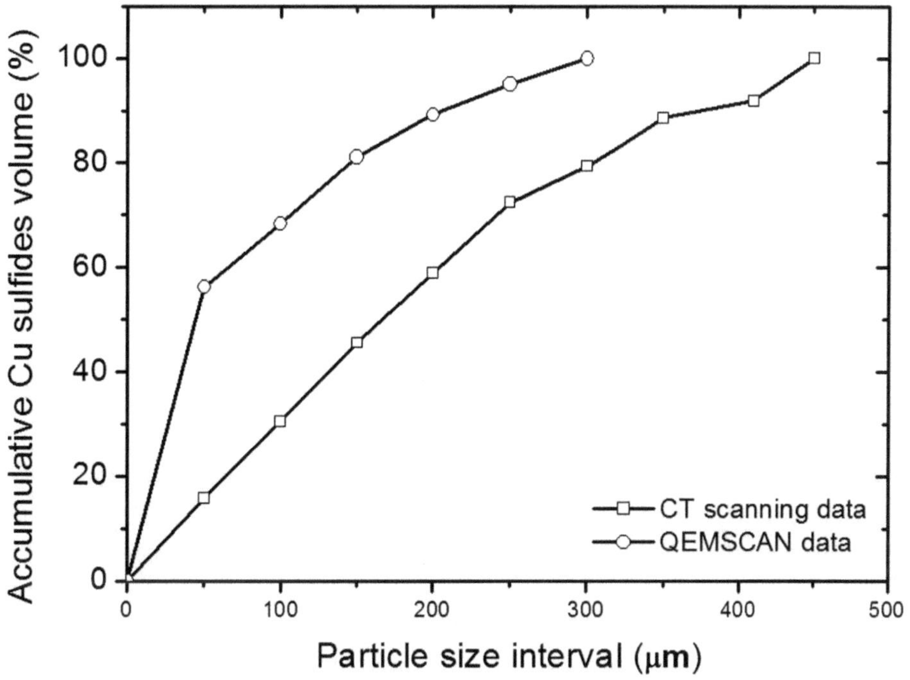

Figure 7. Accumulative copper sulfide volume versus copper bearing particle size plotted with CT scanning and QEMSCAN data.

According to the data obtained from CT scanning, as most of the copper bearing particles are in droplet form, the assumption of spherical shape of copper bearing particles is reasonable. With this assumption, the relevant particle size (diameter) of each volume of individual particles can be calculated. The results are plotted in Figure 8. 100 μm is set to be the threshold between large and small particles in this work since materials are commonly grinded down below 100 μm in beneficiation processes such as flotation. Around 70% of the total copper sulfides appear in particles with size larger than 100 μm. Hence, the majority of the copper value accumulates in the relatively larger particles. The data obtained from QEMSCAN shows different results. No particles larger than 300 μm were found and the amount of copper in smaller (< 100 μm) particles is significant larger (~ 68 %) than that obtained through CT scanning (~ 30 %). As mentioned earlier, this deviation may due to the "section size" nature of the QEMSCAN data. Compared to the data from CT scanning, QEMSCAN data underestimate the volume and size of the large copper bearing particles.

Large copper bearing particles can be liberated easily upon comminution without grinding slag into extra-fine particles. Hence, the energy consumption is largely reduced. Higher liberation rate will also provide higher copper recovery potential in the following slag treating processes, such as copper sulfides flotation, acid copper extraction and magnetic separation. In order to reach the maximum metal recovery, the cooling process of slag can be tailored to generate more large copper bearing particles.

The integration of QEMSCAN and CT scanning can provide the mineralogy and particle size distribution which are crucial metallurgical factors for the recovery of copper from slag. The optimum grinding size should be set according to the size distribution of the copper bearing particles to achieve high liberation of copper sulfide minerals. The following processing conditions, such as flotation reagents and leaching medias, should be selected based on the mineralogy of the copper sulfides.

4. Conclusions

The copper slag studied in this work contains around 0.9 wt% of Cu which is even higher than that of certain ores. This indicates the potential to recover copper from this slag. The slag is fayalite-based and high energy input is required for grinding. Copper majorly exists in the form of bornite, chalcocite and chalcopyrite. Large copper bearing particles are aggregates of various copper sulfides, mainly bornite and chalcocite. Based on the observations, the distribution of copper sulfide phases is random in the horizontal direction of the slag layer but inhomogeneous in the vertical direction of the slag layer. Large copper sulfide particles tend to settle down during solidification. The grain size of copper bearing particles and its distribution are important parameters for the selection of appropriate beneficiation techniques. The integration of QEMSCAN and CT scanning can provide the mineralogy and particle size distribution which are crucial for the design of following copper recovery processes.

Acknowledgements

The authors are grateful to Mr. Matthew D. Dye and Dr. Katharina Pfaff for the QEMSCAN analysis performed in Colorado School of Mines, and Dr. Greet Kerckhofs and the Materials Performance and Non-destructive Testing research group (MTM, K.U.Leuven) for providing CT scanning device and training. Financial and material support from Group Machiels are also acknowledged.

References

1.Gorai B., Jana R.K., and Premchand, Characteristics and utilisation of copper slag—a review. Resources, Conservation and Recycling, 2003. 39(4): p. 299-313.

2.Altundogan H.S., Boyrazli M., and Tumen F., A study on the sulphuric acid leaching of copper converter slag in the presence of dichromate. Minerals Engineering, 2004. 17(3): p. 465-467.

3.Arslan C. and Arslan F., Recovery of Cu, Co & Zn from copper smelter & converter slags. Hydrometallurgy, 2002. 67: p. 1-7.

4.Banza A.N., Gock E., and Kongolo K., Base metals recovery from copper smelter slag by oxidising leaching and solvent extraction. Hydrometallurgy, 2002. 67(1-3): p. 63-69.

5.Carranza F., et al., Biorecovery of copper from converter slags: Slags characterization and exploratory ferric leaching tests. Hydrometallurgy, 2009. 97(1-2): p. 39-45.

6.González C., et al., Reduction of Chilean copper slags a case of waste management project. Scandinavian Journal of Metallurgy 2005. 34: p. 143-149.

7.Li Y.J., Papangelakis V.G., and Perederiy I., High pressure oxidative acid leaching of nickel smelter slag: Characterization of feed and residue. Hydrometallurgy, 2009. 97(3-4): p. 185-193.
8.Maweja K., et al., Effect of annealing treatment on the crystallisation and leaching of dumped base metal smelter slags. Journal of Hazardous Materials, 2010. 183(1-3): p. 294-300.
9.Das B., et al., Characterization and recovery of copper values from discarded slag. Waste Management & Research, 2010. 28(6): p. 561-567.
10.Miller J.D., et al., Liberation-limited grade/recovery curves from X-ray micro CT analysis of feed material for the evaluation of separation efficiency. International Journal of Mineral Processing, 2009. 93(1): p. 48-53.
11.Gbor P.K., Mokri V., and Jia C.Q., Characterization of smelter slags. Journal of Environmental Science and Health Part a-Toxic/Hazardous Substances & Environmental Engineering, 2000. 35(2): p. 147-167.
12.Cardona N., et al., The Physical Chemistry of Copper Smelting Slags and Copper Losses at the Paipote SmelterPart 2 - Characterisation of industrial slags. Canadian Metallurgical Quarterly, 2011. 50(4): p. 330-340.
13.Cardona N., et al., Physical chemistry of copper smelting slags and copper losses at the Paipote smelterPart 1 ‐ Thermodynamic modelling. Canadian Metallurgical Quarterly, 2011. 50(4): p. 318-329.
14.Remeysen K., Combination of Medical Computed Tomography Microfocus Computed Tomography and Petrography in Reservoir Characterization. PhD Thesis. 2006, Katholieke Universiteit Leuven: Leuven, Belgium.
15.Higgins M.D., Measurement of crystal size distributions. American Mineralogist, 2000. 85(9): p. 1105-1116.
16.Videla A.R., Lin C.L., and Miller J.D., 3D characterization of individual multiphase particles in packed particle beds by X-ray microtomography (XMT). International Journal of Mineral Processing, 2007. 84(1–4): p. 321-326.
17.Mertens G. and Elsen J., Use of computer assisted image analysis for the determination of the grain-size distribution of sands used in mortars. Cement and Concrete Research, 2006. 36(8): p. 1453-1459.
18.Ghorbani Y., et al., Use of X-ray computed tomography to investigate crack distribution and mineral dissemination in sphalerite ore particles. Minerals Engineering, 2011. 24(12): p. 1249-1257.
19.Miller J.D. and Lin C.L., Three dimensional analysis of particulates in mineral processing systems by cone beam X-ray microtomography. 2004, Society for Mining, Metallurgy, and Exploration.
20.Dhawan N., et al., Recent advances in the application of X-ray computed tomography in the analysis of heap leaching systems. Minerals Engineering, 2012. 35(0): p. 75-86.
21.Miller J.D., et al., Ultimate recovery in heap leaching operations as established from mineral exposure analysis by X-ray microtomography. International Journal of Mineral Processing, 2003. 72(1–4): p. 331-340.
22.Kutay M.E., et al., Three-dimensional image processing methods to identify and characterise aggregates in compacted asphalt mixtures. International Journal of Pavement Engineering, 2010. 11(6): p. 511-528.
23.Liu Y., et al., Mineral matter–organic matter association characterisation by QEMSCAN and applications in coal utilisation. Fuel, 2005. 84(10): p. 1259-1267.
24.Pascoe R.D., Power M.R., and Simpson B., QEMSCAN analysis as a tool for improved understanding of gravity separator performance. Minerals Engineering, 2007. 20(5): p. 487-495.

25.Farrokhpay S., Ametov I., and Grano S., Improving the recovery of low grade coarse composite particles in porphyry copper ores. Advanced Powder Technology, 2011. 22(4): p. 464-470.
26.Matjie R.H., et al., Behaviour of coal mineral matter in sintering and slagging of ash during the gasification process. Fuel Processing Technology, 2011. 92(8): p. 1426-1433.
27.Lotter N.O., Modern Process Mineralogy: An integrated multi-disciplined approach to flowsheeting. Minerals Engineering, 2011. 24(12): p. 1229-1237.
28.Russ J.C., Practical Stereology 2nd Edition ed. 1999: Plenum Press, New York, NY.
29.D'Hoker N., Valorisation of cooper slag: Characterisation and (ultrasound-assisted) selective acid leaching, in Department of Chemical Engineering. 2011, Katholieke Universiteit Leuven: Leuven. p. 75.
30.Underwood E.E., Quantitative stereology. 1970, Reading, Massachusetts: USA: Addison-Wesley Publ. Co.
31.Davenport W.G., et al., Extractive metallurgy of copper. Fourth edition ed. 2002: Pergamon.
32.IMOA. Mining. 2011 [cited 2011; Available from: http://www.imoa.info/molybdenum/mining_processing/molybdenum_mining.php.
33.Jalkanen H., Vehviläinen J., and Poijärvi J., Copper in solidified copper smelter slags. Scandinavian Journal of Metallurgy, 2003. 32: p. 65-70.
34.Sarrafi A., et al., Recovery of copper from reverberatory furnace slag by flotation. Minerals Engineering, 2004. 17(3): p. 457-459.
35.Pandelaers L., Ferro-alloy dissolution in liquid steel. PhD thesis. 2011, Katholieke Universiteit Leuven: Leuven, Belgium.

REWAS 2013: Enabling Materials Resource Sustainability
Edited by: Anne Kvithyld, Christina Meskers, Randolph Kirchain, Gregory Krumdick, Brajendra Mishra, Markus Reuter, Cong Wang, Mark Schlesinger, Gabrielle Gaustad, Diana Lados, and Jeffrey Spangenberger
TMS (The Minerals, Metals & Materials Society), 2013

Recovery of Zinc and Iron from Steel Mill Dusts by the use of a TBRC: A possible Mini-mill Solution?

Jürgen Antrekowitsch[1], Gerald Schneeberger[1]

[1]Christian Doppler Laboratory for Optimization and Biomass Utilization in Heavy Metal Recycling, University of Leoben,
Franz-Josef-Str. 18, A-8700 Leoben, Austria/Europe
(Corresponding author: juergen.antrekowitsch@unileoben.ac.at)

Keywords: steel mill dusts, zinc, recycling, mini-mill solution, slag optimization

Abstract

The search for an effective mini-mill solution to treat zinc containing residues from steel industry or also cupola furnaces has a long history but has not been very successful yet due to several reasons. Some of them are: Not available technologies, the not well known and not developed product market by the steel producers, still relatively low treatment charges for the landfilling and no effective solution for an additional iron recovery. The present paper describes a process where in two subsequent steps by the use of a Top Blown Rotary Converter technology a simultaneous recovery of zinc, iron and other valuable metals should be possible. Investigations which were done in an advanced technical scale gave satisfying results. These data allowed the calculation of a mass- and energy balance. First economic considerations underline that with such a concept a feasible mini-mill solution could be established for the mentioned residues.

Introduction

During steel production dusts are generated in an amount of 15 to 23 kg /t of steel mainly including iron oxides, zinc oxide, typical slag compounds, lead and halogens. Worldwide two main steel production routes are operating whereof the electric arc furnace route (EAF) with approximately 30 % of the world production generates dusts with high zinc contents up to 40 %, due to the high amount of charged galvanized scrap. The alternative integrated route produces steel in a basic oxygen furnace (BOF) with a comparable low ratio of charged scrap resulting in a dust with low zinc contents normally below 10 %. In some cases the BOF-dust is re-circulated in the furnace and with this reaches zinc values up to 27 %. Thereof 40 to 45 % of the yearly generated electric arc furnace dusts are recycled to recover first of all zinc by different technologies where the waelz kiln is far the dominating concept. Due to various disadvantages like low yields, the recovery of only one metal and the generation of huge amounts of residues there is still the demand for alternatives. Table 1 gives typical compositions of the mentioned dusts. [1], [2]

Table 1: Typical compositions of investigated dusts during the last ten years (mainly Europe)

%	Zn	Pb	SiO_2	CaO	Na	K	F	Cl	Fe
EAF	18-40	2-7	3-5	8-9	1.0-2.0	0.8-1.5	0.2-0.5	2.0-5.5	18-35
BOF	1-15	0-2	2-6	5-15	0-1.0	0.5-1.1	0-0.5	0-2.0	24-45

Further disadvantages regarding the present situation from the view of the steel mills are:

- the necessity to transport the dust to central recycling units often for several 100 km generating high transport costs
- the dependency on price politics of recycling companies
- the payment of treatment fees

Therefore since many years there is the idea to develop treatment concepts which could be installed directly at the steel mill allowing a recycling of the dust under economic conditions also with comparable small throughputs.

Most of the existing processes like the already mentioned waelz kiln as well as others like rotary hearth furnaces or the Primus processes do not offer such a so called mini-mill-solution.

During the investigations at the University of Leoben in the last years following points have to be found as mandatory to create such an efficient mini-mill-solution especially for the treatment of EAF-dusts:

- iron has to be recovered in form of an alloy and for this the necessity of a melting process
- halogens have to be separated successfully to increase the quality of the zinc oxide
- the remaining slag should be low in critical metals like lead and should have an optimum composition and behaviour, so that utilization in road construction or even in cement industry is possible

Also investigations on the treatment of BOF-dusts were done but as a fact of the generally low zinc content compared with the low value of the main component iron, it is not easy to reach an economical solution. This is only possible in the described case where the dust is re-circulated and with this the zinc content increased up to at least 15 % [1, 2].

Investigated process

The investigated concept that should fulfill the above mentioned criteria is based on a two-step process, where the reduction is done by a carbon containing iron bath. Hereby the residues are charged with reducing agents on a liquid metal bath, the oxides become reduced and volatile compounds are evaporated. Also the iron bath contains an amount of 3 % carbon at the beginning of the procedure. Not volatile compounds are partly caught by the metal bath or remain in the slag. Figure 1 shows the process principle of the recycling procedure.

The desired results are as stated a marketable iron alloy, a stabilized slag, which can be used very likely in the cement industry or in road construction and a crude zinc oxide caught from the off-gas stream by a bag house.

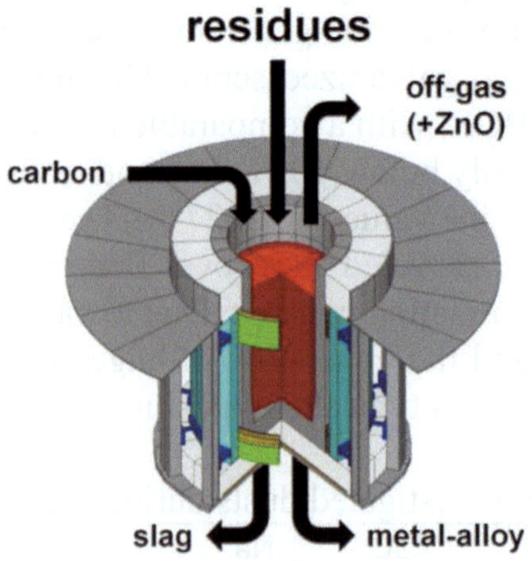

Figure 1. Basic schema of the Iron Bath Process

Nevertheless there are also some problems which have to be considered during the investigation of this process. The iron bath technology has the same disadvantage like all other pyrometallurgical recycling processes for steel mill dusts based on one process step. Unwanted impurities are fumed with the zinc and they are concentrated in the resulting crude zinc oxide. Especially when charging electric arc furnace dusts from carbon steel production directly, the content of lead, chlorine, fluorine, potassium and sodium in the resulting filter dust becomes very high. These contaminations, especially the halogens, are harmful for the possible utilization areas first of all the winning electrolysis where only very low amounts of these elements are allowed but also for others like the zinc sulfate production [3].

In Figure 2 a simple thermodynamic study shows the performance of an electric arc furnace dust under rising temperature with added reducing agents as it would be done on a reducing iron bath. The picture provides data for the behaviour of the dust above 1100 °C for the Zn, Fe and some of the chlorides. Nevertheless, it gives lead in its metalized form which is clear from the thermochemical point of view due to its noble character and therefore the easy reduction of the oxide. This definitely would not happen in the real process where lead oxide or chloride would be evaporated before its reduction, leaving the reaction zone with the off-gas.

This underlines that the zinc oxide product for sure would contain a high amount of impurities, mainly halides and lead oxide.

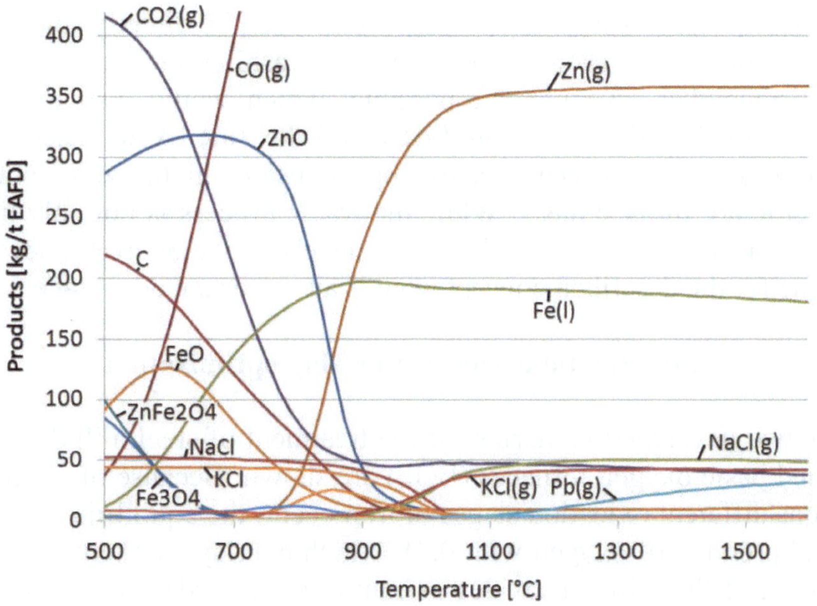

Figure 2. Equilibrium curves of a steel mill dust under reducing conditions

Due to that problem a further process step is taken into account. One option is an additional pyrometallurgical step before the iron bath process. During such a treatment under oxidizing conditions - clinkering - in the field of 1000 – 1200 °C impurities are removed and the cleaned material can be charged after this procedure on a liquid iron bath to reduce the oxides and fume the zinc and reach a high quality zinc oxide.

An important economical factor is the usage of the process energy from the first clinkering step by charging the hot material in the second stage on the liquid iron bath. Figure 3 shows the implementation of the clinkering step onto the iron bath process. A further important point would be to use the energy from the zinc re-oxidation in the off-gas stream in the previous processes.

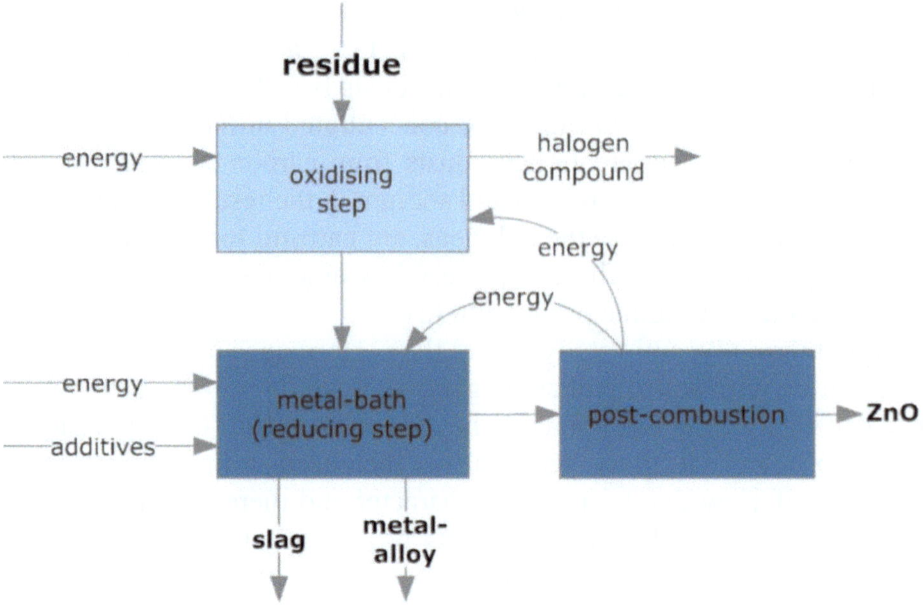

Figure 3: Implementation of a clinkering step in the iron bath process

If the quality of the filter dust is high enough the product can be directly sold to the primary zinc industry without any previous cleaning steps like soda-washing. Or it could probably break into other markets where higher revenues are possible, which would also justify a high energy demand. The "cash cow" of the process is first of all the zinc oxide. Therefore high amounts have to be produced which is easier to achieve by the use of higher zinc containing secondary materials like electric arc furnace dusts. While the waelz process is not only limited by too low zinc contents in the dust but also by too high zinc levels above 30 % for several reasons, such values are welcome for the described concept what is an important advantage.

Investigations concerning slag optimization

In Figures 4 to 6, two developed scenarios for the treatment of steel mill dusts from electric arc furnaces (EAF) and basic oxygen furnaces (BOF) are shown. Because of the problem that under reducing conditions different oxidation stages of iron occur, the equilibrium has to be considered by setting a partial pressure of oxygen with 0.21, which overlays the ternary system and for this allows the formation of the different oxides in equilibrium. Furthermore, different phases aside from iron, calcium and silicon oxides have to be considered.

The optimization by e.g. additives is not a simple procedure which could only be derived from one system. The fact that during a reduction of steel mill dust on a reducing iron bath three different diagrams become relevant led to the challenge of a "cross linked" investigation, which is not easy to handle. To be able to cope with this, it is mandatory to have a computational thermochemical solution which enables to judge whether or not a parameter setting is a possible solution for all relevant ternary systems and how the individual ternary system would change with such a setting.

With this information, trials in the heating microscope were performed verifying the calculated results.

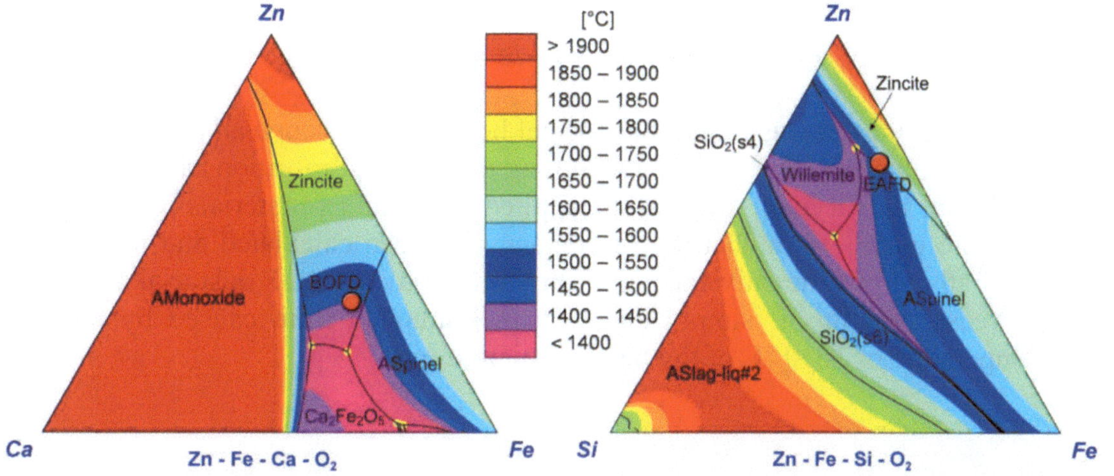

Figure 4: BOFD (left) and EAFD (right) during melting under atmospheric conditions

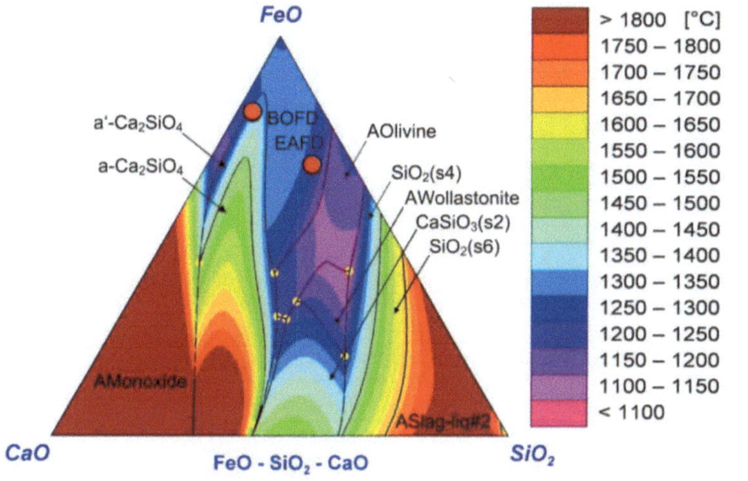

Figure 5: BOFD and EAFD under partial reducing conditions

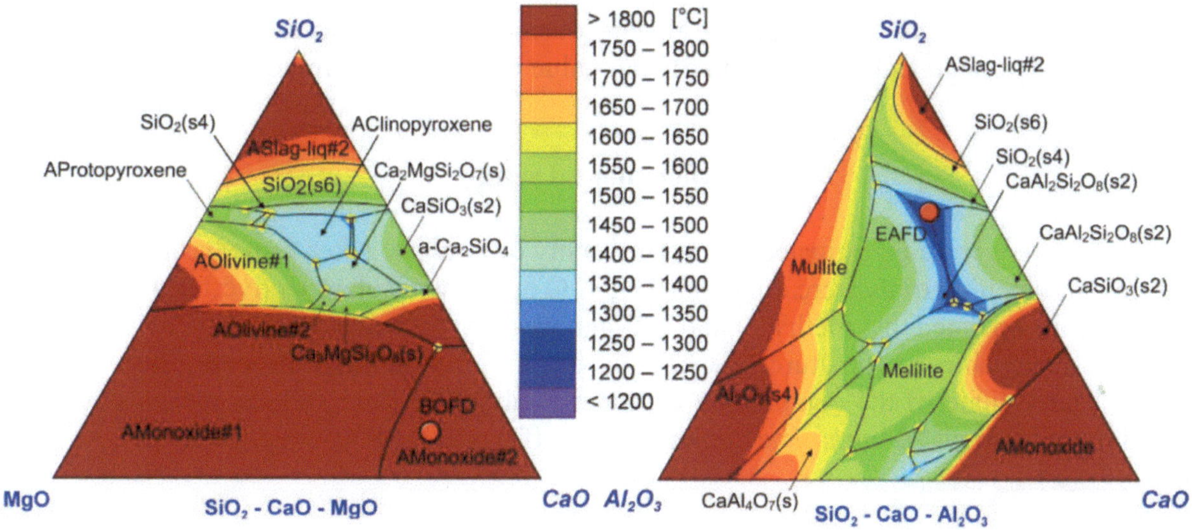

Figure 6: BOFD (right) and EAFD (left) at the end of the process (fully reduced)

The graphs in Figure 7 show the slag regime calculated and determined by the heating microscope for EAF dust and BOF dust as well as an alternative generated by additives (e.g. SiO_2).
From these investigations, it was found that special heating microscope analysis is a reliable tool for the evaluation of materials behaviour concerning melting, optimization trials and taking some boundary conditions into account, can also be applied for complex materials like e.g. electric arc furnace dusts. The graphs show that the discrepancy between calculated and measured (dotted lines) values is rather low. The only exceptions are formed by the BOF dust trial at the end of the process. The reason for this is simply due to a measurement limit caused by the maximum adjustable temperature of 1650 °C of the heating microscope.

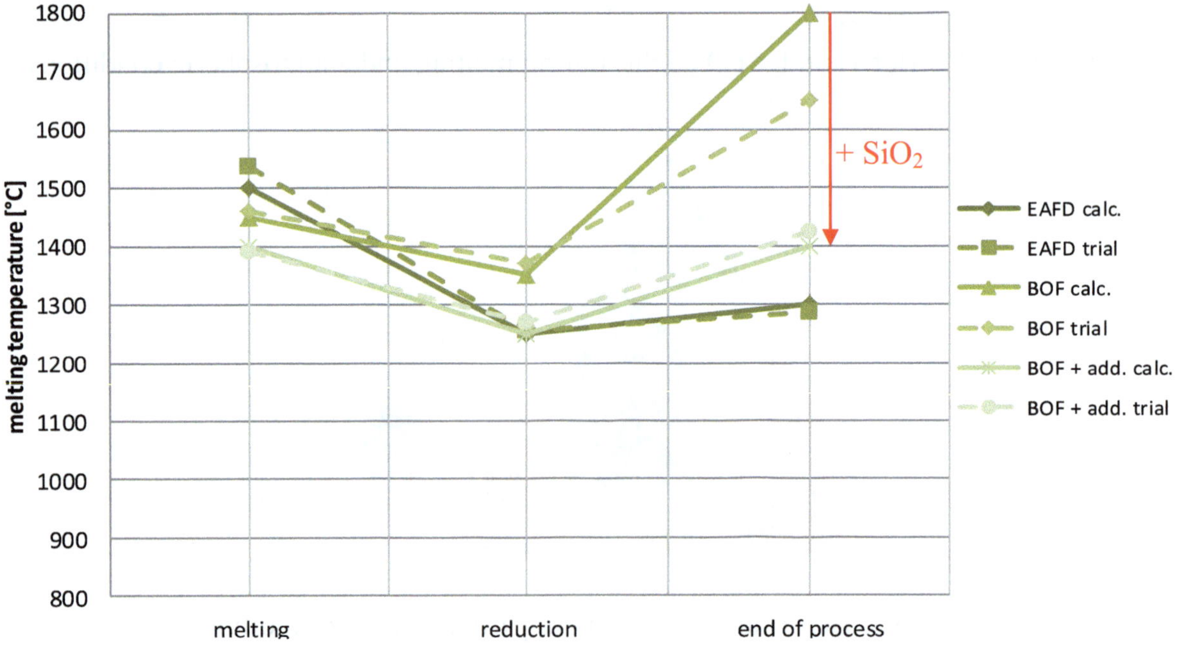

Figure 7: Evolution of melting points of slags during different process steps calculated by Factsage and determined as mean value of three measurements each

Experimental setup

Both process steps were investigated at the University of Leoben. After some basic lab scale tests clinkering and melting trials were done in a TBRC (Top Blown Rotary Converter, Figure 8).
The burner is operated by methane and oxygen with a maximum capacity of 75 kW. The maximum achievable temperature, depending on the investigated process, lies at around 1600 °C. The volume of the combustion chamber is about 80 liters, which means a maximum material input of about 20 liters. The cooling of the off-gas is done by excess air.

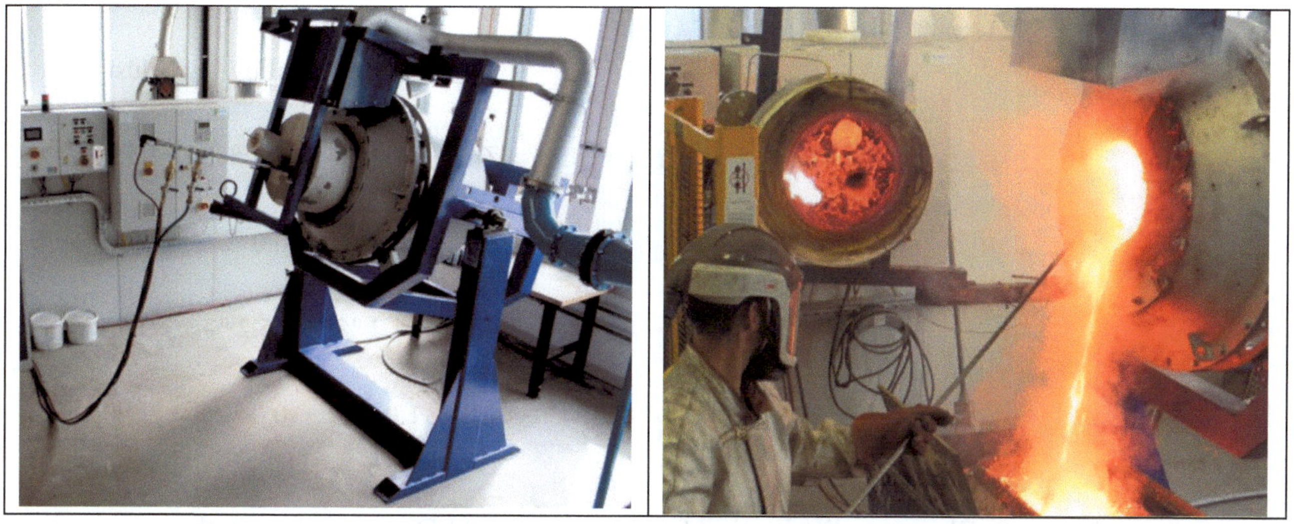

Figure 8: Top Blown Rotary Converter (TBRC)

Results of investigations

Trials in the TBRC for the first clinkering step were carried out with 30 kg of EAFD-pellets which were charged in the 1100 °C preheated furnace and treated for 30 to 60 minutes. The results showed a satisfying volatilization of fluorine, chlorine and lead especially after 60 minutes. Table 1 gives the extraction rates of lead, chlorine and fluorine under different treatment times.

Table 1: Clinkering trials in the TBRC (mean values of 3 trials each)

yield [%]	Pb	Cl	F
EAF-dust (30 min)	76.1	68.4	76.9
EAF-dust (60 min)	87.9	95.1	94.4

Due to some volatile zinc-compounds also some zinc losses resulted but were relatively low with an average amount of 3.5 % in the clinker. With this the zinc amount in the dust pellets increased from 32 % to roughly 35 %.

The clinkered material was then charged in hot stage (800 °C) into the second process step, the reduction. This was also carried out in the TBRC providing a carbonized iron bath at 1400 °C. The mass ratio of iron bath to clinkered pellets was 4:1.
Zinc and iron oxides were reduced and zinc evaporated. Remaining halides and lead oxide went into the off-gas with the zinc. Figure 9 gives a scheme of the elements distribution as well as the mean distribution values of three experiments.

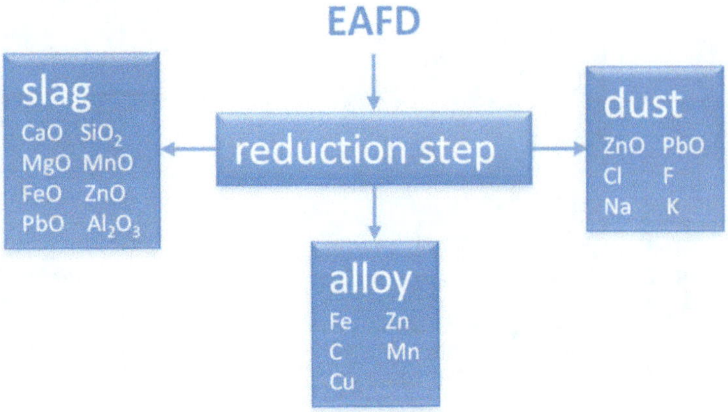

	slag	alloy	dust
Zn	3.0 %	1.5 %	95.5 %
Pb	1.0 %	0.0 %	99.0 %
Fe	14.2 %	85.2 %	0.6 %
Cl	1.0 %	0.0 %	99.0 %

Figure 9: Elements distribution for the system EAFD-carbon-iron bath

The results of the treatment of EAFD on an iron bath gave relatively accurate and detailed information about element distribution and dependence on parameters like temperature, slag composition, etc. Nevertheless, an important field is iron reduction, which was found to be the one with the lowest speed and with this also the lowest yield. Here it is foreseen for the upcoming trials to study the kinetics of this reaction. This will be mainly influenced by the slag composition as well as the temperature and the reaction surface. Two types of slag will be investigated in detail in further trials. Both should lie in the area of 0.9 to 1.2 concerning their basicity to meet the requirements for slag utilization after reduction. The difference would be that one slag is completely based on calcium- and silicon-oxide, while the other one should include a certain amount of wustite. The reactive surface conditions will be changed by using a conductively heated furnace where the relative surface is low and a TBRC where the mixture of charged material, slag and iron bath is expected to be a maximum.

The most interesting product, the recovered zinc oxide, showed the following composition:

Table 3: Composition of the zinc oxide (average values from 3 trials)

%	ZnO	Fe_2O_3	Cl	Pb	F	others
dust	96.8	1.1	0.25	0.3	0.15	1.4

The quality can be seen as similar to that one of a washed waelz oxide or even higher due to the low lead content and also low amounts of carry over. Nevertheless, there is a potential for improvement in further optimization trials.

Table 4 gives the average composition of the produced iron alloy:

Table 4: Composition of the iron alloy (average values from 3 trials)

%	Fe	S	Cu	Pb	C	others
alloy	95.7	0.5	0.7	0.1	2.4	0.6

Of course copper and sulphur are unwanted elements for utilization in the steel mill. Nevertheless, the use for the production of cast iron is one option but also for the steel production if an appropriate dilution could be realized.

Figure 10 gives a general overview of the two step process with the most important elements and compounds in the various products as well as a mass balance.

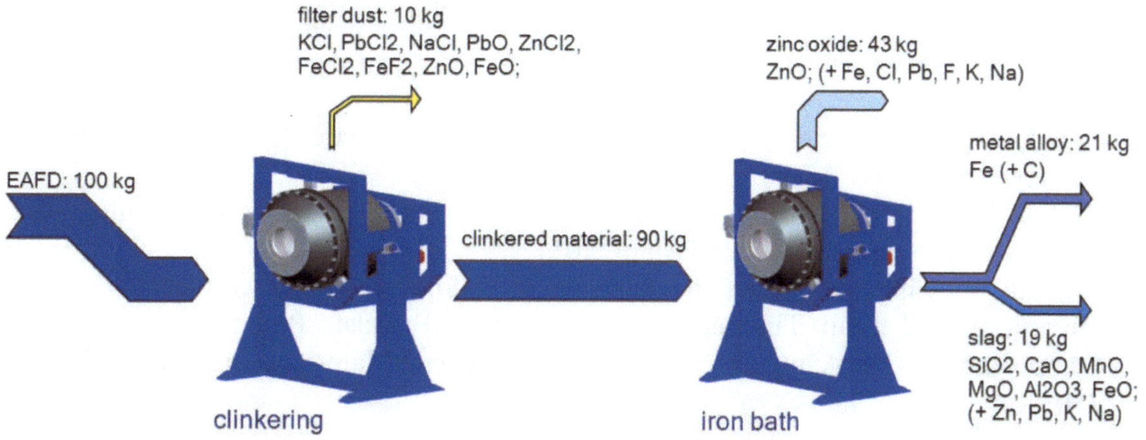

Figure 10: Results of first trials for the EAFD-treatment in the TBRC, off-gas is not considered

Economic considerations

Economic calculations are always difficult to generalize and very dependent on regional business of the individual steel mill. Roughly the revenues from the products could be calculated assuming that the slag goes out for zero costs. The zinc oxide was calculated with 70 % of the LME zinc value, which is based on utilization in primary metallurgy applying the usual concentrate formula. For a LME-value of 1850 USD/t of zinc revenues of 615 USD per ton of charged EAF-dust could be realized. For the iron alloy a low steel scrap price of 300 USD/t was taken as basis for the calculation resulting in revenues generated by the iron alloy of 60 USD per ton of charged EAF-dust. Taking treatment charges into account which are normally paid by the steel mill to the recycler further 50 USD per ton of dust can be added. Summing up overall revenues of 725 USD per ton of EAF-dust should be possible. Based on CAPEX and OPEX calculations this amount should make an economic treatment easily possible down to yearly throughputs of roughly 10000 tons of dust.

With this, the mini-mill solution should be realizable with the additional advantage of an environment friendly concept that only produces very low amounts of newly generated residues.

Conclusion

Recycling of steel mill dusts nowadays is a centralized business which shows various disadvantages for the dust producing steel mills. A compact mini mill solution for a single one or a certain number of small steel mills would be an interesting step forward in recycling of zinc and iron. Investigations at the Christian Doppler Laboratory/University of Leoben developed a two-step process based on a top blown rotary converter which should enable the treatment of comparable low dust amounts under economic conditions. Essential steps are the slag optimization, the pre-treatment in form of a clinkering-process and the reduction on a reducing iron bath. Results from experiments in a technical scale gave satisfying results which of course could be improved by further optimization trials. First economic considerations were also promising due to the flexibility of the chosen aggregate concerning its minimum throughput for economic operations.

References

[1] Gaugl, H.: Alternative Zinc Sources for the Electrolytic Galvanizing. Dissertation, Leoben, 2000

[2] Rütten, J.: Various Concepts for the Recycling of EAFD abd Dust from Integrated Steel Mills, Vernetzung von Zink und Stahl, 2011

[3] Gouzhu, Y.: Characterization and Removal of Halogens in the EAF Dust and Zinc Oxide Fume obtained from Thermal Treatment of EAF Dust. Proc.: Fourth International Symposium on Recycling of Metals and Engineered Materials, Pittsburgh, 2000, 271 - 280

REWAS 2013: Enabling Materials Resource Sustainability
Edited by: Anne Kvithyld, Christina Meskers, Randolph Kirchain, Gregory Krumdick, Brajendra Mishra, Markus Reuter, Cong Wang, Mark Schlesinger, Gabrielle Gaustad, Diana Lados, and Jeffrey Spangenberger
TMS (The Minerals, Metals & Materials Society), 2013

Secondary Processors and Landfills – Partnerships that Work

Prepared for TMS March 3-7, 2013 Conference

By: Ben Brewer
EWS & Recycling Ventures LLC, Nashville, Tn.
ben@elmingtoncapital.com

David J Roth
GPS - Global Solutions, Downingtown PA.
david.roth@global-solutions.us.com

Introduction

Using Best Available Technology is a phase that we often hear when there are environmental discussions on aluminum dross and secondary salt slag processing. The reality is best available technology is a mix between efficient removal of the valuable aluminum, oxides, misc metals and flux from dross and salt cake. This combined with conscientious land fill disposal of those items that finally, at this time, have no economic use is the reality of a company's best available actions. Recycling processes must be looked at with both the economic and environmental benefits weighed for their responsible implementation. This paper will discuss how this is done on a practical basis by Recycling Ventures (a secondary processor) and Environmental Waste Solutions (a Title II landfill), for the aluminum industry.

The Environmental Waste Solutions Story

Environmental Waste Solutions is a company committed to utilizing safer and more sustainable practices for the natural resources that have been given to us. By working smarter and more efficiently, they believe they can continue making our environmental stewardship and performance even better.

They specializes in reclaiming former mine pits, and turning them into carefully designed landfills that are classified for "beneficial use" according to approved civil engineering applications. The proposed site is studied to determine:

- The area of land necessary for the landfill
- The composition of the underlying soil and bedrock
- The flow of surface water over the site
- The impact of the proposed landfill on the local environment and wildlife
- The historical or archaeological value of the proposed site

EWS works to take otherwise unusable land and make it productive during the short term. Our long-term goals include turning the land into wooded preserves.

EWS continually strives to improve it's waste management standards and uphold the highest level of environmental responsibility. Our Camden Tennessee, landfill meets the strictest level of compliance with the standards of the Tennessee Department of Environment and Conservation. EWS also has the accreditation from the US Environmental Protection Agency for the construction and maintenance of our facility. This accreditation was not required, but was pursued to demonstrate our commitment to the highest environmental standards.

Our sites are thoughtfully designed and highly engineered to exceed state requirements and are professionally run to keep the community safe. Additionally, our landfill has undergone rigorous application processes, site surveying and ongoing maintenance, and have exacting plans in place for post-closure procedures to ensure a lifetime of responsibility.

Figure1. Environmental Waste Solutions Site..... Camden, Tennessee.

EWS accepts shipments only by contractual agreement from industrial partners who need to dispose of waste product in an environmentally responsible way. During landfill operations, incoming trucks are weighed on arrival and their load is inspected. Next, the vehicles unload at the tipping point. A warehouse is located on-site, so that if it's raining, covered trucks can come into a closed building and safely offload. This means they can receive shipments 365 days a year, rain or shine. After loads are deposited, bulldozers are used to spread and compact the new material on the working area of the landfill. At the end of the day, the working cell is covered with chert and compressed until it becomes impermeable.

Definition of Class II Industrial landfill

The EWS property is a Tennessee State and Federal Class 2 landfill, which is permitted to be used for industrial waste, including aluminum salt cake and a variety of other materials that are non hazardous waste.

The main material in the EWS landfill is the salt cake. This material is in mono cells. Salt cake placed in the landfill is covered each day with dirt that is compacted with a roller. That minimizes the amount of water that seeps into the landfill. Water that does get in drains to a liner at the bottom, which carries it off to a storage tank, from there it goes to a waste water treatment facility.

How a land fill works.

The basic parts of a landfill are:

- **Bottom liner system** – separates landfill content and subsequent leachate from groundwater. Leachate is liquid that drains downward through the landfill.
- **Individual cells** – where the material is stored within the landfill
- **Storm water drainage system** – collects rainwater that falls on undeveloped portions of the landfill and runs off without contacting waste material.
- **Leachate collection system** – collects rainwater that comes in contact with waste material and collects the water for treatment and proper disposal.
- **Covering or cap** – seals off the top of the landfill

Each of these parts is designed to address specific challenges that are encountered in landfill operation.

- The landfills **bottom liner** system contains the waste so that all potential problems are contained within this liner. The bottom liner prevents the waste from coming in contact with the outside soil, eliminating any potential groundwater problems. EWS uses a 6 - foot wide liner overlapping layered between the ground clay and man-made materials being deposited. The liner is triple layered on the sides and quadruple-layered at the base, far exceeding requirements.
-

Figure 2 ...Cell Preparation

- Landfill content is separated into areas, called **cells.** The material within the cell is compressed by heavy equipment (tractors, bulldozers, rollers and graders) that go over the day's accumulation several times. EWS stores a specific type of waste in each individual cell, hence it is a "monofill" composed of "monocells. This style of landfill use means that in the future, if and when it becomes economically feasible, the landfill can be mined and its contents harvested for recycling and reuse.

- The **storm water drainage** system is built to minimize water discharge. This is accomplished in two ways:

  - Preventing the rainwater from entering site to the extent possible. The cells are covered with a 6" clay layer daily to keep out rainwater.
  - Draining rain water, thru a storm water drainage system. Plastic drainage pipes and storm liners collect water from the landfill and channel it to drainage ditches surrounding the landfill's base. The ditches carry water from collection areas to the site of the storm water collection system.

- The **leachate collection** system is system is designed to exclude any water from a landfill that comes in contact with the waste. This water runoff called, leachate is typically acidic. To collect leachate, perforated pipes run throughout the landfill. These pipes then drain into a leachate pipe, which carries leachate to a leachate collection tank. The leachate is tested and allowed to settle. After settling, all water is professionally treated.

- **Groundwater monitoring** is done at many points around the landfill by groundwater-monitoring stations. These are pipes that are sunk into the ground so water can be sampled and tested for the presence of leachate chemicals. EWS uses three times the number of test wells required, just feet away from the waste stream. More test wells allow us to detect potential situations before they can become a problem.

- A **covering or cap** of six inches of a clay and gravel mixture called chert is placed on top of the waste and compacted each day. This covering seals the compacted waste from the air and prevents water from getting through. When a section of the landfill is finished, it is covered permanently with a polyethylene cap (40 mil). The cap is then covered with a 2-foot layer of compacted soil. The soil is then planted with vegetation to prevent erosion of the soil by rainfall and wind.

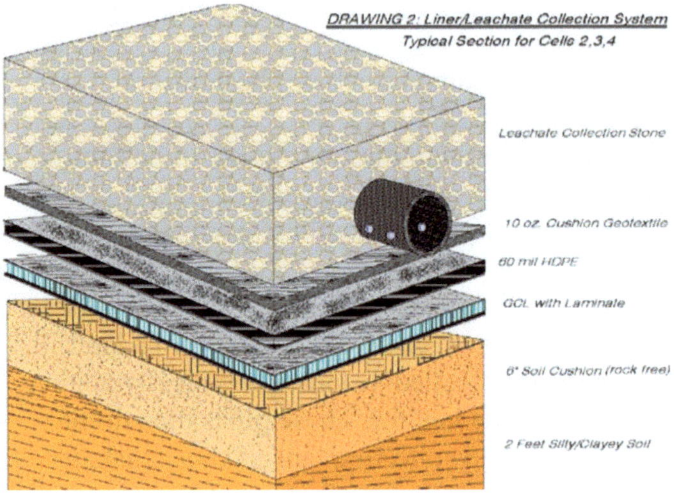

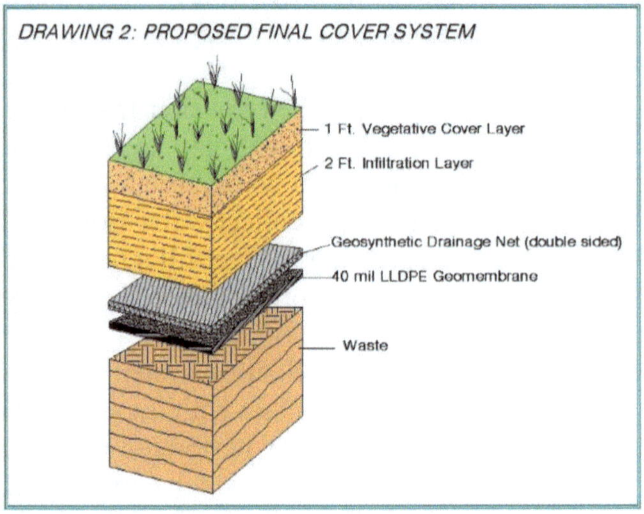

Figure 3 .... Leachate Collection

Figure 4 ...... Final Cover System

The above defines the complicated structure of a properly built and maintained landfill site.

Industry goals and objectives

The aluminum industry has started to focus on sustainability of their product designs and the materials that they use and discard in the production of aluminum. Almost all companies if you go to their web sites discuss increased recyclability and/or the reduction of landfill waste as part of their corporate mission statement or general goals. A few are noted below;

ALCOA ....0% landfill waste by 2030*
Novelis...............80% total recycled content of product by 2020 with 0% land fill waste*
SAPA................74.5% of former waste currently recycled, goal of continuous improvement*

Looking at sustainable solutions for the dross and salt cake are part of the aluminum industries goals and the key concept of the practices sought by EWS /Recycling Ventures model.

The Recycling Ventures Story

Recycling Ventures was created in alliance with EWS to develop economically feasible methods to reduce, and eventually eliminate first salt cake waste and then attack black dross, white dross and other metal slag waste.

In the case of salt cake, Recycling Ventures is an environmentally sensitive company that uses no chemicals to separate the aluminum from the salt cake. Currently Recycling Ventures takes the salt cake waste stream being land filled by EWS and recycles the waste in alignment with current best practices. They are able to reduce the amount of waste by 1-15% through our recycling process.

Additionally, Recycling Ventures has considered the environment first through the design and construction of our facility and stands ready to implement additional technology that shows further waste reduction in the future. Recycling Ventures is also pursuing the future design and implementation of technology that would provide the ability to fully recycle salt cake.

The recycling of black and white drosses in a sustainable manner is in many ways similar to the approach used on salt cake. The goal being to work toward the elimination of as much land fill waste as is feasible and find alternate uses for the oxide , flux and fine aluminum/oxide mixtures.

The RV Process

The DIDION RT Metal/Dross Reclaimer System combined with magnetic and eddy current separation was the first relatively straight forward and simple development that allowed for total salt cake and dross phase separation processing. The system allows for a mechanical concentration of the aluminum and separation of the non aluminum materials such as miscellaneous ferrous, oxides and flux components.

Mechanical separation of the aluminum metal from the flux, oxide and iron bearing materials has several advantages. It lowers the amount of aluminum sent to thermal processing saving the burning of carbon base fuels. Removal of the iron typically increases the value of the metal units produced from the aluminum concentrates that are generates in the process and provides another metal for direct recycling. Separation and sizing of the flux from salt cake and black dross allows for easier flux recycling and development of alternative uses for these materials. Separation of the aluminum/oxides and oxide fines from whites allows for the development of by product markets for these materials. The RV process provides an alternative to the oxides being contaminated with chloride bearing compounds.

Figure 5… DIDION RT 84 TUMBLER SYSYTEM with Magnetic Separation.

The RT Systems as utilized by Recycling Ventures is a very unique piece of equipment used to mechanically process dross, slag and other waste materials in which there is a malleable fraction (aluminum, brass, stainless steel etc.) mixed into more friable fractions (refractories, stone, oxide and flux). This one pieces of equipment allows a land fill with mono cell capabilities many opportunities for alternative recycling of materials.

The system is equipment with all the latest pollution control features to not only control dust generation but also to enable the dust fractions to be potential by products of the process. Bagged fines from the pollution control unit are perfectly packaged for efficient sale or disposal in the landfill if required.

Process flexibility is important in the land fill application and material of almost any size can be charged into the front of the RT unit. Systems are sized for the generators dross/salt cake block size. The oversized materials either come out the front or back ends of the drum. All the fine materials come out the front end of drum.

The equipment can be operated by unskilled labor grades. It is simple to run, low in maintenance and batch operated to keep unlike materials and alloys separate.

Figure 6 – Torit Bag House System

Summary ....Sustainable Solutions for Salt Cake and Dross Processing

Mechanical processing of salt cake and dross is an old approach that is being reinvented to match the current environmental mandate of total process sustainability. Single, relatively inexpensive mechanical separation systems coupled with magnetic and eddy current separation can do what complicated large processing plants were required to do the in the past. Impacting and breaking away the oxides and flux from the metal particles allowing for particle sizing for major advantages in either selling or for final environmentally sound land filling is the philosophy the EWS and RV has taken to minimize waste and maximize today's and the future values of the materials.

Various materials other than aluminum dross can be approached and processed in this same method, such as waste foundry sands, brass along with alloy and stainless steels. Keeping the valuable by products for sale and use, while efficiently land filling for now those items that currently have no viable market but may have in the future.

References*

1. Environmental Waste Solutions Web Site
   http://ewssite.com/how-it-works
2. Alcoa Inc.
   http://www.alcoa.com/sustainability/en/pdfs/2011_Sustainablity_Highlights_Report.pdf
3. Novelis Inc.
   http://www.novelis.com/en.us/Pages/Sustainability.aspx
4. Novelis Inc.
   http://novelis.com/en -us/SiteAssets/Sustainabillty/Novelis%20Sustainability%20Report.pdf
5. SAPA AB
   http://sapagroup.com/pages/488359/Sapa%20Sustainability%20Report2011.pdf
6. D.J.Roth, Mechanical Dross Processing an Approach to Zero Waste from Smelter and Secondary Dross Processing, Prepared for INCAL 2011 – September, 2011
7. D.J.Roth, Primary and Secondary Aluminum Industry and TUMBLER Processing – *February 2010*

REWAS 2013: Enabling Materials Resource Sustainability
Edited by: Anne Kvithyld, Christina Meskers, Randolph Kirchain, Gregory Krumdick, Brajendra Mishra, Markus Reuter, Cong Wang, Mark Schlesinger, Gabrielle Gaustad, Diana Lados, and Jeffrey Spangenberger
TMS (The Minerals, Metals & Materials Society), 2013

MATERIAL AND ENERGY BENEFICIATION OF THE AUTOMOBILE SHREDDER RESIDUES

N. Menad[1], N. Kanari[2], S. Guignot[1], F. Diot[2], L. Filippov[2], F. Thomas[2] and J. Yvon[2]

[1] BRGM – D3E/DMP, 3, avenue Claude Guillemin - BP 36009, 45060 Orléans cedex 02, France.
[2] Laboratoire Environnement et Minéralurgie, UMR 7569 CNRS, Nancy-University, ENSG, BP 40, 54501 Vandœuvre-lès-Nancy, France.

Keywords: automobile shredder residues, thermal beneficiation, iron oxides reduction

Abstract

Although vehicles represent a main key of our modern society, they affect our environment *via* the energy and resource consumption, waste generation during their manufacturing as well as greenhouse gas emissions all along their use. Further, hazardous residues are produced at the end-of-life vehicles "ELV". After collection and dismantling, the remainders of the ELV are directed to shredding operator followed by a series of mechanical and physical separations in order to recover the ferrous and non-ferrous metals. The residue of the shredding process, called automobile shredder residue "ASR" represents about 20-25% of the ELV. The ASR, while toxic enough to be classified as hazardous waste, could be considered as material and energy sources.

The present study deals with the possibility of material and energy beneficiation of the ASR by its use in the metallurgical units. ASR samples from an European automobile shredder company were collected and subjected to the physical separation process followed by a thermodynamic approach and isothermal batch tests to assess the reducing performance and energy capacity of the ASR hydrocarbon matter. Particular attention was devoted to the behavior of several residual and tramp elements (Cl, Pb, Cu, Zn) affecting the metallurgical process and the product quality. Results showed that physical operations (screening, attrition, dry low intensity magnetic separation) lead to a selective extraction of the mineral part of the ASR which can be directed to the blast furnace unit. Direct reduction of hematite by the plastics contained in the ASR was obtained at 1000-1050 °C resulting into multistage steps of Fe_2O_3 converting into metallic iron. Multi-parametric analysis of the results suggests that the purified ASR can partially substitute raw materials used in pig iron and steel production.

Introduction

Vehicles represent a main key of our modern society and their number in use is increasing from year to year, reaching, in 2010, about one billion cars and light trucks on world's roads [1]. However, vehicles affect the environment through their entire life cycle. This environmental impact relates to the energy and resource consumption, waste generation during their manufacturing as well as greenhouse gas emissions all along their use. Further, hazardous residues are produced at the End-of-Life Vehicles "ELV". 11 million vehicles are annually discarded in the European Union, resulting in about 9 Mt of wastes that have to be disposed of [2]. Current European state of practice for spent cars processing falls into a manual depollution followed by the shredding of the cars hulks, which generates 2 to 2.5 million tons of Automotive Shredder Residues "ASR" each year [3]. Without further treatments, this residual stream is mainly landfilled or sent to municipal incinerators [4]. European legislation will impose a recycling rate greater than 95% of the car initial weight by 2015, resulting in less than 5% of waste discharge [5]. Although, the ASR is reactive enough to be classified as hazardous waste, it could be considered as a source of energy and reducing agent since it contains more than 70% in combustible hydrocarbon matter (Figure 1), especially plastics. Note that the shredder companies use, besides ELV, other materials (e.g. plastics from discarded electric and electronic assemblies) in the shredding process leading often to a heterogeneous ASR depending on the input materials. Meeting the legal requirements by addressing the issue of ASR has led to the development of several Post-Shredder industrial lines of treatments [4], in which a fluff fraction is recovered by vacuuming ASR lighter components during shredding. Residual metals and plastic-rich fractions are then extracted from this fluff by means of magnetic, eddy current and sequences of sink-float separations.

Besides being marketed anew, recovered plastics can be valued as a relevant additive to the pulverized coal injected through blast furnace tuyeres [6]. In addition, considering that iron content in ASR can mount up to 18 $(\%)_{wt}$ of the light fraction [7], residual mineral fractions might also be used as part of the ore feedstock in metallurgical furnaces, provided further enrichments are performed to curb specific tramp elements. For instance, upon combustion, materials containing chlorine generate gas which is corrosive for the furnace and the dusts removal post-treatments [8]. K_2O can deposit into coke and sinter porous structures and weakens their Reduction Degradation Index (RDI). Copper is related to surface cracking of steels during hot rolling process. Sb, Sn and As accentuate this negative effect by segregating at grain boundaries during coiling in the hot strip mill, thus reducing the grain cohesion and thereby favoring embrittlement. This segregation is higher in steel containing Ni, Mn and Cr. Lastly, gaseous zinc can condense onto the blast furnaces inner walls, favoring scabs that modify the furnace flow configuration.

Despite such considerations on the chemical composition of plastic and mineral fractions generated from Post-Shredder industrial lines of treatments, few studies consider the design of a separation process which would open the way for a Blast Furnace “BF” injection. The present paper addresses this issue by focusing on 2 plastic fractions and 2 mineral fractions sampled on an industrial Post Shredder line of ASR treatment, so as to dispose of representative composition data. The separation processes associate sieving, gravity separation and low intensity magnetic separation. The separation efficiency is discussed in line with the chemical composition of the various outlets obtained. For the most interesting outlets, their use in a BF is assessed through a mass balance model of the tramp elements inside the furnace, with an attention paid to the effect on the pig iron specifications and the steel quality.

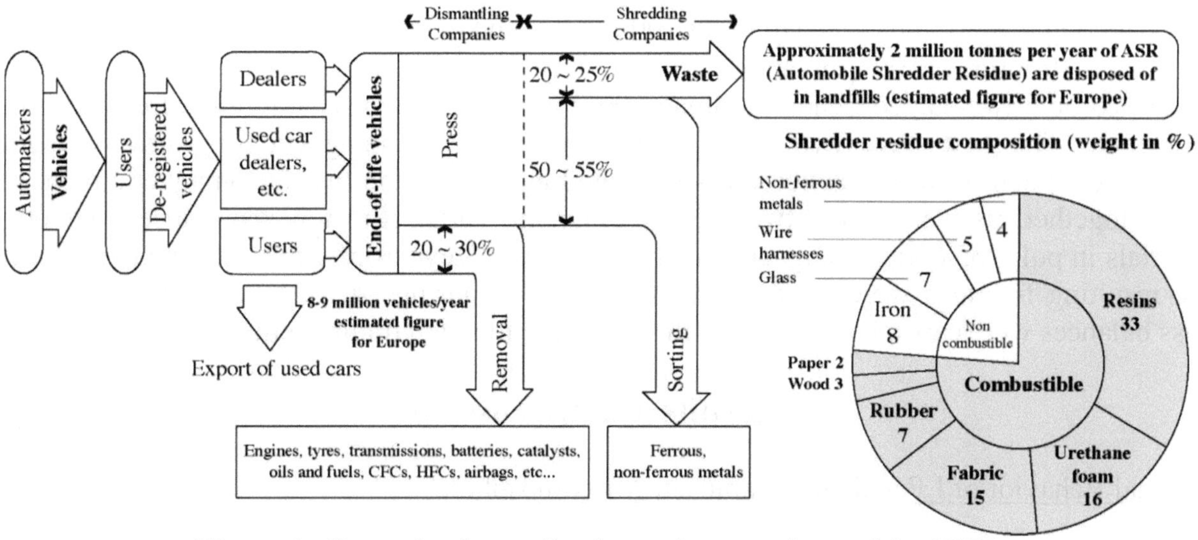

Figure 1: General scheme for the main operations of the ELV.

Materials

Four ASR samples were supplied by an European automobile shredder company and two of them, composed essentially of the wasted plastics. They are separated by sink float process and they are classified as Light Fraction "LF" and Dense Fraction "DF". The carbon content of the LF is about 66.8% whilst that of the DF is 42.1%. Chlorine is almost concentrated in the DF (6.8% Cl) indicating the presence of the PVC and of certain rubbers. Tramp elements (Pb, Zn…) as well as calcium are also found mostly in the DF. The other two samples were sampled from successive dry low intensity magnetic separations and called mineral fractions (MF#1 and MF#1). These fractions contain 36.8% and 19.4% of iron oxides respectively. Chlorine is almost the same in both fractions (0.3%). The carbon content of the MF#1 is approximately 8% whilst that of the MF#2 is 10.5%. The tramp elements (Pb, Zn….) as well as CaO and SiO_2 are found in both fractions.

Methods

Two PVC samples were used for this investigation of the plastic fractions. The first sample is wasted PVC and it is noted as PVC-1. The second sample is pure chemical grade PVC and is called PVC-2. Iron oxide, subjected to reduction, is high purity hematite (Fe_2O_3). Thermodynamic approach using HSC chemistry was considered for selective removal of tramp elements (Pb, Cu, Zn …) *via* chloride formation & volatilization and for hematite reduction. Mixtures of ASR and/or PVC samples with hematite were conditioned as pellets by using kaolin clay as useful bonding and plasticizing agent. Obtained pellets were dried in an oven at 70 °C before thermal treatment and different analysis tests. The thermal behaviors of the individual samples and their mixtures were checked in horizontal experimental set-up including a static electrical furnace. Most of the tests were performed under air atmosphere for the treatment of the ASR and PVC alone. The thermal reduction of hematite containing mixtures was performed without gas circulation and/or under a low flow rate of air. Raw samples and solid reaction products were examined by scanning electron microscopy coupled with energy dispersive spectroscopy (SEM-EDS), X-ray diffraction (XRD) and infrared (DRIFTS) spectroscopy.

The two mineral fractions MF#1 and MF#2 were subjected to physical separations. Sorting operations such as grinding, screening, attrition, and magnetic separation were used in this investigation. Two-stage sieving were performed, first on a rotating, perforated drum loaded with steel bars (cut 2 mm), then manually (cut 100 μm) on the -2 mm fraction, resulting in three sub-fractions: -100 μm, +100 μm and + 2 mm. Dry Low Intensity Magnetic Separation was performed on a two-staged Lenoir Rollmag 500-2 separator. Two separations were successively performed. The first gave two fractions composed of ferromagnetic or weakly magnetic compounds. Each of these fractions was separated again, and the two ferromagnetic parts were mixed together, as well as the two weakly magnetic parts. Attrition was carried out on the materials in pulp (65 $(\%)_{wt}$) using a Cylab-1 (2125 rpm). 5 cycles of 1 min each were performed. The resulting fractions were dried in an oven (105 °C during 24 h). Using composition data, mass balances were carried out at a blast furnace scale for each tramp element.

Results and Discussion

Thermal Behavior of LF and DF Samples in Air Atmosphere

The investigation of reactivity of ASR and PVC samples was performed between 225 °C and 900 °C under air atmosphere. Results are shown in Figure 2 as the evolution of the percent weight loss (%WL) of the samples as a function of the temperature.

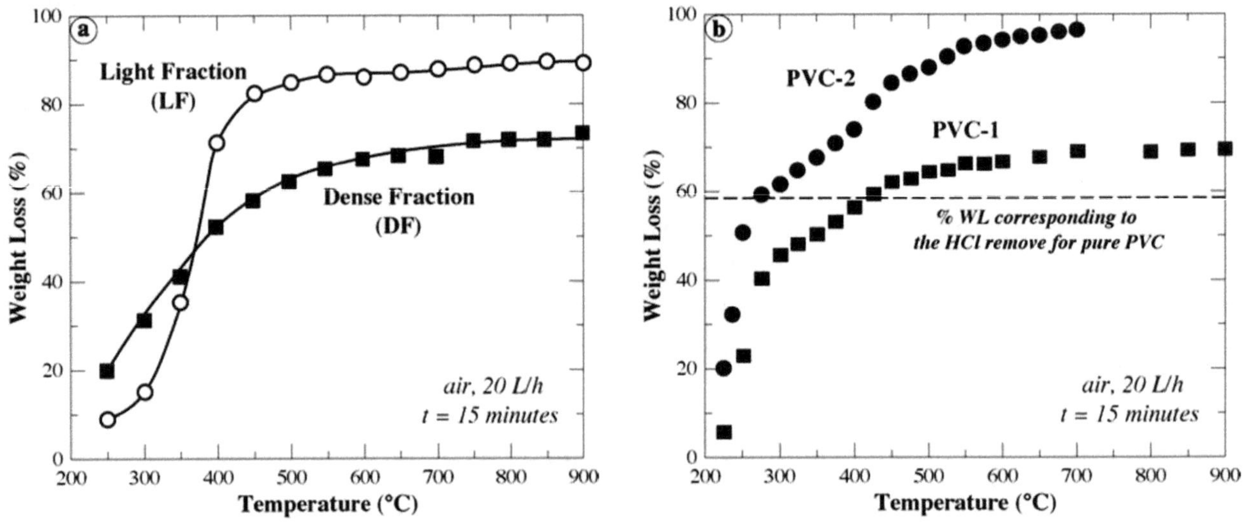

Figure 2: Behavior of ASR (a) and PVC (b) samples in air between 200 and 900°C.

From Figure 2 (a), it can be seen a very rapid weight loss of both plastic fractions (LF and DF) at temperatures lower than 600 °C. This is due to the combustion of the organic matter contained. Scanning electron microscopy (SEM) and X-ray diffraction (XRD) reveal the presence of mineral compounds (Si, Ca, Mg, Al…) in the residues obtained at higher temperatures. A rough examination of the %WL curves for PVC samples (Figure. 2-b) suggests that at least two phenomena should be distinguished at temperatures lower and higher than 300 °C. The first phenomenon could be attributed to a complete (PVC-2) or partial (PVC-1) removal of chlorine. The %WL observed at temperatures higher than 300 °C may indicate the reaction of the hydrocarbons with oxygen giving carbon oxides and water vapor as final reaction products. Detailed analysis for de-chlorination of both PVC samples was given early [9, 10]. However, these results suggest a high reactivity of all studied samples at relatively low temperatures.

The results of the analytical techniques (SEM, XRD, and IR) on treatment products have been at the base of deductions on the thermal behavior of PVC samples. The IR spectra of the PVC-1 treated at different temperatures are grouped in Figure 3. It can be observed that between 2950 and 2840 cm^{-1} some signals characteristics of elongation C-H of CH_x groups. The intensity of these bands decreases with increasing temperature to 'disappear' for T ≥ 600 °C. This is consistent with the data given in Figure 2. The presence of carbonates is observed in all the spectra up to 900 °C. XRD confirms the presence of calcite in the initial PVC and heat treatment products.

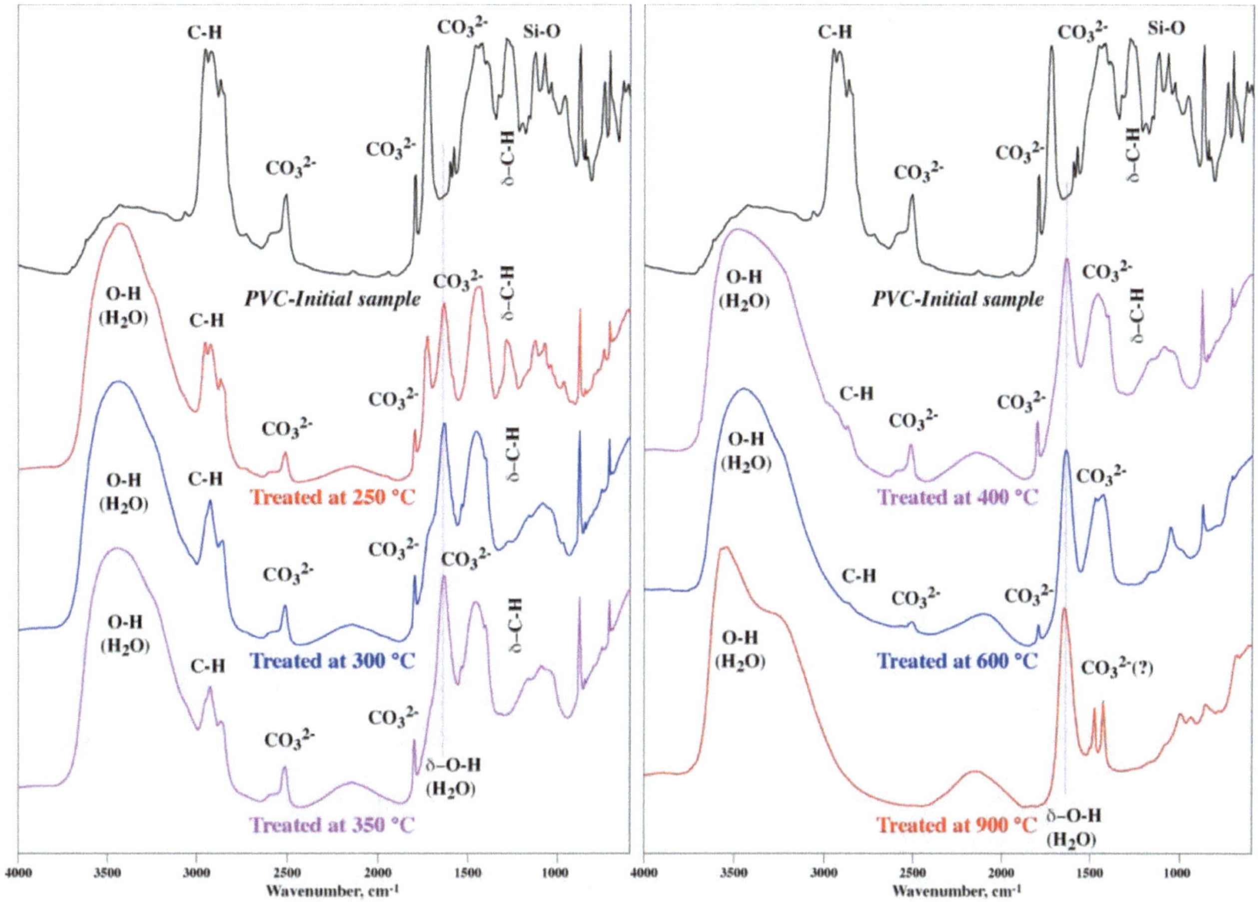

Figure 3 - IR Spectra after treatment of PVC-1 between 250 and 900 °C.

<u>Reduction of Hematite by PVC and ASR</u>

The following mixtures were tested in view of iron oxide reduction

1. 35% LF + 35% hematite + 30% clay,
2. 35% PVC-1 + 35% hematite + 30% clay,
3. 50% PVC-1 + 50% hematite.

The results show that the chlorine was almost removed at about 275°C from PVC, and then, the mixtures were de-chlorinated at 300 °C for 30 minutes. Subsequently, free of chlorine pellets were heated from 600 to 1050 °C for 30 minutes. Several tests were also carried out at 1025 °C with a residence time from 7.5 to 240 minutes. Obtained experimental and analytical results related to the reduction of iron oxides are described below. They are compared with the well-known Chaudron and Boudouard diagrams drawn in Figure 4.

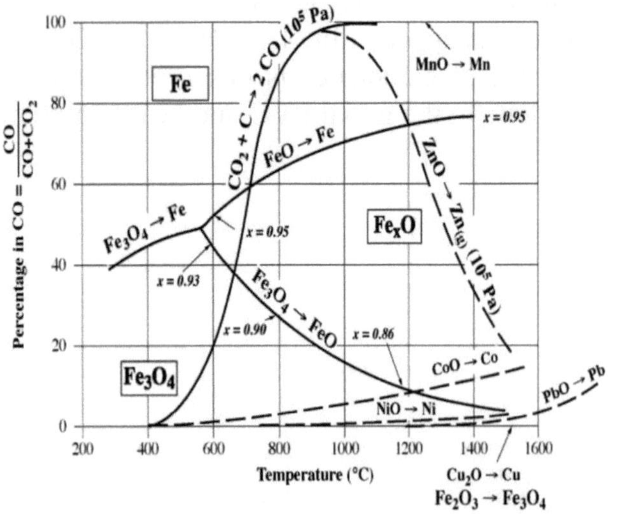

Figure 4: Equilibrium diagram of some metallic oxides reduction by carbon monoxide.

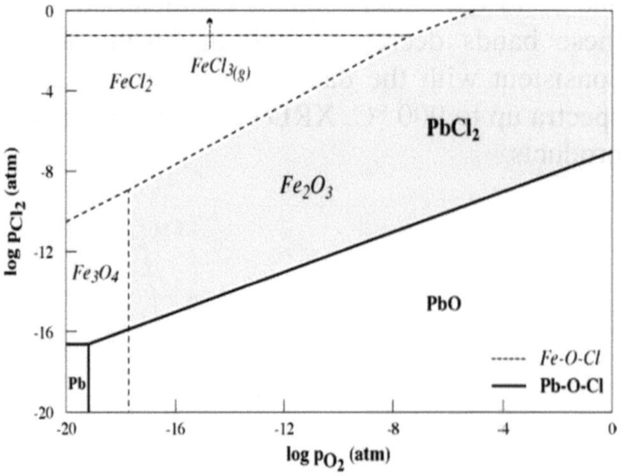

Figure 5: Phase stability diagram of (Fe, Pb)-O-Cl systems at 500 °C.

The following points summarize the results obtained from the reduction of hematite by plastics (PVC and ASR), mechanisms and the reduction steps.

1. The reduction of hematite to magnetite by hydrocarbon of PVC starts at 300 ° C, even in the presence of air (as shown in Figure 4, the reduction of Fe_2O_3 to Fe_3O_4 is possible for low CO concentration).
2. Metallic iron is obtained during the direct reduction of hematite at 1000 °C which is confirmed by XRD and SEM analyses. In addition, the crystallized fayalite (Fe_2SiO_4) is synthesized by the reaction of wüstite with quartz. The presence of the wüstite (Fe_xO) as intermediate phase of magnetite reduction at T > 570 °C is consistent with thermodynamic predictions given in Figure 4. The complete reduction of Fe_2SiO_4 in metallic iron and SiO_2 is obtained at 1050 °C.
3. The pellets reduced at T > 1000 °C contract probably due to the sintering of particles, or to partial melting of some constituents of the pellets. The narrowing of the pellets promotes contact of oxides with reducing agents (C, CO, H_2) improving the reduction process.
4. The reduction of iron oxides, particularly at high temperatures, is carried out by the CO and H_2. The carbon monoxide is generated *in situ* (gasification of carbon by the Boudouard reaction, Figure 4). Hydrogen is also produced during the gasification of carbon by water vapor. The latter comes from the ASR and from the release of structural water from clay during heating.
5. The direct reduction of pellets can be directed using the electric arc furnace for the manufacture of steel.

Considering thermodynamic investigation for a selective extraction of those metals and using an available thermochemical database for the thermodynamic calculation of various Metal-O-Cl systems, a typical example is the predominance area diagram given in Figure 5 for the (Fe, Pb)-O-Cl systems at 500 °C. This diagram clearly shows that lead chloride ($PbCl_2$) is the most stable phase even for a low partial pressure of chlorine and at high partial pressure of oxygen. By analyzing these results, a large zone (greyish zone in Figure 5) of chlorine and oxygen should be chosen in order to transform lead into $PbCl_2$, while hematite remains intact. Similar interpretations are valid for the systems containing zinc and copper. Generally, metal chlorides have lower boiling points compared with their respective sulfides and/or oxides.

As shown in Figure 6, metal chlorides are volatilized at temperatures lower than 1000 °C. This can be suggested that it is possible to achieve a direct reduction of iron oxides into Fe° by ASR and/other plastics and to remove selectively chlorides of tramp elements.

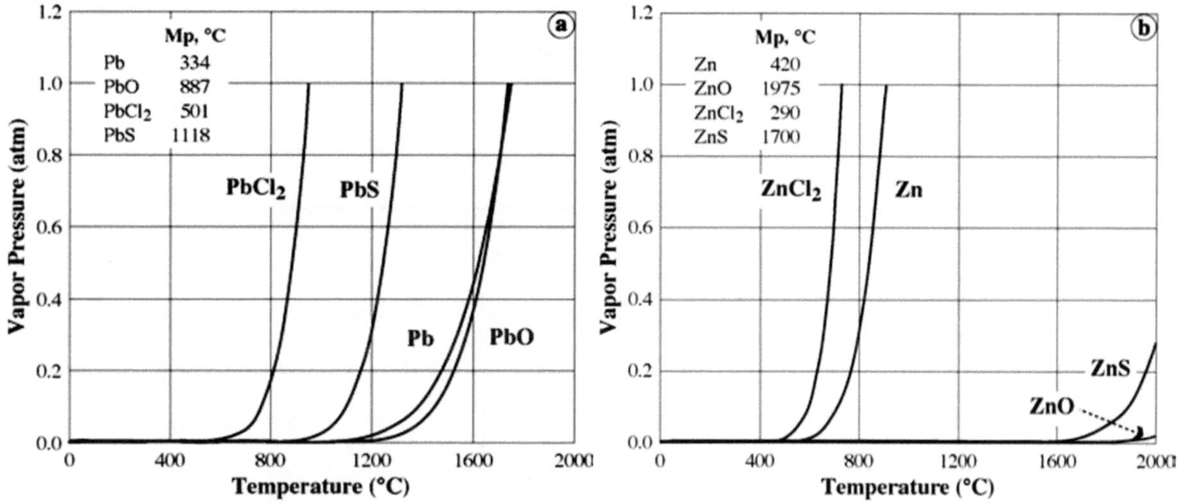

Figure 6: Evolution of vapor pressure of several lead (a) and zinc (b) compounds.

Mineral Fractions (MF#1 and MF#2)

The results of size distribution analysis show that most of the materials in MF#1 and MF#2 are located in -2 mm size classes (Figure 7). Fe and Cu in MF#1 are concentrated in coarser size classes, while Fe is homogeneously distributed throughout the classes of MF #2. Thus, a size cut at 2 mm will enable to purify this sample from Cu and to isolate an iron-rich sub-fraction in MF#1.

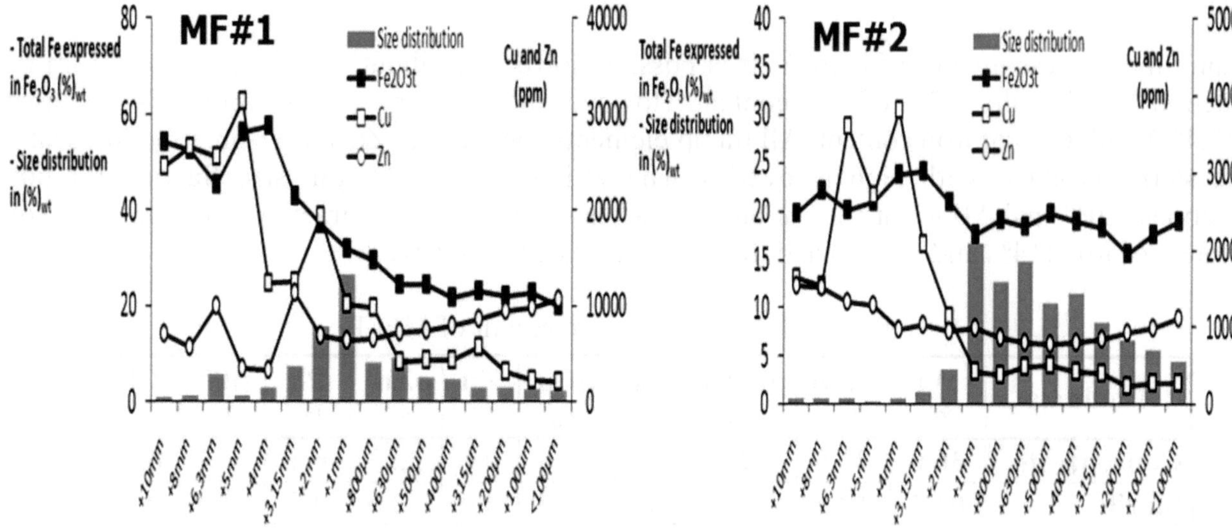

Figure 7: Size distributions and Fe, Cu and Zn repartitions in both mineral fractions.

Afterwards, a low intensity magnetic separation can be performed to recover magnetic metals. Lastly, considering that some of the smallest materials might be embedded in coarser complex matrixes, the following processes of separation is proposed in this study. This process includes sieving step for separate 4 size fraction of -100 µm, -400/100 µm, -2/400 µm, -8/2 µm and 8 µm. Manual magnetic separation to recover metallic scrap was applied on the fraction 8mm. The size fractions of –400/100 and -2/400 µm were treated by wet low intensity magnetic separation. Gravity separation using jig process was performed on the fraction -8/2 mm, the products obtained (light and heavy) were treated by magnetic separation. The size fraction of -100 µm was not treated. The products obtained from each operation were analyzed and the results are shown in tables I and II.

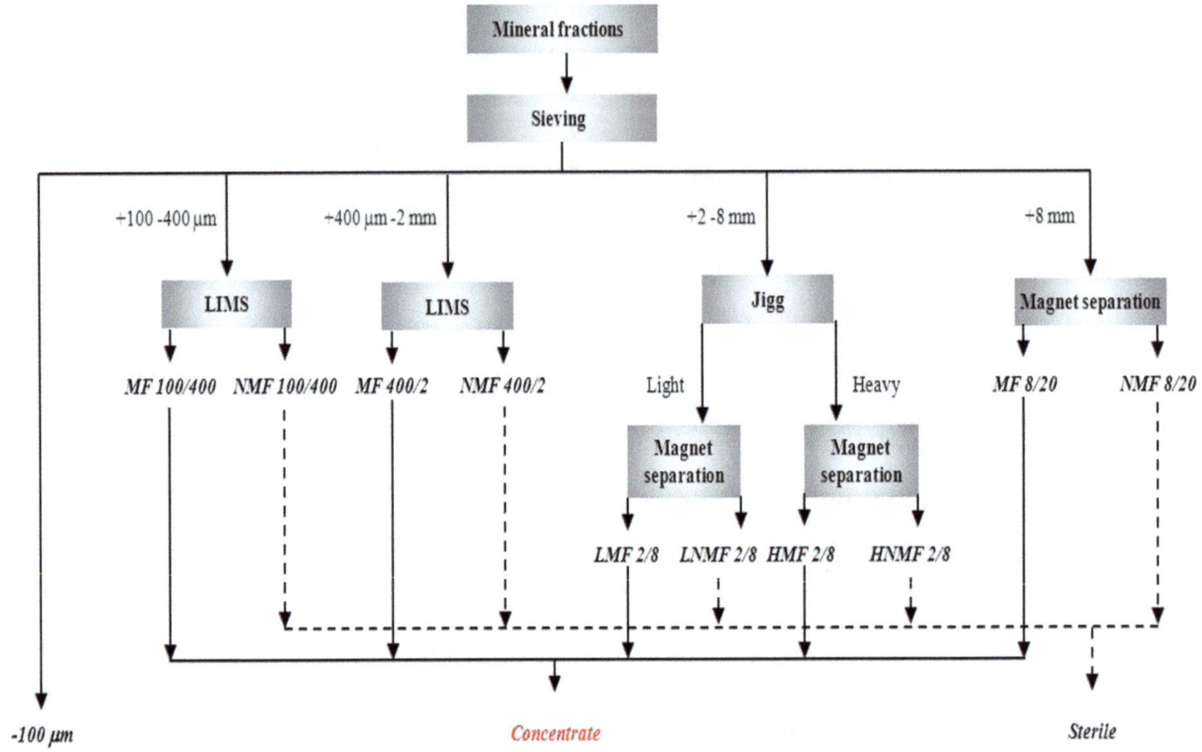

Figure 8: Wet treatment of Mineral fractions (MF#1 and MF#2).

Using the process given in Figure 8, it is possible to recover about 79% of total iron from the fraction MF#1 with 72.5% of iron content. However the recovery of iron from MF#2 is lesser (62.4%) with 62% of iron content. All tramp elements such as Cu, Zn and Pb are more present in the sterile products with high recoveries. The weight % of -100 µm size fraction for both fractions MF# and MF#2 are 2.6% and 5.4% respectively. These fractions contain very low content of iron (14% and 13.3%) and they are not treated in this work.

Table I. Results of wet treatment of MF#1

Products	**Wt %**	**FeOt %**	**ρFeOt %**	**Cu %**	**ρCu %**	**Zn %**	**ρZn %**	**Pb %**	**ρPb %**	**Cl %**	**ρCl %**
Concentrate	36.0	72.5	79.3	0.4	11.6	0.1	22.7	0.1	11.7	0.1	16.6
Sterile	61.4	10.5	19.6	1.6	87.9	3.6	72.6	4.9	85.5	0.3	81.9
-100µm	2.6	14.0	1.1	0.2	0.5	0.2	4.67	0.1	2.8	2.5	1.5
Total	100.0		100.0		100.0		100.0		100.0		100.0
Reconstituted		**32.9**		**1.2**		**0.5**		**0.6**		**0.2**	
Analyzed		**36.8**		**0.94**		**0.6**		**0.5**		**0.3**	

Table II. Results of wet treatment of MF#2

Products	Wt %	FeOt %	ρFeOt %	Cu %	ρCu %	Zn %	ρZn %	Pb %	ρPb %	Cl %	ρCl %
Concentrate	21.4	61.6	62.4	0.3	11.7	0.7	20.8	0.4	16.6	0.1	15.2
Sterile	73.2	8.7	33.7	0.7	86.4	0.7	71.9	0.5	75.6	0.2	79.5
-100μm	5.4	13.3	3.9	0.2	1.9	1.0	7.3	0.7	7.8	0.2	5.3
Total	100.0		100.0		100.0		100.0		100.0		100.0
Reconstituted		**18.8**		**0.6**		**0.7**		**0.5**		**0.2**	
Analysed		**19.3**		**0.5**		**0.9**		**0.6**		**0.3**	

Evaluation of the Investigated Samples Quality

The results of calculations using metallurgical model developed by steelmaking industry in France show that the mineral fraction MF#1 is the closest ore with regards to the iron content. However it is essential to know the nature of the other elements to determine the potential value of this product added to the materials at the agglomeration step before charging into the blast furnaces. Regarding to carbon products, the light plastic fractions (LF) could be more easily valorized in the steel industry, knowing that their major handicap is that they contain volatile carbon while coal or cokes are mainly out of solid carbon. Therefore, valorization channels could focus on:

1. injection into the slag devoted to the electric furnaces to make foam that lead to a better coverage of the electric arc and then improves its thermal performance,
2. addition to the scrap charge to boost the heating of the electric furnace,
3. injection in tuyeres of blast furnaces.

In both the cases (MF#1 and plastic products), it is essential that the process generating these products, used by many shredders, lead to significant tonnages, since the use of coal for foaming is of the order of 5 kg/t of liquid metal for production of electrical steel; what roughly need in France an equivalent of 25 kt of pulverized coal. It must also be considered that the product obtained has a regular composition, what remains to be checked, since the charge of shredders is heterogeneous (ELV, WEEE, other equipment). To achieve this goal, it would certainly be needed to go through a stage of homogenization of the products destined to steelmaking industries.

Conclusions

This work had ambition a multi-parametric analysis of the thermodynamic and experimental use of plastics and automotive shredder residues «ASR» as a substitute of energetic material and reducing agent in metallurgy of iron.

Recycling of ASR is performed at least through two stages: treatment / purification by physical methods followed by an injection of those purified ASR in the furnace. However, the physical methods of extraction of tramp elements (Cl, Pb, Zn, Cu) of ASR strongly depend on their dissemination, their liberation degree as well as their adherence to the matrix that surrounds them. In the absence of a final purification by physical methods, the thermal process developed in this study seems to be quite promising for an alternative process and industrial prospects. We deduced from this thermal approach, that the handicap of coexistence of PVC (source of chlorine) and tramp elements (Pb, Zn, Cu) can be an asset into respective volatile chlorides at low temperatures. In this context, the future directions of the ASR are:

1. light plastic fraction can be introduced into the chain of iron ore agglomeration to substitute combustible / reducing materials,
2. the mineral fraction (MF#1) can also partially substitute for material load for the agglomeration of iron,
3. material partially reduced by plastic in the form of pellets can be injected into the steel basket.

According to metallurgical operations models, there will be three opportunities to recycle the mineral fractions in the steel industry:

1. by assimilation into the iron load intended for the manufacture of the agglomerate produced which is then loaded to the top of the blast furnaces,
2. by direct charge to the topside,
3. by loading to the converter after agglomeration with binders.

The possibilities of valorization of plastic fractions are more numerous, since they can be used as:

1. alternative fuel at the settlement on grid,
2. component of the coke,
3. substitute for pulverized coal injected into the tuyeres of blast furnaces,
4. substitute for pulverized coal injected for frothing slag in electric furnaces,
5. substitute for coal brought in the load of scrap in electric furnaces to reduce oxides and bring energy,
6. thermal dopant in converter to increase the sediment charge.

References

1. Sousanis, J.: "World vehicle population tops 1 billion units, Wards Auto", http://wardsauto.com/ar/world_vehicle_population_110815 (2011). Accessed 25 May 2012.

2. Andersen, F., Larsen, H., Skovgaard, M. "Projection of end-of-life vehicles: development of a projection model and estimates for ELVs for 2005-2030". ETC/RWM working paper 2008/2, Copenhagen (2008).

3. Vermeulen, I., Van Caneghem, J., Block, C., Baeyens, J., Vandecasteele, C. "Automotive shredder residue (ASR): reviewing its production from end-of-life vehicles (ELVs) and its recycling, energy or chemicals valorization". *J. Haz. Mat*, 190 (2011), 8 – 27.

4. GHK/BioIS Intelligence Service. "A study to examine the benefits of the End of Life Vehicles Directive"– (Final report to DG Environment, 2006), from: http://ec.europa.eu/environment/waste/elv_study.htm

5. Directive 2000/53/CE du parlement européen et du conseil du 18 septembre 2000 relative aux véhicules hors d'usage. *J. Officiel des Communautés européennes*, L 269/34 (2000).

6. Menad, N. "Recycling of auto shredder residue". J. *Haz. Mat*, A139 (2007), 481 – 490.

7. Jalkanen, H. "On the direct recycling of automotive shredder residue and electronic scrap in metallurgical industry". *Acta Metall. Slovaca*, 12 (2006), 2625 – 2643.

8. United Nation Environmental Program (UNEP): Converting waste plastics into a resource – compendium of technologies (2010). Available from (accessed on April 10th 2012): http://www.unep.or.jp/Ietc/Publications/spc/WastePlasticsEST_Compendium.pdf.

9. Kanari, N., Menad, N., Diot, F., Filippov, L., Thomas, F., Yvon, J.: "Beneficiation of PVC wastes in iron oxide reduction". Proceedings of the 43rd International October Conference on Mining and Metallurgy, Kladovo, Serbia, 408–411 (2011).

10. Kanari, N., Menad, N., Diot, F., Filippov, L., Thomas, F., Yvon, J. "Valorization of the automobile shredder residues by thermal route" Proceedings of the 4th International Conference on Engineering for Waste and Biomass Valorization. (WasteEng 2012), September 10th – 13th Porto Portugal. A. Nzihou & F. Castro eds. Vol 4, 1085-1090.

REWAS 2013: Enabling Materials Resource Sustainabil
Edited by: Anne Kvithyld, Christina Meskers, Randolph Kirchain, Gregory Krumdick, Brajendra Mish
Markus Reuter, Cong Wang, Mark Schlesinger, Gabrielle Gaustad, Diana Lados, and Jeffrey Spangenberg
TMS (The Minerals, Metals & Materials Society), 20

ISASMELT™ for Recycling of Valuable Elements Contributing to a More Sustainable Society

Gerardo R. F. Alvear F. and Stanko Nikolic

Xstrata Technology
10th Floor, 700 West Pender Street, Vancouver BC V6C 1G8, Canada
isasmelt@xstratatech.com

Keywords: ISASMELT™, TSL, Copper, Recycling, Smelting

Abstract

Metals are essential for modern lifestyles. As standards of living improve and metal consumption increases, it is evident that raw materials are becoming scarcer. The recycling of metals is essential if we are to build a more sustainable society. ISASMELT™ Top Submerged Lance (TSL) technology can enable plant operators to recycle metals efficiently. The implementation of the ISASMELT™ technology for the recycling of valuable metals at Umicore Precious Metals Refining in Belgium and Aurubis AG in Germany are both good examples of how ISASMELT™ technology can be used to recycle a range of metals. Visionary people motivated by the need for a technological breakthrough have developed novel processes for recycling complex materials incorporating the ISASMELT™ technology.

This paper describes how the ISASMELT™ technology is used to recycle a range of materials. It highlights one of the most important strengths of the technology: its versatility and capacity to be incorporated into new, innovative applications.

Introduction

The 96th plenary meeting of the United Nations, held in New York on the 11th of December 1987, presented the outcomes of work done by the World Commission on Environment and Development. One of the main contributions of this commission was the introduction of sustainable development as a concept: “meeting the needs of the present without compromising the ability of future generations to meet their own needs” [1]. The commission's report challenged nations to reorientate their way of drafting national and international policies in order to legislate towards sustainable policies that would contribute to remedy numerous environmental issues and also influence the source of some of the problems in human activity. In particular, the report invited governments to work with non-governmental organizations, industry and the scientific community to support efforts towards sustainable development. This visionary statement was a call for innovation in a variety of fields: social, legal, economic, political and technological. It was a call to challenge the old paradigm of development associated with growth with little regard of the consequences for society and the environment. Instead, it proposed a new one, where innovation was expected to play a central role in helping to create sustainable growth.

In 1987 the non-ferrous industry, and in particular the copper industry, was recovering from the impact of a long economic recession. After four years of low copper prices, the LME price for copper at the end of 1987 was 145.4 U.S.c/lb. of copper, compared with 64.0 U.S.c/lb. of copper at the end of 1983 [2]. Although the copper industry recognized that the higher copper prices were an indication of better days to come, it was also clear that a new wave of technology innovation was required to meet the challenges of the future.

The period between 1983 and 1987 was also when Mount Isa Mines (now part of Xstrata) was demonstrating the application of the Sirosmelt lance in a new technology for lead and copper smelting. This technology was named ISASMELT™ after the mine at which the development occurred. The construction and operation of lead and copper demonstration plants at Mount Isa was the breakthrough in the scale-up of this new technology. Following the successful implementation of both demonstration plants, Mount Isa Mines approved the construction of commercial scale lead and copper plants. The development path from the conceptualization of the idea at the Commonwealth Scientific and Industrial Research Organisation (CSIRO) laboratories in Melbourne to its industrial implementation and successful demonstration required twelve years of fundamental research as well as applied scientific and industrial scale effort [3].

Innovation Takes Time

Twelve years to be able to demonstrate an innovation may appear to be a long time, but in the world of pyrometallurgy technological changes occur slowly. The adoption of new technologies occurs infrequently, as capital investment requirements for new smelters are generally very high and the implementation of new technology results in significant technical challenges. Generally various stages of development are required before a new process can be successfully demonstrated and applied. Consequently, although innovative breakthroughs can create opportunities to position businesses for new frontiers, such revolutions are rare. In most cases companies prefer to select proven technologies to minimize the risk and ensure fast ramp up of their facilities.

In some cases however, proven technologies are not capable of delivering the necessary economic and environmental benefits. In such situations breakthroughs are required and technical innovation plays an essential role in this process. For innovative processes to be successfully implemented many hurdles must be cleared. It is here where a systematic approach to the innovation process is required to mitigate the risk.

To be successful, process innovation requires constant critical analysis of the technology under evaluation. Each component and system requires detailed assessment and confrontation with regard to pre-established goals that the developer expects the process to achieve. Multidisciplinary teams must work collaboratively from a solid scientific foundation to reach the stage where a prototype can be tested. This process of constant experimentation, performance evaluation and product re-engineering, when in the hands of a capable team, becomes the source of more innovative ideas.

This was the case in the use of the Sirosmelt lance for the smelting of lead and copper concentrates in Mount Isa. The innovation evolved from the initial concept of a swirled lance at

CSIRO, to a fully developed process including a robust furnace that was suited to the requirements of Mount Isa Mines. However, this was only the first stage of the innovation of the process. In parallel to the industrial implementation and consolidation of the ISASMELT™ technology in the primary smelting of lead and copper concentrates, the ISASMELT™ concept was also tested and demonstrated for the primary production of other base metals such as nickel and zinc, the recycling of base metals, precious metals (PM) and platinum group metal (PGM) smelting and converting. Table I shows some of the applications investigated for the recycling of Metals. Two common links exist across the applications listed in Table I:

- Applying an innovative lance design to the development of new metallurgical processes; and
- Finding a sustainable process solution to recover valuable metals from metallurgical wastes or discarded end-of-life products; and stabilizing the final waste stream.

Table I. Different Applications of the ISASMELT™ Technology in the Recycling of Metals

Application	Scale	Target
Processing of Roast Leach EW Zinc Plant Residues	- Pilot Scale	- Recovery of Pb, Zn, Ag, Ge and In and reduction of disposable slag
Fuming of Lead Blast Furnace Slags	- Pilot Scale	- Recovery of Zn and Pb and production of dischargeable slag
Processing of Lead Smelter Dross	- Pilot Scale	- Separation and Recovery of Cu and Pb - Production of lead fume and copper matte
Processing of Cu-As residues	- Pilot Scale	- Recovery of Cu as copper matte
Pyrolysis of CCA-treated wood	- Pilot Scale	- Energy recovery - Production of disposable slags - Economic disposal of As
Fuming of EAF dust	- Pilot Scale	- Zn and Pb recovery - Halide recovery - Production of disposable slags
Recycling of Lead Batteries	- Pilot Scale - Industrial Scale	- Pb and Sb recovery - Production of Disposable Slags
Recycling of Copper, Lead, PM and PGM	- Pilot Scale - Industrial Scale	- Pb and Cu recovery - PM and PGM recovery - Production of Disposable Slags
Copper and PM Recycling	- Pilot Scale - Industrial Scale	- Cu and Sn recovery - Zinc recovery - PM recovery - Production of Disposable Slags

These two concepts, metals recovery and waste stabilization, became the key attributes for the successful application of the ISASMELT™ technology for recycling consumer waste and industrial by-products. The new business created by this innovation is now becoming known as "urban mining".

Innovative and Sustainable Metallurgy

Although metallurgy is an ancient discipline it still requires innovation. Sustainable innovative new processes can be developed by going back to the basics and applying the fundamentals of slag chemistry and thermodynamics using modern thermochemical software. The key to this approach is evaluating and considering fundamental principles such as the chemical thermodynamics, slag reaction mechanism, basic chemistry, metal and slag physical properties.

Good examples of how a combination of pyrometallurgical and hydrometallurgical processes can be applied in an innovative manner to recover and recycle base and precious metals are the use of the ISASMELT™ technology by Aurubis AG in the Kayser Recycling System (KRS) in Lünen, Germany [4], and by Umicore at their Base Metals and Precious Metals Operations in Hoboken, Belgium [5].

These two plants operated traditional smelters in the past using technologies such as Peirce-Smith Converters, Blast Furnaces and Hoboken Converters. However they understood that a new sustainable approach was required for them to meet new and more challenging environmental regulations to operate in a profitable and sustainable way in the 21st century [4,5]. The dramatic changes to the existing operations required a willingness to explore how the versatility of an ISASMELT™ furnace could be combined with complementary technologies to produce the targeted metals.

The design of the new operations at Hoboken and Lünen required innovative metallurgical approaches in order to investigate and define suitable operating conditions for a variety of feed materials. Investigations, conducted at lab and pilot scale levels, allowed for the generation of fundamental knowledge that could be combined with state of the art engineering to design, construct and operate both plants. From the fundamental metallurgical design point of view, the aspects listed below are relevant to the conceptualization, design and development of both metallurgical processes:

- Definition of suitable metallurgy to recover the targeted metals.
- Definition of suitable oxygen and sulphur potential conditions required to maximise recovery of the targeted elements.
- Characterization of physical and chemical properties of the molten liquid and gas phases.
- Design of an appropriate slag systems to:
    - Define different operating windows that provide process flexibility for processing complex and variable feeds
    - Optimize distribution of key elements between the slag and metal phases,
    - Contribute to the stabilization of refractory materials under the prevalent operating atmosphere.
- Design of appropriate off-gas handling facilities that allow for adequate post combustion and heat recovery of the generated process gases and dusts.
- Design of complementary pyro- and hydro-metallurgical processes to complement the core designed smelting process to maximise recovery of the targeted elements.

A suitable experimental process, including pilot plant testing for the ISASMELT™ process, was carried out by Umicore and Aurubis in collaboration with Xstrata Technology to define the prevalent operating conditions for their respective processes. Additional efforts in fields such as heat transfer, transport phenomena, combustion engineering, reactor design and finite element calculations were also required to complement the metallurgical design. The results of those investigations generated the required fundamental knowledge that allowed the design and construction engineering teams to complete the design of their respective plants.

Application of Metallurgical Theory for Process Development

Proper design of a sustainable pyrometallurgical process for recovery and recycling of metals requires managing a large number of factors. A key aspect in the recycling of electronic scrap is the diverse range of elements that can be contained in a product targeted for recycling. For example, a mobile phone can contain over 40 elements from the periodic table, including base metals like copper (Cu) and tin (Sn), special metals such as cobalt (Co), indium (In) and antimony (Sb), and precious metals including silver (Ag), gold (Au) and palladium (Pd) [6]. The denoted complexity of only this one part of the feed to an electronic scrap recycling facility shows that for these facilities to run, they not only require the use of state of the art technologies, but also the appropriate use of fundamental theory to maximise the recovery and stabilization rate of all targeted metals in a sustainable manner.

The diversity of minor elements and precious metals in the electronic scrap, along with the content of minor oxides, such as alumina, used in some of the components of the electronic materials are two relevant aspects to consider in the proper design of a metallurgical flowsheet.

Selective Oxidation of Minor Elements and Precious Metals in Recycling of Electronic Scrap

Selective oxidation is a well-known property that can be used as a fundamental tool to design a suitable metallurgical process. Figure 1 shows the oxygen-sulphur potential diagram at 1300 °C for (a) Cu, Pb, Ni, Sn, Zn, and Fe, and for (b) copper smelting processes. The combination of both diagrams can help in providing a better understanding on how to apply selective oxidation in recycling process design.

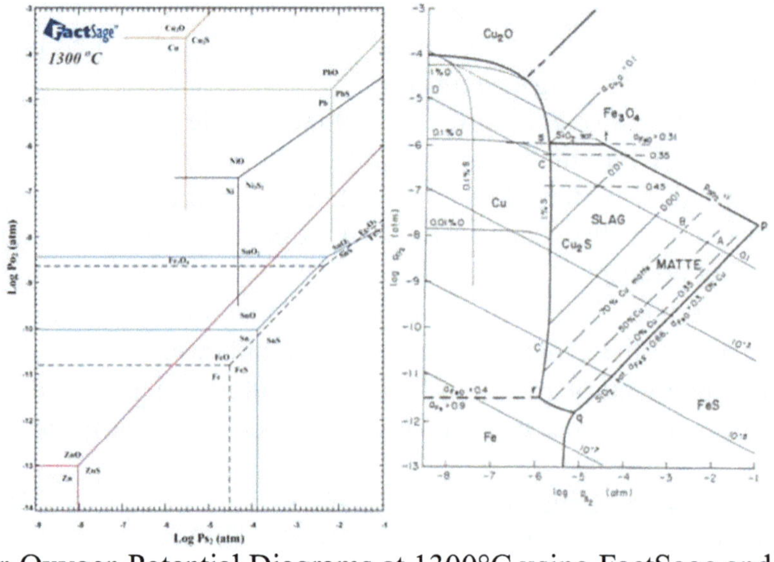

Figure 1 – Sulfur-Oxygen Potential Diagrams at 1300°C using FactSage and from Yazawa[7]

Example 1: Recycling of low copper content scrap with minimum sulphur content

In this case an initial reductive smelting process such as the one in place at the Aurubis ISASMELT™ plant is appropriate. The selection of these conditions will be represented by oxygen potentials below 10^{-9}atm and sulphur potentials below 10^{-4}atm, in the FactSage and Yazawa diagrams shown in Figure 1. Under these conditions base and precious metals will be predominantly deported to the metallic phases while oxide components associated with the recycled materials will form a slag phase. In a subsequent oxidation stage, the oxygen potential is increased to levels between 10^{-7}atm and 10^{-6}atm allowing partial or total oxidation of elements such as Fe, Sn, Pb, As, Sb, Ni, leaving most of the copper and precious metals in the metallic phase. The activity coefficient (affinity) of these elements with the base metal (Cu) play a relevant role in defining the minimum achievable concentration of these elements in the copper phase. The product black copper requires a subsequent pyrorefining stage to produce copper with a suitable quality for electrorefining. The flowsheet for the Aurubis KRS process, including the ISASMELT™ furnace, is shown in Figure 2.

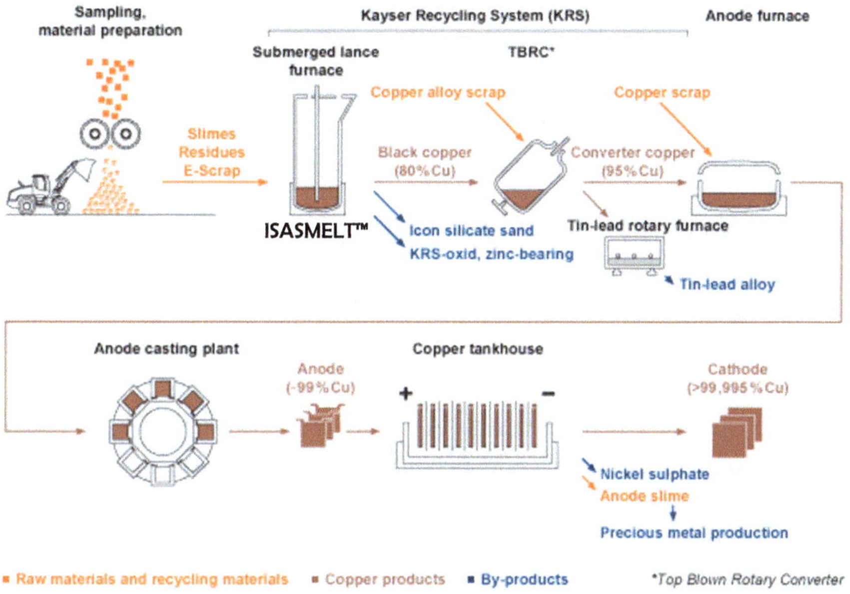

Figure 2 – ISASMELT™ Furnace in the KRS Flowsheet at Aurubis Lünen, Germany [8]

Example 2: Recycling of mixed PGM, copper and lead scrap with higher sulphur content

Due to the relatively high sulfur content in the feed, a copper-lead matte formation stage, as shown in the Yazawa diagram in Figure 1, is required. Sulphur potentials between 10^{-2}atm and 10^{-3}atm and oxygen potential between $10^{-8.5}$atm and 10^{-8}atm represent the matte formation stage of this type of process. After the selected matte grade is reached, a converting stage with further increase in the oxygen and sulphur potentials allows the formation of blister copper. This concept was applied in the design and operation of the Umicore ISASMELT™ at Hoboken shown in the flowsheet in Figure 3.

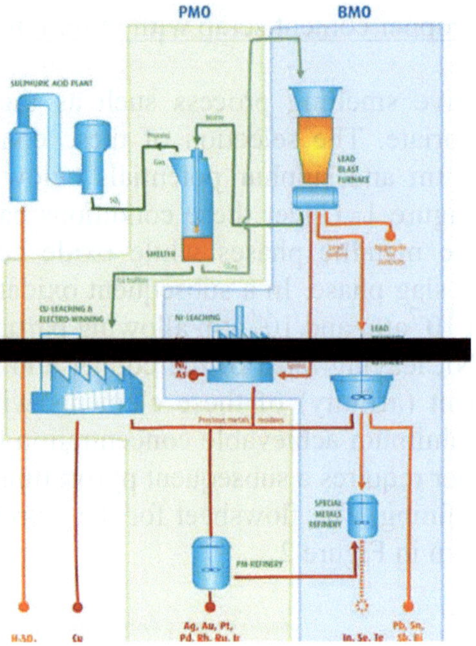

Figure 3 – ISASMELT™ Furnace in the Umicore, Hoboken Plant Operations Flowsheet [5]

Example 3: Refining of copper

Copper produced in an ISASMELT™ furnace needs to be refined. If a pyrometallurgical route is chosen as the first stage, an additional stage to allow oxidation of minor elements is required (levels near $Po_2 = 10^{-5}$atm in Figure 1). Two main aspects are relevant to consider in the design of this process:

- Selective Oxidation of Impurities: Appropriate conditions are required not only to maximize recovery of copper and precious metals but also to recover (in a subsequent reduction process) some valuable elements that will be oxidized in this stage such as tin, lead and copper.

- Selection of Appropriate slag systems to maximise the collection of impurities: There is no single slag system capable of efficiently removing all the impurity elements associated with the generated copper phase. Therefore a combination of selective slagging steps using different fluxes is required to achieve the necessary copper quality.

A different approach for copper refining such as the leaching and electrowinning, applied by Umicore, is also an option to achieve the required quality of copper and recovery of precious metals [9].

Selection of Appropriate Slag Systems for the Smelting Process

As previously mentioned, in addition to the diversity of minor elements and precious metals present in secondary materials such as waste electrical and electronic equipment (WEEE), minor oxides are also present. Therefore, selection of an appropriate slag system is also important to maximize the capacity of slags for the required elements and minor oxides. The selection of the slag system is governed by the chemical and physical properties of the slags, such as liquidus temperature and viscosity. For example, an incremental increase in the alumina content in the slag will decrease the liquidus temperature for given oxygen potential and operating temperature. Figure 4 shows this effect of alumina in a typical fayalite slag a 1200 and 1300 °C at different oxygen potentials, which represent operational conditions from reductive smelting to converting.

As the set of ternary diagrams at 1200 °C show (Figure 4 (a), (c) and (e)), at relatively low oxygen potential conditions (reductive smelting) the increase from 0% alumina to at first 4% and then 8% alumina in the slag increase the liquidus region by pushing the silica saturation boundary towards the silica rich region. However, if fluxing were to follow the path of the higher combined silica and alumina contents, this will play a detrimental role to the slag characteristics due to the higher slag viscosity. This is illustrated by Figure 5 which shows that with higher SiO_2/Fe ratios, namely higher silica concentrations, even though solids precipitation is avoided until over 5 wt% alumina in the slag, the overall slag viscosity is higher if higher silica regions are targeted.

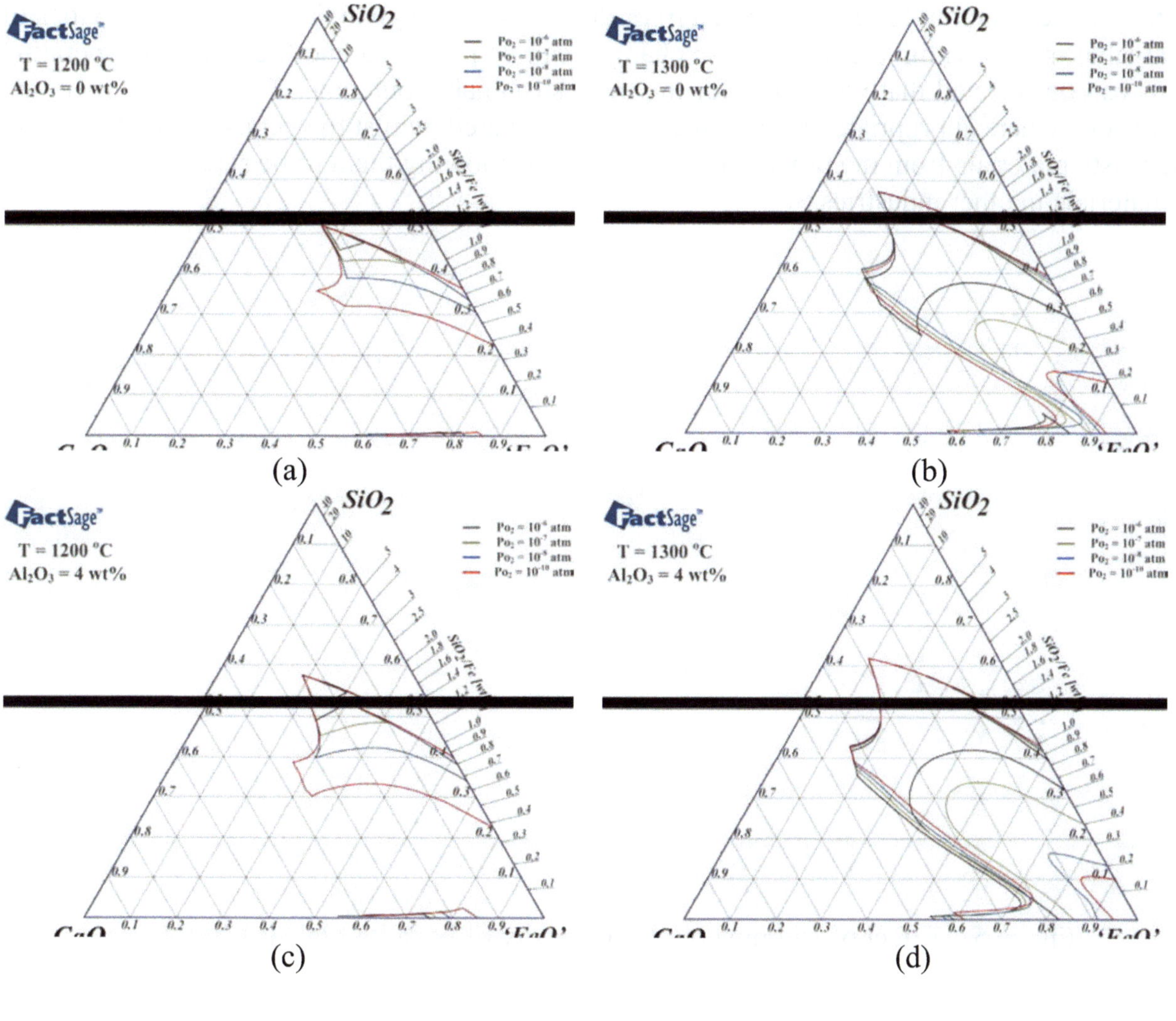

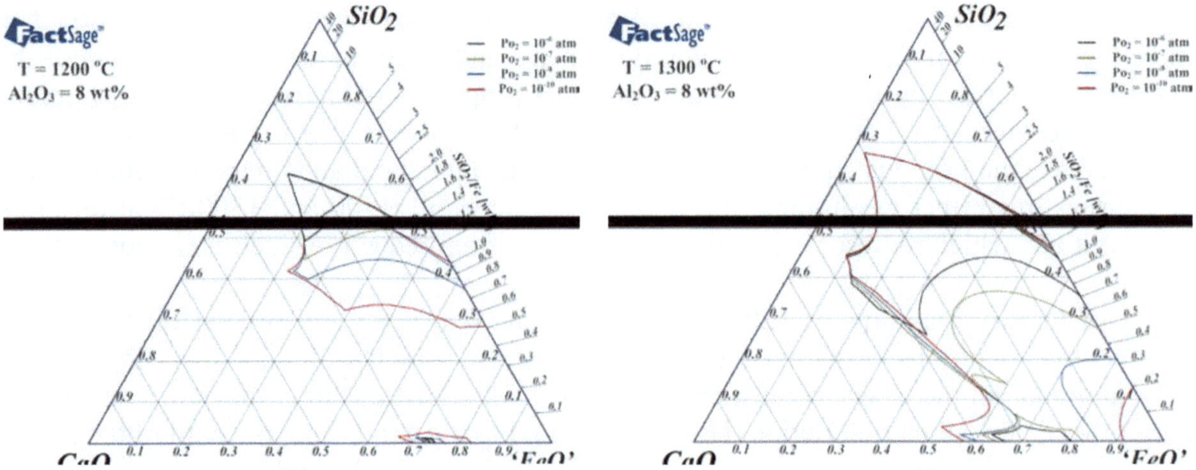

(e) (f)

Figure 4 - Effect of Alumina on Fayalite Slag at 1200 °C and 1300 °C

When considering the effect of temperature on the conditions in Figure 4 we find that at the same oxygen potential conditions (ternary (b), (d) and (f)), the liquidus region will be expanded, pushing the spinel saturation region towards the FeO rich area. The change in operating area by adjusting the temperature, flux content (CaO) or due to oxygen potential changes the activities of the valuable metals within the slag. These changes can be predicted and therefore the operating regions can be selected that will allow for the most sustainable outcome. The optimal operating region is one where appropriate levels of metals are recovered, whilst the versatility and usability of the waste materials from the process is maximised by allowing them to be separated for use as feed materials in other industries.

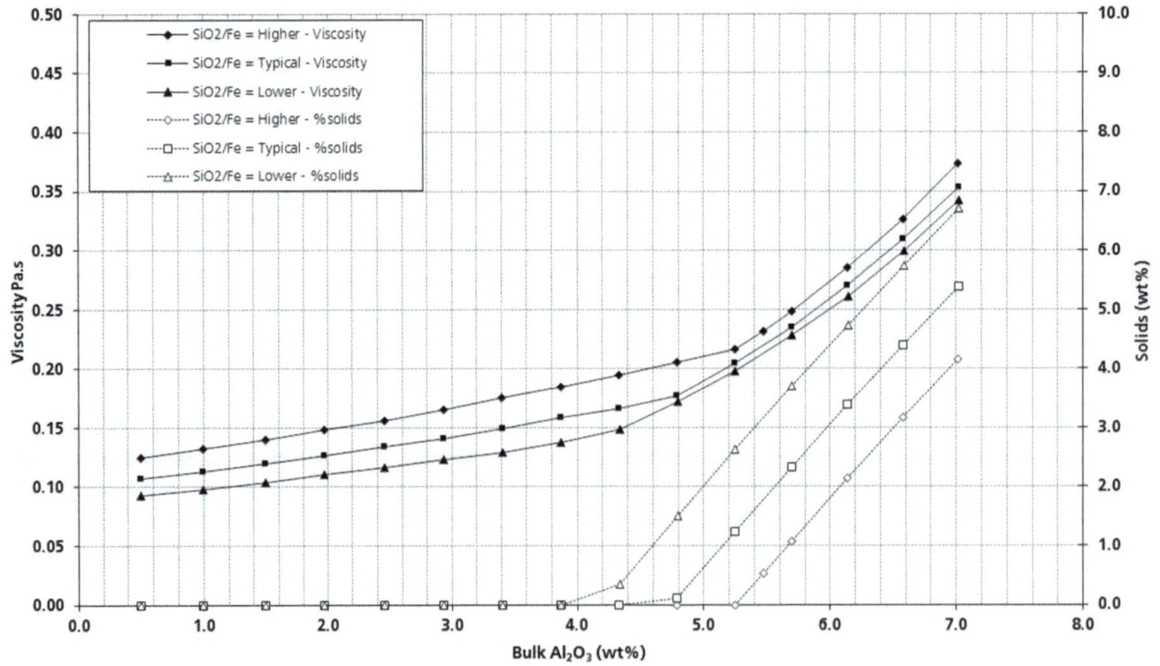

Figure 5 – Effect of Alumina on the Viscosity of Fayalite Slag at 1200 °C

Concluding Remarks

Sustainable innovative new metallurgical processes can be developed by going back to basics and applying the fundamentals of slag chemistry and thermodynamics using modern thermochemical software. Application of basic metallurgical concepts such as oxygen and sulphur potential, selective oxidation, slag capacity, liquidus temperature and viscosity of slags combined with a versatile technology such as ISASMELT™ can be used to design modern, innovative and sustainable recycling plants such as the ones operated by Umicore Precious Metals Refining at Hoboken and Aurubis AG at Lünen. Both companies recognized one of the important strengths of ISASMELT™ technology for their respective processes: its versatility and ability to be incorporated into new, innovative applications, thereby giving these "urban miners" a commercial edge while ensuring they can operate in a sustainable manner.

Acknowledgments

Xstrata Technology gave permission to publish this paper.

Xstrata thanks the staff from Umicore and Aurubis for technical discussions.

References

1. United Nations; "*Report of the World Commission on Environment and Development*"; 96th Plenary Session, New York, USA, 1987.
2. US Congress, Office of Technology Assessment; "*Copper: Technology and Competitiveness*", September 1988, NTIS order #PB89-138887.
3. P. S. Arthur and G. R. F. Alvear F, "*Incremental Scale up of ISASMELT™ - The Key to its Success*", COM 2011, Montreal, Canada, October 2011.
4. J. S. Edwards and G. R. F. Alvear F., "*Converting Using ISASMELT™ Technology*", Copper 2007, Vol 3, pp. 17-28, Toronto, Canada, October 2007.
5. F. Vanbellen and M. Chintinne, "*The Precious Art of Metals Recycling*", Advanced Processing of Metals and Materials, F. Kongoli and R.G. Reddy, Eds., TMS, Warrendale, Vol. 1, 43-52. 2006
6. UNEP; "*Sustainable Innovation and Technology Transfer Industrial Sector Studies, Recycling – From E-waste to Resources*"; Final Report July 2009, pp7.
7. A. Yazawa, "Thermodynamic Considerations of Copper Smelting", Canadian Metallurgical Quarterly, vol. 13, no. 3, 1974, pp. 443-453
8. Aurubis Lunen Website 2011: http://www.aurubis.com/en/our-business/raw-materials/recycling/technology/
9. C. Hagelüken, *"Recycling Of Electronic Scrap At Umicore Precious Metals Refining,"* Acta Metallurgica Slovaca, Vol. 12, 2006, pp.111 – 120,

REWAS 2013
Enabling Materials Resource Sustainability

Enabling Sustainability through Process Design, Modeling & Simulation

Session Chairs

Yongxiang Yang
Juergen Antrekowitsch

MOVING EQUIPMENT AND WORKERS TO MINE CONSTRUCTION SITE AT A LOGISTICALLY CHALLENGED AREA

Laszlo Tikasz[1], Dennis Biroscak[2], Scheale Duvah Pentiah[1], Robert I. McCulloch[1]
[1]BECHTEL Canada Co.; 1981 McGill College Avenue, Montréal, QC, H3A 3A8, Canada
[2]BECHTEL Co.; 50 Beale Street, San Francisco, CA, 94105, USA

Keywords: Safety, Social Awareness, Mine Construction, Traffic Organization, Modeling, Simulation

Abstract

Social sensitivity of habitants, minimal impact on the environment, low-grade infrastructure, high altitude, frequent rock slides combined with expectations for the timely moving of equipment and workers are some of the challenges emerging from the current construction of a mine. Starting with traditional planning, and experiencing issues in the early phase of the construction, a traffic simulator was requested by the Procurement Department in order to validate daily-weekly schedules and predict likely delays or blockages on the long-term.

The now available simulator captures the available routes and the applicable travelling rules. Traffic schedules are inputs to the simulator, which attempts to perform the planned movements of the vehicles of various type and size. The data collected is analyzed and recommendations are made to management thus providing the means of adaptive/responsive planning and dispatching of vehicles and convoys.

The simulator is undergoing real-life tests. It is expected that adaptive traffic planning will result in an improved usage of resources and will help maintain sustainable operation.

Introduction

The ongoing construction of a major, long-life mine is subject to a number of challenges. The pace of the construction work sets the daily-weekly requirements on cargo and workers' movement. Detailed traffic plans are generated for the weeks ahead by traffic and logistics (T&L) experts, carefully matching the forecasted demands and resources. However, the execution of a traffic plan is subject to countless, time-varying restrictions and limitations. These include, among others, the social sensitivity of habitants, a low-grade infrastructure, high altitude and frequent rock slides. It is rather difficult to project the true impact on the overall traffic performance and size the necessary countermeasures. The mine construction traffic was initially drawn on traditional planning, and T&L Specialists experienced issues in its early phases. As a result, a traffic simulator was requested by the Procurement Department in order to validate daily-weekly schedules and predict likely delays or blockages on the long-term.

To the mine construction site, the road crosses through the Andean region, at an elevation of 3,700 – 4,600 m above sea level, illustrated on Figure 1. The road is technically the only available road. It was intended to serve the local communities only: getting basic commodities to habitants and connecting neighbors and neighboring settlements. Then, industrial operations started in the region and recently the mine construction was started. The construction of the mine commenced around September 2011 with approximately 480 workers onsite. At peak construction, around October 2013, there will be approximately 6685 workers onsite. Towards the end of the construction phase, around December 2014, there will be approximately 2085 workers onsite. Once the mine is operational there will be approximately 1,350 permanent jobs.

Figure 1. Road to Mine Construction Site

This major development should meet high safety and health standards, handle human resource issues, manage community relations and limit environmental impact to assure long-term sustainability to the stakeholders. It was stated by the mine owner, that regarding workplace safety, health and risk management, the most sensitive issue was road traffic and helicopter operation. In this paper, attempts are made to underline some sustainability aspects related to road traffic planning and operation.

Sustainability Related to Road Traffic

Sustainability, as often applied to industrial activities, consists in balancing long-term economic, environmental and social considerations when managing the business and ensuring viability for all participants. Moving equipment and workers to mine construction over a long period of time is indeed a world in a drop of water when sustainability is addressed. Specifically, safety risk management and handling social/ community issues are particularly applicable to road traffic.

Managing safety and risk

Workplace safety and health starts with preventive screening of workers/ drivers for high altitude. A policy is implemented on internal authorization to drive vehicles, maintenance and vehicle age control, safe driving, en route control. Traffic organization is based on a *Road Management Plan*, what defines, among others, the followings:

- Operating times (normally daylight traffic is allowed only)
- Prohibitions under low visibility
- Minimal distances between vehicles
- Defining solo vehicles and grouping vehicles to convoys
- Zones for passing/ zones prohibited for passing
- Obligatory stopovers for slow moving vehicles

Travelling conditions – in general – are bad. Road width, curvature, slope and pavement impose constant risk on drivers; road improvement is a continuous support activity to road traffic. Building a new branch of road is planned to handle a great part of the traffic over a longer distance towards the mine. In Figure 2, the narrow bend of the road is under extension/ leveling by a bulldozer. It is obvious that the hut is "at a wrong place". Even after the bend is extended, it will stay extremely close to the heavy traffic. Relocating it seems plausible – but not doable in most cases, subject to the given circumstances, due to social and community issues.

Figure 2. Road Extension at a Community

Respecting social sensitivity

The road was built to connect settlements, then settlements developed and further houses were built right beside the road. As of today, passing traffic is extremely close to residential buildings at certain settlements. Great effort is done to widen roads, add extra lanes, build bypasses or purchase properties and relocate them to a safe spot nearby.

The lodging, catering and health facilities are rather limited at the settlements. A rock slide, suddenly blocking the road at a random place, might order a convoy of buses to unplanned halt. Providing the basic needs for 30-40 extra people at settlements far exceeds local resources and disturbs the community; this requires immediate mitigation from T&L service.

A registry of events was compiled for holidays, fairs, ceremonial events and regular local activities in the involved settlements. The *Traffic Forecast Plan* considers these temporary traffic limitations (speed control, one-way traffic or no passing during the events).

Further, otherwise minor events, unrelated to mine construction or any above, might have serious effect on timely moving equipment and workers. Closing the pedestrians' crossing in front of a local school can completely paralyze traffic for hours. In Figure 3, local habitants are working on a building next to the road. First, the mine construction traffic was restricted to a "one-way" traffic, then ordered to complete halt, to provide time and access for residents to complete their work in a safe manner.

Figure 3. A Temporary Traffic Restriction

On the Road

The mine construction is ongoing. Road maps and settlement maps are available and it was possible to travel the road and collect data on vehicle movement (logging arrival/ departure at settlements, collecting GPS data etc.). A traffic model and simulator was requested by the Procurement management to aid traffic scheduling. During model testing and validation, model results were regularly compared to real, observed situations. To start with, the following Figure 4 and 5 illustrate - in pairs - the real conditions and situations and the simulated ones. The details of the modeling work are addressed in the subsequent sections of the paper.

Figure 4 shows a general traffic situation, vehicles pass through a regular road sector. In this case, the vehicles are regular-sized trucks, organized into convoys. The convoy is led by an escorting vehicle, with a board mounted to its top, marking the fact that a convoy is coming behind it, containing two (2) members (trucks). These are important pieces of information to counter traffic members, flagmen and road repair workers, equally. The simulated twin-pair shows a fraction of a map, road marked in red and the small boxes represent the vehicles.

Figure 5 shows a dedicated "rest site", real and simulated, respectively: medium-sized busses, organized to convoys, made a temporary halt at the roadside. The roads are generally barely wide enough for single-lane traffic; passing is prohibited at many road sectors. Parking, resting is allowed only at dedicated sectors.

Figure 4. Roads – General Traffic

Figure 5. Roads - Rest Site for Convoys

Travelling restrictions are categorized by the T&L experts as permanent and temporary restrictions. These restrictions and the applicable traffic rules are summarized in Table 1 and Table 2. Real and simulated situations at travel restrictions are shown in Figure 6 - Figure 8.

Table 1. Permanent Traffic Restrictions

Type	Time Duration	Traffic Rules
Bridges Road narrows Steep roads Settlements - roads	Permanent or regular, limited to certain periods of the day	Single-lane, alternating traffic speed limitation keeping convoys together

Table 2. Temporary Traffic Restrictions

Type	Time Duration	Traffic Rules
Road problems		single-lane, alternating traffic
Weather problems	limited to certain periods of a day	speed limitation
Settlements - events		keeping convoy together
Rock slides	several days	road blockage

Figure 6. Permanent Traffic Restrictions - Bridges

Figure 7. Permanent Traffic Restrictions - Settlements

Figure 8. Temporary Traffic Restrictions - Rockslides

Discrete-Event Traffic Modeling

There is a recurring need at Bechtel Mining and Metals (M&M) for modeling and simulation activities to complement various pre-feasibility and feasibility studies, design alternatives as well as ongoing plant/ mine transformation, demolition, construction and operation activities. The scope of the activities spans over industries, among others, addresses:

- Aluminium production
- Copper ore mining and processing
- Iron ore mining and concentrate producing
- Coal mining and transportation

Part of the processes and most of the complementary people, product and material movement that raised questions turned to be logistic-type problems; discrete event models (DEM) and process simulation were used to address the issues. During the years, the ever-growing M&M Center of Excellence (CoE) Process Model Library has been extended with Traffic Model components. These include:

- Routes, composed of sectors, determined by and set to the travelling characteristics of the real roads
- Parking/staying sites, bridges, narrows and intersections
- Settlements, with individual traffic restrictions (daily restrictions, holidays, events etc.)
- Parking/staying sites, bridges, narrows and intersections with rules for intended usage
- Road blockages
- Operation-related traffic, (scheduled, defined)
- Third-party traffic (random, estimated)

Model prototypes are built from these library modules. Then, the parameters of the model components are tuned to design values and specific traffic scheduling is applied (pick-ups, regular trucks, heavy trucks, buses etc.). Also, the simple, default blocks of vehicles could be transformed into 3D shapes better resembling the vehicles applied at a particular site. At this stage, the model starts to "work", and step-by-step model verification could begin.

The 3D visualization is beneficial to demonstrate the forming traffic and to spot issues. The usage of the simulator is organized in three steps:

- Compilation of input data – vehicle parameters, traffic restrictions, traffic scheduling; data arrangement and pre-processing
- Simulation - performing scenarios, visual demonstration of the forming traffic; data collection
- Evaluation of output data – traffic analysis, vehicle utilization, progress with mine construction; post-processing and chart generation.

The DEM simulation shell is from Flexsim. The available development environment is shown in Figure 9. A road map serves as background; the real road is marked with red, the red lines mark the modeled road. Vehicles are color-coded (shown as blue, grey and yellow boxes). Poles (green and yellow ones) mark important segments of the road (speed limit zone, narrows etc.). A bridge is marked with a purple block. For ease of use, pre- and post-processing were developed in Microsoft Excel. Post-processing was kept open-ended, allowing further analysis on behalf of the T&L Specialists.

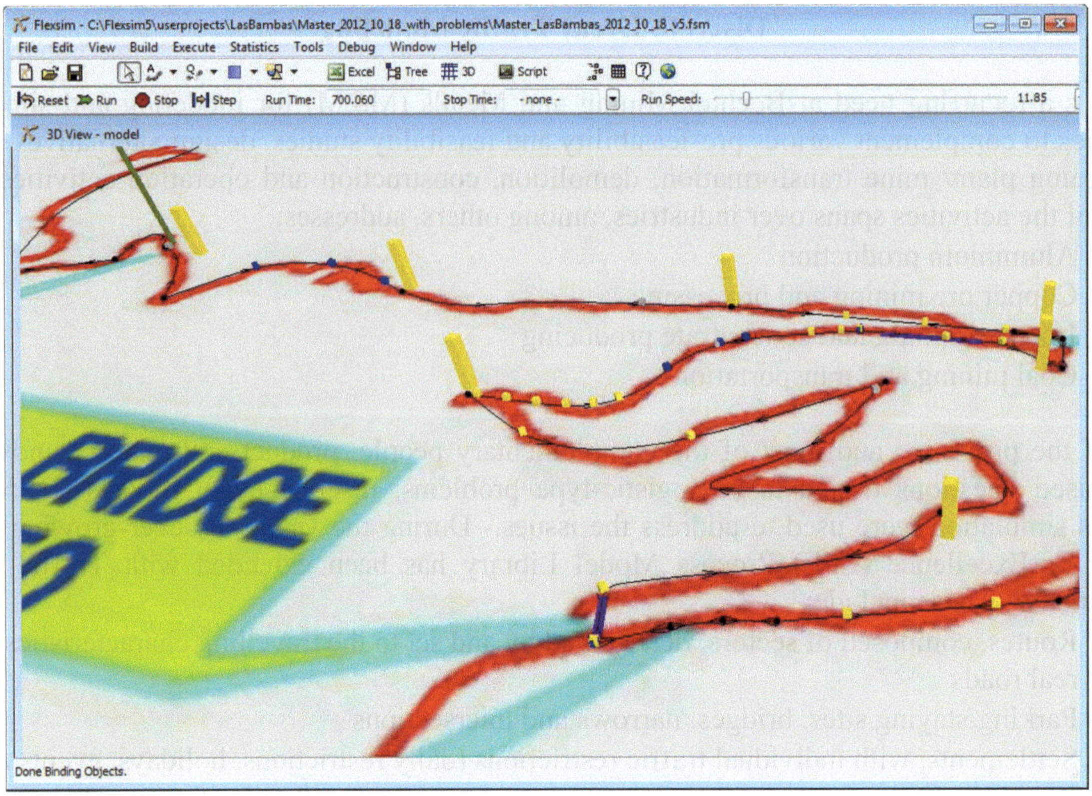

Figure 9. Flexsim DEM Environment

T&L experts measure utilization of access points, routes, intersections, staging areas, and control movement of vehicles on the road. The model allows having an "overview" of the ongoing traffic. As the road is about 200 km long, the blocks, representing the vehicles are not visible. The black lines, "electronic leashes" in Figure 10 are to spot vehicles and group of vehicles as they move.

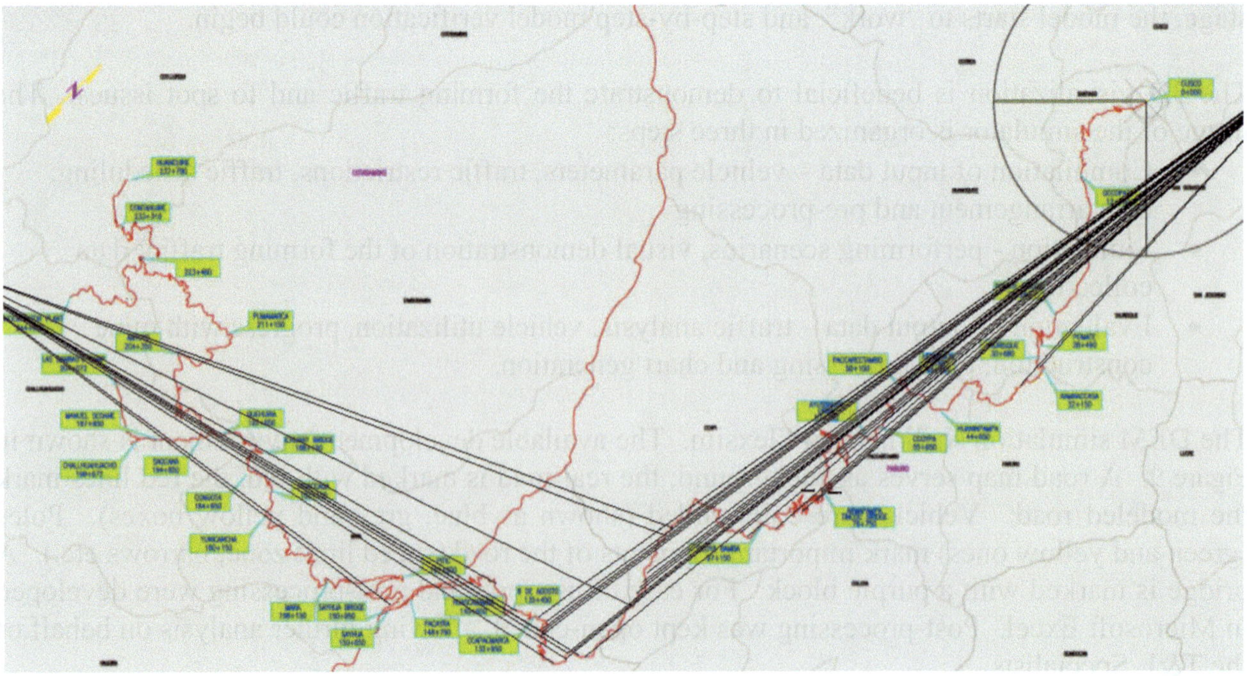

Figure 10. Vehicle Movement Overview

Model–Based Traffic Planning and Validation

During the construction of the mine, special care was taken in minimizing impact on the life of local habitants and the environment. As a result, a number of issues were captured in the model, such as limiting or even banning traffic on local and national holidays, allowing traffic only during daylight (adapting to sunset/sunrise times and to seasonal weather conditions). It was understood that sustainable operation during mine construction was primordial towards mine operation for the years to come.

The traffic simulation is based on the continuously updated *Traffic Forecast Plan*. When a time period of interest is set, a schedule is generated for that period (typically days, or a week). Then, using the traffic model, T&L experts could simulate combined effects of schedules and projected traffic restrictions. Figure 11 shows waiting times at a bridge, over a week-long period, where alternating, single-lane traffic is ordered. Figure 12 shows the travelling times for the groups of vehicles.

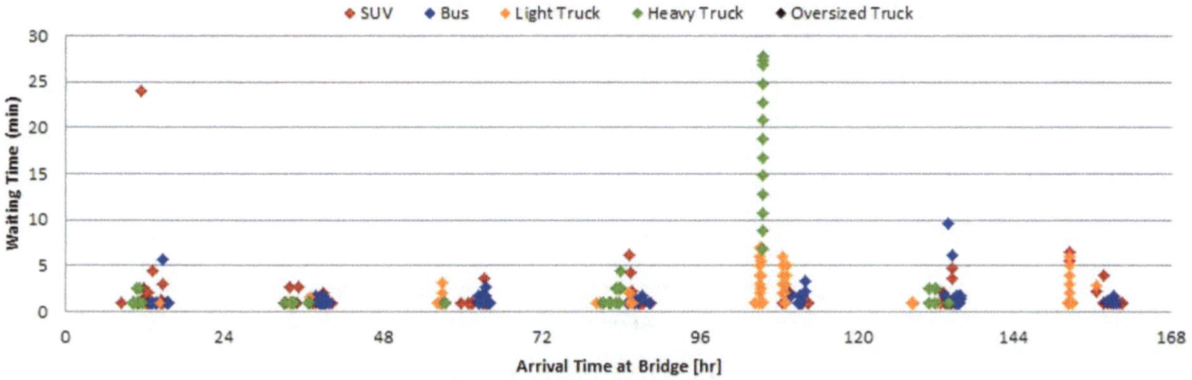

Figure 11. Vehicles' Waiting Time at a Bridge

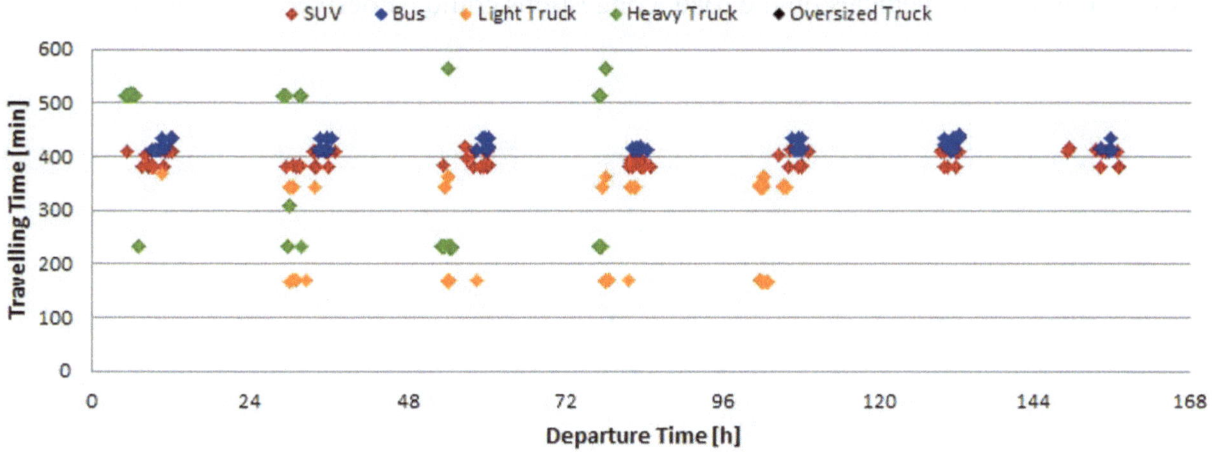

Figure 12. Vehicles' Travelling Time

In general, the model execution is fast. It could be up to 100 times faster than real time, depending on model granularity as well as animation and reporting overload. It was found, that results for projected one-week traffic could be generated within minutes, i.e. still in real-time with the ongoing process. This means that in critical situations, a T&L Specialist, weighing options, first could apply the possible changes to a model, then run the model and compare the outcomes. It could be, indeed, a model-based "calculator" to traffic planning and validation.

The simulator could be extended with data collecting/ recording utilities. Those data then could serve for formal traffic density and road capacity analysis. A time-space diagram is given in Figure 13. Trips of selected vehicles (color-coded lines for vehicle #1, 7, ... 65) are shown. Traffic stream properties, density (number of vehicles over a unit length of the road) and traffic flow (number of vehicles passing a reference point per time unit) can be derived from the curves.

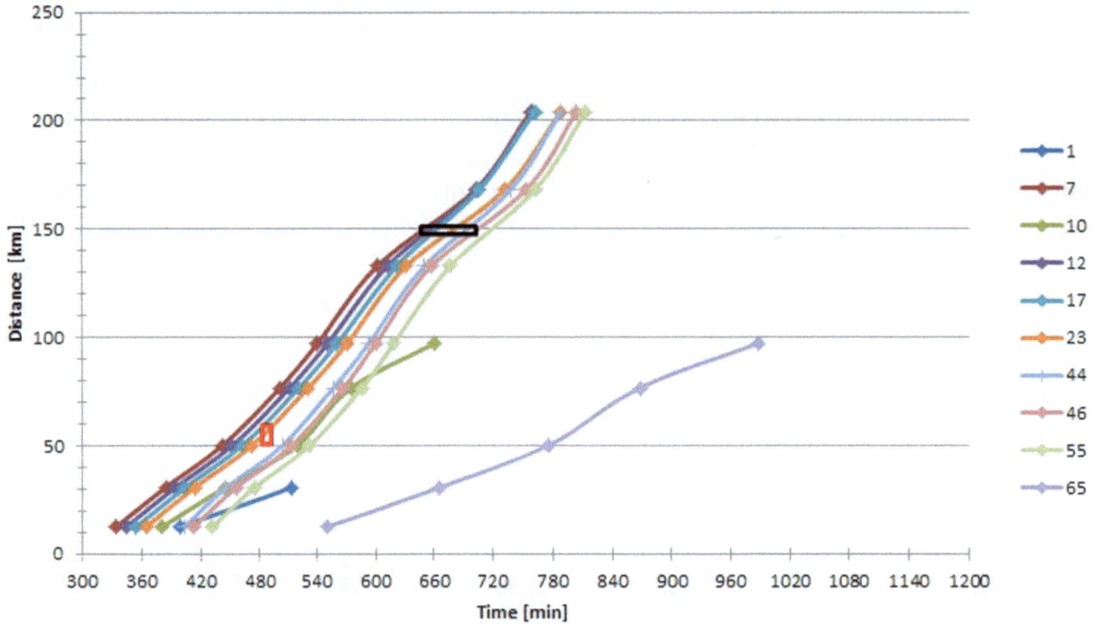

Figure 13. Time-Space Diagram

N-cumulative curves are shown in Figure 14. At selected sites of the road (coded as 100, 200,... 800), number of passing vehicles were counted over the day. Travelling times between sites, delays, total number of vehicles on the road, congestions could be derived from the curves.

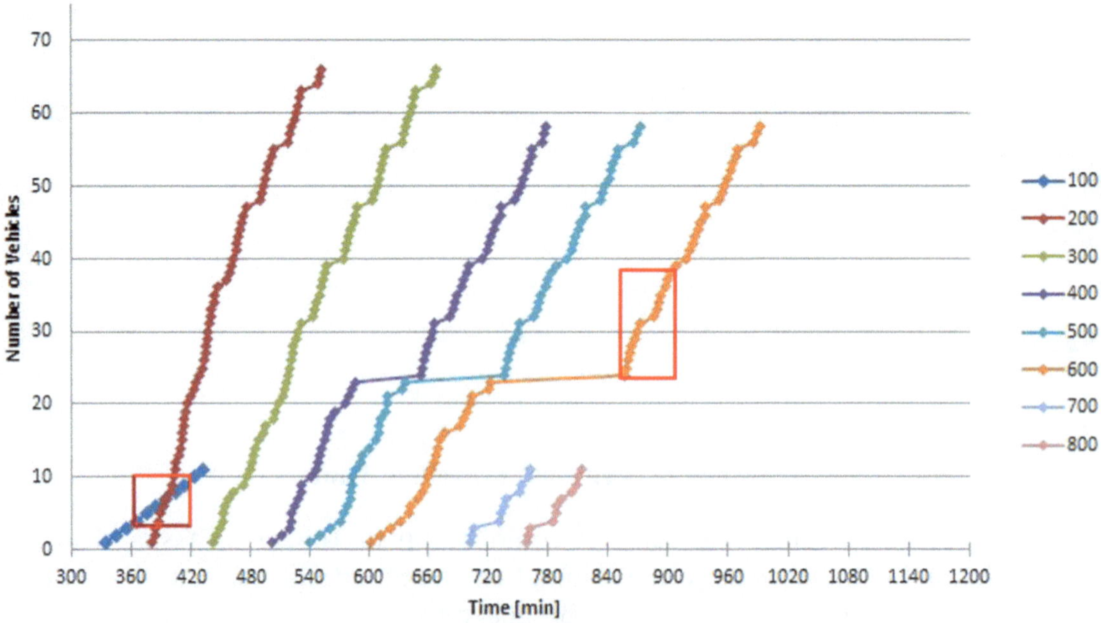

Figure 14. N-Cumulative Curves

Conclusions

Sustainability is key measure for an ongoing mine construction project. A traffic simulator is provided to aid planning/ validating movement of workers and cargo.

Testing, maintaining, extending and improving the simulator are ongoing. As the mine construction progresses, roads will be altered and new roads will be added to the existing ones. Traffic rules will be added and adapted to evolving site conditions. The modular structure of the model/ simulator provides more flexibility such as the capability of adding elements to the model to represent new scenarios.

T&L experts are gradually including the traffic modeling approach to their daily traffic planning routine. Their comments on data preparation (planning), traffic analysis (data collection, reporting) and interfacing with other applications (GPS-based vehicle monitoring) will be captured to the new model release in the coming months. Further results are expected to be published at coming TMS conferences.

Trademarks

Flexsim is registered trademark of Flexsim Group. Excel is registered trademark of Microsoft.

Acknowledgements

Special thanks are due to our team member, John Maxwell, T&L Specialist at Bechtel M&M Santiago Office, for providing on-site photos and interpretation of the real traffic challenges.

The authors thank Bechtel for permitting the publication of this study.

References

1. Laszlo Tikasz, C. Mark Read, Robert Baxter, Rafael L. Pires, Robert I. McCulloch; "Safe and Efficient Traffic Flow for Aluminium Smelters"; Paper presented at TMS Light Metals 2010.

2. Responsible Mining for the Benefit of All, Sustainability Report, 2011. www.xstrata.copper.com

3. Laszlo Tikasz, Philippe Toutoungi, Robert I. McCulloch, "Ore Hauling Alternatives from Mine to Site", Paper submitted to the 23rd World Mining Conference 2013.

REWAS 2013: Enabling Materials Resource Sustainabil
Edited by: Anne Kvithyld, Christina Meskers, Randolph Kirchain, Gregory Krumdick, Brajendra Mish
Markus Reuter, Cong Wang, Mark Schlesinger, Gabrielle Gaustad, Diana Lados, and Jeffrey Spangenberg
TMS (The Minerals, Metals & Materials Society), 20

PREPARATION AND CHARACTERIZATION OF FIBROUS COPPER POWDER USED FOR CONDUCTIVE FILLER

Youqi Fan[1], Yongxiang Yang[2], Yanping Xiao[2], Zhuo Zhao[1]

[1]School of Metallurgy and Resource, Anhui University of Technology, Ma'anshan, Anhui, 243002, PR China

[2]Department of Materials Science and Engineering, Delft University of Technology, Mekelweg 2, 2628 CD, Delft, The Netherlands

Keywords: Fibrous copper powder, Thermodynamic analysis, Coordination precipitation, Thermal decomposition, Conductive Filler

Abstract

A novel two-stage process is investigated for preparing the fibrous copper powder used for conductive filler in conductive paste. Based on thermodynamic simulation of Cu(II) - $C_2O_4^{2-}$ - NH_3 - NH_4^+ - H_2O system, the rod-shape copper oxalate salt is synthesized firstly with coordination precipitation method, using ammonium oxalate and a purified copper salt solution which could be from either primary or secondary copper-bearing resources. According to the results of XRD, element analysis and X-ray single crystal diffraction, it can be inferred that the composition of the copper oxalate salt is $[Cu(NH_3)_3]C_2O_4 \cdot xH_2O$, and the NH_3 plays a significant role in the formation of rod-shape crystal. Secondly, the copper oxalate salt is decomposed to fibrous copper powder at 350℃ in inert atmosphere. The thermo-decomposition procedure is characterized by TG-DSC-FTIR, XRD and SEM. It is found that the phase of copper oxalate salt transforms as following: $[Cu(NH_3)_3]C_2O_4 \cdot xH_2O \rightarrow [Cu(NH_3)_3]C_2O_4 \rightarrow CuC_2O_4 \rightarrow Cu$.

Introduction

The conductive paste usually comprises burning infiltration-type conductive paste and curing conductive paste. A certain quantity of conductive filler is required as a conductive medium, generally using lower resistivity metal powders such as Ag, Cu, Au, Ni, etc., wherein Ag is the most extensive.

The conductive properties of the conductive paste mainly depends on three factors [1]: 1) the volume fraction of the conductive filler; 2) its own conductivity of conductive filler; 3) the dispersion state of the filler in a polymer matrix. The study showed that the morphology of the electrically conductive filler particles is one of the most important parameters of the many factors affecting the conductivity. Comparative study of the effect of conductive filler morphology (spherical, flake, fibers, etc.) was undertaken. It's found that the more irregular of the filler particles (such as metal fibers, metalized glass fibers, carbon nanotubes, etc.), the lower of seepage critical concentration φ, and the less specific consumption of filler required to achieve the same conductivity[2-5]. In copper conductive paste, , the fibrous copper powder filler with a length / diameter ratio (referred to as the "aspect ratio") will help to improve the conductive properties of the paste, and to reduce the specific consumption of copper powder. Accordingly, the development of a new process for preparing fibrous copper powder will contribute to improving the performance of the conductive paste, as well as reduce the cost significantly.

The traditional methods of copper powder preparation mainly include the electrolytic process, atomization method and liquid-phase reduction method, etc. [6], wherein the copper powder morphologys refer to dendritic, spherical, flake shape, except fibrous shape. In recent years, the preparation of nanowires and nanorod of copper or copper oxide is reported in literatures. The methods can be divided into two categories: chemical vapor deposition(CVD) and physical vapor deposition(PVD) [7]; the template method [8, 9], but all these methods have some shortages, such as harsh preparation conditions or complicated process, or high cost, etc.

In the literatures [10], oxalate liquid precipitation method is reported to be an effective way to prepare special morphology of powder. Currently, the researches mainly focus on preparation of copper oxalate salt by this way, whereas rarely on the preparation of fibrous copper powder , except few reports on porous copper oxide powder synthesis[11, 12]. N.Jongen [13] studied the effect of the hydroxypropyl methyl cellulose (HPMC) on crystal growth and obtained square-shape and short rod-shape copper oxalate powder; T.Ahmad [14] prepared nickel oxalate and copper oxalate nanorods using reversed micelles method; M.Y.Li [15] synthesized copper oxalate nanowires in the ionic liquid of ethanol and BMImBF4 with copper acetate and oxalic acid as raw material, by controlling of the reaction temperature 180 ℃.

In this study, we propose a new process for preparing fibrous copper powder, which is divided into two steps:

1) The rod-shape copper oxalate salt is synthesized firstly with coordination precipitation method, using ammonium oxalate and a purified copper salt aqueous solution which could be from either primary or secondary copper-bearing resources, just like the copper sulfate electrolyte in the electrolytic refining of copper, copper-bearing leaching solution in anode slime treatment or in hydrometallurgy of copper.

2) copper oxalate salt is decomposed to fibrous copper powder at certain temperature in inert atmosphere.

Therefore, thermodynamic analysis of Cu(II) - $C_2O_4^{2-}$ - NH_3 - NH_4^+ - H_2O system is to be undertaken first in the research. and copper oxalate salt will be synthesized on the base of thermodynamic calculation results. The chemical compostion, phase, morphology, and thermostability of the powder will be analyzed.

Thermodynamic analysis

Pourbaix diagram (Eh-pH) of Cu-$C_2O_4^{2-}$-H_2O and Cu-NH_3-H_2O systems

Considering the complexity of the aqueous solution, Pourbaix diagram (Eh-pH) is first introduced to analyze the stability conditions of species. The Eh-pH diagram is calculated and plotted by the "EpH" module of software Factsage 6.3 [16, 17]. Since of the functional restriction of the program, in which the number of inputed non-metals is no more than three, so the Eh-pH diagrams of Cu-$C_2O_4^{2-}$-H_2O and Cu-NH_3-H_2O systems are respectively plotted at the conditions of T=298K, $C_{Cu2+} = C_{C2O42-} = C_{NH3} = 1$mol/L.

Thermodynamic equilibrium calculation of Cu^{2+}-$C_2O_4^{2-}$-NH_3-NH_4^+-H_2O System

Cu^{2+}-$C_2O_4^{2-}$-NH_3-NH_4^+-H_2O solution is a complex system, which contains not only multi-level coordination reactions of Cu^{2+} and NH_3, but also precipitation reaction of Cu^{2+} and $C_2O_4^{2-}$, Cu^{2+} and OH^-. In the system, the main species including: Cu^{2+}, $Cu(NH_3)^{2+}$, $Cu(NH_3)_2^{2+}$, $Cu(NH_3)_3^{2+}$, $Cu(NH_3)_4^{2+}$, $Cu(NH_3)_5^{2+}$, $Cu(OH)^+$, $Cu(OH)_{2(aq)}$, $Cu(OH)_3^-$, $Cu(OH)_4^{2-}$, $CuC_2O_{4(aq)}$, $Cu(C_2O_4)_2^{2-}$, $CuHC_2O_4^+$, $CuC_2O_{4(s)}$, $Cu(OH)_{2(s)}$, $H_2C_2O_4$, $HC_2O_4^-$, $C_2O_4^{2-}$, OH^-, H^+, NH_4^+, NH_3, Cl^-.

When the solution system is in equilibrium, based on the principle of simultaneous balance, the respective chemical reactions are in equilibrium and the equilibrium concentration of each species in the system are simultaneously meet thermodynamic balance equation. Further, according to the principle of mass conservation, the balance system of the same element and the initial element of the quality of the quality and equal, ignoring the volume changes before and after the solution in the equilibrium reaction, the sum of each kind of element equilibrium concentration is equal to the initial concentration. Based on the above two principles, the model of the thermodynamic equilibrium of the system can be built, which is shown in our previous paper[18].

Experimental

Synthesis of copper oxalate salt

Certain concentration of copper chloride solution and ammonium oxalate solution were prepared based on the requirements of stoichiometric ratio. All the solutions ware elaborated by deionized water. The ammonium oxalate solution was injected into the bulk copper chloride solution with peristaltic pump, controlling reaction temperature at 30~70℃. Meanwhile, ammonia water was added to adjust and stabilize the solution pH around a certain value. After a duration of agitating and aging, the precipitate was filtered, washed with deionized water and ethanol, then dried in a vacuum drier at 100~120℃ for more than 12 hours. Finally, the copper oxalate salt powder was obtained.

Thermal Decomposition of copper oxalate salt

Some copper oxalate salt powder was put into an alundum boat, and then was set in electric tube furnace. By controlling an appropriate atmosphere, temperature and time, the copper powder was obtained finally.

Characterization of powder

The structure and composition of as-prepared powder was measured with X-ray diffraction (Bruker AXS D8 Advance) and X-ray single crystal diffraction (Bruker SMART APEXⅡ). The chemical compositions of copper oxalate salt was obtained by elemental analyser (VARIO EL III) and chemical analysis method. The morphology, particle size and dispersion were characterized with scanning electron microscope(JSM-6490LV) . The TG-DSC-FTIR analysis of powder was conducted with NETZSCH STA 449C thermal analyzer.

Results and Discussion

Analysis of Pourbaix diagram

The Eh-pH diagrams of Cu-$C_2O_4^{2-}$-H_2O and Cu-NH_3-H_2O are shown in Fig.1. It can be found in Fig.1(a) that by adding $C_2O_4^{2-}$, Cu^{2+} in aqueous solution is precipitated as CuC_2O_4(s) when pH less than 8.0 , while precipitated as $Cu(OH)_2$(s) when pH value more than 8.0. From Fig.1(b), it's obvious that before adding $C_2O_4^{2-}$, the copper mainly exists as $Cu(NH_3)_4^{2+}$ in solution when pH value less than 9.0. Based on the above analysis, it's supposed that when adding oxalate into the Cu-NH_3-H_2O system, the copper ion can be precipitated as CuC_2O_4. However, in the more complex Cu-NH_3-$C_2O_4^{2-}$-H_2O system, the stability region of speces will differ from the situations in Fig.1, which needs further thermodynamic equilibrium calculation.

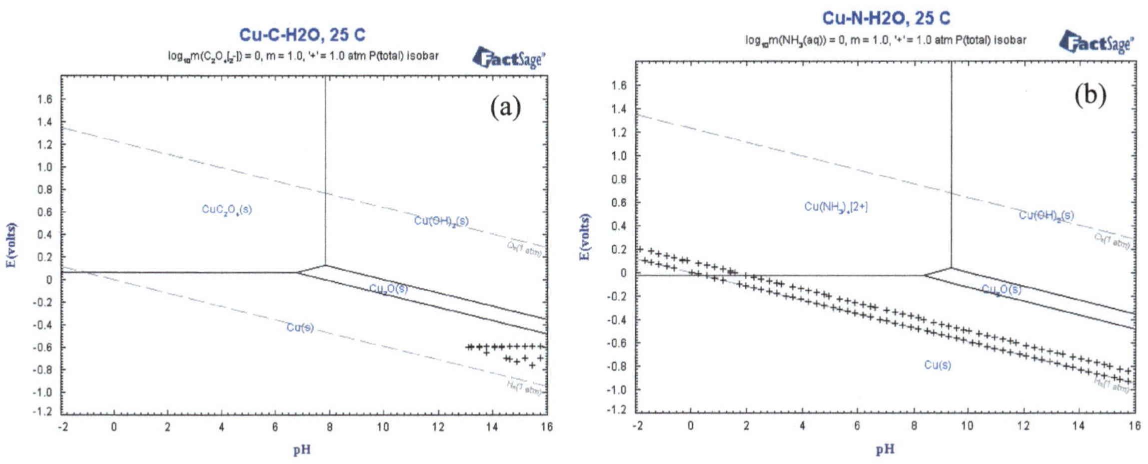

Fig.1 Eh-pH diagrams of Cu-$C_2O_4^{2-}$-H_2O and Cu-NH_3-H_2O system

Thermodynamic equilibrium calculation of Cu^{2+}-$C_2O_4^{2-}$-NH_3-NH_4^+-H_2O System

The 3D surface of lg$[Cu]_T$-pH- $[N]_T$ in solution is shown in Fig.2 when $[C]_T$ is 0.4mol/L. As can be seen from Fig.2, The pH value affects greater on the total concentration of copper ions $[Cu]_T$ in the solution. when the pH value is gradually increasing, the $[Cu]_T$ has a trend of decreasing - increasing - decreasing. Comparatively, the total ammonia concentration $[N]_T$ has a relatively small impact of change on the $[Cu]_T$. It can be seen clearly from Fig.3. when $[N]_T$ changes between 0 ~ 4 mol / L, the variation of $[Cu]_T$ with pH value turns to be consistent: when pH less than 4.0, $[Cu]_T$ decreases with the increasing of pH, and the $[N]_T$ almost has no impact on $[Cu]_T$; $[Cu]_T$ reaches the minimum value when the pH value is between 4.0 to 5.0, and the effect of $[N]_T$ on $[Cu]_T$ become notable. The larger is $[N]_T$, the more is $[Cu]_T$ in solution. when the pH value is elevated to between 5.0 to 8.0, with the increase of the pH value, $[Cu]_T$ sharply increases. This is mainly due to the free NH_3 concentration in the solution increasing rapidly within the pH range, so that more free copper ion coordinate with NH_3 to form a complex ion of Cu (NH3) x (x = 1,2, ..., 6) . When the pH value is up to 8.0 ~10.5, $[Cu]_T$ reaches its maximum value, and remains substantially stable; when the pH continues to rise, due to the hydrolysis of the copper ions Cu $(OH)_2$, $[Cu]_T$ in the solution decline rapidly.

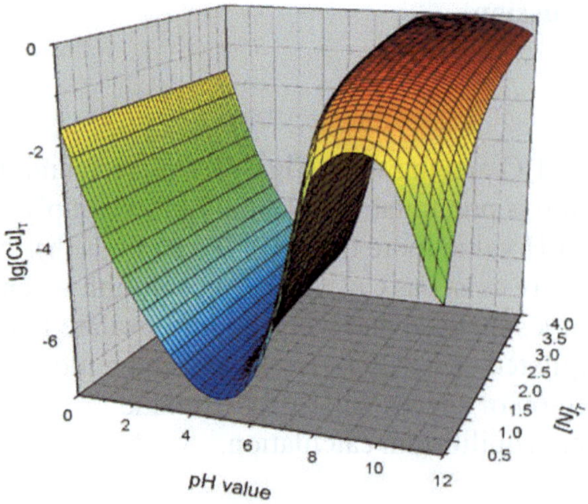

Fig.2 3D surface of lg$[Cu]_T$-pH- $[N]_T$ in solution when $[C]_T$ = 0.4mol/L

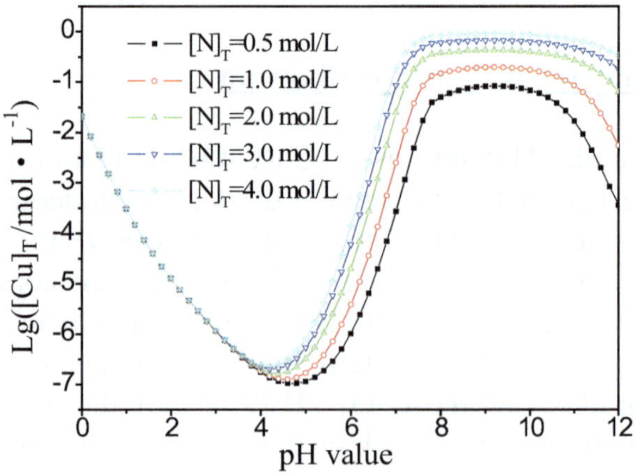

Fig.3 relationship curve of lg$[Cu]_T$ and pH value at various total ammonia concentration

The 3D surface of lg$[Cu]_T$-pH-$[C]_T$ in solution when $[N]_T$ = 2.0 mol/L is shown in Fig.4. As can be seen from the Figure, the total concentration of oxalate $[C]_T$ affects slightly on the $[Cu]_T$.- Fig.5 shows that $[Cu]_T$ value has a down-up-down trend with the increasing of pH as $[C]_T$ changes between 0~1 mol/L, which is similar to Fig.3. However, the effect of $[C]_T$ on $[Cu]_T$ is much more obvious when pH less than 8.0. The higher is $[C]_T$, the lower is $[Cu]_T$, since more copper is precipitated as copper oxalate.

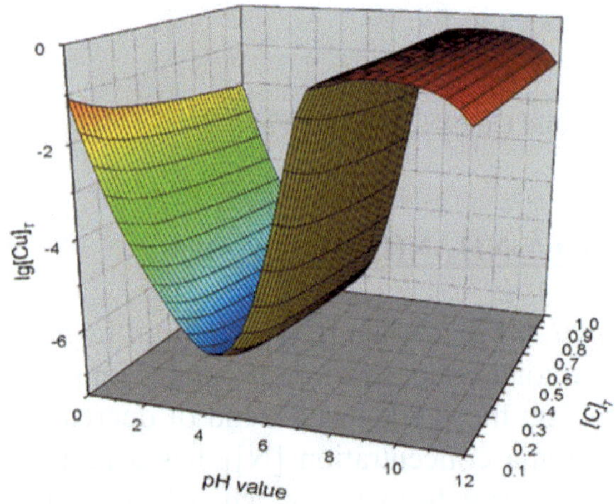

Fig. 4 3D surface of lg$[Cu]_T$-pH-$[C]_T$ in solution when $[N]_T$ = 2.0 mol/L

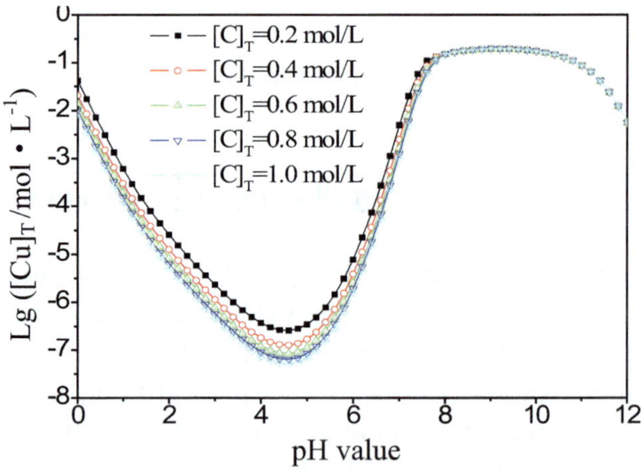

Fig. 5 relationship curve of lg$[Cu]_T$ and pH value at various total ammonia concentration

Fig.6 shows the concentration distribution of copper-bearing species in solution. It's seen from the figure, when the pH value is less than 4.0, the copper exists entirely in the form of free copper ions, and the copper ions precipitates with oxalate directly. now the reaction rate is relatively fast, so aggregation growth have priority in the precipitation process, which is difficult to control and obtain a special morphology particle; With the pH value increased, copper ion coordinates mainly with NH_3. Due to the decreasing of precipitation reaction rate, it's easier to systhesize powders with a certain crystal growth habit.

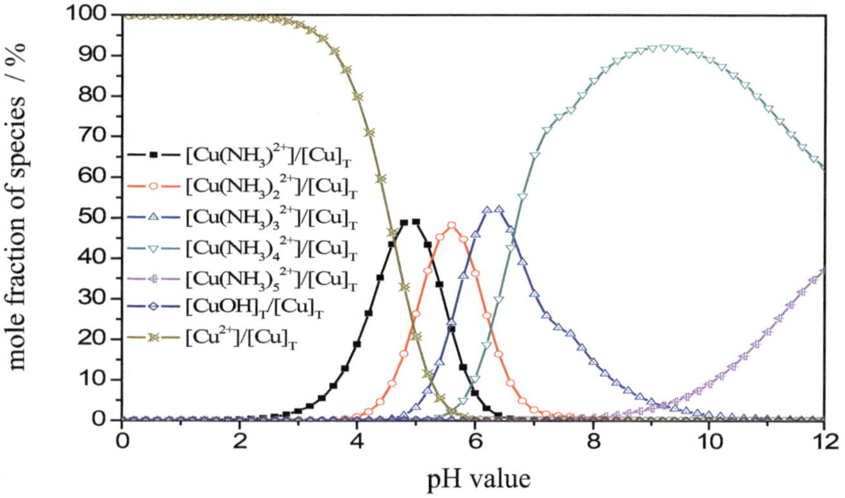

Fig. 6 Concentration percentage curves for copper-contained species under $[N]_T = 2.0$ mol/L and $[C]_T = 0.2$ mol/L conditions (T = 298 K)

Characterization of the copper oxalate salt

Copper oxalate salt was synthesized at various experiment conditions. the XRD patterns and SEM photos are shown in Fig.7. It is found that pH value has a significant effect on phase and morphology of copper oxalate salt. Three different kinds of products were obtained, namely pie shape (particle size 5μm), flower shape(particle size 50μm) and rod shape (particle diameter 1μm, length >20μm). By analyzing the corresponding XRD patterns, the phases differ from each other. Compared with standard PDF card, the powder prepared under the condition of pH 5.0 is proved to be $CuC_2O_4{\bullet}H_2O$; however, the other four patterns don't show any $CuC_2O_4{\bullet}xH_2O$ peaks, and also accord with none of the standard PDF cards at present, which suggests they are complicated copper oxalate instead of $CuC_2O_4{\bullet}xH_2O$.

The chemical composition and structure of them were analyzed by chemical analysis, element analyser and X-ray single crystal diffraction, the results are show in Table 1 and Fig.8. From the data in Table 1, the molecular formula of sample Fig.7(b) and Fig.7(c) is inferred as $Cu(NH_3)_2C_2O_4 \cdot xH_2O$ （$0<x<1$） and $Cu(NH_3)_3C_2O_4 \cdot xH_2O$ （$0<x<1$） respectively. Fig.8 shows that the molecular chain of rod-shape powder is linear, and the functional group NH_3 is perpendicular to the molecular chain plane, which plays a significant role in the formation of rod-shape crystal. The cell parameters in Table 2 indicate that the crystal stucture belongs to triclinic system.

Table 1 Chemical composition of copper oxalate salt synthesized at pH 7.0 and 8.0

Sample \ Contents	*Cu(wt.%)*	*N（wt.%）*	*C（wt.%）*	*H（wt.%）*
Fig.7(b)	36.90	15.33	13.28	3.17
Fig.7(c)	30.05	21.66	11.43	4.22

Fig.7 XRD patterns and corresponding SEM photos of copper oxalate salts synthesized at various pH values: (a) 5.0; (b) 7.0; (c) 8.0

[redacted]

[redacted]

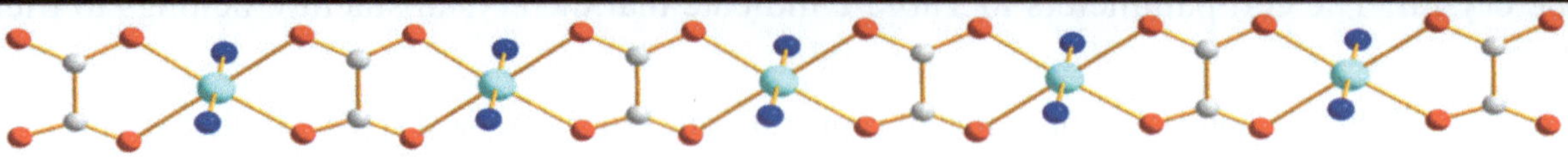

Fig.8 molecular chain of fibrous copper oxalate salt synthesized at pH 8.0

Table 2 the cell parameters of of fibrous copper oxalate salt synthesized at pH 8.0

a（Å）	b（Å）	c（Å）	α	β	γ
5.5046（16）	5.7125（17）	6.2887（18）	75.11°	79.05°	78.35°

Procedure of thermal decompostion of the rod-shape copper oxalate salt

The thermostability of rod-shape copper oxalate salt was investigated by TG-DCS-FTIR analyser.The result is shown in Fig.9. The FTIR spectrums of offgas at the points of time a, b, c, d are plotted in Fig.10. Compared with standard spectrum database, they are proved to be (a) NH_3, (b) NH_3 with little CO_2, (c) CO_2 and NH_3, (d) CO_2. Due to the hysteretic quality of detection and interference of offgas, it is reasonable to consider the three peaks ahead to be NH_3, and the last peak to be CO_2.

From the TG curve of Fig.9, it can be seen that the copper oxalate salt powder is continually losing weight as the temperature rise. There are three endothermic peaks respectively around 450K, 500K, 520K in the DSC curve, meaning the gradually elimination of the NH_3 from the $Cu(NH_3)_3C_2O_4{\cdot}xH_2O$ according to the FTIR spectrums in Fig.10; In addition, there exists a relatively weak exothermic peak around 580K, indicating the decompostion of CuC_2O_4 with emission of CO_2.

According to the above analysis, the procedure of rod-shape copper oxalate salt is as following:

$$Cu(NH_3)_3C_2O_4{\cdot}xH_2O \xrightarrow{\text{dehydration}} Cu(NH_3)_3C_2O_4 \xrightarrow{\text{deamination}} CuC_2O_4$$

$$\xrightarrow{\text{decarbonation}} Cu$$

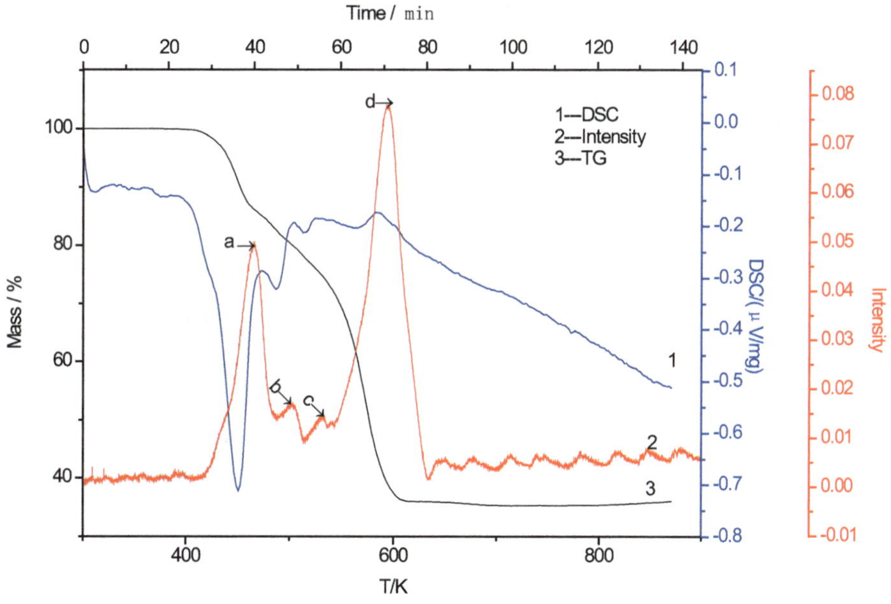

Fig.9 TG-DSC-FTIR curves of rod-shape copper oxalate at heating rate 10K/min in Ar atmorsphere

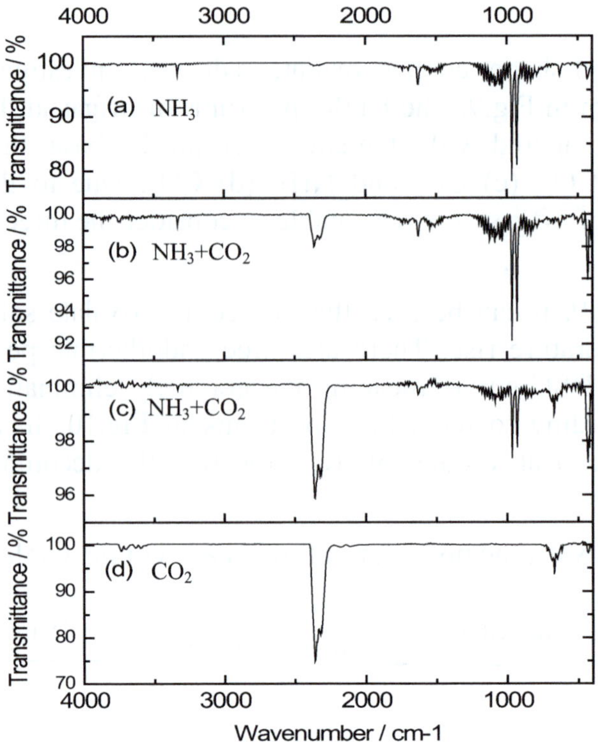

Fig.10 FTIR spectrums of offgas at the points of time a, b, c, d in Fig.9

Characterization of thermal decomposition product

Fig.11 shows the phase and morphology of the final product after thermal decompostion of rod-shape copper oxalate salt at the temperature of 623 K in nitrogen atmosphere. Compared with standard PDF card, the product is proved to be copper powder. And the morphology of particle performs to fibrous shape.

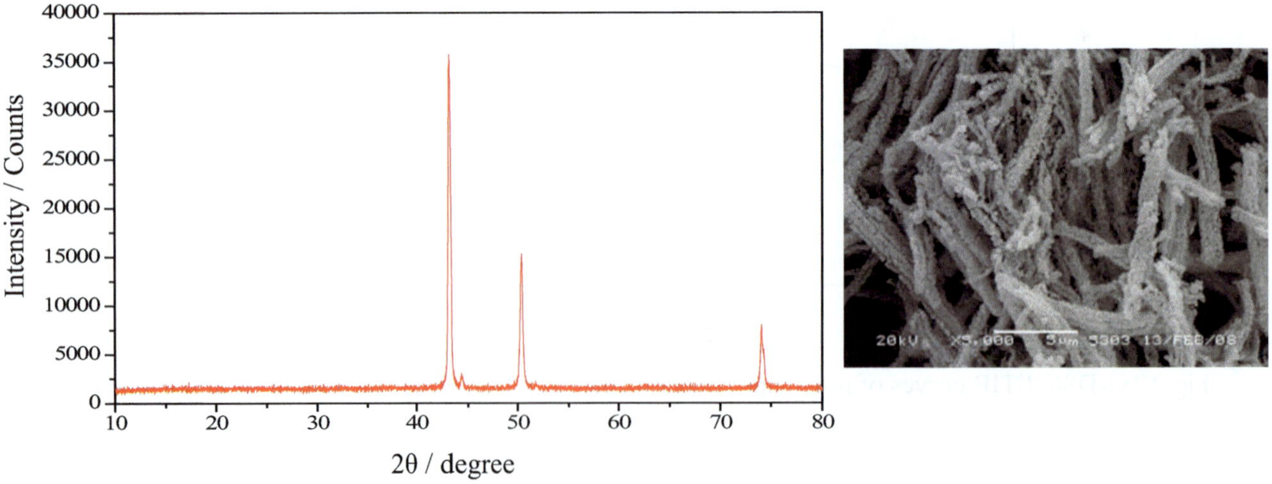

Fig.11 The X-ray diffraction pattern and SEM photoe of the thermal decompostion product of rod-shape copper oxalate salt

Conclusions

The research proposed a novel two-stage process for preparing the fibrous copper powder used for conductive filler in conductive paste. Firstly, thermodynamic calculation was carried out for the aqueous system and proved the possibility of process. meanwhile the effect of technical parameters on the equilibrium were discussed theoretically, wherein the pH value affect the distribution of species in solution to a large extent.

Copper oxalate salt with various morphologies——pie shape, flower shape and rod shape—— were synthesized by controlling different experimental conditions, especially the pH value. The results of characterization indicated that the molecular formula of rod-shape powder is $Cu(NH_3)_3C_2O_4 \cdot xH_2O$ ($0<x<1$) and the NH_3 plays significant role in the formation of rod shape. After thermal decomposition, the fibrous copper powder was prepared with the decompostion procedure: $Cu(NH_3)_3C_2O_4 \cdot xH_2O \rightarrow Cu(NH_3)_3C_2O_4 \rightarrow CuC_2O_4 \rightarrow Cu$

Acknowledgments

The work was supported by The Nature Science Research Key Project of Anhui Province of China (KJ2012A042), and The Young Talents Fund of Anhui Province in China (2012SQRL033ZD).

References

[1] Boudenne. Handbook of Multiphase Polymer Systems[M]. John Wiley and Sons, 2011: 427-431
[2] H.Koshikawa, H.Usui, Y.Maekawa. Thermally stable and anisotropically conducting membranes consisting of sub-micron copper wires in polyimide ion track membranes[J]. Journal of Membrane Science, 2009, 327(1-2): 182-187
[3] D.J.Amarasekera. Conductive plastics for electrical and electronic applications[J]. Reinforced Plastics, 2005, 49(8): 38-41
[4] J.E. Mark. Some Novel Polymeric Nanocomposites[J]. Acc. Chem. Res., 2006, 39 (12): 881-888
[5] H.M. Ma, X.L. Gao. A three-dimensional Monte Carlo model for electrically conductive polymer matrix composites filled with curved fibers[J]. Polymer, 2008, 49(19): 4230-4238
[6] O.D.Neikov, S.S.Naboychenko, I.V.Murashova, et al. Handbook of Non-Ferrous Metal Powders - Technologies and Applications[M]. Elsevier, 2009: 331-367
[7] E. Comini, G. Faglia, M. Ferroni, et al. Physical Vapor Deposition of Copper Oxide Nanowires[J]. Procedia Engineering, 2010, 5: 1051-1054
[8] Y.Konishi, M.Motoyama, H.Matsushima, et al. Electrodeposition of Cu nanowire arrays with a template[J]. Journal of Electroanalytical Chemistry, 2003, 559: 149-153
[9] X.B. Cao , F. Yu, L.Y. Li, et al. Copper nanorod junctions templated by a novel polymer-surfactant aggregate[J]. Journal of Crystal Growth, 2003, 254(1-2): 164-168
[10] T.Ahmad, A.Ganguly, J.Ahmed, et al. Nanorods of transition metal oxalates: A versatile route to the oxide nanoparticles[J]. Arabian Journal of Chemistry, 2011, 4(2):125-134
[11] Z.G. Jia, L.H.Yue, Y.F. Zheng, et al. The convenient preparation of porous CuO via copper oxalate precursor[J]. Materials Research Bulletin, 2008, 43(8-9): 2434-2440
[12] X.J. Zhang, D.G. Zhang, X.M. Ni, et al. Optical and electrochemical properties of nanosized CuO via thermal decomposition of copper oxalate[J]. Solid-State Electronics, 2008, 52(2): 245-248
[13] N.Jongen, P.Bowen, J.Lemaitre, et al. Precipitation of Self-Organized Copper Oxalate Polycrystalline Particles in the Presence of Hydroxypropylmethylcellulose (HPMC): Control of Morphology[J]. Journal of Colloid and Interface Science, 2000, 226(2): 189-198
[14] T.Ahmad, R.Chopra, R.V.Kandalam, et al. Nanorods of copper and nickel oxalates synthesized by the reverse micellar route[J]. Journal of Nanoscience and Nanotechnology, 2005, 5(11): 1840-1845

[15] M.Y. Li, W.S. Dong, C.L.Liu, et al. Ionic liquid-assisted synthesis of copper oxalate nanowires and their conversion to copper oxide nanowires[J]. Journal of Crystal Growth, 2008, 310(21): 4628-4634
[16] C.W. Bale, P. Chartrand, S.A. Degterov,et al. FactSage Thermochemical Software and Databases [J]. Calphad, 2002, 26(2): 189-228
[17] C.W. Bale, E. Bélisle, P. Chartrand, et al. FactSage Thermochemical Software and Databases – Recent Developments [J]. Calphad, 2009, 33(2): 295-311
[18] Y.Q. Fan, C.F. Zhang, J. Zhan, et al. Thermodynamic equilibrium calculation on preparation of copper oxalate precursor powder. Trans. Nonferrous Met. Soc. China, 2008, 18(2): 454-458

REWAS 2013: Enabling Materials Resource Sustainability
Edited by: Anne Kvithyld, Christina Meskers, Randolph Kirchain, Gregory Krumdick, Brajendra Mishra, Markus Reuter, Cong Wang, Mark Schlesinger, Gabrielle Gaustad, Diana Lados, and Jeffrey Spangenberger
TMS (The Minerals, Metals & Materials Society), 2013

Silver Selenide Thermodynamics for Copper Anode Slime Refining

Dawei Feng, Pekka Taskinen*

Department of Materials Science and Engineering, Aalto University
Vuorimiehentie 2K, Espoo, FI-00076 Aalto, Finland

Keywords: Silver selenide, Thermodynamic stability, Solid state galvanic cell.

Abstract

Copper anode slimes are a by-product of the electrolytic refining of anode copper, which contains significant amounts of silver, selenium, copper, and gold. Slimes are usually smelted to recover silver and gold. The thermodynamics of the smelting of such selenium-rich materials have received only little attention. In this work, the numerical values on the standard thermodynamic functions of Ag_2Se (Naumannite) were determined by the electromotive force (EMF) method in a solid-state galvanic cell with superionic conductor $RbAg4I_5$ as the solid electrolyte. Ag_2Se was synthesized from pure elements in evacuated quartz glass ampoules and examined to be homogenous by SEM and EDS. According to the experimental data on the EMF versus temperature, the analytical equations were obtained for the polymorphic forms of Ag_2Se. The temperature of phase transformation from α-Ag_2Se to β-Ag_2Se is determined experimentally to be 407.7 K by interpolation of the EMF vs. T data, and the enthalpy of phase transformation is 6.06 kJ•mol^{-1}. The Gibbs energy of formation for Ag_2Se is given by

$$\Delta\overline{G}_{\alpha\text{-}Ag_2Se}, J = -(40869.14 \pm 0.58129) - (27.94759 \pm 1.53034) \cdot T, (350 < T/K < 408),$$

$$\Delta\overline{G}_{\beta\text{-}Ag_2Se}, J = -(35062.17 \pm 0.09895) - (42.17847 \pm 0.21827) \cdot T, (408 < T/K < 500).$$

Introduction

Anode slimes are a by-product of the electrolytic refining of copper anodes, containing significant amounts of silver, gold, the platinum group metals, and selenium. Slimes typically contain from 5 to 53 pct copper, 5 to 20 pct silver, up to 1 pct gold, 1 to 45 pct selenium, and 2 to 30 pct lead, as well as small amounts of tellurium, sulfur, arsenic, bismuth, antimony, and nickel. Silver is found partly in the metallic state, associated with selenium as silver selenide (Ag_2Se) and silver copper selenide (AgCuSe); copper is present in the metallic state, as the sulfate or the oxide and associated with selenium as copper selenide (Cu_2Se) and AgCuSe; lead is present mainly as lead sulfate ($PbSO_4$).

The primary aim of the processing of anode slimes is to remove copper, selenium, tellurium, etc., and to leave silver, gold, and the platinum group metals, as an alloy called ‘‘dore′.’’ However, the thermodynamics of anode slime smelting have received little attention. The chemical potential diagram for the Ag-Se-O and Cu-Se-O systems and a simple computer model were presented by D. Swinbourne, A. Yazawa, and G. Barbante [1].

The objective of the present study is to determine the thermodynamic stability of Naumannite experimentally, by using electromotive force (EMF) measurements in solid-state galvanic cells at low temperatures. Ag_2Se(cr) is the only intermediate phase in the Ag-Se system and it exists in two polymorphs, α and β. The α-polymorph (low naumannite) is stable below 406 K. The temperature range from 350 to 500 K was chosen in order to obtain thermodynamic data for both α-Ag_2Se (stable below 406 K) and β-Ag_2Se (stable above 406 K) in equilibrium with pure selenium.

2. Experimental

2.1 Materials synthesis

Silver powder from Alfa Aesar (99.99 % in purity) and selenium powder from Koch-Light Laboratories Ltd (99.999 % in purity) were used for synthesis of the intermetallic Ag_2Se phase. The Ag_2Se intermetallic compound was prepared by a direct synthesis from the elements. The Ag and Se powder were mixed in mole ration 2:1 together and sealed in an evacuated quartz glass ampoule. Then the ampoule was heated up from room temperature to 673 K at a rate of 4 K/min and kept for 2 days. After that it was heated to 873 K and then kept at that temperature for 2 days. Finally, it was heated to 1173 K (the melting point of Ag_2Se) and kept there for one hour, and then cooled down to room temperature.

The resulting phase mixture from the synthesis was examined to be homogenous by SEM and EDS. The compounds were ground with 1% atomic excess of selenium and pressed uniaxially to obtain a pellet of 6 mm in diameter and 2 mm in thickness, under pressure of about 0.1G Pa.

The solid electrolyte, $RbAg_4I_5$, was synthesized according to the method described by Owens and Argue [2]. Weighing of 0.8 mole fraction of silver iodide from Alfa Aesar (99.9 % in purity) and 0.2 mole fraction of rubidium iodide from Alfa Aesar (99.8 % in purity) was followed by mixing. The mixture was sealed in a glass tube under vacuum, and heated at 220 °C for 2 hr. Then after being cooled down and maintained at 160 °C for 15 hours, thereby $RbAg_4I_5$ was obtained.

2.2 Temperature and EMF Measurements

Constant temperature profile inside a Lenton tube furnace type LTF 12/50/610 was located by measurements using a moveable thermo-resistance. Schematic illustration of the experimental furnace for carrying out measurements with the solid electrolytes is shown in Fig 1. During the EMF-measurements, temperatures on both ends of the galvanic cell were measured using two PT100 sensors (platinum resistance thermometers, PRT). The PRTs, with tolerance class B 1/10 DIN, i.e. ± 0.03 °C variation at 0 °C, according to the manufacturer, were calibrated in a mixture of ice and water at 0 °C. The obtained resistance values above 100Ω (Ro = 100.026 Ω and Ro = 100.03 Ω, respectively) were added to a program based on LabVIEW software code from National Instruments that records the temperature values from each PRT. The accuracy of temperature and EMF readings reach 0.0001 K and 0.001 mV, respectively.

The identical platinum wires for the EMF-measurements were used and the lead wires from the PT100 sensors for temperature measurements were connected to a KEITHLEY-6517B/electrometer/high resistance meter and a KEITHLEY-2000-multimeter, respectively. Input impedance of the electrometer for EMF-measurements was 2 10^{14} Ω, which allows the cells to function in a reversible way. The measured EMF-values and temperatures were simultaneously transferred to a computer through a IEE-488-GPIB-cable and a KEITHLEY-KUSB-488A USB-to-GPIB interface adapter, and the readings were recorded by the software giving two measured values every 5 seconds.

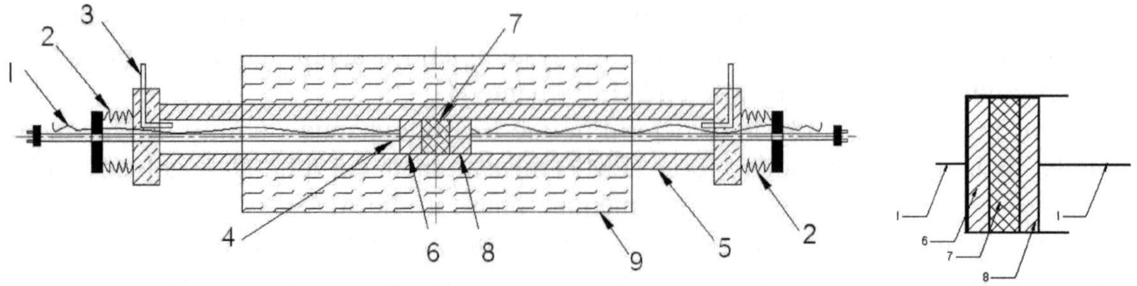

Fig.1. Schematic illustration of a experimental furnace for carrying out measurements with the solid electrolytes (1) platinum wire, (2) spring, (3) inert gas inlet/outlet, (4) PRT, (5) furnace tube of quartz glass, (6) sample system, (7) solid electrolyte, (8) reference system, (9) Lenton tube furnace

Most measurements were performed by heating and cooling the cell in steps from 1 to 10 °C. To reach steady state EMF reading, it took from few hours up to two weeks. The equilibriums were considered to be reached when the EMF-values were constant, or their variations were not significant (< 0.1 mV) and they were oscillating about a certain value for several hours. Temperature differences between the two electrodes of the EMF cell were controlled to be much less than 1 °C (0.1 < T(°C) < 0.8), by manually adjusting the horizontal position of the galvanic cells and observing the real-time temperature readings over the cell from the highly accurate PRTs. The uncertainty of temperature and EMF was established to be ± 0.5 K and ±0.1 mV, respectively. Thus, the possible thermoelectric effect generated in the cell EMF by the temperature gradients is negligible.

Functionality of the experimental electrochemical system was tested by measuring the EMF of the symmetric cell Ag | Ag^+ | Ag, which theoretically should not result in any measurable EMF or electric potential difference. The equilibriums in this study were considered reproducible when the EMF readings in heating and cooling coincided, and another galvanic cell with the same chemical composition resulted in the same EMF-values.

In this particular cell design, the resistance thermometers are connected to each end of the cell, in order to record the temperature exactly on the anode and cathode, thus eliminating the temperature errors of measurements.

Gas flow of dry argon (99.999% in purity) to the EMF cell was purified before introduction to the cell, by passing though an auxiliary furnace with titanium wire at 900 K for removing oxygen traces. Owing to these improvements, the accuracy of EMF measurement and its stability as a function of time was very good, allowing also very long measuring campaigns for each experimental cell.

2.3 EMF cell

The measurements were performed on the solid state galvanic cell when one mole silver selenide is formed from solid pure silver and selenium, and its virtual cell reaction is given by (A).

The virtual reaction of the electrochemical cell is

$$2\,Ag(s) + Se(s) = Ag_2Se(s) \qquad (A)$$

The EMF of the cell

(-) Pt |Ag(s) | $RbAg_4I_5$(s) | Ag_2Se(s), Se(s, l) | Pt (+)

was measured in a temperature range of 350 K-500 K, under the ambient atmospheric pressure.

The Gibbs energy change of the cell reaction, except for the work of volume expansion, is related to the reversible EMF of reaction (A) by the Nernst equation

$$\Delta\overline{G}_{Ag_2Se} = -zFE \qquad (1)$$

where E is the electromotive force produced by the cell and F is the Faraday constant (96 485 $C \cdot mol^{-1}$) and the number of elementary changes transferring in the virtual cell reaction (A) is z =2.

The other fundamental thermodynamic properties of Ag_2Se at selenium saturation can be derived from the general properties of Gibbs energy and its functional relationships with temperature [3]:

$$\Delta\overline{S}_{Ag_2Se} = zF\left(\frac{\partial E}{\partial T}\right)_P \qquad (2)$$

and

$$\Delta\overline{H}_{Ag_2Se} = zF\left[T\cdot\left(\frac{\partial E}{\partial T}\right) - E\right]_P \qquad (3)$$

3. Results

The observed values of EMF (E, mV) at different temperatures obtained in this study are shown in Table 1, where temperature is the mean value of both end of electrodes, the E_{calc} column gives the EMF values according to linear equations obtained in the present study by the least squares fitting.

Table 1. Experimental EMF values

T, K	E, mV	E_{calc}, mV	E_{meas} -E_{calc}, mV
465.33	283.61	283.44	0.18
455.51	281.41	281.29	0.12
445.77	279.22	279.16	0.06
426.24	274.93	274.89	0.04
406.89	270.70	270.50	0.21
404.02	270.06	270.10	-0.04
402.06	269.63	269.82	-0.19
392.52	268.08	268.47	-0.39
368.51	264.85	265.09	-0.24
359.09	263.74	263.76	-0.02
349.65	262.41	262.43	-0.02
350.41	262.05	262.58	-0.53
360.01	264.03	263.96	0.07
369.64	265.85	265.34	0.51
379.32	267.06	266.73	0.33
388.95	267.98	268.11	-0.12
398.71	269.13	269.51	-0.38
408.44	271.02	270.98	0.04
418.20	273.12	273.12	0.00
427.91	275.22	275.24	-0.02
437.67	277.32	277.37	-0.06
447.47	279.43	279.52	-0.09
457.33	281.59	281.67	-0.09
467.17	283.80	283.82	-0.02
476.99	286.12	285.97	0.15
486.85	288.24	288.13	0.12
496.77	290.13	290.29	-0.16

The analytical equations were obtained by the least square method using the linear relationship $E = a + b \cdot T$ as:

$$E, \mathrm{mV} = (211.79 \pm 3.01) + (0.1448 \pm 0.0079) \cdot T, (350 < T/K < 408) \quad (4)$$

$$E, \mathrm{mV} = (181.70 \pm 0.51) + (0.2186 \pm 0.0011) \cdot T, (408 < T/K < 500) \quad (5)$$

The Gibbs energy change of reaction (A), according to equation (1), representing the two polymorphic forms of Ag_2Se over their stability rages in these measurements, are given by equations:

$$\Delta \overline{G}_{\alpha\text{-}Ag_2Se}, \mathrm{J} = -(40869.14 \pm 0.581) - (27.95 \pm 1.53) \cdot T, (350 < T/K < 408) \quad (6)$$

$$\Delta \overline{G}_{\beta\text{-}Ag_2Se}, \mathrm{J} = -(35062.17 \pm 0.099) - (42.18 \pm 0.22) \cdot T, (408 < T/K < 500) \quad (7)$$

As the EMF of the cell was measured as a function of temperature, the molar entropy and enthalpy formation of α-Ag_2Se and β-Ag_2Se at selenium saturation can be derived from equations (2) - (3), as following:

$$\Delta\overline{S}_{\beta\text{-}Ag_2Se} = z\cdot F\cdot\left(\frac{\partial E}{\partial T}\right) = (42.18\pm0.22)\ \text{J}\cdot\text{K}^{-1}\text{mol}^{-1} \quad (8)$$

$$\Delta\overline{S}_{\alpha\text{-}Ag_2Se} = z\cdot F\cdot\left(\frac{\partial E}{\partial T}\right) = (27.95\pm1.53)\ \text{J}\cdot\text{K}^{-1}\text{mol}^{-1} \quad (9)$$

$$\Delta\overline{H}_{\alpha\text{-}Ag_2Se} = z\cdot F\cdot\left[T\cdot\left(\frac{\partial E}{\partial T}\right) - E\right] = -(40.87\pm0.58)\ \text{kJ/mol} \quad (10)$$

$$\Delta\overline{H}_{\beta\text{-}Ag_2Se} = z\cdot F\cdot\left[T\cdot\left(\frac{\partial E}{\partial T}\right) - E\right] = -(35.06\pm0.099)\ \text{kJ/mol} \quad (11)$$

4. Discussion

Figure 2 shows a comparison of the EMF measurements obtained by various authors according to the literature sources and the data wherein. Obviously, our data agrees pretty well with Oehsen & Schmalzried [5] by solid coulometric titration. Moreover, the data obtained in this study confirm the results given by Osadchii & Voronin [6] at lower temperatures, and with Kiukkola & Wagner [7] at higher temperatures above the polymorphic transformation temperature. However, Yamamoto & Takahashi's data below the polymorphic transformation temperature are clearly higher than all the other author's values. At 373.15 K, the EMF value difference with the other authors is as large as 5.2 mV. This may result from the non-equilibrium in the measurement, due to lack of sufficient equilibration time in the observations, which was limited for a few hours only. In fact, the equilibrium of reaction (A) was typically reached in this work after 1-2 weeks in the present experimental condition, resulting from slow diffusion of silver ions at low temperature [8].

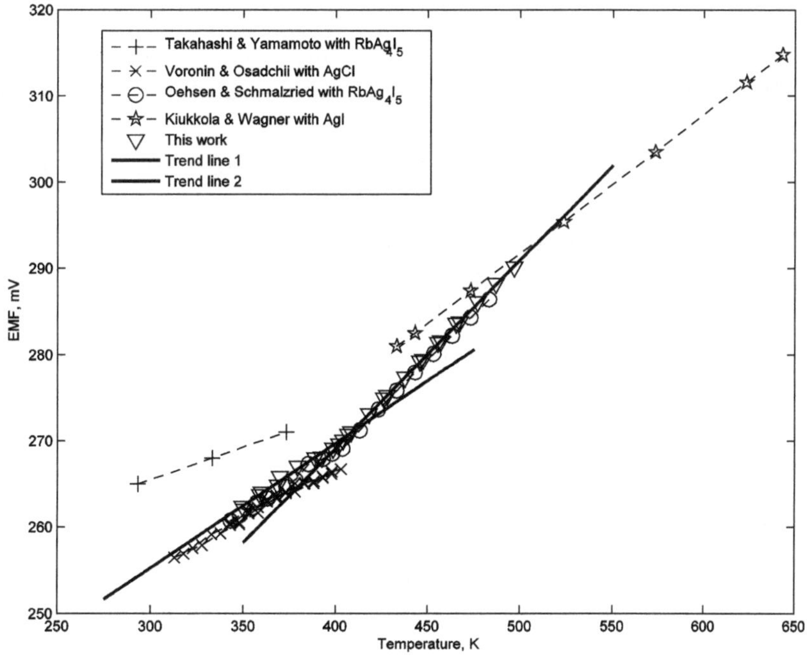

Fig.2. Variation of EMF of Ag_2Se with temperature according to various authors and the present work Takahashi & Yamamoto [4], Oehsen & Schmalzried [5], Voronin & Osadchii [6], Kiukkola & Wagner [7]

Using the fundamental equations of thermodynamics, the thermodynamic features of the crystallographic modifications of naumannite under constant atmospheric pressure were calculated. The results are shown in Table 2 and compared to the literature data.

Table 2. A comparison of values of the standard thermodynamic properties of Ag_2Se

Compound	$-\Delta\overline{G}$, kJ/mol	$\Delta\overline{S}$, $J\cdot K^{-1}mol^{-1}$	$-\Delta\overline{H}$, kJ/mol	References
α- Ag_2Se	49.47±0.13	144.99±0.56	42.73±0.29	Osadchii & Echmaeva [9]
α- Ag_2Se	48.90±1.0	148.20	42.70	Nasar & Shamsuddin [10]
α- Ag_2Se	49.59	149.20	43.09	Voronin & Osadchii [6]
α- Ag_2Se	49.24±0.46	154.60±0.22	41.12±0.58	This work
β - Ag_2Se	47.43±0.29	169.01±0.78	35.02±0.48	Osadchii & Echmaeva [9]
β - Ag_2Se	48.87	148.20	42.70	Nasar [10]
β - Ag_2Se	47.58	169.44	35.04	Voronin & Osadchii [6]
β - Ag_2Se	47.64±0.07	169.57±1.53	35.06±0.099	This work

Table 2 shows the standard thermodynamic functions calculated for the Gibbs energy of formation of Ag_2Se using the data obtained in this study and reported in the literature. Phase transition of the low temperature naumannite (α- Ag_2Se) to its high temperature modification (β - Ag_2Se) occurs at an equilibrium temperature of 407.7 K, and its enthalpy of transformation calculated from the observations of this work is 6.06 $kJ\cdot mol^{-1}$. However, data varies somehow even for the same author. In different studies, Osadchii & Echmaeva [9] in 2007 reported the stable polymorphic transformation temperature T_{trs}= 405.4 K, whereas Voronin & Osadchii in 2011 reported that T_{trs}= 397.5 K [6,6]. This difference may result from their electrolyte selections, where AgCl + KCl was used in the later experiment instead of $RbAg_4I_5$. This is supported by the fact that ionic transfer numbers affect accuracy of the measurements [11].

5. Conclusion

In this study, new experimental data on the stability of naumannite, Ag_2Se in equilibrium with pure selenium, were obtained over a temperature range of 300-500 K. The results obtained in this study agree very well with the previous experimental data, and provide a higher accuracy based on the advanced EMF cell design and direct control of temperature difference over the solid electrolyte. At lower temperatures, the results obtained were well compatible with the data provided by Osadchii & Echmaeva [9], while they tend to confirm the experimental observations by Kiukkola & Wagner [7] at higher temperatures. Furthermore, our data agrees very well with Oehsen & Schmalzried [5] by solid coulometric titration.

The equilibrium temperature of polymorphic phase transformation from α- Ag_2Se to β - Ag_2Se is determined in this work experimentally to be 407.7 K, by interpolation of the EMF vs. Temperature data. The transformation temperature obtained is slightly higher than the earlier reported value of 406.3 K [7]. The enthalpy of phase transformation is 6.06 $kJ\cdot mol^{-1}$, which is slightly lower than the value 6.824 $kJ\cdot mol^{-1}$ given in the literature [9]. The results obtained in this study clearly indicate that $RbAg_4I_5$ is a suitable solid electrolyte in the low temperature EMF methods.

Acknowledgements

The authors are grateful to Sönke Schmachtel for helping analyze of the data errors, Markus Aspiala for helping set up of the furnace, and Marko Järvenpää for SEM and EDS analyze. This work was made as a sub task of ISS project of Elemet Program (Fimecc Oy), supported financially also by Boliden Harjavalta Oy, Boliden Kokkola Oy, Norilsk Nickel Finland Oy and Outotec Finland Oy.

Reference

[1] D. Swinbourne, A. Yazawa, G. Barbante, Metall. Mater. Trans. B-Proc. Metall. Mater. Proc. Sci. 28 (1997) 811-819.

[2] B.B. Owens, G.R. Argue, Science. 157 (1967) 308-310.

[3] H. Ipser, A. Mikula, I. Katayama, Calphad. 34 (2010) 271-278.

[4] T. Takahashi, O. Yamamoto, J. Electrochem. Soc. 117 (1970) 1-5.

[5] U. VONOEHSEN, H. SCHMALZRIED, Berichte Der Bunsen-Gesellschaft-Physical Chemistry Chemical Physics. 85 (1981) 7-14.

[6] M. Voronin, E. Osadchii, Russian J. Electrochem. 47 (2011) 420-426.

[7] K. Kiukkola, C. Wagner, J. Electrochem. Soc. 104 (1957) 379-387.

[8] I. Rom, W. Sitte, Solid State Ionics. 101 (1997) 381-386.

[9] E.G. Osadchii, E.A. Echmaeva, Am. Mineral. 92 (2007) 640-647.

[10] A. Nasar, M. Shamsuddin, Metallurgical and Materials Transactions B. 28 (1997) 519-522.

[11] K. KIUKKOLA, C. WAGNER, J. Electrochem. Soc. 104 (1957) 308-316.

REWAS 2013: Enabling Materials Resource Sustainabil
Edited by: Anne Kvithyld, Christina Meskers, Randolph Kirchain, Gregory Krumdick, Brajendra Mish
Markus Reuter, Cong Wang, Mark Schlesinger, Gabrielle Gaustad, Diana Lados, and Jeffrey Spangenber
TMS (The Minerals, Metals & Materials Society), 20

MEASUREMENT OF THERMODYNAMIC PROPERTIES OF TELLURIUM IN MOLTEN IRON BY TRANSPIRATION METHOD

Shumpei Suzuki[1], Takeshi Yoshikawa[1], Takayuki Nishi[2] and Kazuki Morita[1]

[1]Institute of Industrial Science, The University of Tokyo;
Rm. Fw-402, 4-6-1, Komaba, Meguro-ku, Tokyo, 153-8505, Japan
[2]Technical Research & Development Bureau, Nippon Steel & Sumitomo Metal Corporation;
Sunayama 16-1, Kamisu City, Ibaraki, 314-0255 Japan

Keywords: Steelmaking, Thermodynamics, Tellurium, Transpiration method

Abstract

Tellurium is added to sulfur free-cutting steel in order to improve its machinability by inclusion control. However, the thermodynamic properties of the steel and inclusions have not yet been studied. In the present work, the dissolution of tellurium into steel was investigated. First, the equilibrium Te_2(g) partial pressure of solid tellurium was determined at 663–693 K by the transpiration method. The equilibration of molten iron under a controlled tellurium partial pressure, prepared by the transpiration method, was then performed at 1823 K. An approximately linear relationship between the tellurium content of the molten iron and the partial pressure of Te(g) was found at mass% Te < 0.01.

Introduction

Lead-added free-cutting steel is widely used as a steel with excellent machinability. Currently, environmental restrictions affect lead usage, and the general trend is to minimize or eliminate lead in free-cutting steels. Although sulfur free-cutting steel is also a major free-cutting steel, the anisotropy of its mechanical properties is regarded as a serious issue. MnS inclusions form in sulfur free-cutting steel and contribute to the stress concentration factor to provide machinability in cutting processes. However, such inclusions elongate after hot-rolling processes, resulting in anisotropic mechanical properties of the steel. Tellurium is known to be an effective additive element for modifying elongation by forming an MnS–MnTe liquid phase around MnS inclusions in steels, and preventing elongation of the MnS inclusions by a sacrificial effect [1]. In controlling tellurium addition and inclusion formation, the thermodynamic properties of tellurium in the steel and inclusions are crucial, but the necessary thermodynamic information is not available.

Tellurium is added to free-cutting steels in amounts of less than 0.05 mass%; therefore precise addition control is indispensable for obtaining the required steel properties. In the present work, the thermodynamic properties of tellurium in molten iron were investigated at 1823 K. First, the equilibrium Te_2(g) partial pressure of solid tellurium was measured at 663–693 K. Then, the equilibration of molten iron under a controlled tellurium partial pressure was conducted, using the transpiration method, at 1823 K.

Experimental

Measurement of Tellurium Partial Pressure

Experiments were conducted using an experimental apparatus consisting of a horizontal Kantal furnace equipped with a quartz tube (shown in Figure 1). The temperature was maintained at 663–693 K with an accuracy of ±1 K, using a proportional integral differential (PID) controller and a chromel–alumel thermocouple. Granular tellurium (99.99% purity, ~1 mm diameter) was charged in the tube and placed in the hot zone of the furnace. Argon gas was flown through the tube at a constant rate of 9.17×10^{-4} to 3.67×10^{-3} m^3/s. The gas was cooled in region 5 in Figure 1, and solid tellurium deposited. The deposited tellurium was dissolved in acid, and the amount was determined using inductively coupled plasma emission spectroscopy (ICP-AES). Although tellurium exists in several molecular forms, the majority is present as $Te_2(g)$ below the melting point of tellurium (722 K). The tellurium partial pressure was therefore determined as follows:

$$P_{Te_2} = \frac{mRT}{2MFt} \quad (1)$$

Here, m (g) is the amount of tellurium deposited, M (g/mol) is the atomic weight of tellurium, and F (m^3/h) and t (h) specify the flow rate of argon gas and experimental time, respectively.

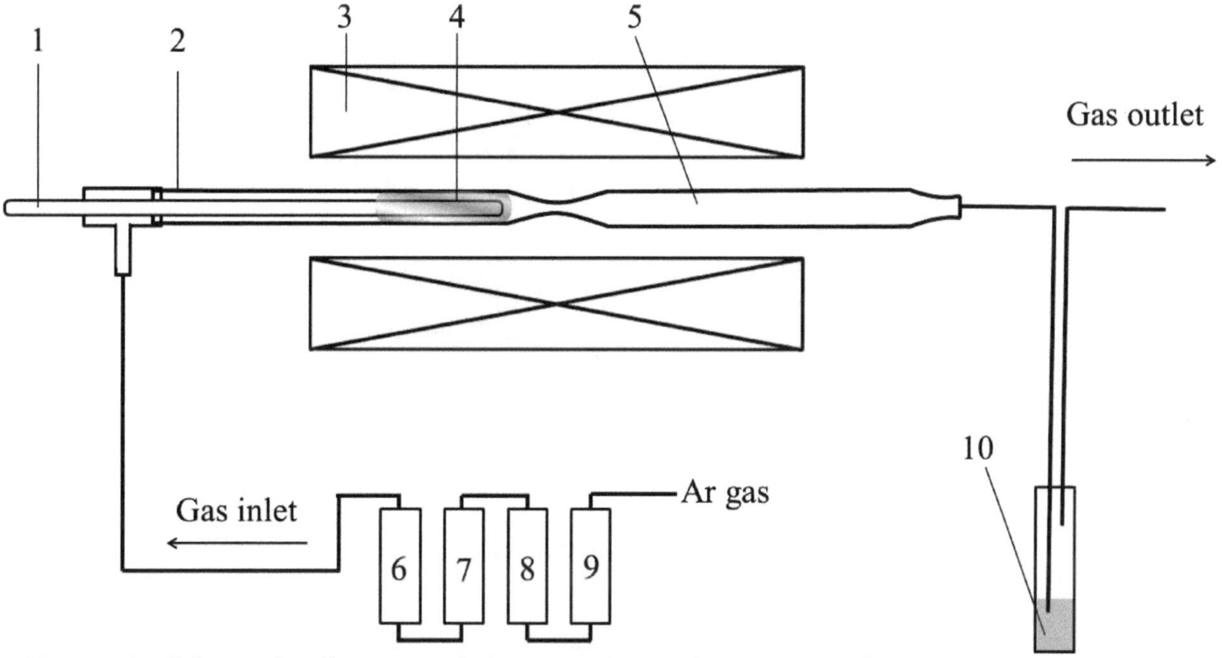

Figure 1. Schematic diagram of the experimental apparatus for measurements: 1 K-type thermocouple, 2 quartz reaction tube, 3 Kantal furnace, 4 granular tellurium, 5 space for deposition of tellurium, 6 mass flow controller, 7 $Mg(ClO_4)_2$ + soda lime, 8 silica gel, 9 Mg deoxidation furnace, and 10 bubbler (mixed acid).

Equilibration of Molten Iron Under Controlled Tellurium Partial Pressure

The experimental apparatus consisted of a vertical $MoSi_2$ resistance furnace equipped with a mullite reaction tube (Figure 2). The temperature was controlled at 1823 K with an accuracy of ±1 K using a PID controller and a Pt–6%Rh/Pt–30%Rh thermocouple. The granular tellurium (99.99% purity, ~1 mm diameter) was charged in the quartz tube and placed in the hot zone of the upper furnace and heated at 663–693 K. By introducing argon gas at a constant rate (3.67×10^{-3} m^3/s) into the tube, saturated tellurium vapor was generated. Iron (4 g; >99.9% purity) was placed in a high-purity alumina crucible, which was hung at the end of the gas lance with tungsten wires and placed in the hot zone of the lower furnace. The iron was melted under a controlled tellurium partial pressure at 1823 K for a period of 16 h. It was then pulled up and quenched in the furnace. The concentrations of tellurium in the iron were determined by ICP-AES.

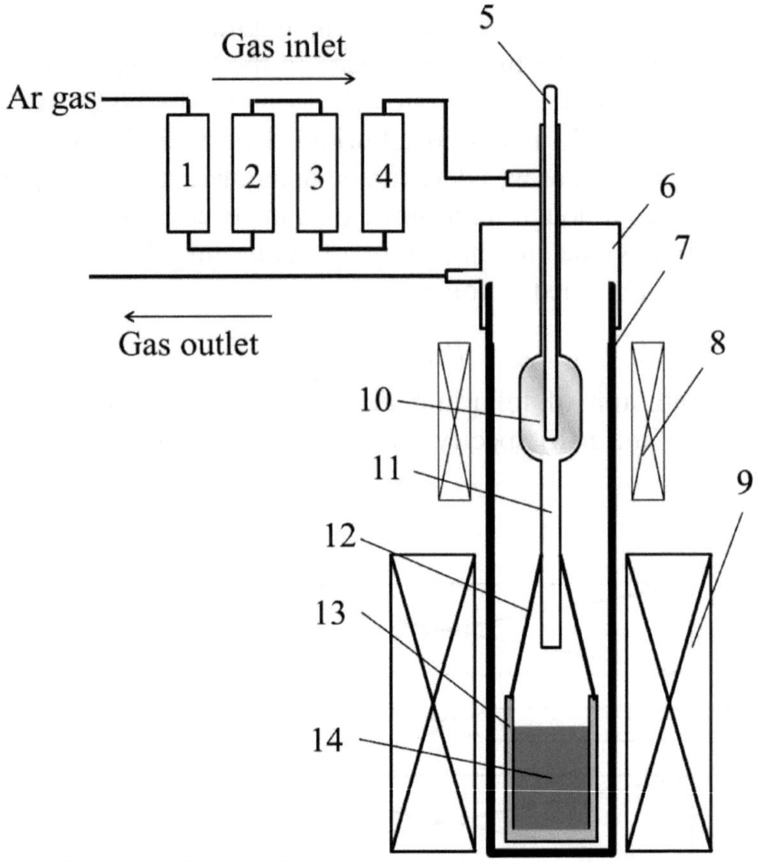

Figure 2. Schematic diagram of experimental apparatus: 1 silica gel, 2 $Mg(ClO_4)_2$ + soda lime, 3 Mg deoxidation furnace, 4 mass flow controller, 5 B-type thermocouple, 6 stainless-steel flange, 7 mullite reaction tube, 8 Kantal furnace, 9 $MoSi_2$ resistance furnace, 10 granular tellurium, 11 alumina gas lance, 12 tungsten wire, 13 alumina crucible, and 14 molten iron.

Results and Discussion

Measurement of Tellurium Partial Pressure

The region for cooling the gas after the experiments is shown in Figure 3. Needle-shaped deposits appear in the high-temperature region, whereas a coating on the quartz tube is seen in the low-temperature region. As the tellurium content of the acid in the gas bubbler placed after the furnace was negligible, the evaporated tellurium was completely collected as these deposits. The vapor pressures of $Te_2(g)$, calculated from the measured amounts of tellurium in the deposits, are plotted against the flow rate in Figure 4. The vapor pressures were observed to be constant, even when the flow rate of the argon gas was varied in the range 9.17×10^{-4} to 3.67×10^{-3} m^3/s. The temperature dependence of the vapor pressures of $Te_2(g)$ is shown in Figure 5, together with the reported data [2, 3]. The data in the present study are in good agreement with those reported. We therefore concluded that the transpiration method is suitable for generating stable tellurium vapor, and adopted these partial pressures in the equilibrium experiments. $P_{Te_2(g)}$ can be written as Eq. (2) in the temperature range used.

$$\mathrm{Log} \cdot P_{Te_2(g)} = -8413.2 \times \frac{1}{T} + 8.0573 \qquad (2)$$

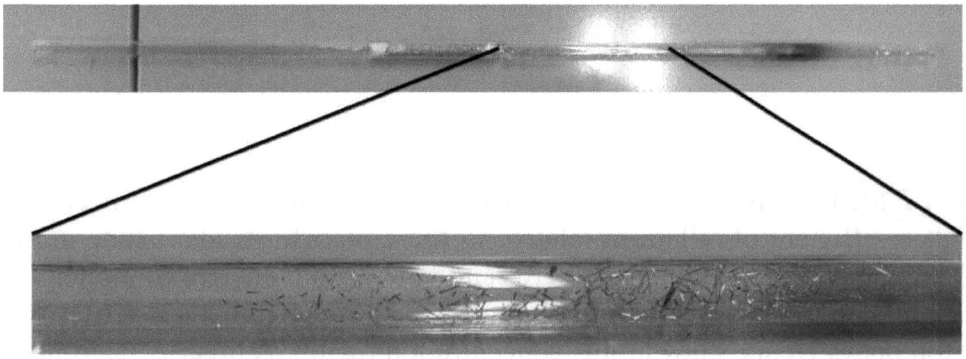

Figure 3. Deposited tellurium in quartz tube.

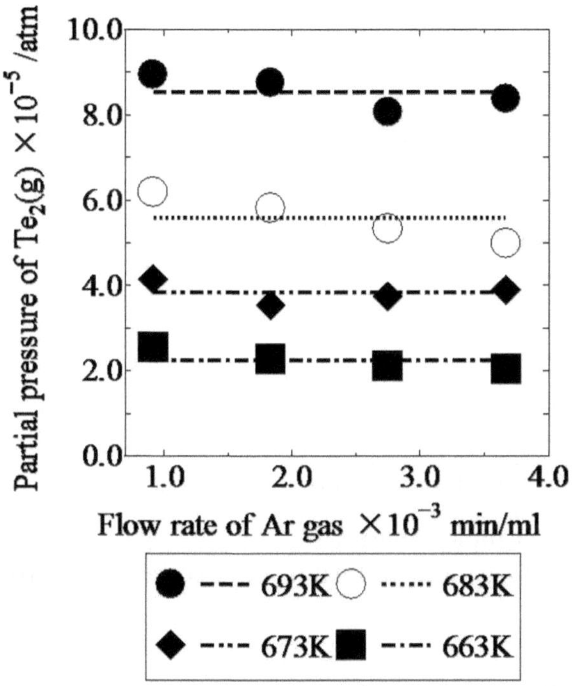

Figure 4. Measured partial pressure of Te_2(g) against gas flow.

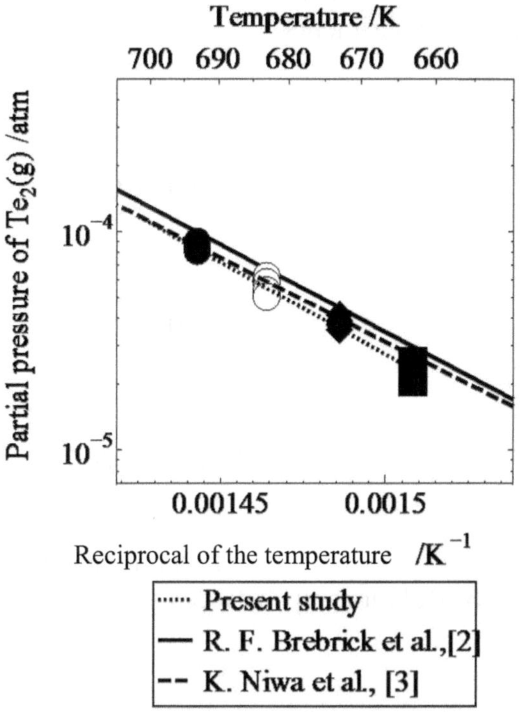

Figure 5. Temperature dependence of partial pressure of Te_2(g).

Equilibration of Molten Iron Under Controlled Tellurium Partial Pressure

Te_2(g), which was generated by the transpiration method at 663–693 K, dissociates at high temperature. The standard Gibbs energy change of the disassociation reaction in Eq. (3) is expressed by Eq. (4).

$$\mathrm{Te_2(g)} = 2\mathrm{Te(g)} \tag{3}$$

The equilibrium constant in Eq. (4) is extremely small at 1823 K, and the generated Te_2(g) is considered to dissociate completely.

$$\frac{P_{\mathrm{Te_2(g)}}}{P_{\mathrm{Te(g)}}^2} = \exp\left(-\frac{\Delta G_1^0}{RT}\right) \tag{4}$$

The concentrations of tellurium in molten iron are plotted against the partial pressure of Te(g) in Figure 6. It is clear that the concentration of tellurium in iron increases with increasing partial pressure of Te(g). The dissolution reaction of tellurium into molten iron is given by Eq. (5), and the standard Gibbs energy change for the reaction, ΔG_2^0, is given by Eq. (6).

$$Te(g) = \underline{Te}(1\text{mass\%, in Fe}) \tag{5}$$

$$\Delta G_2^0 = -RT \ln \frac{f_{Te} \cdot [\text{mass\%Te}]\text{in Fe}}{P_{Te}} \tag{6}$$

Here, R and T are the gas constant and the experimental temperature, respectively, and f_{Te} is the activity coefficient of tellurium relative to the 1 mass% standard state in molten iron. As an approximately linear relationship between the tellurium content of molten iron and the partial pressure of Te(g) is observed in Figure 6, it is speculated that the activity of tellurium in molten iron obeys Henry's law in this concentration region. A precise determination of the standard Gibbs energy change for the dissolution of tellurium will be made after conducting additional experiments.

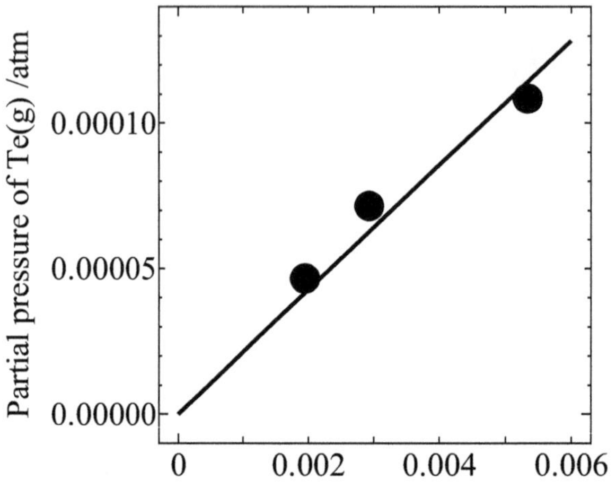

Figure 6. Relationship between partial pressure of Te(g) and tellurium concentration in iron samples.

Conclusions

In order to determine thermodynamic properties of tellurium in iron, dissolution of tellurium vapor, generated by transpiration method, into molten iron was investigated and obtained the findings as below.

(i) The equilibrium Te_2(g) partial pressure of solid tellurium was determined at 663–693 K by the transpiration method. Since the data obtained in the present study are in good agreement with those reported, the transpiration method was found to be suitable for generating equilibrated tellurium vapor.

(ii) The equilibration of molten iron under a controlled tellurium partial pressure, prepared by the transpiration method, was performed at 1823 K. An approximately linear relationship between the tellurium content of the molten iron and the partial pressure of Te(g) was found when Te content was less than 0.01 mass%.

References

[1] T. Kato *et al.*, *Electric Furnace Steel*, **53**(2) (1982), 195–202.

[2] R.F. Brebrick *et al.*, *J. Phys. Chem.*, **72**(5) (1968), 1032–1036.

[3] K. Niwa *et al.*, *J. Chem. Soc. Jpn.*, **61** (1940), 667–670.

REWAS 2013: Enabling Materials Resource Sustainability
Edited by: Anne Kvithyld, Christina Meskers, Randolph Kirchain, Gregory Krumdick, Brajendra Mishra, Markus Reuter, Cong Wang, Mark Schlesinger, Gabrielle Gaustad, Diana Lados, and Jeffrey Spangenberger
TMS (The Minerals, Metals & Materials Society), 2013

THERMODYNAMIC MODEL FOR ACIDIC METAL SULFATE FROM SOLUBILITY DATA

Petri Kobylin[1], Hannu Sippola[1,2], Pekka Taskinen[1]

[1]Department of Materials Science and Engineering, Aalto University; FI-00076 Aalto; Finland
[2]Permanent address FCG Design and Engineering; Osmontie 34; Helsinki, FI-00601, Finland

Keywords: thermodynamic modeling, Pitzer model, acid mine drainage

Abstract

Acidic metal sulfate solutions are generated in a large scale in the hydro- and pyrometallurgical industries. Acid mine drainage has long been a significant environmental problem in coal and metal mining. Acidic metal sulfate solutions are also produced in steel industry. The demand of recycling and reuse of materials has increased significantly especially in EU. Dumping a neutralized deposit is not an option any more. Thus, several techniques of recycling and reuse of sulfuric acid and/or metal sulfates from the side streams are needed.

When developing alternative solutions a better understanding of the thermodynamic behavior of $MnSO_4$–H_2SO_4–H_2O system is needed. In this study a thermodynamic model of this system is developed using Gibbs energy minimisation, Pitzer model and CALPHAD method to yield thermodynamically consistent set of values for solubility of metal sulfate in a wide temperature and concentration range.

Introduction

The water – manganese sulfate – sulfuric acid system has been studied in the literature due to its key importance in many hydrometallurgical processes, which typically operate at temperatures between 50 and 300 °C. Hydrometallurgical processes such as stainless steel pickling acid regeneration, manganese ore leaching as well as acid mine drainage from tailings ponds need internally consistent thermodynamic databases to improve, develop and deeper understand the systems and phenomena in the aqueous process solutions and environments.

Manganese has oxidation states of +2, +3, +4, +6 and +7 but in this work only the most stable +2 state was considered. In aqueous sulfuric acid solutions, manganese sulfate forms hydrates with 1, 2, 4, 5 and 7 molecules of crystalline water. Thermodynamics of the $MnSO_4$-H_2SO_4-H_2O system have been modelled earlier by Przepiera [1] using the Pitzer model. The Pitzer parameters used in Przepiera's model have also been reviewed critically in this work.

Przepiera [1] assessed the $MnSO_4$-H_2SO_4-H_2O system at 0-100°C and up to 30 mol/kg of sulfuric acid using the Harvie's modification [2, 3] of the Pitzer model for describing activity coefficients. Przerpiera included both enthalpy of solution and solubility data in his assessment, but the paper is lacking some thermodynamic data and that is why Przepiera results are not recalculated in this work. Przepiera tabulated solubility data at 20, 40, 70 and 90 °C and they are included in figures 1-4 for comparison up to 10 mol/kg of H_2SO_4. The concentration range of Przepiera seems to be too high for Pitzer model, with 30 mol/kg of sulfuric acid.

The new improved thermodynamic models of the binary systems $MnSO_4$-H_2O and H_2SO_4-H_2O have been published in separate papers [4, 5] by the authors. In the $MnSO_4$-H_2O paper $MnSO_4 \cdot 7H_2O(s)$, $MnSO_4 \cdot 5H_2O(s)$ and $MnSO_4 \cdot H_2O(s)$ were found to be stable phases with the peritectic transition temperatures at 9.2 and 23.9 °C, respectively. Adding sulfuric acid to the system will decrease this temperature due to lowering the activity of water so that there is a peritectic point with different compositions at each temperature from 0-90 °C. The $MnSO_4$-H_2O system was successfully assessed from -11 to 175 °C from pure water up to solubility limit of manganese sulfate 4.26 mol/kg. The H_2SO_4-H_2O system has been assessed by Sippola [5] with experimental EMF cell and osmotic coefficient data only, and it is valid up to 6.1 mol/kg and over a temperature interval of 0-55 °C. Sippola found out that four different K_2 equations for the dissociation of HSO_4^- are equally suitable for presenting the H_2SO_4-H_2O system. The K_2 equation of Matsushima and Okuwaki [6] was chosen in this work since it has been found out by the authors to be able to describe well the $FeSO_4$-H_2SO_4-H_2O and $NiSO_4$-H_2SO_4-H_2O systems well [7, 8].

The aim of this study is to compile and reassess the experimental observations of the system $MnSO_4$-H_2O-H_2SO_4 at 0-90 °C and H_2SO_4 concentration range up to 10 mol/kg in order to validate our previous $MnSO_4$-H_2O and H_2SO_4-H_2O binary models [4, 5] with this ternary system. All experimental data used in the modelling were taken from the literature. The resulting thermodynamic model was obtained using the thermodynamic equilibrium software package MTDATA® (www.mtdata-software.com), which uses a global Gibbs energy minimisation routine and includes the Pitzer activity coefficient model for the excess Gibbs energy of the aqueous solutions. The CALPHAD (CALculation of PHAse Diagrams) method [9] was used in the modelling, to ensure internal consistency of the thermodynamic database.

Theory

The Pitzer model is one of the most used activity coefficient models for the aqueous solutions. The original approach assumes that the aqueous solution consists only of ions, and no ion complexes are formed. Details of the Pitzer model used are available in [10-12]. Later, Harvie and Weare [3] and Harvie et al. [2] included unsymmetrical electrostatic mixing terms in their modification of the Pitzer model, which has been shown to improve the fit in multicomponent systems. The values for the internal constant parameters of the Harvie's modification of the Pitzer equation used in this work are shown in Table I. All the necessary Pitzer model equations, variables and parameters have been explained in our previous paper [5, 13].

Table I. Internal parameters (b = 1.2) of the Pitzer model used in this work

parameter	1-1, 1-2, 1-3 and 1-4 electrolyte	2-2 electrolyte
α_1(kg/mol)$^{1/2}$	2.0	1.4
α_2(kg/mol)$^{1/2}$	---	12

The consistent concentration unit in aqueous solutions is molality of $MnSO_4$ and H_2SO_4 (mol/kg of water), which is used throughout this paper. The temperature dependency equation in MTDATA® for heat capacity of a species has the following form:

$$C_p = A + B\left(\frac{T}{K}\right) + C\left(\frac{T}{K}\right)^2 + D\left(\frac{T}{K}\right)^{-2} \tag{1}$$

and thus Gibbs energy has a temperature dependent form

$$G(T) = A_G + B_G\left(\frac{T}{K}\right) + C_G\left(\frac{T}{K}\right)\ln\left(\frac{T}{K}\right) + D_G\left(\frac{T}{K}\right)^2 + E_G\left(\frac{T}{K}\right)^3 + F_G\left(\frac{T}{K}\right)^{-1}. \tag{2}$$

The general temperature dependency available in MTDATA® for the Pitzer equation parameter (p) is

$$p = A_{Pitz} + B_{Pitz}\left(\frac{T}{K}\right) + C_{Pitz}\left(\frac{T}{K}\right)\ln\left(\frac{T}{K}\right) + D_{Pitz}\left(\frac{T}{K}\right)^2 + E_{Pitz}\left(\frac{T}{K}\right)^3 + F_{Pitz}\left(\frac{T}{K}\right)^{-1} \tag{3}$$

Parameter optimizations

Evaluation of the thermodynamic properties of the aqueous phase as well as the condensed manganese sulfate hydrates was carried out using the MTDATA® assessment module, version 4.81 and MTDATA Studio, version 5.03. The assessment module minimises the weighted sum of squares of errors between the measured and fitted values, according to equation (4). Thus, the objective function (OF) to be minimised in the optimisation can be written as

$$OF = \sum_{i=1}^{n} W_i\left(\frac{C_i - E_i}{U_i}\right)^2 \tag{4}$$

where n is the number of properties (data items) to be reproduced, C_i and E_i are the calculated and experimental values of property i, U_i is the uncertainty associated with value E_i and W_i is the weight assigned to property (data item) i.

Experimental phase equilibrium data used in the optimisation

The solubility measurements have been made at temperatures ranging from 0 to 90 °C [14, 15]. The solubilities have been reviewed by Linke and Seidell [16]. The experimental solubility data used in the optimisation at temperature range of 40-80 °C are shown in Table II. The H_2SO_4 concentration upper limit was 10 mol/kg. Details of the experimental data used have been added as supplemental material including uncertainties of each experiment.

All weights for experimental data were set to 1, with the exception of rejected values, where W = 0 was used.

Table II. The experimental data used in the assessment of H_2O-$MnSO_4$-H_2SO_4 ternary system. H_2SO_4 cut-off limit was 10 mol/kg

		Data points		
Experiment	Temperature / °C	accepted	total	
Solubility of $MnSO_4 \cdot H_2O(s)$	40-80	12	12	[14]

Thermodynamic data used in the optimisation

The simplified HKF model [17, 18] was used for the ions (except $HSO_4^-(aq)$), see Appendix 1. Thermodynamic data for $HSO_4^-(aq)$ were calculated from $SO_4^{2-}(aq)$ data and the sulfuric acid second dissociation ($HSO_4^-(aq) = SO_4^{2-}(aq) + H^+(aq)$) constant K_2 value of Matsushima and Okuwaki [6], from equation (5).

$$\log_{10} K_2(T,K) = 577.214 - 246.01\log_{10} T - \frac{12717}{T} + 0.283133T - 1.37566 \cdot 10^{-4} T^2 \quad (5)$$

C_p function of the H_2O was fitted to experimental data from the literature; see details in Kobylin et al. [13].

The thermodynamic values $\Delta_f H°_{298.15}$, $S°_{298.15}$ and C_p for ions and C_p for $MnSO_4 \cdot 7H_2O(s)$, $MnSO_4 \cdot 5H_2O(s)$ and $MnSO_4 \cdot H_2O(s)$ were taken from Hämäläinen et al. [19] and DeKock [20], respectivelyΔ $_fH°_{298.15}$, $S°_{298.15}$ of $MnSO_4 \cdot 7H_2O(s)$, $MnSO_4 \cdot 5H_2O(s)$ and $MnSO_4 \cdot H_2O(s)$ at 25 °C were optimised with H_2O-$MnSO_4$ binary system published earlier by the authors [4]. The gas phase was assumed to be ideal.

Results and discussion

The temperature dependencies of the Pitzer parameters $\beta^{(0)}$, $\beta^{(1)}$ and C^{ϕ} for Mn^{2+}-HSO_4^- binary interaction were optimised in this work with temperature dependency of A_{Pitz} + F_{Pitz}/T. The optimised Pitzer parameters are shown in Table III. The interaction parameters used for H_2SO_4-H_2O binary from [5] are also shown in Table III and $MnSO_4$-H_2O binary parameters are published in [4].

Table III. Assessed Pitzer parameters used in this work. $MnSO_4$-H_2O binary parameters from [4] were used.

	A_{Pitz}	F_{Pitz}		
$\beta^{(0)}$	0.00000	100.6319	Mn^{2+}(aq)-HSO_4^-(aq)	this work
$\beta^{(1)}$	5.26731	0.0000	Mn^{2+}(aq)-HSO_4^-(aq)	this work
C^{ϕ}	-0.01528	0.0000	Mn^{2+}(aq)-HSO_4^-(aq)	this work
$\beta^{(0)}$	-0.04083	20.4876	H^+(aq)-SO_4^{2-}(aq)	[5][a]
C^{ϕ}	0.18522	-42.794	H^+(aq)-SO_4^{2-}(aq)	[5][a]
$\beta^{(0)}$	0.02808	54.141	H^+(aq)-HSO_4^-(aq)	[5][a]
$\beta^{(1)}$	-0.00516	147.759	H^+(aq)-HSO_4^-(aq)	[5][a]

[a] "Okuwaki set A" from reference [5] was used in this work.

The solubility of $MnSO_4$ in the aqueous sulfuric acid solution was calculated at temperatures from 0 to 90 °C, using the optimised properties of this work from 40 to 80 °C. Calculated solubilities at 20 and 90 °C are thus extrapolated. Figures 1-4 show the solubility results together with the experimental points at 20, 40, 70 and 90 °C, respectively. Data from the earlier thermodynamic modelling study by Przepiera [1] have been superimposed in the figures. At 20 °C, the experimental data are quite consistent; see Figure 1. At 40 and 70 °C, both models are working reasonably well. At 90 °C this work is not able to extrapolate the experimental solubilities well. The $MnSO_4$ solubility to pure water is somewhat different at 90 °C than the experimental observations. Mean activity coefficients of both H_2SO_4 and $MnSO_4$ were calculated at 20 and 90 °C to show the non-ideal behaviour of the solution; see figure 5.

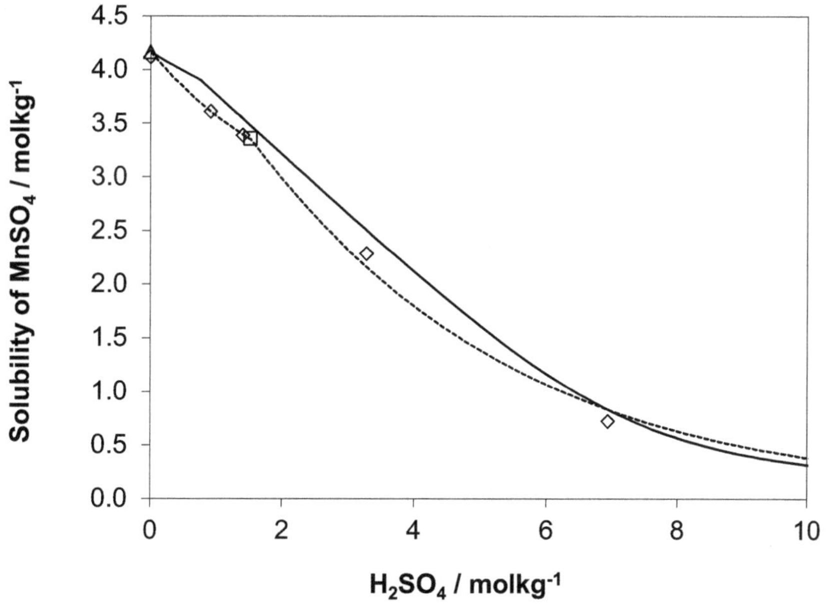

Figure 1. The assessed and experimental solubility data of the system at 20 °C. —this work (extrapolated); ---Przepiera [1]; △Krepelka [21]; ◇Petlicka [14]; □ Taylor [15].

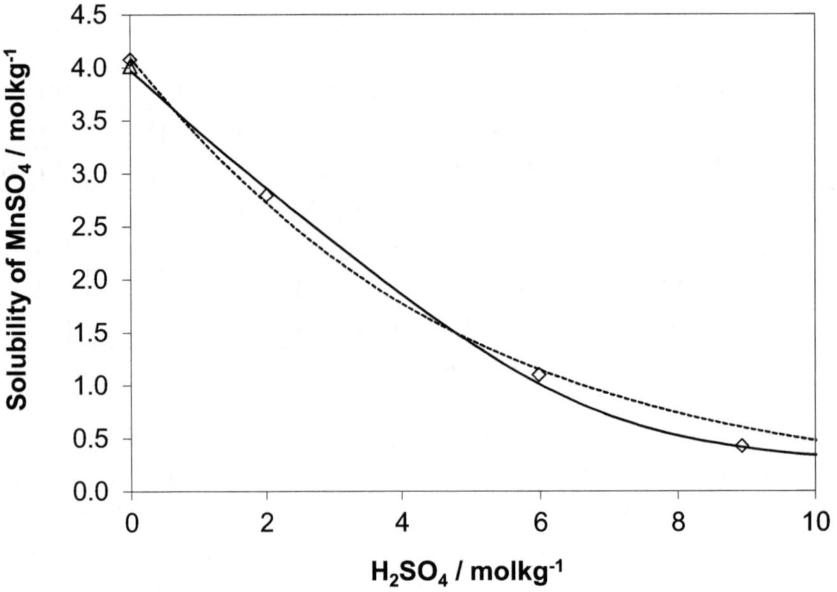

Figure 2. The assessed and experimental solubility data of the system at 40 °C. —this work; ---Przepiera [1]; △Rohmer [22]; ◇Petlicka [14].

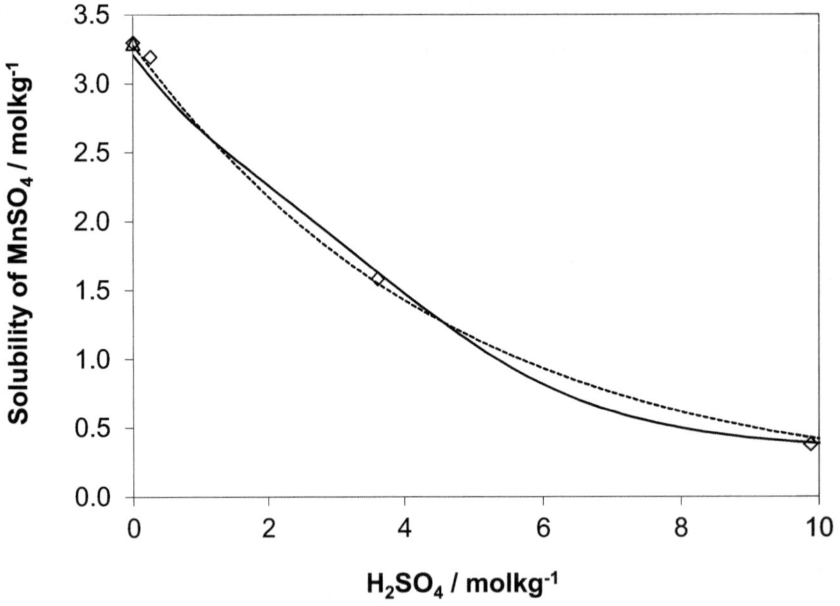

Figure 3. The assessed and experimental solubility data of the system at 70 °C. —this work; ---Przepiera [1]; △Krepelka [21]; ◇Petlicka [14].

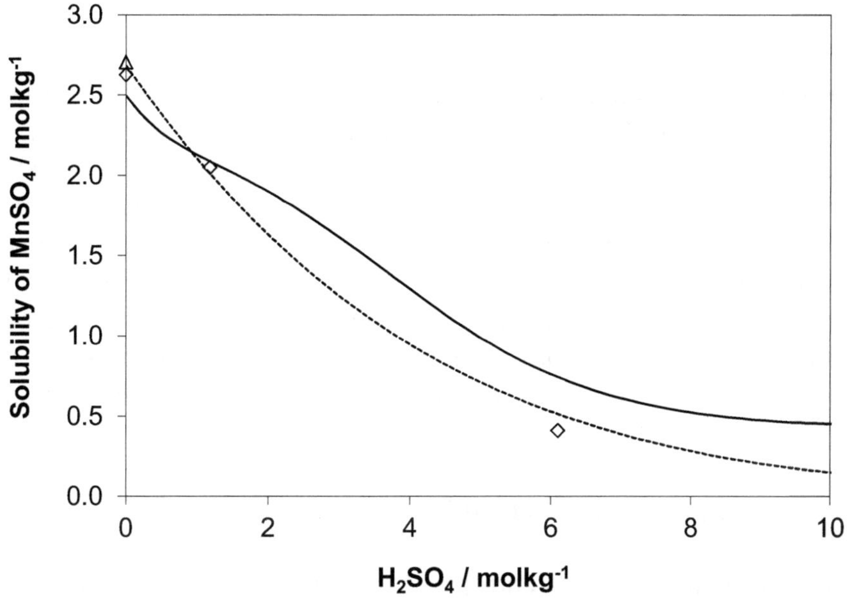

Figure 4. The assessed and experimental solubility data of the system at 90 °C. —this work (extrapolated); ---Przepiera [1]; △Krepelka [21]; ◇Petlicka [14].

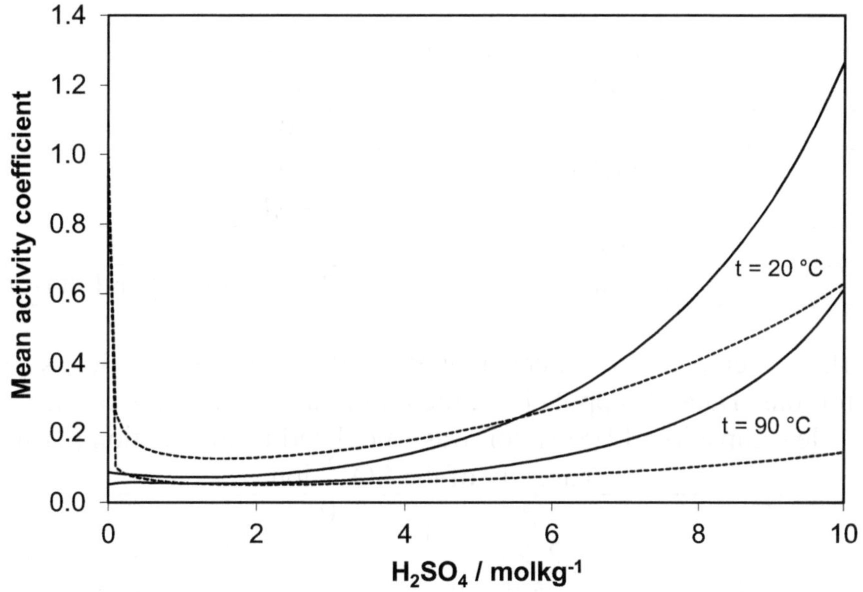

Figure 5. Calculated mean activity coefficients at 20 and 90 °C as a function of H_2SO_4 additions. Molality of $MnSO_4$ is 0.3 molkg^{-1}. —$\gamma_{\pm}(MnSO_4)$; ---$\gamma_{\pm}(H_2SO_4)$.

Conclusions

In this work, the earlier model was carefully compared using the available solubility data in the literature. The current model presents the experimental data available with a good accuracy and consistently up to 90 °C, and to sulfuric acid concentrations up to 10 mol/kg.

Due to the lack of experimental data, the heat capacity of crystalline $MnSO_4 \cdot 7H_2O(s)$, $MnSO_4 \cdot 5H_2O(s)$ and $MnSO_4 \cdot H_2O(s)$ should be measured on a wide temperature interval, preferably from 0-500 K. More solubility measurements of $MnSO_4$ in sulfuric acid solutions at higher temperatures, above 90 °C, are also needed to ensure the correct solubilities. There is also a need to make water activity and vapour pressure measurements at moderate to high temperatures to improve the current model in the areas of industrial processes. Since thermodynamic properties of crystalline phases evaluated in [13] are related to the chosen values of manganese ion it would be more convenient to use $\Delta_f H°_{298.15}$, $S°_{298.15}$ and C_p values of Mn^{2+} ion that are well evaluated and generally accepted.

Appendix 1. Thermodynamic properties of ions

Enthalpy of formation and standard entropy of the ions were taken from the literature; see Table A.I.

The heat capacities of the ions were estimated using a simplified HKF model, explained in detail in our previous paper [13]; see Table A.II.

Table A.I. The thermodynamic properties of ions at 25 °C

Ion	$\Delta_f H°_{298.15}$ / $J \cdot mol^{-1}$	$S°_{298.15}$ / $J \cdot mol^{-1} \cdot K^{-1}$	
$H^+(aq)$	0.0	0.00	by definition
$OH^-(aq)$	-230015.0	-10.90	[23]
$Mn^{2+}(aq)$	-220597.2	-73.64	[24]
$SO_4^{2-}(aq)$	-909340.0	18.50	[23]
$HSO_4^-(aq)$	-885200.0	137.50	Eq (5) and $SO_4^{2-}(aq)$ data

Table A.II. The temperature dependency of $C_p(J/(mol \cdot K)) = A + B(T/K) + C(T/K)^2 + D(T/K)^{-2}$ for ions. Data of Sippola [25]; the minimum temperature of the C_p function is 283.15 K. Heat capacity of HSO_4^- ion was calculated from equation (5) and the heat capacity of the SO_4^{2-} ion.

Ion	T_{max}, K	A	B	$C \cdot 10^{-3}$	$D \cdot 10^5$
$OH^-(aq)$	343.15	21245.30	-84.1929	93.5145	-4083.14
	448.15	-5250.28	20.4720	-23.0421	985.66
$Mn^{2+}(aq)$	323.15	13234.90	-53.9590	61.8370	-2365.61
	398.15	-414.06	2.8850	-4.8000	-27.90
	448.15	2737.52	-4.9020	0.0000	-1314.44
$SO_4^{2-}(aq)$	328.15	46200.60	-186.8004	211.9290	-8546.29
	403.15	1080.77	-0.7188	-3.9917	-676.58
	448.15	5857.78	-10.7722	0.0000	-2907.89
$HSO_4^-(aq)$	328.15	48246.05	-197.6415	227.7310	-8546.29
	403.15	3126.22	-11.5599	11.8103	-676.58
	448.15	7903.23	-21.6133	15.8020	-2907.89

References

[1] A. Przepiera, "Solubility equilibria of salts in electrolyte solutions MeSO4-H2SO4-H2O Systems," (Report (in Polish) 551/35, Prace Naukowe Politechniki Szczecinskiej, Institut Inzynierii Chemicznej I Chemii Fizycznej, 1999).

[2] C.E. Harvie, N. Møller, J.H. Weare, "The Prediction of Mineral Solubilities in Natural Waters: The Na-K-mg-Ca-H-Cl-SO4-OH-HCO3-CO3-CO2-H2O System to High Ionic Strengths at 25°C," *Geochimica et Cosmochimica Acta,* 48 (1984), 723-751.

[3] C.E. Harvie, J.H. Weare, "The Prediction of Mineral Solubilities in Natural Waters: The Na-K-mg-Ca-Cl-SO4-H2O System from Zero to High Concentration at 25° C," *Geochimica et Cosmochimica Acta,* 44 (1980), 981-997.

[4] P.M. Kobylin, P.A. Taskinen, "Thermodynamic Modelling of Aqueous Mn(II) Sulfate Solutions," *CALPHAD,* 38 (2012), 146-154.

[5] H. Sippola, "Critical Evaluation of the 2nd Dissociation Constants for Aqueous Sulfuric Acid," *Thermochimica Acta,* 532 (2012), 65-77.

[6] Y. Matsushima, A. Okuwaki, "The Second Dissociation Constant of Sulfuric Acid at Elevated Temperatures from Potentiometric Measurements," *Bulletin of the Chemical Society of Japan,* 61 (1988), 3344-3346.

[7] P.M. Kobylin, H. Sippola, P.A. Taskinen, "Thermodynamic Model for Acidic Fe(II) Sulphate from Solubility Data," *CALPHAD,* 38 (2012), 185-193.

[8] P.M. Kobylin, H. Sippola, P.A. Taskinen, "Thermodynamic model for acidic Ni(II) sulfate from solubility data," *CALPHAD,* (accepted 2012).

[9] P.J. Spencer, "A Brief History of CALPHAD," *CALPHAD,* 32 (2008), 1-8.

[10] K.S. Pitzer, "Thermodynamics of Electrolytes. I. Theoretical Basis and General Equations," *The Journal of Physical Chemistry,* 77 (1973), 268-277.

[11] K.S. Pitzer, G. Mayorga, "Thermodynamics of Electrolytes. II. Activity and Osmotic Coefficients for Strong Electrolytes with One Or both Ions Univalent," *The Journal of Physical Chemistry,* 77 (1973), 2300-2308.

[12] K.S. Pitzer, G. Mayorga, "Thermodynamics of Electrolytes. III. Activity and Osmotic Coefficients for 2–2 Electrolytes," *Journal of Solution Chemistry,* 3 (1974), 539-546.

[13] P.M. Kobylin, H. Sippola, P.A. Taskinen, "Thermodynamic Modelling of Aqueous Fe(II) Sulfate Solutions," *CALPHAD,* 35 (2011), 499-511.

[14] J. Petlicka, "Solubility of the Manganese Sulfate-Sulfuric Acid-Water System," *Hutnicke Listy,* 25 (1970), 727-731.

[15] D. Taylor, "The System Manganese Sulfate-Sulfuric Acid-Water," *Journal of the Chemical Society,*(1952), 2370-2375.

[16] William F. Linke, Atherton Seidell, *Solubilities, inorganic and metal organic compounds; a compilation of solubility data from the periodical literature* (4th edition, vol. 2, Princeton, N.J.: Van Nostrand, 1965), 1-1914.

[17] H.C. Helgeson, D.H. Kirkham, "Theoretical Prediction of the Thermodynamic Behavior of Aqueous Electrolytes at High Pressures and Temperatures I," *American Journal of Science,* 274 (1974), 1089-1198.

[18] H.C. Helgeson, D.H. Kirkham, G.C. Flowers, "Theoretical Prediction of the Thermodynamic Behavior of Aqueous Electrolytes by High Pressures and Temperatures IV," *American Journal of Science,* 281 (1981), 1249-1516.

[19] M. Hämäläinen, H. Rannikko, H. Sippola, "Vesiliuossysteemien termodynaaminen mallitus," (Report (in Finnish) TKK-V-C108, Department of Materials Science and Engineering, 1991).

[20] C.W. DeKock, "Thermodynamic properties of selected transition metal sulfates and their hydrates," (Report IC 8910, U.S. Department of the Interior, Bureau of Mines, 1982).

[21] J.H. Krepelka, B. Rejha, "Solubility of the Hydrates of Manganous Sulfate," *Collection of Czechoslovak Chemical Communications,* 5 (1933), 67-75.

[22] R. Rohmer, "Dehydration of Manganous Sulfate Heptahydrate by the Aqueous Method," *Comptes Rendus Hebdomadaires des Seances de l'Academie des Sciences,* 209 (1939), 315-317.

[23] J. D. Cox, Donald D. Wagman, V. A. Medvedev, *CODATA key values for thermodynamics* (New York, Hemisphere Publishing Corporation, 1989), 1-271.

[24] E.L. Shock, H.C. Helgeson, "Calculation of the Thermodynamic and Transport Properties of Aqueous Species at High Pressures and Temperatures: Correlation Algorithms for Ionic Species and Equation of State Predictions to 5 Kb and 1000°C," *Geochimica et Cosmochimica Acta,* 52 (1988), 2009-2036.

[25] H. Sippola, "Solubility of ferrous sulphate in sulphuric acid—a thermodynamic model," (Licentiate's Thesis (in Finnish), Helsinki University of Technology, 1992), 1-96.

REWAS 2013: Enabling Materials Resource Sustainability
Edited by: Anne Kvithyld, Christina Meskers, Randolph Kirchain, Gregory Krumdick, Brajendra Mishra, Markus Reuter, Cong Wang, Mark Schlesinger, Gabrielle Gaustad, Diana Lados, and Jeffrey Spangenberger
TMS (The Minerals, Metals & Materials Society), 2013

Practical thermodynamic model for acidic sulfate solutions

Hannu Sippola[1,2], Petri Kobylin[1]and Pekka A. Taskinen[1]

[1]Department of Materials Science and Engineering, Aalto University, FI-00076 Aalto, Finland
[2]Permanent address FCG Design and Engineering, Osmontie 34, FI-00601 Helsinki Finland

Keywords: "Sulfuric acid, iron sulfate, thermodynamics"

Abstract

Practicable thermodynamic description of sulfuric acid - water system is essential when modeling of acidic sulfate solutions required in process optimization and waste management. Traditional Pitzer model is limited up to 6 mol/kg (35w%) sulfuric acid solutions. Local composition models such as electrolyte NRTL can deal H_2SO_4-H_2O systems up to high concentrations but their use is practically limited to vapor-liquid equilibrium. Mole fraction based version of Pitzer equation is capable to describe H_2SO_4-H_2O system up to 40 mol/kg (80w%) acid concentration at temperatures range from 328 K (55°C) down to 200 K but the number parameters is quite extensive for practical purposes. Recently developed NPL Pitzer model has proved to be capable to describe sulfuric acid -water system in temperature range 0-55 °C using only five parameters with simple temperature dependency of a+b/T. The capabilities of NPL Pitzer model is demonstrated here in wide temperature range up to 40 mol/kg (80w%) solutions with H_2SO_4-H_2O and H_2SO_4-$FeSO_4$-H_2O systems.

Introduction

Sulfuric acid is an important chemical and widely used in industry such as phosphate fertilizer and pigment production, steel pickling and many hydrometallurgical applications, production of organic chemical, explosives, etc. [1]. Quite often spent acidic sulfate solutions, such as acid main drainage, are generated as undesired by-product.

Thermodynamic modeling of sulfuric acid is complicated due to the incomplete dissociation of sulfuric acid

$$HSO_4^- = SO_4^{2-} + H^+ \qquad K_2 = a(SO_4^{2-})\, a(H^+) / a(HSO_4^-) \qquad (1)$$

Both Pitzer equation and its modifications as well as several NTRL models has been used to describe the behavior of sulfuric acid-water system. The number of used parameters and terms within has increases during the years generating more precise descriptions and extended temperatures for the sulfuric acid-water system (table I). However, the complexity of the models has made them unusable in practical purposes especially for higher order systems.

Table I. Pitzer interaction models for aqueous sulfuric acid.

Author(s)	Year	Modifications[a]	Temperature range	Number of parameters	Total number of terms
Pitzer et al. [2]	1977		5-55°C	4	8
Reardon and Beckie [3]	1987	U	5-55°C	4	9
Sippola [4]	1992	U	5-55°C	4	8
Holmes and Mesmer [5]	1992		25-200°C	5	17
Clegg et al. [6]	1994	UA	0 -55°C	9	32
Clegg and Brimblecombe [7]	1995	M	-70 -55°C	10	40
Knopf et al. [8]	2003	A	-90 -200°C	10	34
Christov and Moller [9]	2004	U	0-200°C	5	20
Friese and Ebel [10]	2010	M	-70 -55°C	11	66

[a] U=unsymmetrical mixing terms [11]; A=Archer extension [12]; M=mole fraction scale

Most of the local composition models for H_2SO_4-H_2O system (Table II) are focused on vapor-liquid equilibrium and the speciation of sulfuric acid. Electrochemical cell measurements are generally omitted in model parameter estimation. Thus, the vapor-liquid equilibrium is presented quite well but they are not very successful in describing the behavior of aqueous sulphuric acid at concentrations below 1 mol/kg, i.e., where the dissociation of bisulfate ion practically occurs. Some models even neglect this equilibrium completely.

Table II. Local composition models for aqueous sulfuric acid.

Author(s)	Year	Model	Temperature range	Concentration range (mol/kg)	Number of parameters (terms)
Liu and Gren [a] [13]	1991	Liu-Harvey-Prausnizt	25°C	1 - 76	2
Rosen and Engels [a] [14]	1998	NRTL	0-240°C	0 – 96 w%	10 (17)
Messnaoui and Bounahmidi [15]	2006	eNRTL	25-75°C	1 – 7	4(6)
Bollas et al. [16]	2010	eNRTL (refined)	25°C	0-50 (65)	10
Simonin et al. [17]	2006	MSA-NRTL	25°C	0.1-6.0	6
Simonin et al.[b] [18]	2004	MSA (BIMSA)	25°C	0.1-27	6
Campos et al. [19]	2006	UNIQUAC	0-150°C	0-100 w%	7(15)

a) Complete dissociation of sulfuric acid is assumed.
b) Bisulfate ion is considered via equilibrium only. No ion specific parameters used.

In 2010 Sippola [20], later published in ref [21] modeled the aqueous sulfuric acid with original Pitzer model and found out that only 4 Pitzer parameters with simple temperature dependency of (a+b/T) is sufficient to present the stoichiometric osmotic and activity coefficients in temperature range 0-55°C up to 6 molal sulfuric solution equally well as the more complicated models by Clegg et al. [6] and Clegg and Brimblecombe [7]. Six different K_2-equations were tested and the differences in predicted values between K_2 equations were found well below experimental accuracy.

Recently, a modified Pitzer equation for concentrated aqueous solutions was developed in National Physical Laboratory (UK) [22]. The NPL modification of Pitzer equation (NPL Pitzer) was found to be able to model sulfuric acid solutions well and to the high concentrations. The temperature and concentration range used in the assessment were 0–55°C and 0–15 mol

H_2SO_4/kg H_2O, respectively. One of 14 studied parameter sets (set K) was found to have excellent extrapolation capabilities to higher temperatures and concentrations. [23].

Theory

Original Pitzer equation

Pitzer [24] introduced an interaction model for excess Gibbs energy of aqueous solutions which is based on virial coefficients similar to gas phase. After recombining virial coefficients and changing moles to molalities (m) the excess Gibbs energy divided by gas constant R, absolute temperature (T) and mass of solvent (w_W) is:

$$\frac{G^{E,m}}{w_w RT} = f(I) + 2\sum_c \sum_a m_c m_a \left[B_{ca} + \left(\sum_c m_c z_c \right) C_{ca} \right] + \sum_{c<c'} \sum m_c m_{c'} \left[2\Phi_{cc'} + \sum_a m_a \psi_{cc'a} \right] + \sum_{a<a'} \sum m_a m_{a'} \left[2\Phi_{aa'} + \sum_c m_c \psi_{aa'c} \right] \tag{2}$$

where c refers to a cation, a to an anion and n to a neutral species. B_{ca}, C_{ca} are interaction parameters for the cation c and the anion a, $\Phi_{cc'}$ and $\Phi_{aa'}$ for two different cations and anions, respectively. z_i is the charge of the ion. Parameters $\psi_{cc'a}$ $\psi_{aa'c}$ are for interactions between three ions. Function f(I) describes the long range forces between ions and is dependent on temperature and ionic strength defined as:

$$I = 1/2 \sum_i m_i z_i^2 \tag{3}$$

Assuming ψ and C to be independent on concentration, the equations for activity and osmotic coefficients are obtained with appropriate differentiation from equation (2) and summarized in equations (4) – (7)

$$\ln \gamma_M = z_M^2 f^{\gamma}(I) + \sum_a m_a \left[2B_{Ma} + ZC_{Ma} \right] + z_M \sum_c \sum_a m_c m_a C_{ca} + z_M^2 \sum_c \sum_a m_c m_a B'_{ca} + \sum_c m_c \left[2\Phi_{Mc} + \sum_a m_a \psi_{Mca} \right] + z_M^2 \sum_{c<c'} \sum m_c m_{c'} \Phi'_{cc'} + z_M^2 \sum_{a<a'} \sum m_a m_{a'} \Phi'_{aa'} + \sum_{a<a'} \sum m_a m_{a'} \left[m_M \psi_{Maa'} \right] \tag{4}$$

$$\ln \gamma_X = z_X^2 f^{\gamma}(I) + \sum_c m_c \left[2B_{cX} + ZC_{cX} \right] + |z_X| \sum_c \sum_a m_c m_a C_{ca} + z_X^2 \sum_c \sum_a m_c m_a B'_{ca} + \sum_a m_a \left[2\Phi_{Xa} + \sum_c m_c \psi_{Xac} \right] + z_X^2 \sum_{c<c'} \sum m_c m_{c'} \Phi'_{cc'} + z_X^2 \sum_{a<a'} \sum m_a m_{a'} \Phi'_{aa'} + \sum_{c<c'} \sum m_c m_{c'} \left[m_X \psi_{cc'X} \right] \tag{5}$$

$$(\phi - 1) = \left(\frac{2}{\sum_i m_i} \right) \left\{ \begin{aligned} & I f^{\phi}(I) + \sum_c \sum_a m_c m_a \left[B^{\phi}_{ca} + ZC_{ca} \right] + \sum_{c<c'} \sum m_c m_{c'} \left[\Phi^{\phi}_{cc'} + \sum_a m_a \psi_{cc'a} \right] \\ & + \sum_{a<a'} \sum m_a m_{a'} \left[\Phi^{\phi}_{aa'} + \sum_c m_c \psi_{aa'c} \right] \end{aligned} \right\} \tag{6}$$

Osmotic coefficient (ϕ) is generally used in aqueous systems instead of activity of water and is defined as:

$$\phi = -\left(\frac{1000}{M_w \sum m_i}\right) \ln a_w \tag{7}$$

where M_w is the molecular weight of water. More detailed description of the Pitzer model can be found elsewhere [24-27].

<u>NPL Pitzer equation</u>

NPL Pitzer equation is obtained by replacing the reduced molality and ion strength defined as

$$m_r = \frac{n_i}{\left(n_w + \sum_j n_j\right) M_w} \quad \text{and} \quad I_r = \frac{1}{2}\sum_i \frac{n_i z_i^2}{\left(n_w + \sum_j n_j\right) M_w} \tag{8}$$

with their counterparts in equations 4-6 [22].

The reduced molalities will converge to molalities in dilute solutions. Thus, the model coincides with the original Pitzer model at infinite dilution.

Experimental data and methods

<u>Sulfuric acid – water -system</u>

The experimental data used for the evaluation were chosen to be the most accurate considered ones and are listed in table III. The detailed discussion of the used experimental data can be found in refs [21, 23, 28]

Table III The experimental data used for the sulfuric acid sulfate – water –system

Experiment	Measurements	Temperature range (°C)	m (H_2SO_4) (mol/kg)	Excluded molalities
Cell A [9]	Pt,H_2 | H_2SO_4 |$PbSO_4$,Pb,Hg	0-50	0.005 - 0.02	All data <0.005
Cell B [10]	Pt,H_2 | H_2SO_4 |Hg_2SO_4,Hg	5-55	0.1003 - 8.0	5.767 (5°C)
Cell C [11]	Pt,H_2 | H_2SO_4 |PbO_2,$PbSO_4$,Pt	5-55	0.1 - 7.2	
Isopiestic data [12]	Critical evaluation of several isopiestic measurements	25	0.2 – 15.0	0.1 (25°C) All data > 15
Isopiestic data [13]	Isopiestic measurements	110-200	0.5 - 5.6	

<u>Iron sulfate – water -system</u>

The experimental data used for the parameter assessment are the same used previously to fit parameters for the original Pitzer model [29]. The detailed discussion of the used experimental data can be found in the same reference.

Table IV The experimental data used for the iron sulfate – water –system.

Experiment	Temperature range (°C)	$m(FeSO_4)$ (mol/kg)
Freezing point depression [30]	-1.819 to -0.172	0.067-0.978
Eutectic point [30]	-1.821	0.982
Solubility of melanterite [31]	0-55	1.036-3.483
Solubility of szomolnokite [32]	65-100	1.666-3.098
Solubility of szomolnokite[b] [33][34]	120-220	0.022-1.481
Peritectic point[a] [32]	56.6	3.59
$a(H_2O(l))$ aqueous solution [35]	25	0.1-1.9641
$a(H_2O(l))$ saturated solution [36],[35],[37]	18.6-31.9	Saturation
Enthalpy of solution [38]	19-21	0.036-1.696
Heat Capacity of solution [38][39]	9-91.6	0.176-3.545
Dissociation pressure [40][41]		

a) This peritectic point between melanterite and szomolnokite is not directly measured.
b) Bruhn et al. [33]120 and 150 °C and Hasegawa et al. [34] 130 and 150 °C values not included.

Iron sulfate – sulfuric acid – water - system

The experimental data used for the evaluation were chosen to be the most accurate considered solubility data [32, 42-44]. The Okuwaki and Matsushima's equation [45] for K_2 was chosen to retain consistency with the earlier assessment with original Pitzer model [29].

Computer software

MTDATA version 4.81 was used for parameter fitting [46].

Results

Fitted parameters

The fitted binary parameters and their temperature dependencies are shown in table V.

Binary interactions	H^+	Fe^{2+}	HSO_4^-
SO_4^{2-}	C^ϕ : a + b/T	$\beta^{(0)}$: a + bT + cT^2 + d/T $\beta^{(1)}$: a + bT + cT^2 + d/T $\beta^{(2)}$: a C^ϕ : a + b/T	θ : a+b/T
HSO_4^-	$\beta^{(0)}$: a + b/T $\beta^{(1)}$: a + b/T C^ϕ : a + b/T	$\beta^{(0)}$: a + b/T $\beta^{(1)}$: a + b/T C^ϕ : a + b/T	
Ternary interactions	Fe^{2+} - SO_4^{2-} - HSO_4^-	H^+ - Fe^{2+} - SO_4^{2-}	
	ψ : a + b/T	ψ : a	

Thus, the total number of parameters for ferrous sulfate – sulfuric acid – water system is 14 including 29 fitted terms.

Sulfuric acid – water – system

The predicted osmotic coefficients at temperatures 25, ≈110, ≈140, ≈170 and ≈200°C are shown in figures 1 and 2. The comparison with more complicated model by Clegg and Brimblecombe [7] is shown in figure 3. Please note that the maximum concentration used in the parameter estimation was 15 mol/kg at 25°C and ≤ 8 mol/kg elsewhere.

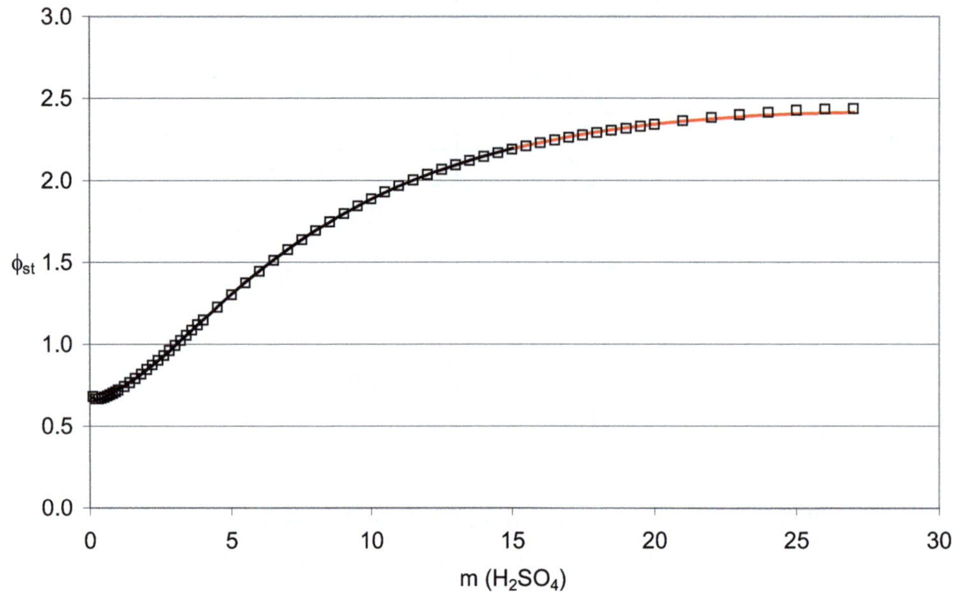

Figure 1. Predicted stoichiometric osmotic coefficients of aqueous sulfuric acid at 25°C versus experimental data [12]. The values above 15 mol/kg are extrapolated.

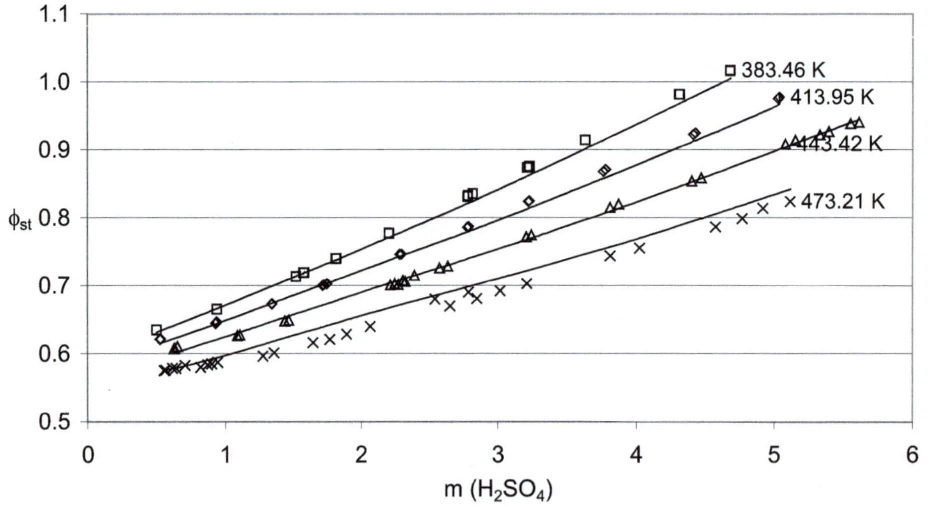

Figure 2. The experimental [13] and predicted stoichiometric osmotic coefficient at around 110, 140, 170 and 200 °C.

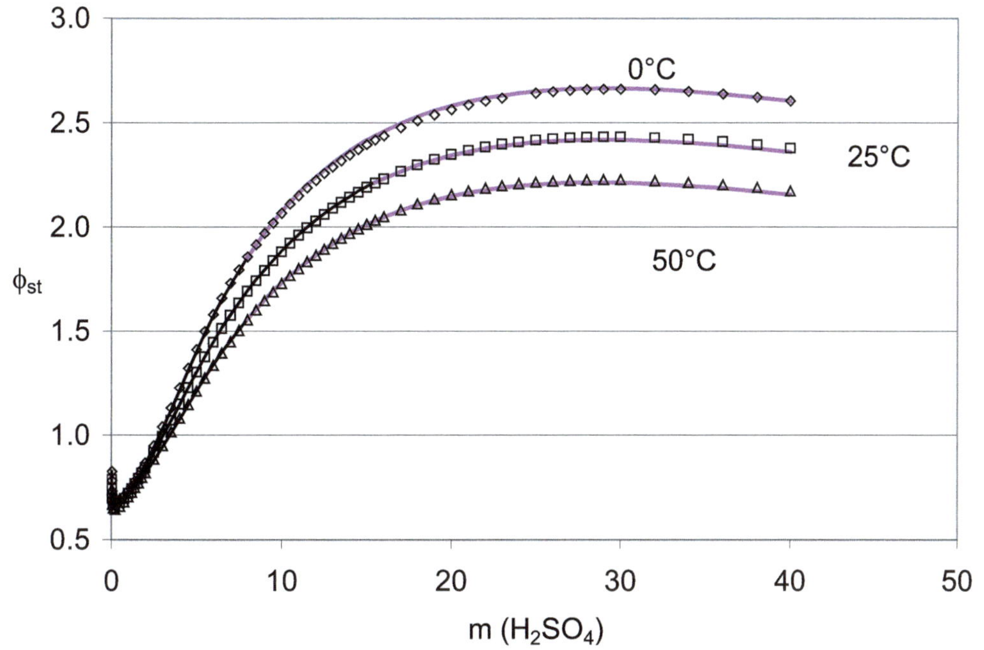

Figure 3. Stoichiometric osmotic coefficient data at 0, 25 and 50 °C compared to values from Clegg and Brimblecombe's [14] model (dots). These values were not included in the assessment.

<u>Iron sulfate – water – system</u>

The predicted and measured solubility of iron sulfate are shown in figure 4.

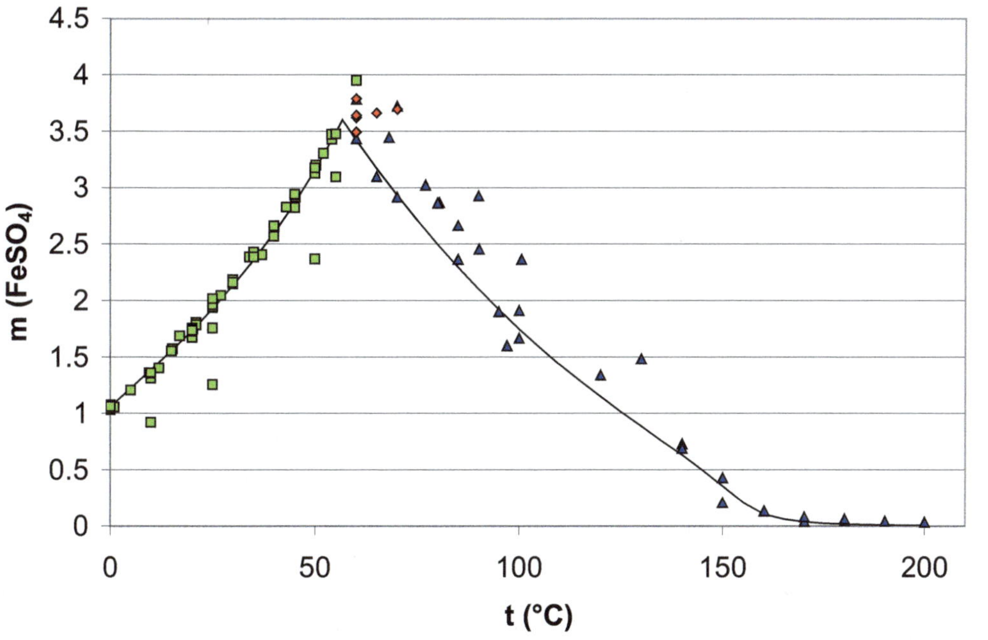

Figure 4. A comparison of the predicted solubility of iron sulfate (solid line) with experimental data (dots, see ref [29]) in temperature range 0-200°C. Squares present heptahydrate, diamonds tetrahydrate (metastable) and triangles monohydrate.

Iron sulfate – sulfuric acid – water – system

The predicted solubility of ferrous sulfate at temperatures 25, 50 and 90°C are shown in figures 5-7.

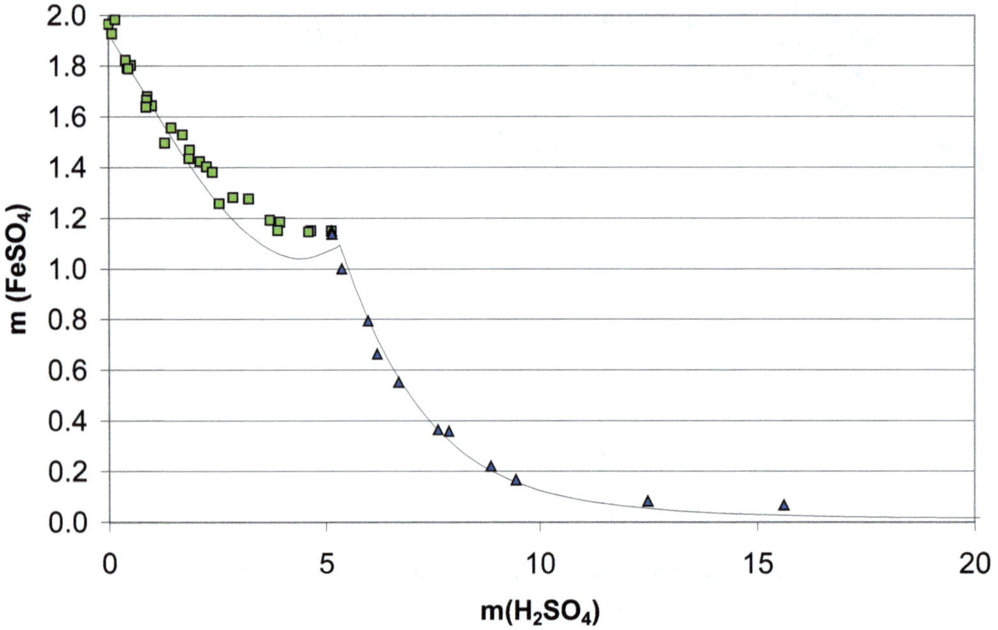

Figure 5. A comparison of the predicted solubility of iron sulfate in aqueous sulfuric acid (solid line) with experimental data [32, 43, 44, 47] at 25°C. Squares present heptahydrate and triangles monohydrate.

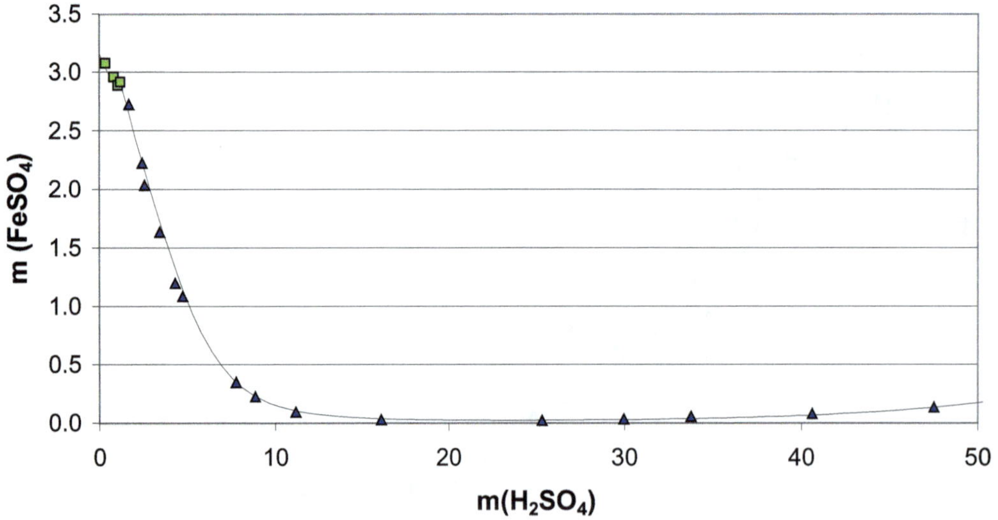

Figure 6. A comparison of the predicted solubility of iron sulfate in aqueous sulfuric acid (solid line) with experimental data [42] at 50°C up to 50 mol/kg. Squares present heptahydrate and triangles monohydrate.

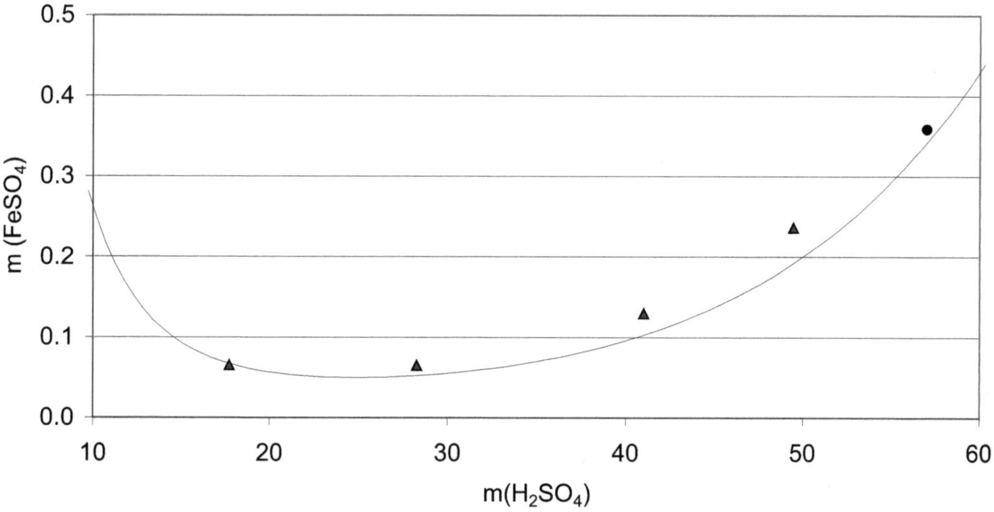

Figure 7. A comparison of the predicted solubility of iron sulfate in aqueous sulfuric acid (solid line) with experimental data [42] at 90°C between 10 - 50 mol/kg. Triangles present monohydrate and filled circle transition point from monohydrate to anhydrous iron sulfate.

As can been seen in figure 7, the predicted values are quite close to the experimental ones up to the transition point to anhydrous ferrous sulfate. Since the anhydrous $FeSO_4$ is not included in the model, it cannot be extrapolated beyond this transition point, .i.e. beyond 50 molal sulfuric acid solution. The transition point varies slightly with temperature.

Conclusions

As can been seen figures 4 - 7 the NPL modification of Pitzer equation is capable to describe acidic ferrous sulfate solution quite well with reasonable number of parameters and terms within. Moreover, the extrapolating capability of the sulfuric acid water system is excellent (figure 3).

References

[1] H. Müller, *"Sulfuric Acid and Sulfur Trioxide,"* Ullmann's Encyclopedia of Industrial Chemistry, Wiley-VCH Verlag GmbH & Co. KGaA, 2000), 1-71.

[2] K.S. Pitzer, R.N. Roy and L.F. Silvester, "Thermodynamics of Electrolytes 7. Sulfuric acid.", *J Am Chem Soc,*99(1977),4930-4936.

[3] E.J. Reardon and R.D. Beckie, "Modeling chemical equilibria of acid mine-drainage: The $FeSO_4$-H_2SO_4-H_2O system", *Geochim Cosmochim Acta,*51(1987),2355-2368.

[4] H. Sippola, *"Solubility of Ferrous Sulphate in Sulphuric Acid - A Thermodynamic Model"* (Licentiate thesis, Helsinki University of Technology, 1992), .

[5] H.F. Holmes and R.E. Mesmer, "Isopiestic studies of H_2SO_4(aq) at elevated temperatures: Thermodynamic properties", *J Chem Thermodyn,*24(1992),317-328.

[6] S.L. Clegg, J.A. Rard and K.S. Pitzer, "Thermodynamic properties of 0-6 mol kg^{-1} Aqueous Sulfuric Acid from 273.15 to 328.15 K", *J Chem Soc Faraday Trans,*90(1994),1875-1894.

[7] S.L. Clegg and P. Brimblecombe, "Application of Multicomponent Thermodynamic Model to Activities and Thermal Properties of 0-40 mol kg^{-1} Aqueous Sulfuric Acid from <200 to 328 K", *J Chem Eng Data,*40(1995),43-64.

[8] D.A. Knopf, B.P. Luo, U.K. Krieger, et al, "Thermodynamic Dissociation Constant of the Bisulfate Ion from Raman and Ion Interaction Modeling Studies of Aqueous Sulfuric Acid at Low Temperatures", *J Phys Chem A,*107(2003),4322-4332.

[9] C. Christov and N. Moller, "Chemical equilibrium model of solution behavior in the H-Na-K-OH-Cl-HSO_4-SO_4-H_2O system to high concentration and temperature", *Geochim Cosmochim Acta,*68(2004),1309-1331.

[10] E. Friese and A. Ebel, "Temperature Dependent Thermodynamic Model of the System H^+–NH_4^+–Na^+–SO_4^{2-}–NO_3^-–Cl^-–H_2O", *J Phys Chem A,*114(2010),11595-11631.

[11] C.E. Harvie and J.H. Weare, "The Prediction of Mineral Solubilities in Natural Waters: The Na-K-Mg-Ca-Cl-SO_4-H_2O System from Zero to High Concentrations at 25 °C", *Geochim Cosmochim Acta,*44(1980),981-997.

[12] D.G. Archer, "Thermodynamic Properties of the NaCI + H_2O System
II. Thermodynamic Properties of NaCI(aq), NaCI·$2H_20$(cr), and Phase Equilibria", *J Phys Chem Ref Data,*21(1992),793-829.

[13] Y. Liu and U. Grén, "Simultaneous correlation of activity coefficients for 55 aqueous electrolytes using a model with ion specific parameters", *Chemical Engineering Science,*46(1991),1815-1821.

[14] A. Bosen and H. Engels, "Description of the phase equilibrium of sulfuric acid with the NRTL equation and a solvation model in a wide concentration and temperature range", *Fluid Phase Equilib,*43(1988),213-230.

[15] B. Messnaoui and T. Bounahmidi, "On the modeling of calcium sulfate solubility in aqueous solutions", *Fluid Phase Equilib,*244(2006),117-127.

[16] G.M. Bollas, C.C. Chen and P.I. Barton, "Refined electrolyte-NRTL model: Activity coefficient expressions for application to multi-electrolyte systems", *AIChE J,*54(2008),1608-1624.

[17] J. Simonin, S. Krebs and W. Kunz, "Inclusion of Ionic Hydration and Association in the MSA-NRTL Model for a Description of the Thermodynamic Properties of Aqueous Ionic Solutions:â€‰ Application to Solutions of Associating Acids", *Ind Eng Chem Res,*45(2006),4345-4354.

[18] T. Vilarino, O. Bernard and J. Simonin, "Ionic Solutions in the Binding Mean Spherical Approximation. Thermodynamics of Associating Electrolytes up to Very High Concentrations", *J Phys Chem B,*108(2004),5763-5770.

[19] F.L.P. Pessoa, C.E.P. Siqueira Campos and A.M.C. Uller, "Calculation of vapor–liquid equilibria in aqueous sulfuric acid solutions using the UNIQUAC equation in the whole concentration range", *Chemical Engineering Science,*61(2006),5170-5175.

[20] H. Sippola, "Critical Evaluation of the 2nd Dissociation Constants for Aqueous Sulphuric Acid", *Poster presented at ICCT-2010, Tsukuba Science City, Japan, August 1-6,*(2010),.

[21] H. Sippola, "Critical Evaluation of the 2nd Dissociation Constants for Aqueous Sulphuric Acid", *Thermochim Acta,*532(2012),65-77.

[22] J. Pihlasalo, H. Davies, P.A. Taskinen, *Validation of a new Pitzer type model and database for aqueous solutions with Outotec HydroCopper^{TM} process data.(Presented at* CALPHAD, June 16-20 2008).

[23] H. Sippola, "Thermodynamic modelling of concentrated sulfuric acid solutions", *Calphad,*38(2012),168-176.

[24] K.S. Pitzer, "Thermodynamics of Electrolytes. I: Theoretical Basis and General Equations", *J Phys Chem,*77(1973),268-277.

[25] C.E. Harvie, N. Møller and J.H. Weare, "The prediction of mineral solubilities in natural waters: The Na-K-Mg-Ca-H-Cl-SO_4-OH-HCO_3-CO_3-CO_2-H_2O system to high ionic strengths at 25°C", *Geochim Cosmochim Acta,*48(1984),723-751.

[26] K.S. Pitzer, *"Semi-empirical Equations for Pure and Mixed Electrolytes,"* Thermodynamics(Singapore, McGraw-Hill, 1995), 290-321.

[27] K.S. Pitzer, *"Ion Interaction Approach: Theory and Data Correlation,"* Activity Coefficients in Electrolyte Solutionsed. Pitzer K.S. (Florida, USA, CRC Press, 2000), 75-154.

[28] H.V. Sippola, *Critical evaluation of the sulfuric acid - water system in wide concentration and temperature range(Presented at* CALPHAD XL, May 22 - 27, 2011 2011).

[29] P.M. Kobylin, H. Sippola and P.A. Taskinen, "Thermodynamic modelling of aqueous Fe(II) sulfate solutions", *Calphad,*35(2011),499-511.

[30] F. Fraenckel, "Über die Existenzgebiete der Ferrosulfat-Hydrate", *Zeitschrift für anorganische Chemie,*55(1907),223-232.

[31] A. Seidell, W.F. Linke, *Solubilities of inorganic and metal organic compounds. Vol. 1(* New York: Van Nostrand, 1958), 1487 s.

[32] W. Bullough, T.A. Canning and M.I. Strawbridge, "The solubility of ferrous sulphate in aqueous solutions of sulphuric acid", *Journal of Applied Chemistry,*2(1952),703-707.

[33] G. Bruhn, J. Gerlach and F. Pawlek, "Untersuchungen über die Löslichkeiten von Salzen und Gasen in Wasser und wäßrigen Lösungen bei Temperaturen oberhalb 100°C", *Zeitschrift für anorganische und allgemeine Chemie,*337(1965),68-79.

[34] F. Hasegawa, K. Tozawa and T. Nishimura, "Solubility of Ferrous Sulfate in Aqueous Solutions at High Temperatures", *Shigen-to-Soza,*112(1996),879-884.

[35] T.G. Oykova and C. Balarew, "Thermodynamic study of magnesium sulfate-ferrosulfate water system at 25°C", *Comptes Rendus de l'Academie Bulgare des Sciences,*27(1974),1211-1214.

[36] T. Ojkova, "Die Veränderung der mittleren Ionenaktivitätskoeffizienten der Komponenten in Dreistoffsystemen mit Doppelsalzbildung", *Z. Phys. Chem (Leipzig),*263(1982),1153-1161.

[37] M. Diesnis, "Détermination des États hygrometriques critiques", *Ann. Chim. Appl.,*7(1937),5-69.

[38] J. Perreu, "Sur la calorimétrie des solutions aqueuses de borax, de sulfate ferreux, de nitrate cuivrique et de nitrate de magnesium", *Comp Rend,*213(1941),286-289.

[39] K.A. Kobe and E.J. Couch, "Enthalpy-Concentration Diagram for System Ferrous Sulfate Water", *Ind Eng Chem,*46(1954),377-381.

[40] A.A. Malinin, S.I. Drakin and A.G. Ankudimov, "Study of equilibrium pressure of dehydration of some crystal hydrates of salts", *Russ. J. Phys. Chem.,*53(1979),755.

[41] I.-. Chou, R.R. Seal II and B.S. Hemingway, "Determination of melanterite-rozenite and chalcanthite-bonattite equilibria by humidity measurements at 0.1 MPa", *Am. Mineral.,*87(2002),108-114.

[42] A. Belopol'skii and S. Shpunt, "System $FeSO_4$–H_2SO_4–H_2O at temperatures 50–90°", *Zh Prikl Khim,*14(1941),716-733.

[43] A.P. Belopol'skii and V.V. Urusov, "System $FeSO_4$-H_2SO_4-H_2O II. Solubilities and densities of solutions at temperatures from −20° to +25°", *Zh Prikl Khim,*21(1948),781-793.

[44] A.P. Belopol'skii, V.N. Kolycheva and S.Y. Shpunt, "System $FeSO_4$-H_2SO_4-H_2O III.The solubility of $FeSO_4.7H_2O$ in aqueous solutions of sulfuric acid at temperatures from 10° to 50°", *Zh Prikl Khim,*21(1948),794-801.

[45] Y. Matsushima and A. Okuwaki, "The Second Dissociation Constant of Sulfuric Acid at Elevated Temperatures from Potentiometric Measurements", *Bull Chem Soc Jpn,*61(1988),3344-3346.

[46] R.H. Davies, A.T. Dinsdale, J.A. Gisby, et al, "MTDATA - thermodynamic and phase equilibrium software from the national physical laboratory", *Calphad,*26(2002),229-271.

[47] F.K. Cameron, "The Solubility of Ferrous Sulphate", *J Phys Chem,*34(1929),692-710.

REWAS 2013: Enabling Materials Resource Sustainability
Edited by: Anne Kvithyld, Christina Meskers, Randolph Kirchain, Gregory Krumdick, Brajendra Mishra, Markus Reuter, Cong Wang, Mark Schlesinger, Gabrielle Gaustad, Diana Lados, and Jeffrey Spangenberger
TMS (The Minerals, Metals & Materials Society), 2013

THERMODYNAMIC ANALYSIS OF LEAD-FLUORIDE ION-WATER SYSTEM

Jiayuan Li [1,2], Tianzu Yang [1], Lin Chen[1], Weifeng Liu[1]

[1] School of Metallurgical Science and Engineering, Central South University, Changsha 410083, China
[2] Department of Chemical Engineering and Life Science, Xiangnan University, Chenzhou 423000, China

Key words: Lead fluoride, Lead-fluoride ion-water system, Thermodynamics

Abstract

Based on thermodynamic data, this paper analyzed the feasibility of removing fluoride in the waste acid solution using lead ions. Lead-fluoride ion-water system E-log[F^-] diagram and the solubility curve of lead fluoride at 25℃ were drawn. It was found that insoluble PbF_2 only exists stably in a certain concentration range of fluoride ion. With the increase of lead ion concentration, the stable area for solid lead fluoride expands and the total fluoride ion concentration decreases. Therefore, it is possible to remove fluoride ion in the waste acid solution lead ion.

Introduction

In recent years, with the sharp increase of the production capacity of non-ferrous smelting industry, a large number of fluoride, chloride, and heavy metal ions transferred to the smelting flue gas in the process of roasting, smelting and blowing.
Acid making with flue gas process will produce a lot of waste acid in smelters. The waste acid contains large amounts of fluoride, chloride, sulfate and heavy metal ions, which has been recognized as one of the major problems smelting industry. This imposes a serious threat to human health and environmental issues. Thus, a renewed interest in the ions removal from industrial wastewaters has been greatly increased [1, 2, 3, 4, 5, 6]. Removal of fluorine is an unavoidable problem to cyclic utilization and standard discharge of waste acid. At present, lime precipitation treatment is still widely used as an important method [7, 8, 9]. Additionally, electrodialysis [10, 11, 12], sorption [13, 14, 15, 16], ion exchange resins [17, 18, 19, 20] and membrane technology [21, 22, 23, 24, 25] have been reported.

Lime method is low cost, but it will generate a lot of waste residue, inconvenience of subsequent disposal and results in second pollution. Electric flocculation and ion exchange method are difficult to deal with the high concentration of fluoride and result in high energy consumption, more processing costs. According to the above finding and requirements of the handling of waste acid in smelter, we make use of materials containing rich lead oxide which come from smelter production process as defluoridation reagents. This paper mainly makes a theoretical analysis of the fluoride removal from waste acid based on smelter rich lead oxide materials, which is of great significance for the smelter waste acid treatment.

Equilibrium Equations in Solution

The divalent lead ion can react with fluoride ion to create a series of complex ions in the fluoride solution. The complex reactions and equilibrium constants are shown as follows:

$$H^+ + iF^- = HF_i^{1-i}, \quad K_i = \frac{[HF_i^{1-i}]}{[H^+]\cdot[F^-]^i}, \quad i = 1,2 \tag{1}$$

$$Pb^{2+} + jF^- = PbF_j^{2-j}, \quad K_j = \frac{[PbF_j^{2-j}]}{[Pb^{2+}]\cdot[F^-]^j}, j = 1,2,3,4 \tag{2}$$

$$Pb^{2+} + kOH^- = Pb(OH)_k^{2-k}, \quad K_k = \frac{[Pb(OH)_k^{2-k}]}{[Pb^{2+}]\cdot[OH^-]^k}, k = 1,2,3,4 \tag{3}$$

$$H^+ + OH^- = H_2O, \quad K = 10^{-14} \tag{4}$$

Where K_i , K_j and K_k are cumulative stability constants of complex ion at all levels. The stability constants [26] of ions are shown in Table 1.

Table 1 Stability constants of ions in Pb-F-H_2O solution at 25℃

Complex	lgK_1	lgK_2	lgK_3	lgK_4
Pb^{2+}-F^-	1.44	2.54	3.42	3.08
Pb^{2+}-OH^-	6.3	10.3	13.3	7.6
H^+-F^-	3.176	0.591		

The total concentrations of lead, $[Pb]_T$, and fluoride, $[F]_T$, in the solution can be calculated by Eqs. (4)- (6):

$$[Pb]_T = (1+\sum_{j=1}^{4} K_j[F^-]^j + \sum_{k=1}^{4} K_k[OH^-]^k) \cdot [Pb^{2+}] \tag{5}$$

$$[F]_T = [F^-] + [H^+]\cdot\sum_{i=1}^{2} iK_i[F^-]^i + [Pb^{2+}]\cdot\sum_{j=1}^{4} jK_j[F^-]^j \tag{6}$$

Where $[M]_T$ denotes the total concentration of M and [M] is the concentration of free M.

When the solution coexists with insoluble fluoride (PbF_2), the relationship between [Pb^{2+}] and [F^-] is determined by the solubility product (K_{sp}) of lead fluoride:

$$K_{sp} = [\mathrm{Pb}^{2+}]\cdot[\mathrm{F}^-]^2 = 2.7\times10^{-8} \tag{7}$$

$[Pb]_T$, $[F]_T$ and [H^+] in Eqs.(4)-(6) can be controlled during practical work. The concentrations of free [Pb^{2+}] and [F^-] can be calculated at the three given values.

Moreover, when the solution is in an equilibrium condition, only one electrode potential exists based on the principle of simultaneous equilibrium:

$$E_{\mathrm{Pb^{2+}/Pb}} = E_{\mathrm{PbF^{+}/Pb}} = E_{\mathrm{PbF_2(s)/Pb}} = E_{\mathrm{PbF_3^-/Pb}} = E_{\mathrm{PbF_4^{2-}/Pb}} \tag{8}$$

The above value can be calculated by the following equation:

$$E_{\mathrm{Pb^{2+}/Pb}} = E^o_{\mathrm{PbF^{+}/Pb}} + \frac{RT}{nF}\ln[\mathrm{Pb}^{2+}] \tag{9}$$

Using the total concentration of lead, $[Pb]_T$, to take the place of the concentration of free lead ion, [Pb^{2+}], in Eq.(9), Eq.(10) is formed. And the relationship between the lead electrode potential and [F^-] at given total concentration of lead, $[Pb]_T$, is represented by Eq.(11):

$$E_{\mathrm{Pb^{2+}/Pb}} = E^o_{\mathrm{Pb^{2+}/Pb}} + \frac{RT}{nF}\ln\frac{[\mathrm{Pb}]_\mathrm{T}}{1+\sum_{j=1}^{4}K_j[\mathrm{F}^-]^j} \tag{10}$$

$$E_{\mathrm{PbF_2(s)/Pb}} = E^o_{\mathrm{PbF_2(s)/Pb}} - \frac{RT}{nF}\ln[\mathrm{F}^-] \tag{11}$$

Then, E-lg$[F]_T$ diagram at a given temperature and a total lead concentration is plotted based on the relationship between [F^-] and $[F]_T$, Eq.(11). In order to demonstrate the effect of $[Pb]_T$ on lead electrode potential, E-lg$[F]_T$ diagram of lead is plotted.

In E-lg$[F]_T$ diagram, two intersection points of the two curves are ascertained by Eqs.(10) and (11), which are marked as α and β, respectively. And the range of $[F]_T$ corresponding to the two points is the concentration range of total fluoride, in which the precipitation PbF_2 exists stably.

The calculation of points α and β could be educed based on the relationship among the point potentials, $E_{Pb^{2+}/Pb}$ and $E_{PbF_2(s)/Pb}$. Also, they could be calculated by Eq.(12) based on the principle of solubility product:

$$[Pb^{2+}]\cdot[F^-]^2 = \frac{[Pb]_T \cdot [F^-]^2}{1+\sum_{j=1}^{4} K_j [F^-]^j} = K_{sp}(PbF_2) \tag{12}$$

It is concluded that $[F^-]$ and the corresponding $[F]_T$ are gained with the help of Eq.(12) and Eq.(6), respectively. The total fluoride concentration corresponding to α and β changes as the total lead concentration changes. So, the practical solubility curves of solid PbF_2 at different temperatures are gained according to $[Pb]_T$–$[F]_T$ diagram.

Thermodynamic Analysis of Pb-F-H$_2$O System

Relationships between Distribution of Fluoride and pH Value

The distribution fraction of fluoride and lead in the aqueous solution at different pH value is shown in Fig.1 and Fig.2.

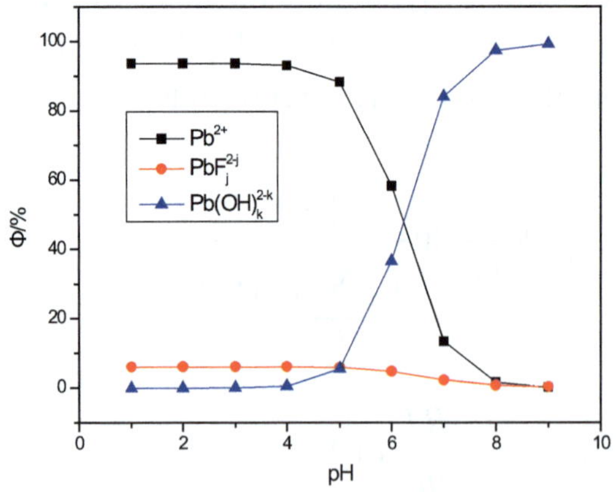

Fig.1 Relationships between distribution of lead and pH value

Fig.1 shows the distribution fraction of lead ion, lead hydrogen complexes and lead fluoride complexes (δ) diagram at different pH. When pH is less than 5, there are mainly lead ions in the solution. Once the pH is greater than 7, there are mainly lead hydrogen complexes in the solution.

Fig.2 shows the distribution fraction of fluoride ion and hydrogen fluoride (δ) diagram at different pH. When the solutions have pH values greater than 4, hydrogen fluoride were less than

10%, and the fluoride ion were more than 80%. We attempt to study the equilibrium in near-neutral pH condition (pH=5～6), and the fluoride ions are more than hydroxide ions, so the complex of lead with hydroxide ion can be ignored.

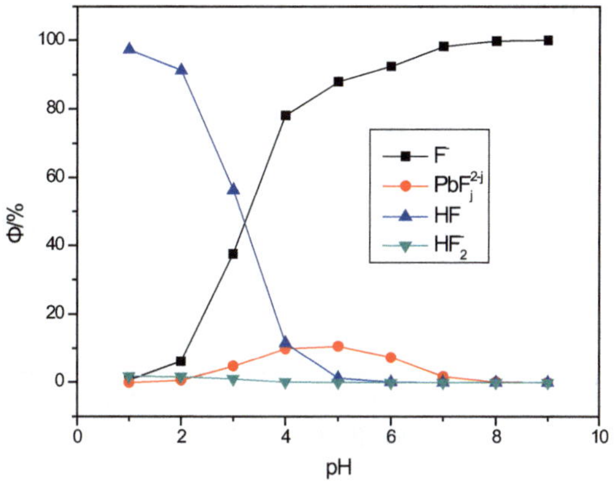

Fig.2 Relationships between distribution of fluoride and pH value

Relationship between Solubility Product and $\lg[F]_T$ at Different $[Pb]_T$

To elucidate the reaction mechanism of lead and fluoride in Pb-F-H_2O system, the relationship between solubility product and $\lg[F]_T$ at different $[Pb]_T$ is studied. Fig.3 shows the curves of $[Pb^{2+}][F^-]^2$ vs $\lg[F]_T$ in Pb-F-H_2O system at different $[Pb]_T$. PbF_2 precipitation is just observed at the limiting point (α), corresponding to the value of $[Pb^{2+}][F^-]_2$ in the solution larger than the solubility of the PbF_2. However, when $[F]_T$ reaches the limiting point β, the precipitated PbF2 will dissolve into the solution again because the value of $[Pb^{2+}][F^-]_2$ in the solution is smaller than the solubility of PbF_2. That is to say, the insoluble salt PbF2 will only exist stably in a certain concentration range of the chlorine ion (between α and β). The points α move to left and β move to right with increasing $[Pb]_T$ in the system and the concentration range of the fluoride ion expands significantly. The total fluoride concentration is usually less than 1 mol/L in waste acid from sulfuric acid production process. So, if we want to use lead ions to remove the fluoride ion in solution, the bigger lead ion concentration the better. Thus, using lead ions to deep purification fluoride becomes feasible because the insoluble salt PbF_2 will exist stably in the certain concentration range of lead ion.

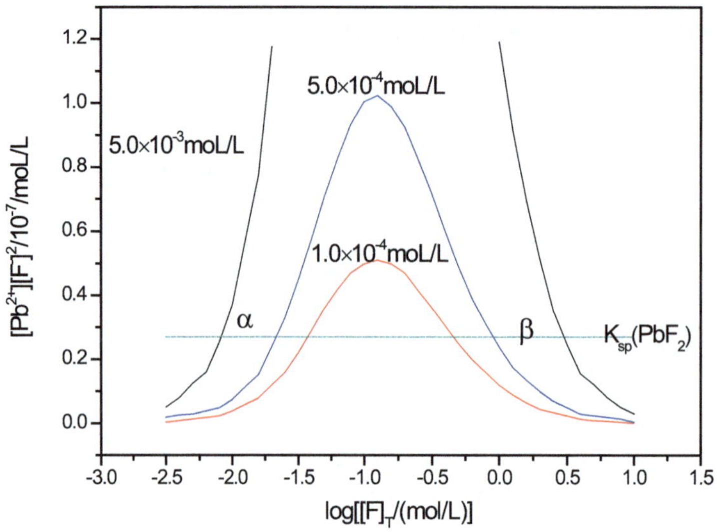

Fig.3 $[Pb^{2+}][F^-]^2$ as function of $\lg[F]_T$ in Pb-F-H_2O system at 25℃

E—lg$[F]_T$ Diagram at Different $[Pb]_T$

To elucidate more clearly, the changes of E with $[F]_T$ at different $[Pb]_T$ are studied as well. Fig.4 shows the E—lg$[F]_T$ diagram at different $[Pb]_T$. It is clear that the equilibrium concentration ranges of fluoride ion ascertained by points α and β are similar to those in Fig.3. The areas ascertained by points α and β in which solid lead fluoride exists stably at given conditions expand a little with increasing the total lead concentration. Point α moves to top left corner while point β moves to bottom right corner, thus the ascertained areas in which metal lead exists stably also expand a little with increasing the total lead concentration. Moreover, the corresponding potentials $E_{Pb^{2+}/Pb}$ and $E_{PbF_2(s)/Pb}$ increase with increasing the total lead concentration. In summary, when the acidity of system are given, $[F]_T$ corresponding to points α and β changes with increasing the total lead concentration, as well as changes with increasing the lead solubility in the solution, demonstrating easy removal fluoride at a higher $[Pb^{2+}]$.

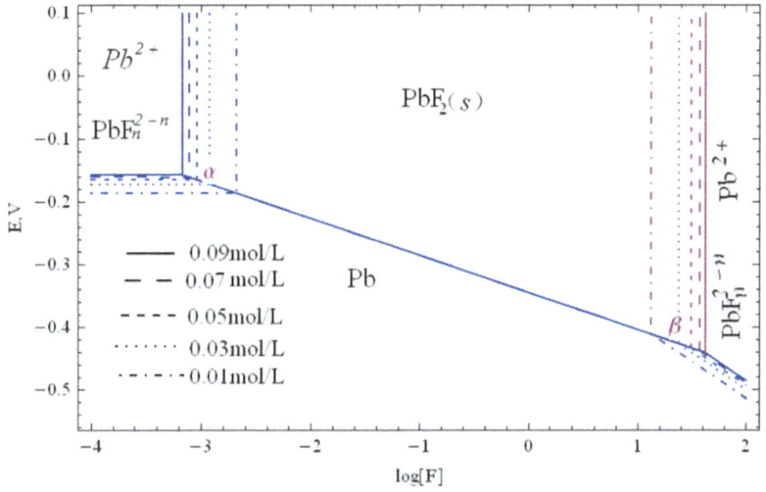

Fig.4 E-log$[F]_T$ diagram of lead-fluoride ion-water system at different $[Pb]_T$ (t=25℃)

Solubility Curve of $PbF_2(s)$

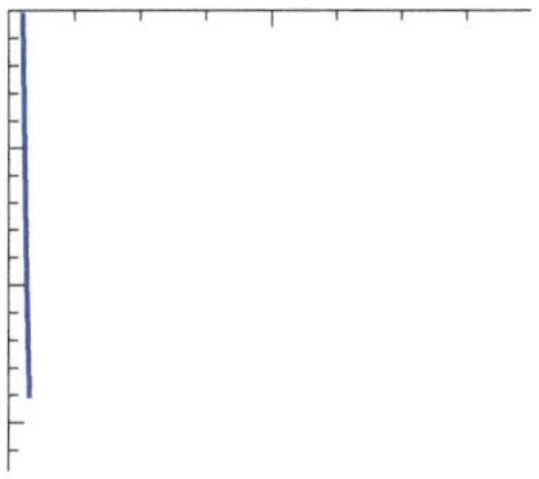

Fig.5 Relationship of $[Pb]_T$ and $[F^-]$ at 25℃

Fig.5 shows the solubility of $PbF_2(s)$ as a function of $[F]_T$ in Pb-F-H_2O system at 25℃. It is clear that the $[F]_T$ concentration reaches a low value in solution with $[Pb]_T$ concentration reaching 120 mg/L in the system. That is to say, when the total lead concentration is more than 120 mg/L, the fluoride ions can be removed from the solution. However, when the total lead concentration is less than 120 mg/L, the solubility of fluoride rises sharply because less lead can form soluble lead fluoride complexes (PbF_i^{2-i}, where i=1,2,3,4) with fluoride. The same phenomena were reported by other researchers.

Thus, using lead ions to deep purification fluoride becomes feasible and the bigger lead ion concentration the better, because the insoluble salt PbF_2 will exist stably in the certain concentration range of lead ion.

Conclusions

1) The thermodynamic analysis of Pb-F-H_2O system is introduced here based on the principle of simultaneous equilibrium.

2) The E-lg$[F]_T$ diagram, $[Pb^{2+}][F^-]^2$-lg$[F]_T$ diagram and the solubility curves of lead fluoride are plotted, respectively.

3) The insoluble salt PbF_2 only exists stably in a certain concentration range of fluoride ion. The solid lead fluoride exists stably would be expanded and the total fluoride ion concentration would decrease with the increase of lead ion concentration. Therefore, using lead ion removal of fluorine ion is possible.

Acknowledgements

This work is supported financially by Major national science and technology programs of China (No. 2010ZX07212-008), Scientific and Technological commission of Hunan Province of China (No. 2012SK3112).

References

1. A. Bhatnagar, E. Kumar, M. Sillanpää, "Fluoride Removal from Water by Adsorption—a Review,"*Chem. Eng. J.*, 171 (2011), 811-840.

2. J. Dron, A. Dodi, "Comparison of Adsorption Equilibrium Models for the Study of CL^-, NO_3^- and SO_4^{2-} Removal from Aqueous Solutions by an Anion Exchange Resin,"*J. Hazard. Mater.*, 190 (2011), 300-307.

3. C.T. Benatti, C.R.G. Tavares, E. Lenzi,"Sulfate Removal from Waste Chemicals by Precipitation," *J. Environ. Manage.*, 90 (2009) , 504-511.

4. R. Malaisamy, A. Talla-Nwafo,K.L. Jones, "Polyelectrolyte Modification of Nanofiltration Membrane for Selective Removal of Monovalent Anions,"*Sep. Purif. Technol.*, 77 (2011), 367-374.

5. Y.C. Leea, E. J. Kimb, H.J. Shinc, M. Choia, J.W. Yanga, "Removal of F^-, NO_3^-, and PO_4^{3-} Ions from Aqueous Solution by Aminoclays," *J. Ind. Eng.Chem.*, 18 (2012) , 871-875.

6. B.D. Turner, P. Binning, S.L.S. Stipp, "Fluoride Removal by Calcite: Evidence for Fluorite Precipitation and Surface Adsorption," *Environ. Sci. Technol,* 39 (2005), 9561-9568.

7. Esin Alkan, Esengul Kır, Lutfi Oksuz,"Plasma Modification of the Anion-Exchange Membrane and its Influence on Fluoride Removal from Water," *Sep. Purif. Technol.*, 61 (2008), 455-460.

8. Nash C. Lu, J.C. Liu, "Removal of Phosphate and Fluoride from Wastewater by a Hybrid Precipitation–Microfiltration Process," *Sep.Purif. Technol.* 74 (2010), 329-335.

9. Suresh K. Nath, Shreemoyee Bordoloi, Robin K. Dutta, "Effect of Acid on Morphology of Calcite during Acid Enhanced Defluoridation,"*J. Fluorine Chem.,* 132 (2011), 19-26.

10. S.K. Adhikary, U.K. Tipnis, W.P. Harkare, K.P. Govindan,"Defluoridation during Desalination of Brackish Water by Electrodialysis," *Desalination*, 71(1989), 301-312.

11. M. Behbahani, M.R. Alavi Moghaddam, M. Arami, "Techno-Economical Evaluation of Fluoride Removal by Electrocoagulation Process: Optimization through Response Surface Methodology," *Desalination*, 271 (2011), 209-218.

12. Mohammad M. Emamjomeh, Muttucumaru Sivakumar,"Fluoride Removal by a Continuous Flow Electrocoagulation Reactor," *J. Environ. Manage.,* 90(2009) , 1204-1212.

13. H. Lounici, L. Addour, D. Belhocine, H. Grib, S. Nicolas, B. Bariou, "Study of a New Technique for Fluoride Removal from Water,"*Desalination*, 114 (1997), 241-251.

14. T. Hiemstra, W.H.V. Riemsdijk , "Fluoride Adsorption on Goethite in Relation to Different Types of Surface Sites," *J. Colloid Interface Sci.*, 225 (2000) , 94-104.

15. Y.L. Tang, X.H. Guan, J.M. Wang, N. Gao, M.R. Mcphail, C.C. Chusuei, "Fluoride Adsorption onto Granular Ferric Hydroxide Eflects of Ionic Strength, pH, Surface Loading, and Major Co-existing Anions," *J. Hazard. Mater.,* 171 (2009) , 774-782.

16. M. Mohapatra, K. Rout, P. Singh, S. Anand, S. Layek, H.C. Verma, B.K. Mishra, "Fluoride Adsorption Studies on Mixed-Phase Nano Iron Oxides Prepared by Surfactant Mediation-precipitation Technique," *J. Hazard. Mater.,* 186 (2011) , 1751-1757.

17. M.F.A. Taleb, G.A. Mahmoud, S.M. Elsigeny, E.-S.A. Hegazy, "Adsorption and Desorption of Phosphate and Nitrate Ions using Quaternary (polypropylene-g-N, N-dimethylamino ethylmethacrylate) Graft Copolymer," *J. Hazard. Mater,* 159 (2008), 372-379.

18. Imam Bakhsh Solangi, Shahabuddin Memon, M.I. Bhanger, "Removal of Fluoride from Aqueous Environment by Modified Amberlite Resin," J. Hazard. Mater., 171 (2009) , 815-819.

19. S. Meenakshi, Natrayasamy Viswanathan,"Identification of Selective Ion-exchange Resin for Fluoride Sorption," *J. Colloid Interface Sci.* 308 (2007), 438-450.

20. Kaisa Vaaramaa, Jukka Lehto, "Removal of Metals and Anions from Drinking Water by Ion Exchange,"*Desalination*, 155 (2003) , 157-170.

21. Guanghui Zhang, Yong Gao, Ying Zhang, Ping Gu, "Removal of Fluoride from Drinking Water by a Membrane Coagulation Reactor (MCR),"*Desalination,* 177 (2005) , 143-155.

22. Peter Sehn, "Fluoride Removal with Extra Low Energy Reverse Osmosis Membranes: Three years of Large Scale Field Experience in Finland," *Desalination,* 223 (2008) , 73-84.

23. Ali Tor.,"Removal of Fluoride from Water Using Anion-exchange Membrane under Donnan Dialysis Condition," *J. Hazard. Mater.,* 141 (2007) , 814-818.

24. Kang Hu, James M. Dickson,"Nanofiltration Membrane Performance on Fluoride Removal from Water,"*J. Membrane Sci.*, 279 (2006), 529-538.

25. Esin Alkan, Esengul Kır, Lutfi Oksuz,"Plasma Modification of the Anion-exchange Membrane and its Influence on Fluoride Removal from Water," *Sep. Purif. Technol.*, 61 (2008), 455-460.

26. Central South Institute of Mining and Metallurgy, *Chemical Analytic Handbook*, (Beijing, Science Press, 1998) ,126-127.

REWAS 2013

Enabling Materials Resource Sustainability

Enabling Sustainability through Life Cycle Management, LCA and Industrial Ecology

Session Chairs

Gabrielle Gaustad

Jeffrey Spangenberger

REWAS 2013: Enabling Materials Resource Sustainabi
*Edited by: Anne Kvithyld, Christina Meskers, Randolph Kirchain, Gregory Krumdick, Brajendra Mish
Markus Reuter, Cong Wang, Mark Schlesinger, Gabrielle Gaustad, Diana Lados, and Jeffrey Spangenber*
TMS (The Minerals, Metals & Materials Society), 20

Abstract for REWAS 2013: Enabling Materials Resource Sustainability

Stock dynamics and emission pathways of the global aluminum cycle

Daniel B. Müller[1], Gang Liu[1], and Colton Bangs[2]

[1] Norwegian University of Science and Technology
[2] Umicore

Key Words: Climate change mitigation, stock dynamics, recycling, socio-economic metabolism, material flow analysis (MFA)

Climate change mitigation in the materials sector faces a twin challenge: satisfying rapidly rising global demand for materials while significantly curbing greenhouse-gas emissions. Process efficiency improvement and recycling can contribute to reducing emissions per material output; however, long-term material demand and scrap availability for recycling depend fundamentally on the dynamics of societies' stocks of products in use, an issue that has been largely neglected in climate science. Here, we show that aluminium in-use stock patterns set essential boundary conditions for future emission pathways, which has significant implications for mitigation priority setting. If developing countries follow industrialized countries in their aluminium stock patterns, a 50% emission reduction by 2050 below 2000 levels cannot be reached even under very optimistic recycling and technology assumptions. The target can be reached only if future global per-capita aluminium stocks saturate at a level much lower than that in present major industrialized countries. As long as global in-use stocks are growing rapidly, radical new technologies in primary production (for example, inert anode and carbon capture and storage) have the greatest impact in emission reduction; however, their window of opportunity is closing once the stocks begin to saturate and the largest reduction potential shifts to post-consumer scrap recycling.

The full paper is available in the following journal publication:

Liu G., C. Bangs, and D.B. Müller (2012): Stock dynamics and emission pathways of the global aluminium cycle. Nature Climate Change. Published online 07 October 2012.

REWAS 2013
Enabling Materials
Resource
Sustainability

Enabling Sustainability through Systems Modelling and Design, Life Cycle Management, LCA and Industrial Ecology

Session Chairs

William Rankin
Diana Lados

REWAS 2013: Enabling Materials Resource Sustainabi
*Edited by: Anne Kvithyld, Christina Meskers, Randolph Kirchain, Gregory Krumdick, Brajendra Mish
Markus Reuter, Cong Wang, Mark Schlesinger, Gabrielle Gaustad, Diana Lados, and Jeffrey Spangenber*
TMS (The Minerals, Metals & Materials Society), 20

A Green Urban Mobility System Solution from the EU Ingrid project

Fabrizio D'Errico[1], Adamo Screnci[2], Marco Romeo[3]

[1]Politecnico di Milano, Dept. of Mechanical Engineering, Via La Masa 34, 20156, Milan, Italy
[2]Mc Phy Energy SA, 40 Rue Des Berges, 38000 Grenoble, France
[3]CiaoTech-PNO Group B.V.,Via A. Pestalozza 31, 20131 Milan, Italy

Keywords: magnesium , hydrogen , energy storage , sustainable mobility , e-mobility

Abstract

With a mandate to reach 20/20/20 targets, new strategies are now focusing on the increased use of electricity to power transportation. Particularly in major urban areas of the EU, capillary use of electric vehicles are being encouraged, however, as these vehicles will be powered by the grid, there is always the risk that load peaks will occur. This work is just one of several being developed as part of the 23.9 MLN Euros INGRID European project started in July 2012, which combines solid-state high-density hydrogen storage systems with advanced ICT technologies for distribution grids. One possible solution which has been designed, is an off-grid utility to store renewable electricity captured from wind/solar sources and a re-charging point for full battery electric cars. This work shows the preliminary financial assessment of two business models for the *Park-for-Recharging* concept to promote green e-mobility as a more convenient and economical means of by-car transport.

Introduction

With a mandate to reach 20/20/20 targets, new strategies are now focusing on the increased use of electricity to power transportation. Particularly in major urban areas of the EU, capillary use of electric vehicles are being encouraged. By the year 2020, European countries expect to see up to one million electric cars on the road.

As these cars will be powered by a grid, there is a real risk that load peaks will occur causing dips in the power supply as the power network that exists today does not have the capability to handle a large increase in demand. But if the power supply and demand for that power could be more closely matched, it would allow hundreds of drivers to recharge car batteries simultaneously at random points throughout a city, even while attending a soccer game or leaving their cars in long term airport parking facilities.

To avoid the frequency at which alternating current is transmitted deviate from the stipulated 50 Hz, falling when there is an excessive demand, and rising in the case of oversupply, the introduction of a large number of e-vehicles must be carefully planned. For example, let's assume that 20.000 battery powered electric city-cars, each with a 20 k WH battery capacity of 50 km and an approximate fuel efficiency of 0,11 kWh/km, would need to be re-charged daily, and that each car would require around 5,5 kWh per day. If 50% of the fleet, 10,000 cars, would simultaneously tap into the grid, the required output would be approximately 200 MW, the amount of electrical power a medium size centralized power storage station could provide in an instant. One possible solution which has already been designed is an off-grid utility to store renewable electricity captured from wind/solar sources and a re-charging point for full battery electric cars. This work addresses one of several issues which will be developed as part of the INGRID European project, which began in July 2012. Seven European partners were granted a financial contribution of 13,8 MLN Euros by the European Commission's Seventh Framework

Programme for European Research. The project will be coordinated by: *Engineering Ingegneria Informatica*, the largest private-owned Italian ICT technology provider (active as ICT specialist for the Smart Grid market); the *Agenzia per la Tecnologia e l'Innovazione* (ARTI), which represents the operational arm of the Puglia regional authority for the innovation and technology transfer in Puglia Region (Italy); *Enel Distribuzione*, which is Italy's largest electricity distribution company; *Hydrogenics* (Belgium), a leader in water electrolyzer and fuel cell units; the French *McPhy Energy SA* company, a leader in the area of innovative technologies for solid hydrogen and safe storage; and the research institutions *Fundacion Tecnalia Research & Innovation* (TECNALIA, Spain); and *Ricerca sul Sistema Energetico* (RSE, Italy).

The core innovation of the INGRID project will consist of combining solid-state high-density hydrogen storage systems with advanced ICT technologies for smart distribution grids which will monitor and control a large number of renewable energy sources in order to balance power supply and demand.

The seven company consortium will design, build, deploy and operate a 39 MWh energy storage facility using McPhy hydrogen-based solid state storage and Hydrogenics electrolysis technology and fuel cell power systems in the Puglia region of Italy where over 3.500 MW of solar, wind, and biomass have already been installed. The hydrogen energy storage installation, including a novel fast responding 1.2 MW hydrogen generator, will provide effective smart grid-balancing support.

The goal of the INGRID project is to illustrate (fig.1) the main concept of hydrogen-based storage through Mg-hydrides for a true "zero-emission" supply-chain that can deliver and dispatch renewable electricity (green-electricity).

Furthermore, this project would lead in future large modular power storage concept that can provide power for refueling large urban electric e-vehicle fleets with real zero-emissions over the well-to-wheel path. Based on plug-in battery electric vehicles (BEVs) like citycars, electric compact minibuses and electric public taxis, a Green Urban Mobility System would be developed according to the Parking-for-Recharging (or P4R) concept detailed in fig.2.

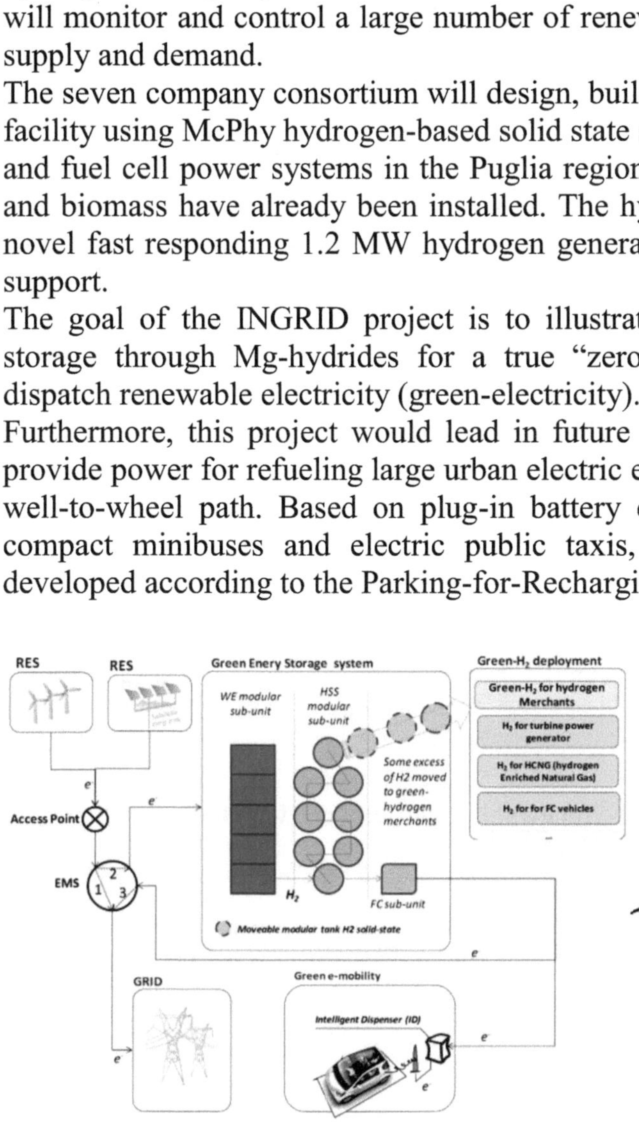

Fig. 1 - The overall concept of European INGRID project

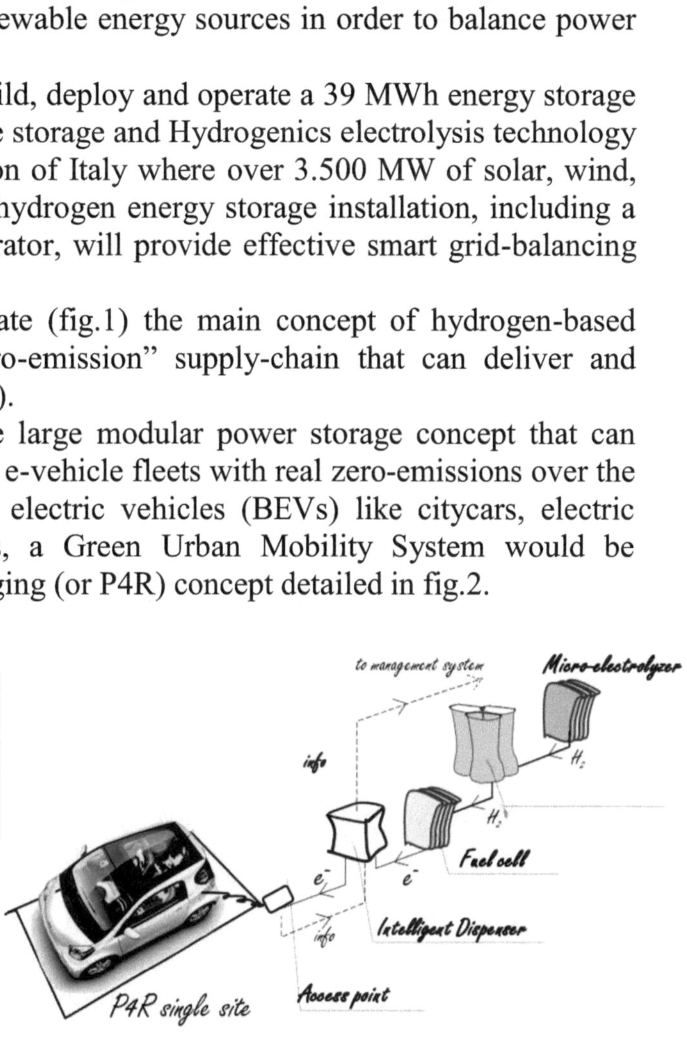

Figure 2 – The scheme of the *P4R* concept to be realized as demonstrative show-case.

This work aims to illustrate results from the preliminary financial assessment of a possible business for the P4R concept studied to promote green e-mobility as a more convenient and economical means of by-car transport.

The Project Scope and Background

The GUMS concept concerns on fully integrated intelligent monitoring, control and communication solution which can be one natural deployment of large amount of RES captured, stored and delivered by INGRID modular systems. The promising result is offered by an inexhaustible magnesium-based material which has been produced and sold by Mc Phy Energy Company since 2009[1]. Today, the storage of a large quantity of GWh at an affordable cost can represent a key-solution to solving the balance of a renewable energy source and a large intermittent electricity supply, in order to accommodate variable electrical demands. Hydrogen can be stored in solid form by several materials; among these materials magnesium metal, under certain conditions, acts as "sponge" absorbing and desorbing hydrogen gas just "locking" (i.e. storing) H_2 molecules by MgH_2 solid phase compound formation.

The innovative solution Mc Phy Energy has brought into marketplace consists in impressive increase in the kinetics of absorption and desorption of H_2 in magnesium-based materials. This fact allows such material to be a practical solution for commercial scopes where reduced time for storing and delivering of hydrogen is really mandatory. An important consequence of practical solid-state storing of hydrogen is new solutions to overcome historical infrastructural barriers, which up to now have prevented hydrogen deployment for power generation, transmission and conversion into electricity for fueling electric devices. The final result of this process is an industrial application of a real "zero-emission" supply-chain, as shown in fig.3: a water electrolyzer, device capable to slit water into hydrogen and oxygen, is fed by RES meanwhile hydrogen can be indefinitely stored safely in solid state mode until demand side requests it.

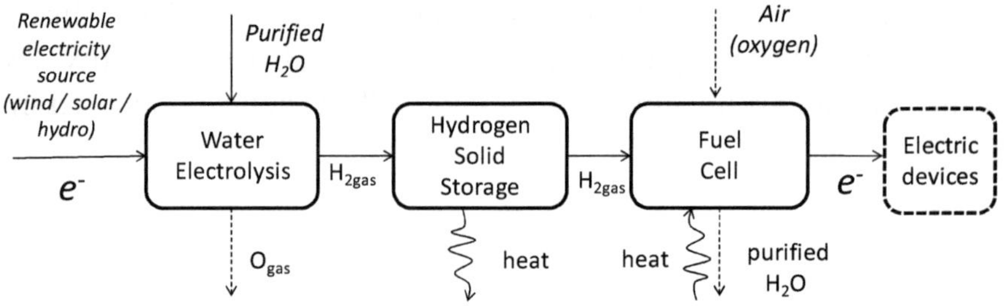

Figure 3 – Layout of the regenerative system for the renewable electricity production realized by a conventional open-loop.

More specifically, the P4R concept studied is aimed to set and demonstrate a key-solution for building a market-ready high sustainable supply-chain for delivering renewable electricity (green-electricity) to feed large urban mobility systems, based on plugged-in battery electric vehicles (BEVs) like citycars, electric compact minibuses and electric taxis.

---

[1] There are several ways of producing magnesium, with the electrolytic and thermal reduction processes (Pidgeon) most widely used. These processes vary significantly with respect to their raw materials, energy carriers, country of production, etc. Therefore, their carbon footprints and energy demand vary too. Studies shows that the electrolytic method has the lowest carbon footprint - ranging from 14.4 kg CO_2 eq./kg Mg in Canada to 24.5 kg CO_2 eq./kg Mg ingot in Australia [D, E]. The thermal Pidgeon process, used mainly in China, generates the highest amount of GHG emissions - ranging from 27-42 kg CO_2 eq./kg Mg ingot, due to a high energy consumption and use of coal for process heat and electricity [2, 3, 4, 5]. For last years a novel lo impacting magnesium extraction process has been developing by Gossan Resources, a Canadian company that lead a magnesium large extraction project in Manitoba. Primary magnesium produced using the Gossan-Zuliani Process is claimed to achieve a Global Warming Potential of only 9.1 kg CO2 eq./kg Mg ingot [6].

While technical feasibility is highly assured by several applications previously exploited by the main partners active in green-power generation and exploitation of fuel cell systems, such mature technologies today can be successfully coupled with a more recent advanced and cost effective energy storage system. Large quantify of GWh can therefore be stored into solid state hydrogen and distributed at different times and places, and in differed modes. The main aim of the supply-chain proposed is to bridge the source availability and the power demand, addressing future amounts of electric power needed to feed capillary, fully electric mobility systems, which will be truly 100% emission-free over the entire supply-chain. Such a safe storing technique of hydrogen can function as a cost-effective solution for an easy integration and development of a Green-Fuel Distribution System (GFDS) for urban centers, namely a ready-solution to build capillary "on-demand" electricity distributed generation (DG) access points for fueling large fleets of battery electric citycars.

Expected Results

The GUMS concept pursues the following broad objectives[2]:

- to realize a true zero-emission supply-chain to feed "green-electricity" to urban mobility systems, which were previously blocked by infrastructure barriers which prevented the exploitation of hydrogen (primarily, the initiation of a cost-effective "Hydrogen Economy");
- to assess the Parking for Recharging (P4R) concept, the decentralized delivering points which will feed 100% green-electricity to electric citycars.
- to exploit a "dual-side" communication system between vehicle and supply point by introducing an "intelligent dispenser" (as shown fig.3) which allows the collection of information on vehicle plugged-in (battery charge status, GPS location of the plugged-in vehicle, time to complete charging, actual average travel distance capacity, etc.);
- to control and manage (in the real-time) the uninterrupted supply of green-electricity in an on-demand mode;
- to develop an ICT communication platform to effectively transfer relevant information acquired by the intelligent dispenser to the consumer, thus allowing rapid and economically effective access to vehicles at P4R sites (fig.4);

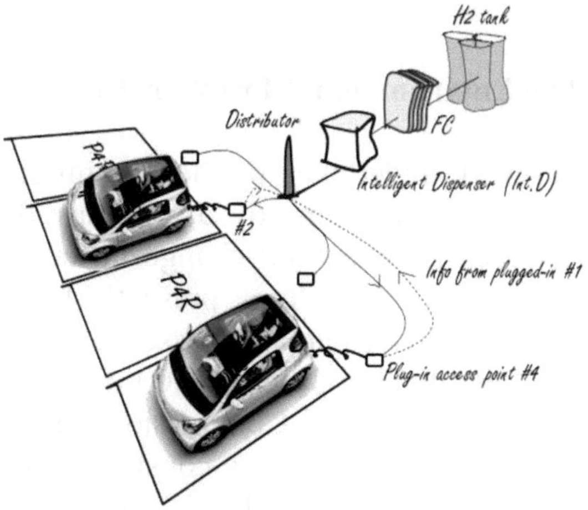

Figure 4 – The P4R cluster area as demonstrative show-case.

[2] In agreement with the "Strategic research agenda for Europe's electricity networks of the futures", European Commission, source: www.smartgrids.eu, latest check on 30 October 2012.

Environmental and financial assessments were prepared during the design stage, and relevant results are shown below. A more studied assessment could be done by gathering data from the construction and testing stage after the installation of the pilot. A sustainable mobility regional/local scenario has been created, (fig.5) and analyzed in terms of environmental (net reduction of global warming potential, $KgCO_2/eq$ cut by alternative scenarios) and economic impact (profitability, return of investments from shareholders, etc.), using a EU town which introduces P4Rs as the most efficient means of mobility compared to private, conventional unclean and expensive cars.

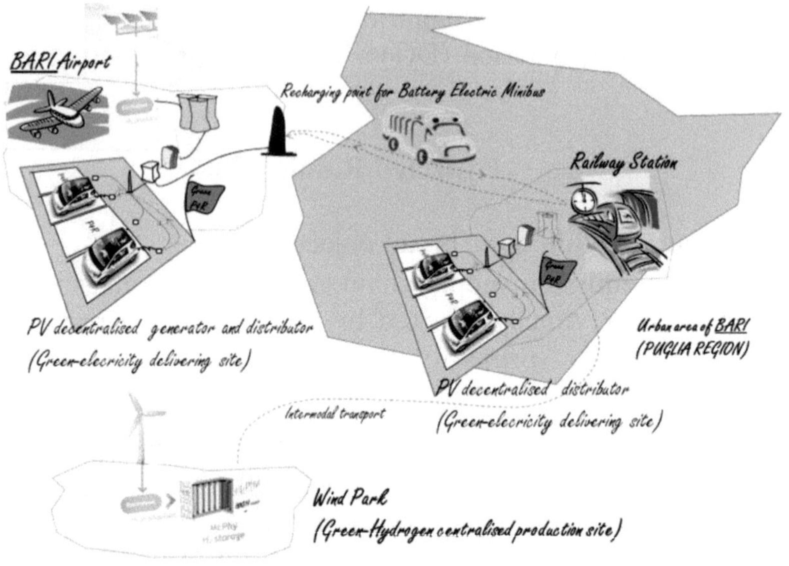

Figure 5 - A vision of the Green Urban Mobility System concept applied to an EU town as presented to Puglia regional authority. Green-electricity is produced and stored either in wind or solar farms nearby towns (main cope of EU-INGRID project) or employing distributing generation (DG) of green-electricity deploying large spaces for photovoltaic system (airports, railway station, shopping centers, etc). Thanks the magnesium hydrides solution, green-hydrogen - produced by electrolizers fed by green-electricity - can be safely stored and dispatched; green- hydrogen stocks dispatched to delivering point, the P4R, feed e-urban vehicles available for people at distributed access points in towns.

The Environmental Problem Targeted

An in-depth ex-post evaluation work undertaken by the Commission [7], has shown that, while several features of the transport system have improved in the last decade - notably its efficiency, safety and security - there has been no structural change in the way the system operates. The inability to modify the current transport paradigm, presently based on fossil fuels and on the dominance of road transport for both freight and passengers, is one of the main causes of unsustainable trends: growing Green House Gases (GHG) emissions, persistent oil dependency and mounting congestion. In other words, EU Commission stated that mobility of people and businesses today is not sustainable and needs to be re-considered structurally [7].
Today transport accounts for around 25% of EU CO_2 emissions and CO_2 emissions from transport have been growing over the last 20 years. Although most attention has focused on CO_2, there are several other (GHG), some of which on a per-kilogram emitted basis have a much greater warming potential than does CO_2. Currently, it is used to generally refer to CO_2 as the transport-related emissions from road transport [7]. Secondly, transport emissions are assumed to include not only direct emissions from the combustion of fuel by vehicles, but they also should

take into account further emissions associated with the production and distribution of transport fuels – i.e. the "well-to-wheel" emissions (ratio between the final energy transmitted to the wheels divided by the primary energy at the source) - by a complete and more significant Life Cycle Assessment (LCA).

The renewed sustainable development strategy of the European Union adopted by the European Council in June 2006 defines a sustainable transport system as the one that "meets society's economic, social and environmental needs whilst minimizing its undesirable impacts on the economy, society and the environment"[3]. As EU Commission stated, transport is a complex system that is based on the interaction of many components all of which need to evolve together: vehicles, infrastructure, behavior etc. Sustainable transport implies for EU to find a proper balance between (current and future) environmental, social and economic sustainability goals. For example, among very active EU State Members, Federal German Government in 2011 declared intention to put into force a national plan on electro-mobility. One million full-electric cars (or battery electric vehicles, BEVs) for example, would take up approximately 0.3% of total power demand. This order of magnitude will certainly not be exceeded in the medium term (till 2020), but a considerably large fleet of electrical cars is envisaged in the long run. Replacing a third of today's total car traffic with electric operation (not significantly before 2020) will require about 5% of present gross electricity demand.

Based on these key-issues, the GUMS concept seeks to demonstrate a "zero-emission" green supply-chain for feeding BEVs as a new approach to implement a public car-sharing service. The proposed P4R station pilot will work fully off-grid, producing "green-fuel" for charging BEVs, as described in Fig.1 by using the method shown in fig.2 which employs "green-fuel" stored in solid form hydrogen as it can be produced: a) locally by solar energy captured by PV panels covering large floor surfaces (public and private common spaces) areas, in airports, railway station, shopping centers, etc. (fig.5); b) available in access point production and storage sites according to the scheme in fig.1. In each case, the *P4R* conceptual station is completed by the Intelligent Dispenser (ID) that manages the charging phase while having the capability to pick-up relevant data in real-time from plugged-in vehicles such as the GPS position and the travel distance allowed by current battery charge levels. The ID processes the data and transfers it in an user-friendly form to mobile devices. Users are then able to access the GUMS services, via an applet on their own mobile devices, to check location and mileage allowed by citycars connected in all the P4Rs available in their own area.

The main benefit of the widespread application of the GUMS proposal in the EU the realization of 100% green urban mobility. If electro-mobility is preferable to fossil fuel mobility in terms of CO2 emission, would it not be true that an electric vehicle can be considered a zero-emission vehicle. In referring to 'well-to-wheel' CO2 emissions, all emissions generated at a power generation site should be taken considered since the e-city car is recharged via the grid, and electricity sourced grid to vehicle generates indirect emissions (emission not directly produced by travelling) that strictly depend on the electricity mix of the EU countries.

To reliably assess CO_2 emission potentially saved by GUMS pathway, it is necessary to create a comparison scenario, this the "as-is" scenario for current urban mobility system made of diesel or petrol cars.

[3]Microclimates in urban centers and urban sprawl today are heavily impaired by exhaust, fine-dust and noise emissions from traffic. The need to reduce such emission and keen competition among European municipalities and regions as sustainable places where to live and work are leading to a wider acceptance for zero-emission mobility in urban areas.

A) "Well-to-wheel" of combustion engine vehicle

To assess the "well-to-wheel" emissions it is necessary to assess emissions produced over the "well-to-tank" pathway (extraction of crude oil, refining process and final delivery to refueling station) thus adding the "tank-to-wheel" emissions produced by vehicle during transport. According to a 2006 study research of the European Commission Joint Research Center Institute for Energy (EJRC-IE) it has been possible to calculate the total efficiency of the petrol consumption for a vehicle as it stated that around 22% of the energy contained in diesel (18% of that contained in petrol) is actually transmitted to the wheels, in ideal conditions (meanwhile the rest is lost as heat). Since gasoline's energy content accounts for about 34.3MJ/l, thus 42MJ/l (=34.3MJ/l/81.7%) are energy expressed in MJ of crude oil needed to produce 1 liter of gasoline at the gas pump.

Referring to the most efficient gasoline cars, rated at around 23 km per liter of gasoline, implies efficiency of around 23km/l/42MJ/l=0.54km/MJ. Finally, considering that CO2 content of crude oil is 73.0 gCO_2 per MJ, verisimilarly a modern efficient petrol powered car could emit around $135gCO_2/km$.

B) "Well-to-wheel" of BEV vehicle connected to the grid

According to the EU JRC-IE, the "tank-to-wheel" efficiency of a lead acid BEV averages 60% (72% if BEV is equipped with lithium batteries). As a BEV averagely consumes about 0.11 kWh of electricity from the battery to drive 1 km, an average of 0.18 kWh/km (=0.11kWh/km/60%) of electrical energy is required tank-to-wheel equipped with lead acid batteries, considering the normal urban driving condition. Considering the production and distribution, 1kWh of electricity generates, with the 2006 average EU energy mix, around 0.443kg of CO_2, consequently a BEV fuelled by EU grid electricity emits averagely around 79g CO_2/km (=0.443 $kgCO_2/kWh \cdot 0.18kWh/km$). A BEV fuelled by EU grid electricity emits averagely around 79 gCO_2/km.

Concerning overall efficiency of the electro-mobility, EU-JRC IE calculated that around 40% of the primary energy arriving at the power plant is converted into electricity, while the energy efficiency of electricity distribution is around 92.5%. The "well-to-tank" energy efficiency is therefore estimated at around 37% (=40%·92.5%), and the total average "well-to-tank" efficiency for a plugged-in to grid is therefore accounted around 22% (=60%·37%) for BEV lead acid batteries, 26% (=72%·37%) for BEV equipped with lithium batteries.

Therefore, comparing the pathway A) and B) in terms of total emissions and efficient consumption of power source, we can note that:

- gasoline fuelled vehicles are 18-22% energy efficient with average emissions of $130gCO_2$ /km;
- BEV are 22-26% energy efficient with average emissions of 0.443 $kgCO_2/km$ when fueled by the EU electricity mix.

C) "Well-to-wheel" of BEV vehicle sourced by "green-fuel" stored via solid state hydrogen

The "well-to-tank" conversion path shown in fig.2, consisting of electricity provided hypothetically by solar power, converted into hydrogen by electrolizers, stored in solid form as magnesium hydrides and then reconverted on-demand by a fuel cell is 27% efficient (= 70%·95%·42%, where 70% is the average efficiency

of electrolizers, 95% the efficiency of hydrogen solid storage and 42% efficiency of fuel cells). To such value we have to add the 75% "tank-to-wheel" efficiency for a BEV (see above).

A BEV vehicle travelling through GUMS concept has an average 20,2% (=27%·75%) overall "well-to-wheel" efficiency, close to a low efficiency of gasoline vehicle, but with zero emissions.

Therefore, consider a medium-size EU city that deploys GUMS concept substituting 100 gas fuelled vehicles that travel averagely 50 km per day, each car can reduce 0.675 tonCO2/day (=100 vehicle·50 km/day·0.135 kgCO_2/km).

Financial assessment of the concept: business model and a market solution for a public use

Today grid is not able to interact with the next scenarios that consider large fleets of BEVs to plug-in. The main problem is the difficulty of balancing peak loads caused by thousands of BEVs connecting that foreseen or managed by SMART grids[4].

This fact assumes that in a transient regime, (i.e. few BEVs plugged-in to urban grid) the demand for electricity by local electric citycars can be managed through the SMART grid "pushing" users to recharge at night or by managing and manipulating peak request loads that are possible at any point of the local grid by "transferring" electricity from BEVs which are "predicted" (or "known") to extend their own connections from other points where large number of BEVs would also connected at the same time, (i.e. the early morning when many e-citycars move from home and re-connect to other recharging points in business centers and downtown).

Therefore, preparing the transmission infrastructure for the distribution of electricity based entirely on SMART grids, it becomes crucial that the grids have the capability to manage thousands of BEVs in order to prevent bottlenecks (as we have today) if sustainable electro-mobility is to grow.

Actually this problem is quite similar to the *chicken and egg problem*[5] suffered by the hydrogen economy that has resulted in a long-standing debate in the United States: hydrogen deployment could create attractive markets, but such markets can't be easily developed because of the large infrastructure investment needed to allow this novel technology solution of hydrogen deployment to effectively compete in the marketplace.

Also problematic is the possibility that BEV owners willingly accept that the car battery is "used" to re-balance the network energy since they would expect that the level of the charge would be increased according to the amount of time the car has been plugged-in. This is not so much an economic or a technical issue, obviously, but only a habit that could create unexpected conditions that could be undervalued[6] today.

An insight of a future sustainable urban mobility in the EU has to consider that BEVs shall be fed by totally green electricity produced from renewable energy sources (RES), as clarified in the previous section.

Starting from these initial assumptions, we can predict large opportunities for the exploitation of electricity that can be produced locally (or stored then delivered locally) and dispatched as green-fuel to P4R sites. As shown, the *P4Rs* are different from conventional pubic car parks that are

[4] Most probably the problem of monitoring and satisfying large BEV fleet in urban center could be solved by next generation of SMART grids that could be fully integrated with adequate and efficient back-up systems that allow to get demand and supply of power properly balanced. This is main scope of the large collaborative INGRID project.

[5] When they refer to a "chicken and egg" problem, philosophers and scientists intend to point out a circular problem, unsolvable since it is hard to identifying how solve the problem because cause and effect are not distinguishable. Solving a circular problem needs to get out from circular consequences, for example introducing a new element that has not already been considered in the circular consequence

[6] Authors disagree with a certain school of thought that claim that management of variability of power generated by RESs (especially those ones that characterized by an extreme variability, as wind source) can be solved introducing an another high variable "system", as it must be considered plugging-in of masses of BEVs which decide to connect to the network when they want or need.

based on a business model of "*parking for staying*". The battery charging system adopted is the pattern established in recent years consisting of a charger pillar. The largest advantage of P4R concept consists in the ability to produce 'green fuel" in a completely off-grid mode.

One of the biggest problem of storing and delivering electricity through the hydrogen double conversion is the low energy efficiency of the loop, namely the low "round trip efficiency" (RTE) of the system water electrolyzer-hydrogen storage-fuel cell. More specifically, the RTE of the loop is greatly affected by the low efficiency of (today) fuel cells as hardly achieves values higher than 30-40%. As stated in the previous section, the overall "well-to-tank" efficiency of the C) scenario, namely BEV fueled by green-electricity provided by P4R concept, has about the same overall efficiency of today efficient gasoline cars[7], with the substantial differences being that no emissions are produced.

In the following we discuss two scenarios. The calculations were made on the technical features of current commercial systems like water electrolizers (WE), hydrogen solid storage systems (HSS), fuel cells (FC), small electric city-cars, photovoltaic panels, wind turbines, etc. available today. But it is worth stating that especially for the loop WE→HSS→FC, the calculations took into consideration separate system unit data which is not fully and effectively integrated. For that reason, the preliminary results may be overly conservative in some aspects.

A more refined financial assessment of the proposed business model will be done. In fact the P4R pilot is expected to "generate" a useful database to be used to refine preliminary assessment based, as explained above, on literature and commercial data of separate system units.

Stage I - Defining the reference "systems" and boundaries

The system under discussion is made up of 4 main subunits that complete the entire supply-chain (i.e. well-to-tank) from the production of the "well", the green-electricity, to the final upload of the BEVs' "tank": a) a PV system of proper size; b) water electrolysis (WE) units to convert green-electricity produced by PV system into green-hydrogen; c) a back-up energy unit to indefinitely store – without losses – a high efficiency hydrogen via-solid state in form of magnesium hydrides (HSS); d) a fuel cell unit (FC) to dispatch electricity from hydrogen reconversion to electric cars as needed.

Stage II – Sizing the reference "system"

Early dimensioning has been made for all the subsystems. Size was chosen to target the recharging of a small BEV fleet of 166 vehicles equipped with 15 kWh battery capacity that can travel approximately 70 km per day. Considering 0.11 kWh/km efficiency of a BEV, every BEV of the fleet consumes about 7.7 kWh per day in an normal urban setting. An approach to conservatively size subsystems being careful to:

a) employ reliable data, especially the unit costs of sub-systems;
b) not underestimate costs of equipment (main capex), operating and maintenance plant costs (O&M),
c) include capex and opex for a small fleet of BEV.

The basic calculation was made using the daily maximum power consumed by a BEV of the sample fleet to travel approximately 70 km per day, allowing for recharging. Considering basic relationships of the in and out of material flow over the supply-chain subunits (refer to Table I),

---

[7] Some detractors claims that calculation of "well-to-tank" efficiency of electric cars sourced by RES electricity, shall consider also low efficiency of solar panels that convert solar energy into electricity. We consider this a misjudged issue, since it should be the same – to make a correct comparison –for the petroleum natural resources, thus calculating how much if efficiency of the original transformation of ancient fossilized organic materials into fossil fuel that originated energy storing into crude oil.

the average hydrogen produced per day by the WE subsystem was calculated and therefore the power (electricity) needed by WE to satisfy a production was about 150,5 kgH_2 per day.

Table I – Summary of inlet-outlet for each subsystem of supply-chain from WELL (left side of table) to TANK (right side).

PV	WE		Storage system		FC (electricity re-conversion)		Supply-Chain WE-ST-FC	BEV user
Efficiency	Efficiency		Efficiency		Efficiency		Efficiency	Fuel efficiency (kWh / km)
12%	62%		98%		50%		28%	11%
FROM WELL: power produced (kWh)	inlet (kWh)	outlet (kg H2)	inlet/outlet (kg H2)	waste heating power (kWh / kg H2)	inlet (kg H2)	outlet (kWh)	outlet (kWh)	**TO TANK:**inlet (kWh - per kg H2 supplied by WE-ST-FC)
55,65	55,65	1,04	1,04	1,68	1,04	15,66	15,66	15,66

This brought the kWh needed per day at the dispatching points (the TANK side) to feed the entire 15 kWh battery capacity for the fleet, to 2'265 kWh. The demand of the kWh necessary for the operation of WE requires a PV subsystem (average data from commercial literature) as set forth below in Table II.

Table II - PV subsystem sizing and cost necessary to fulfill the kWh required for the normal operation of the unit WE.

Total required energy per year	3.392.640,00	kWh
Yearly functioning	3100	hours
Functioning factor per year	0,95	
Total irradiation	1900	kWh / anno/ mq
Energy consumption	3.392.640,00	kWh
Plant efficiency	0,127	
Required surface	14.059,84	Mq
Power per module	120	Watt
Power capacity per module	80	Watt/mq
Module surface	1,5	Mq
Number of modules	9.374,00	
Total surface	14.061,00	Mq
Peak power of plant	1.124,88	kW
Total energy produced by plant	3.382.514,16	kWh
Unit cost	5.000,00	Euro/kWp
Total cost (Capex)	5.624.400,00	Euro

Dimension of HSS subsystem was conceived to satisfy a total storage capacity of - at least - 300kg of hydrogen.

The preliminary analysis on a "well-to-tank" system was completed considering dispatching points equipped with distributed fuel cells at P4R access points that have to operate in stationary mode for - at least - 12 hours per day (an average FC subsystem of 208 kW working averagely 12 hours per day satisfy 2'495 kWh per day).

Based on this assumptions, it was possible to calculate costs of all the listed subsystems, both capital cost (Capex) and operating costs predictable (O & M). To align whole system to an

average duration of photovoltaic system when it is well-maintained, the lifetime (i.e. period time for assessing investments, revenues and costs) was set at 20 years but it is worth to notice that such a prolonged "lifetime" take into account periodic heavy maintenance / substitution / revenue of all system units. In other words, the duration of the current systems WE, HSS storage and FC is certainly not extendable to twenty years but especially for these three subsystems it has been assumed that, after a certain number of years subsystems must be renewed and / or regenerated. Whereas, a new capital costs resulted from these interventions.
Finally a tentative price of kWh dispatched has been defined conservatively targeting break-even point at the end of twenty years, but considering in the O&M costs also profits of not still specifiable investors.

Define and analyze the sustainability of the system based on an intelligent public car-sharing business model.
Such a scenario considers that one or more investors cover purchasing, maintenance and periodic renewal of the fleet of BEV that are get available in urban centers accordingly with scheme of Fig.1 and Fig.2 and provided to push a *green "intelligent" car-sharing* public use upon payment of a monthly fee that grants access sure to BEV, with unlimited mileage and 24 hour day. By "intelligent" we mean that users have real-time access to relevant information for every vehicle plugged-in to *P4R* point, e.g. battery charge status (travel distance allowed) and moreover location of that vehicle. Such information are provided to users on a virtual map on electronic device (e.g. smart phone, tablet, website access mode) to instantaneously locate vehicle to pick-up.
In our simulation it was stated the number of users of "e-car sharing" service considering the availability of 15 P4R sites: an average turnover of 7 users for the 166 vehicles has been considered.

Table III – Summary data results from financial assessment simulation of "public intelligent e-car sharing" scenario.

NPV	€ 462.179,46
IRR	0,51%
Break-Even time (years)	20
Break even price (EUR/kWh supplied)	€ 1,00
Break even price (EUR/ kg H2 delivered)	€ 0,06

Table IV – Calculated service costs for citizen for a profitable P4R service.

Card holders (number)	1164
Card holders per car	7
Yearly card strike price	€ 786,86
Strike Revenue 2012	€ 916.284
Mark-up	110,00%
Yearly "card" price	€ 1.652,40
Monthly "card" price	€ 137,70

Conclusion

The EU Commission ROADMAP 2050 – Practical guide to a prosperous, low-carbon Europe (April 2010) states that the energy transition towards a decarbonized economy has benefits, that reach beyond climate change mitigation. Over the past decade a combination of high growth in demand for energy, slowing growth in oil supply and growing concern about climate change have been driving the case for renewable energy and energy diversification. This ROADMAP has developed a reference projection of the energy transition to test different scenarios for technology and climate policies in the next half-century. Among those ones, lower energy costs per unit of output, more stable and predictable energy prices, an increased security of energy supply for much more economic stability, a more stable sustainable energy. In such context, the GUMS concept can contribute in a concrete and fundamental portion in sustaining the European vision for this pathway.

The basic concept of a *P4R* plant has been developed in order to building the pre-requisite conditions for the setting up of the EU "green energy economy from RES" in the near term period. Managing and deploying energy from RES for BEVs recharging help securing EU Countries from energy dependency on fossil fuel.

Acknowledgements

This work makes use of preliminary study that brought large Consortium of INGRID project, co-funded by the European Commission within the FP7 Framework Programme. Authors thank European Commission for supporting any project dissemination activities. This work reflects only the authors' views. The Community is not liable for any use that may be made of the information contained therein.

References

1. IPCC Fourth Assessment Report: Climate Change 2007 (AR4)
2. M. Hakamada, T. Furuta, Y. Chino, Y. Chen, H. Kusuda, M. Mabuchi, Life cycle inventory study on magnesium alloy substitution in vehicles, Energy, Vol. 32 (2007) pp. 1352-1360.
3. J. Du, W. Han, Y. Peng, Life cycle greenhouse gases, energy and cost assessment of automobiles using magnesium from Chinese Pidgeon process. Journal of Cleaner Production, Vol. 18 (2010) pp. 112-119.
4. S. Ramakrishnan, P. Koltun, Global warming impact of the magnesium produced in China suing the Pidgeon process. Resources, Conservation and Recycling, Vol. 42 (2004) pp. 49-64.
5. PE International, 2012. Gabi 5 LCA software. Available at: http://www.gabisoftware.com/software/gabi-software/gabi-5/.
6. J. Zuliani, D. Reeson, "Making Magnesium a More Cost and Environmentally Competitive Option", 9th International Conference on Magnesium Alloys and their Applications, July 8-12 (2012), Vancouver, Canada
7. White Paper Roadmap to a Single European Transport Area, "Towards a competitive and resource efficient transport system", IMPACT ASSESSMENT report, EU Commission, 2011 (source: http://ec.europa.eu/transport/strategies/2011_white_paper_en.htm ; latest check on website 30 October 2012).

REWAS 2013: Enabling Materials Resource Sustainabil
*Edited by: Anne Kvithyld, Christina Meskers, Randolph Kirchain, Gregory Krumdick, Brajendra Mish
Markus Reuter, Cong Wang, Mark Schlesinger, Gabrielle Gaustad, Diana Lados, and Jeffrey Spangenberg*
TMS (The Minerals, Metals & Materials Society), 20

RECYCLING-ORIENTED PRODUCT CHARACTERIZATION FOR ELECTRIC AND ELECTRONIC EQUIPMENT AS A TOOL TO ENABLE RECYCLING OF CRITICAL METALS

Prof. Dr.-Ing. Vera Susanne Rotter, Dr.-Ing. Perrine Chancerel,
Dipl.-Ing. Maximilian Ueberschaar,
Technische Universität Berlin, Berlin, Germany

Key Words (3): Recycling, WEEE, critical metals

Abstract

To establish a knowledge base for new recycling processes of critical elements, recycling-orientated product characterization for Electric and Electronic Equipment (EEE) can be used as a tool. This paper focuses on necessary data and procedures for a successful characterization and provides information about existing scientific work. The usage of this tool is illustrated for two application: Hard Disk Drives (HDD) and Liquid Crystal Display (LCD) panels. In the first case it could be shown that Neodymium and other Rare Earth Elements are concentrated in magnets (25% by weight) and contribute largely to the end demand of Neodymium. Nevertheless, recycling is limited by the difficult liberation and competing other target metals contained in HDD. In the second case it could be shown that also for this application the usage of Indium is concentrated in LCDs, but unlike in magnets the concentration is lower (200 ppm). The design of LCDs with two glued glass layers and the Indium-Tin-Oxide layer in between make the Indium inaccessible for hydro-metallurgical recovery, the glass content puts energetic limitations on pyro-metallurgical processes. For the future technical development of recycling infrastructure we need an in depth understanding of product design and recycling relevant parameters for product characterization focusing on new target metals. This product-centered approach allows also re-think traditional "design for recycling" approaches.

Introduction & Background

Current and future innovations in the information, telecommunication and entertainment sector base increasingly on the use of technology metals like e.g. Gallium, Indium, Tantalum and Rare Earth Elements (REE), which provide specific chemical and physical properties. The rising contents of these high-functional and strategic metals in WEEE build up an important source for secondary raw materials. Despite the fact that the primary resources for these metals are finite, no relevant recycling is applied yet.

Based on estimations of future supply situation and the impact on the economy, the EU Commission declared a number of metals as critical. Following this study, Antimony, Beryllium, Gallium, Germanium, Cobalt, Indium, Magnesium, Niobium, Tantalum, Tungsten, Platinum Group Metals (PGM) and Rare Earth Elements (REE) have to be considered as critical metals. [1] A raw material is "critical" when the risks of supply shortage and their impacts on the economy are higher compared to other raw materials. The indicator "supply risk" takes into account the political and economic stability of the producing countries, the level of concentration of production, the potential for substitution and as well the recycling rate. Therefore, restraints through the currently applied recycling technologies are one of the key arguments for classifying metals as critical. Recovery of critical metals from waste electronic equipment (WEEE) is very limited in current practice due to a lack of incentives under present socio-economic and technical boundary conditions. Any effort in *a priori*

planning of new infrastructure for recycling has to be based on a good understanding of historic and future usage of metals and detailed product characterization. For bulk materials like plastics, steel, aluminum, copper, and precious metals exist well-established methods and a profound knowledge base that enable detailed process modeling.
This paper aims at introducing 'recycling-oriented product characterization' for WEEE as a tool to enable recyclers on each step of the value chain to identify the composition and recycling potential for materials in general, and for critical metals in processed mass flows in particular. This information provides the basis for a multi-dimensional optimization of the recovery of these materials in treatment processes. As an example, two case studies will present results for Indium in LCD flat screens and REE in hard disk drives. In addition, questions on information storage and accessibility for third parties have to be addressed if 'recycling-oriented characterization' is used for holistic optimization of metal recovery along the value chain. Finally yet importantly "recycling-oriented product characterization" can link the recycling industry with equipment manufacturers by addressing 'design for recycling'.

Recycling-oriented product characterization

Recycling-oriented characterization is defined as a systematic approach to support the design and the operation of recycling processes [2]. It provides and requires a deep understanding of the needs for recycling processes, specifically focusing on the prevention of avoidable material mixing and dilution while also providing relevant information for the design of new EEE [2]. The characterization of WEEE is structured using following levels:

1. General data on the equipment: equipment type, producer, model name or number and year of production
2. Physical and mechanical properties: weight and size of the whole equipment, material fractions, assemblies, components, connection types
3. Chemical analysis of the assemblies and components: e.g. polymer characterization and element analysis

The data analysis (for example estimate of the content of metal in a specific equipment mix) is facilitated if the data are recorded and organized systematically in a database. Information on the uncertainties, measured for example through a standard deviation, should accompany all datasets.

The focus on critical metals affects for instance level 3, in which the content of critical metals will be measured. However, recyclers report that not only the metal concentration is relevant for the process design and operation, but also:

- Mechanical properties of the metal-containing assembly (for example: Tantalum-containing capacitor): What is the particle size? Which other materials are connected, and how are they connected? Are the other materials abrasive?
- Can assemblies or components that contain critical metals be identified visually?

The analysis of data available in the literature shows that the scientific community focuses either on specific metals, or on specific equipment types. Even though some investigations report the results of own analyses, most research uses secondary data and assumptions to calculate the content of critical metals.

Oguchi et al. [3] report the results of measurements of the concentration of tin, silver, gold, palladium, cobalt, gallium, tantalum and other metals in printed circuit boards of 21 different equipment types including televisions, computers, mobile phones and digital cameras. Other publications focusing on specific equipment types investigated the content of critical metals of mobile phones [4], mobile phones and computers [5], personal computers [6] and LED products [7]. Further research focuses were the content of indium in flat display [8], and the composition of electronic components [9]. Buchert et al. [10] combined literature data to estimate the content of cobalt, gallium, germanium, indium, platinum-group metals, rare earth elements and tantalum in flat displays, notebook computers, Smartphone and LED lighting products.

To conclude, even though some equipment types like mobile phones and computers were analyzed in several investigations, the available data do not provide with sufficient information for further research, for example to estimate the potential of WEEE as a source of raw materials. The data gaps concern all detail levels, from the detailed analysis of the metal content in relevant assemblies and components (like tantalum in capacitors, rare earth elements in magnets and gallium in ICs), to batch analysis of critical metals in complex equipment mixes received for treatment by the recycling industry.

Example: Indium & REE in WEEE

The characterization of end-of-life products reveals critical metals containing components in WEEE. With this knowledge, further treatment steps can be adapted such that certain assemblies can be separated upfront. To assess the recycling potential of critical elements in end-of-life products, in first instance devices with high theoretical contents and a high proportion of collected WEEE will be investigated.

74% of primary Indium is used for the production of flat panel Liquid Chrystal Displays (LCD) [11]. Returned LCD monitors are therefore a potential source for secondary Indium. Approximately 25% of the worldwide production of rare earth elements is Neodymium [12]. 13% of this share goes into the production of Neodymium-Iron-Boron (NdFeB) permanent magnets [1] and hence in the production of Hard Disk Drives (HDD). Current estimates place the rate of magnets in circulation worldwide applied in personal computers at 34% [13]. Both metals show examples for high recycling potential on the basis of the concentrated use in defined applications
With the basis of the two cases it will be illustrated how further the characterization procedures help assessing "whether" and "how" critical metals can be recovered.

<u>REE in NdFeB – magnets</u>
Due to the limited number of available of literature data an experimental investigation was carried out with discarded HDDs sampled at several German recycling plants undergoing following steps

- Assessment of product disassembly
- Material and component composition
- Chemical analysis of separated magnets

The disassembly of devices shows the complexity in product design and gives information about adequate liberation strategies. It is possible to calculate the effort for an economic recycling. Through the manual disassembly of devices and components also the constitutional heterogeneity expressed by the material composition and its variation can be described. Furthermore, the by analyzing single components in particular trace elements can be located and quantified. Figure 1 shows the average composition of HDD. HDDs consist of a variety of components and materials like Aluminum, Ferrous or Copper. Focusing on REEs, NdFeB magnets are of great interest, but also the installed printed circuit boards (PCBs) contain traces of Scandium, Europium, Lanthanum and Cerium. Nevertheless, printed circuit boards have in the same time high content of precious metals like Gold which is recovered in highly efficient integrated smelters where REE end up in the slag. [14]

Figure 1: Statistical evaluation of the composition of HDDs. Red [absolute weight per unit): Median; grey: upper (80) and lower (20) quantile; max, min as error bars

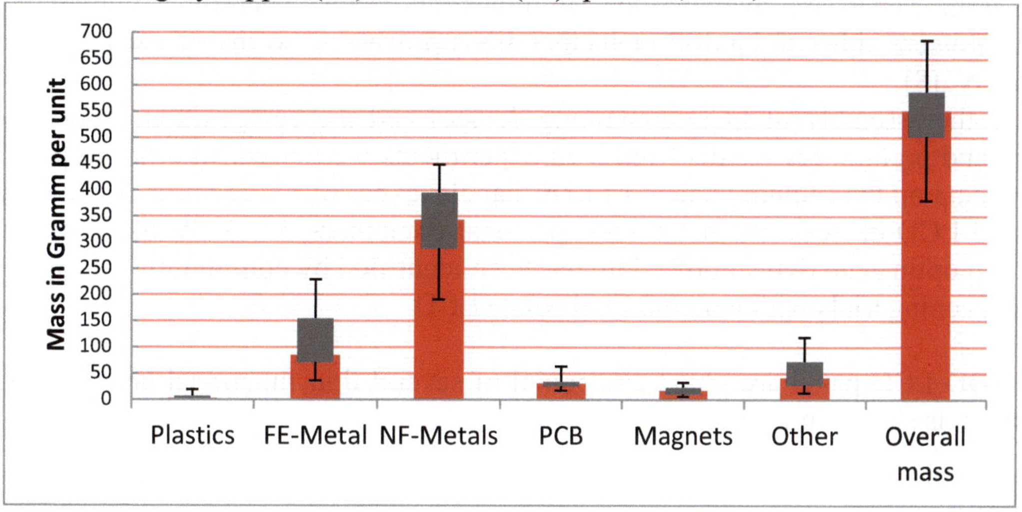

Although a few chemical analyses of NdFeB magnets have been published, specific, separated investigations of the two types of installed permanent magnets could not be found in literature. The contents of REEs and possible alloy differences were examined. The comparison with literature data is shown in Table 2.

Table 2: REE content in NdFeB magnets according to selected authors and own experiments

Unit %	Xu et al.; 2000	Du, Graedel; 2006	Zakotnik et al.; 2008	Lyman, Palmer; 1993	Own analysis* - hard disk drives (n=3) (preliminary results)	
Sample	Mix	Mix	Mix	Mix	Voice-Call Motor	Spindle Motor
Nd	18	20	13.78	26.3	23.3	25.4
Pr	1.82	5	n/a	1.85	3.9	0.3
Dy	5.3	5	0.66	n/a	1.6	0.06
Tb	n/a	1	n/a	n/a	0.01	0.02
Value type	exp.	theo.	exp.	exp.	exp.	

* The digestion methods aqua regia and nitric acid-water (difference in result 1-2%) showed the highest solubility relative to sulfuric acid. The REE content was determined by a first screening with X-ray fluorescence analysis (XRF) and the adapted chemical analyses an ICP-OES (Inductively Coupled Plasma - Optical Emission Spectrometry)

An assessment of recycling strategies of critical elements in NdFeB magnets based on this dataset is possible. In the first place the magnets can be recycled as alloys by reinterring. This requires a predictable composition not only in terms of the Neodymium content but also of the alloying elements. While many studies generalize the composition for NdFeB magnets the detailed own results show that there is a distinct difference of alloying element between two motor magnet types both used in HDD, the voice call motor and the spindle motor. Only manual disassembly would allow a distinction between these two magnet types after liberation which is necessary for alloy-specific recycling. Other processes like hydrometallurgical separation requires only the general liberation and separation of magnets from the other HDD material, in particular the PCBs due to the competing recovery of various elements (precious metals & copper vs. REE).

Indium in LCD

LCD panels in computer monitors consist of a high percentage of Indium and represent the currently leading technology in flat panel display production. Still, no recovery processes for Indium from LCDs are available on an industrial scale. Therefore, Indium containing fractions are sent either to landfill or incinerations processes, so that the scarce metal is lost for recovery. [8]

To assess the assembly of LCD containing devices and the composition, information from literature and preliminary results from own experiments were used. In order to provide more profound data on the Indium content of LDC panels experiments focused on the disassembly of selected LCD devices like notebooks, personal computers, LCD TV devices, and mobile phones. An optimal analytical procedure has been identified by comminuting, heating, burring and acid leaching trials with material of obsolete LCD.

Unlike HDD LCD panel are glued compound of several thin functional layers. The material composition is shown in figure 1.

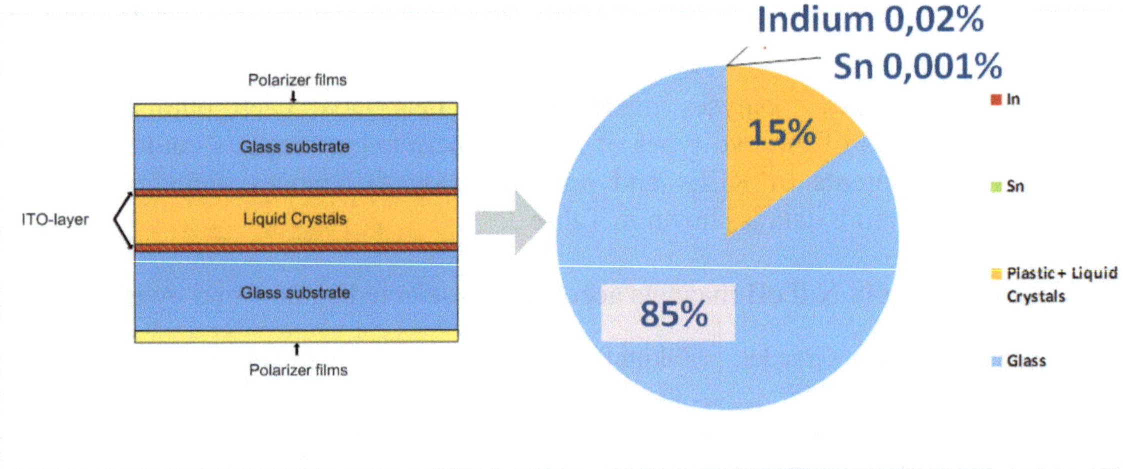

Figure 1: Composition of LCD panels [Source: Götze, Rotter 2012]

Results of own chemical experiments and the comparison with selected literature are shown in Table 3.

Table 3: Indium content in LCD panels according to selected authors and own experiments

	Angerer et al., 2009	Böni et al., 2012	Takahashi et al., 2009	IUTA & FEM, 2011	Own analysis**	
Sample	Mix	Mix	Mobile Phones	Computer, TVs, Laptops	Notebook/PC (n=23)	Mobile phone (n=11)
mg/cm2	4000	234	4408*	696*	722±300	710±570
mg/kg	1000*	58.5*	1102	174	175±60	320±160
Value type	theo.	theo.	?	exp.	exp.	

* own recalculation using a specific weight of the LCD panel 4 kg/m2

** LCD pieces of ca. 1 cm² were digested with 10 ml nitric acid solution (65% HNO_3) and heated up to a temperature of 180° C with 20-bar pressure. The Indium concentration was determined by using an atom absorption spectroscopy analysis device (AAS).

The recovery of Indium from LCD panels is challenging, due to the low concentration of the critical metal in the whole device. The design of this type of monitor makes the dismantling of the device and the disassembly of the flat panel itself very hard.
Currently used shredding technologies dissipate the Indium in non-recyclable concentrations. To avoid this, a complete disassembly of the LCD monitor is necessary. Further mechanical processing steps have to be done in a sealed atmosphere to prevent losses by dust discharge.

Concluding from these results, the need but also potential strategies for upgrading the Indium concentration though pre-processing can be identified (Figure 2). Starting from Indium distribution and Indium content in LCD devices, available recycling and Indium recovery processes were investigated from scientific literature and interviews with selected companies [8]. Each shown pathway of Indium recovery from waste LCD equipment requires verification in terms of economical and thermodynamic limits.

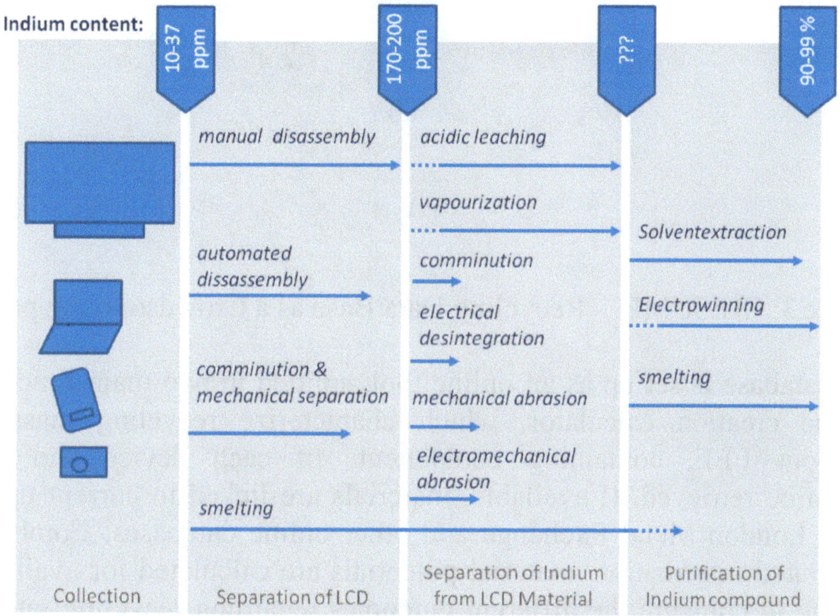

Figure 2: Possible recycling paths for Indium from waste LCD equipment [Source: Götze, Rotter 2012]

Recycling-oriented characterization and data management

Data about contents of critical metals in WEEE, recycling potential and possibilities of liberalization is important for target-orientated recycling approaches. Many research teams worldwide characterize EEE products in order to generate knowledge on this topic. Subsequently, in practice working recyclers are frequently confronted with non-filtered information out of scientific community and try to implement this partially into their running recycling processes. On the other side, recyclers themselves generate an extensive knowledge about their processed materials flows. Nevertheless, this information is sparely shared with others, especially competitors. This can result in reduced overall recycling rates for specific metals and missed business opportunities e.g. less profitable division of labor between different market actors. [15]
One approach, to face this problem is to build up a commonly generated database (Bottom-Up) to combine best practices, individual strategies and general experience as common knowledge basis of single market actors within the recycling market. The fundamental idea of

this database is Crowdsourcing[1], which means the availability for every company connected to the recycling industry and the possibility to read and insert data. Compared to out-sourcing where a product or service is purchased from a specific business partner, Crowdsourcing tasks are announced to a generally large group of people in the form of an open call [16]. In this case, companies from recycling industry shall constitute the group, contributing with more or less detailed information about recycling strategies, material composition or demand and supply of waste material. This tool can handle each recycling step starting at the waste collection and ending with the production of new products. Figure 3 shows the relevant stakeholders for a successful database.

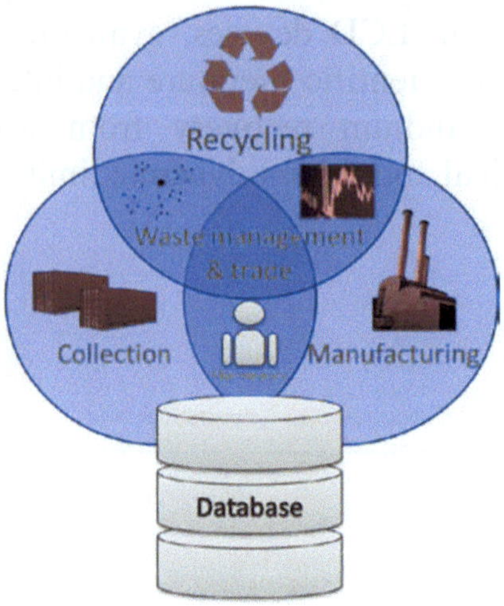

Figure 3: 'RESUME' - Recycling Data Base as a Crowdsourcing product

The recycling database is set up as an online tool, splitted in two main functional units. The first is a value creation calculator, which characterize recycling mass flows. Stored information about EEE, containing components of each device and the elementary composition can be retrieved. If available, materials are linked to current up-to-date market prices from the London Metal Exchange and other online databases. Combining the online data and contributed information, material potentials are calculated for available ways of re-cycling, allowing to compare the different outcomes regarding recycling rate and ecological attractiveness.

The other functional unit is a market place, where materials can be traded directly. Like on an online market place like eBay, each material shall be offered or requested. The platform is realized as PHP web page with backend SQL database.

Linkage to manufacturers – Design for Recycling

The product design essentially determines the materials used in the product, the material combinations and the type of connections (screws, glue etc.), and considerably influences the efficiency of the recycling process. A concrete example of design-for-recycling relates to the accessibility of printed circuit boards. The manual removal of printed circuit boards would partially prevent the losses of precious metals in side-streams [19]. How easy the manual removal is, depends on product design. Design-for-recycling models to link product design to

[1] **Crowdsourcing** is a process that involves outsourcing tasks to a distributed group of people. The difference between Crowdsourcing and ordinary outsourcing is that a task or problem is outsourced to an undefined public rather than a specific body. Crowdsourcing is related to, but not the same as, human-based computation, which refers to the ways in which humans and computers can work together to solve problems. These two methods can be used together to accomplish tasks. [17, 18]

the amount and composition of the output fractions of mechanical pre-processing and to the products of the subsequent material recovery processes were developed by Reuter and Van Schaik [20].

The example of Indium in LCD shows that an early separation of the LCD through disassembling conditions the recovery of the Indium. In this case, design for recycling should focus on making a quick removal of the LCD possible. In the case of the Rare Earth Elements in NdFeB–magnets, even though the separation into a separate fraction also conditions the recovery of the REE, it may be possible to separate the NdFeB-magnets later in the pre-processing facility, for example after size reduction. In this case, the connection of the magnets to other materials (for example casing) should be able to break easily to make the magnets easily sortable, also in automated processes. Two examples – two different recommendations of the designers!

Recycling-oriented product characterization plays a central role to link product design and design of recycling systems. As explained before, it gives the recyclers the possibility to get and analyze data on the characteristics of products to adapt their processes. As a feedback, they can communicate to manufacturers or associations of manufacturers which product characteristics affect positively or negatively the efficiency of recycling, and formulate recommendations for an improved product design. From a manufacturer's perspective, it provides the designers with concrete basic ideas on which product characteristics are keys for the recyclers. Therefore, recycling-oriented product characterization can be seen as a tool for an improved dialogue among actors of the product life cycle.

However, even though recycling-friendly designed, products have an improved recyclability; it does not mean that the materials are actually recycled. Recyclable materials can only be recycled if adequate infrastructure, consumer attitude and appropriate business models are implemented to collect and treat the materials [21].

Conclusions

The following conclusions can be drawn

- A recycling-focused approach towards product characterization enables *a priori* planning of new technology and infrastructure for the recovery of metals in general.
- The present knowledge base of recycling-relevant product characteristics has to extended to tackle challenges of recycling critical metals.
- Beyond the information of the recycling potential expressed as 'concentration' or 'grade' information on liberation behavior, content of alloying elements and impurities is relevant to design recycling systems, as shown for LCD panels and HDD.
- 'Design-for-recycling' recommendation are metal and product specific and cannot be generalized.
- We have to rethink traditional 'design-for-recycling' recommendation if it comes to new recycling challenges of critical metals

GEFÖRDERT VOM
Bundesministerium für Bildung und Forschung

The project on which this publication is based is funded by the German Federal Ministry of Education and Research (BMBF) (UPgrade - Project reference number 033R087A).

References

1. Ad-hoc Working Group, *Critical raw materials for the EU* (Brüssel, Belgien: Raw Materials Supply Group, European Commission - Enterprise and Industry, 2010)
2. P. Chancerel and S. Rotter, "Recycling-oriented characterization of small waste electrical and electronic equipment," *Waste management*, vol. 29 (2009), no. 8, 2336–52.
3. M. Oguchi, S. Murakami, H. Sakanakura, A. Kida, and T. Kameya, "A preliminary categorization of end-of-life electrical and electronic equipment as secondary metal resources," *Waste management*, vol. 31 (2011), no. 9–10, 2150–60.
4. J. Huisman, "QWERTY and Eco-Efficiency analysis on cellular phone treatment in Sweden," (2004).
5. C. Hagelüken and M. Buchert, "The mine above ground – opportunities & challenges to recover scarce and valuable metals from EOL electronic devices," in *(International Electronics Recycling Congress IERC,* 2008).
6. Microelectronics and Computer Technology Corporation (MCC), "Composition of a Desktop Personal Computer," (Electronics Industry Environmental Roadmap, Austin, Texas, 1996).
7. cycLED, Cycling resources embedded in systems containing Light Emitting Diodes, (research supported by FP7 Eco-innovation Program, webpage www.cyc-led.eu/Publications.html, 2012)
8. R. Götze and V. S. Rotter, "*Challenges for the Recovery of Critical Metals from Waste Electronic Equipment - A Case Study of Indium in LCD Panels*," Proceedings of the Electronics goes green 2012+, (Berlin: Fraunhofer Verlag, 2012).
9. H. Wichmann, R. Sprenger, F.T. Dettmer, C. Schmidt-Nädler and M. Bahadir, *Wert- und Schadstoffpotentiale von elektronischen Bauteilen*, (Düsseldorf: VDI Verlag GmbH, 2002).
10. M. Buchert, A. Manhart, D. Bleher, and D. Pingel, "Recycling kritischer Rohstoffe aus Elektronik-Altgeräten," (LANUV-Fachbericht 38, *Landesamt für Natur, Umwelt und Verbraucherschutz Nordrhein-Westfalen*, www.oeko.de/aktuelles/dok/1330.php, 2012).
11. European Commission, Enterprise and Industry, "Annex V to the Report of the Ad-hoc Working Group defining critical raw materials", Online: http://ec.europa.eu/enterprise/policies/raw-materials/files/docs/annex-v_en.pdf; (July 2010)
12. A. Van Schaik, "Kwantificering kritische (grond)stoffen in E-Waste producten", *MARAS – Material Recycling and Sustainability,* (Den Haag, Nederland, 2011)
13. X. Du, T. E. Graedel, "Global Rare Earth In-Use Stocks in NdFeB Permanent Magnets", *Journal of Industrial Ecology*, (1530-9290) 15 (2011), 836–843
14. P. Chancerel, "Substance flow analysis of Remondis Electrorecycling" (Fachgebiet Abfallwirtschaft, TU Berlin, 2008)
15. S. Heyer, P. Krämer, R. Götze, M. Ueberschaar, G. Walter, S. Flamme, S. Rotter, G. Seliger, "Bottom-up organized Recycling Networks as Strategies for Corporate Sustainability", *Proceedings of the Electronics goes green 2012+*, (Berlin: Fraunhofer Verlag, 2012).

16. P. Sloane, “A Guide to Open Innovation and Crowdsourcing: A Compendium of Best Practice, Advice and Case Studies from Leading Thinkers, Commentators and Practitioners”, *Kogan Page Ltd*, (2011)
17. J. Howe "The Rise of Crowdsourcing", *Wired*, (2006)
18. A. J. Quinn, B. B. Bederson"Human Computation: A Survey and Taxonomy of a Growing Field", *University of Maryland, College Park*, (2010)
19. P. Chancerel, C. E. M. Meskers, C. Hagelüken, and V. S. Rotter, “Assessment of Precious Metal Flows During Preprocessing of Waste Electrical and Electronic Equipment,” *Journal of Industrial Ecology*, vol. 13 (2009), no. 5, 791–810.
20. M. Reuter and A. van Schaik, “Opportunities and limits of recycling: A dynamic-model-based analysis,” *MRS Bulletin*, vol. 37 (2012), no. 04, 339–347.
21. P. Chancerel, “Substance flow analysis of the recycling of small waste electrical and electronic equipment - An assessment of the recovery of gold and palladium,” (Thesis, Technische Universität Berlin, 2010).

REWAS 2013: Enabling Materials Resource Sustainabi
*Edited by: Anne Kvithyld, Christina Meskers, Randolph Kirchain, Gregory Krumdick, Brajendra Mish
Markus Reuter, Cong Wang, Mark Schlesinger, Gabrielle Gaustad, Diana Lados, and Jeffrey Spangenber*
TMS (The Minerals, Metals & Materials Society), 20

CRITICAL ANALYSIS OF EXISTING RECYCLABILITY ASSESSMENT METHODS FOR NEW PRODUCTS IN ORDER TO DEFINE A REFERENCE METHOD

E. Maris[1] D. Froelich[1]

[1] Laboratoire Conception Production Innovante; Institut Arts et Métiers ParisTech de Chambéry ; 4, Rue lac Majeur Savoie Technolac, Le Bourget du Lac, 73375 Cedex, France

Keywords: Recyclability, standard, ecodesign.

Abstract

The designers of products subject to the European regulations on waste have an obligation to improve the recyclability of their products from the very first design stages. The statutory texts refer to ISO standard 22 628, which proposes a method to calculate vehicle recyclability. There are several scientific studies that propose other calculation methods as well. Yet the feedback from the CREER club, a group of manufacturers and suppliers expert in ecodesign and recycling, is that the product recyclability calculation method proposed in this standard is not satisfactory, since only a mass indicator is used, the calculation scope is not clearly defined, and common data on the recycling industry does not exist to allow comparable calculations to be made for different products. For these reasons, it is difficult for manufacturers to have access to a method and common data for calculation purposes.

A critical analysis of the standard and the various calculation methods identified in scientific journals was performed. An initial discussion brought to light several possible scopes of calculation. Additional indicators, such as quality loss or economic value loss, would be complementary to the mass indicator. Case studies were performed to compare these different methods. A new method and its scope of calculation are proposed in order to develop a reference method.

Glossary

ELV: End-of-life vehicle

WEEE: Waste Electrical and Electronic Equipment

PRM: Primary raw material

SM: Secondary material

SRM: Secondary raw material

Introduction

The European directives on end-of-life product treatment are now in application. The assessment of recycling networks in terms of the quantities collected and the efficiency of the treatment and recycling is one of the objectives of the statutory texts. Recycling efficiency will be improving in the near future thanks to the use of increasingly successful sorting and transformation technologies. But this is not sufficient. Companies must have an ecodesign strategy for their new products that integrates the constraints of recycling in order to improve the efficiency of end-of-life product treatment.

This strategy must also integrate the use of recycled materials, provided that there are important and long-lasting sources as well as good quality materials; this is the case today for certain materials, but not all.
The European directives contain a recyclability calculation reference based on standard 22 628 [1] for automobiles, while proposals of standards for WEEE and earth-movers are still under discussion. However, a report drafted by industrial members of the CREER club indicated that the product recyclability calculation method used in this standard is not satisfactory, for the following reasons:

- The scope of calculation is not defined,
- The data on treatment networks does not exist or is not capitalized in a common data base,
- The mass recyclability indicator does not suffice to estimate the quality of the materials produced.

For these reasons, it is difficult for industrialists to have access to a method and common data for calculation purposes.

The objective of this study is to examine the state-of-the-art recyclability calculation methods for new products, with the aim to define a reference method for designers.
The study is exploratory and includes 4 sections:

- Critical Analysis of the definitions of product recyclability and principle proposal for the calculation of recyclability
- Proposal of the indicators of mass, properties and economic value
- A case study

Critical Analysis of the definitions of product recyclability and principle proposal for the calculation of recyclability

Critical analysis of definition of definitions in the European directives and standards

The 22 628 standard on recyclability calculation and vehicle recovery defines recyclability as the capacity of components, materials or both to be removed from the end-of-life flow to be recycled. The recyclability rate is the percentage of the mass of a new vehicle that can be recycled, potentially re-used or both. The recyclability rate thus includes recycled material and re-used components. Recycling is defined as the recovery of waste from products, materials or substances for the same purposes as their initial function or for other purposes, excluding energy recovery.

Recyclability rate	
Re-use of component parts	Recycling of materials

The standard does not indicate whether the calculation should be made in relation to a regulated reference network. A product should be considered as recycled when there is a representative recycling network that is economically viable at the national or European level. Regulated networks exist in Europe and are financed by consumers through eco-taxes on products that are redistributed in eco-friendly waste management networks. These networks accept several types of products. ELVs are accepted and represent up to 40% of the incoming products for shredding facilities. Certain products such as furniture and high voltage WEEE are not subject to regulations. It is therefore difficult to estimate their recyclability in a regulated network.

There are at least two differences between the notions of recyclability and recycling.
The first difference is that the recycling rate estimates a network treating a set of products while the recyclability rate estimates the ecodesign of a product when the latter is treated in a network.
The second difference is the product lifespan; for example, small electric equipment products will arrive at their end of life after 5 years on average, versus 15 years for vehicles. The recyclability of a product is the efficiency of a reference network with regard to the design of a product that will be treated at its end of life, with the end of life varying according to product.

The notion of recyclability is used in Directive 2005 / 64 / CE concerning vehicle re-use, recycling and recovery possibilities. It is not referred to in all the directives and is not harmonized in the various texts.
The definitions of the recycling rate also differ according to directive. For example, Directive 94 / 62 / CE concerning packaging and packaging waste defines the recycling rate as the total quantity of waste from recycled packaging divided by the total quantity of packaging waste produced. It does not indicate in which recycling step the material is considered to be recycled, incoming or outgoing?

A product designer has several scopes (see Figure 1) to take into account in order to calculate the recyclability of his product, and several questions to ask:

- How can one integrate a product lifespan if one does not know the efficiency of the network at the product's end of life?
- Is it necessary to integrate the re-used components?
- What scope of calculation should be used? Is it necessary to stop at the SM or the SRM step, i.e. components or components restored to their original level?
- What type of outlet exists for the recycled material?

The design of a product has effects on the mass efficiency and the purity rate of its separate fractions as well as on the outlets for the materials. It is necessary to define more reference treatment scenarios for every product, i.e. typical networks, and more successful indicators to help designers to integrate the end-of-life recycling constraints of their products.

Conclusions:
It is thus necessary to define:

- Recyclability indicators estimating the preservation of the mass and quality of the secondary raw material with regard to the primary raw material
- A reference network for a product
- A scope of calculation

We can say that a product's recyclability is the capacity of a product and a reference network to restore the materials, the technical properties and the economic value close to those of its origin when a product arrives at its end of life.

Principle proposal for the calculation of product recyclabilty

Proposal of indicators

The relevant criteria for network efficiency are thus:

- The preservation of the mass of the primary raw material (PRM) with regard to the mass of SRM
- The preservation of quality, which is translated as the preservation of properties and the preservation of the economic value of the SRM with regard to the PRM.

Proposal of a reference network

Reference networks are regulated networks in which products are collected. For example, household electrical appliances such as vacuum cleaners, coffeemakers etc. are collected and treated all together.

Regulated networks exist for:

- ELV
- WEEE
- Packaging
- Boats,
- Airplanes,
- Construction and demolition waste
- etc.

There are two possible scopes and three ways to calculate the mass recyclability rate.

Proposal of a calculation scope

The first scope complies with the vision of the 22 628 standard, and there are two methods for its calculation.

First method:

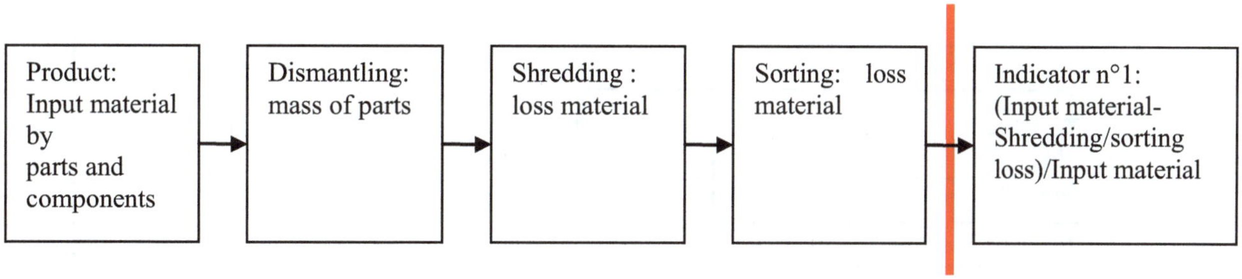

Comments:

Indicator n°1, which expresses the quotient SM/input material, differs from the recycling rate measured in the field. By not taking into account the rates of impurities in the SM, in certain cases mass indicators are superior to 1 because both the impurities and the main material are counted together.

Indicator 1 does not reflect the quality losses due to the impurities of the SM and the mass losses during the recycling by melting of the SM to produce MPS (ingots). These losses depend on product design. For example, if the designer chooses to make a massive part from aluminum or aluminum sheets. The association of materials resulting from design choices that generates these losses is not taken into account.

In the calculation of the mass of dismantled parts, market demand and the waste generated during part renovation are not taken into account.

It is easy to calculate this indicator from data supplied by professionals, yet it is not a very precise support tool for designers with product recycling objectives.

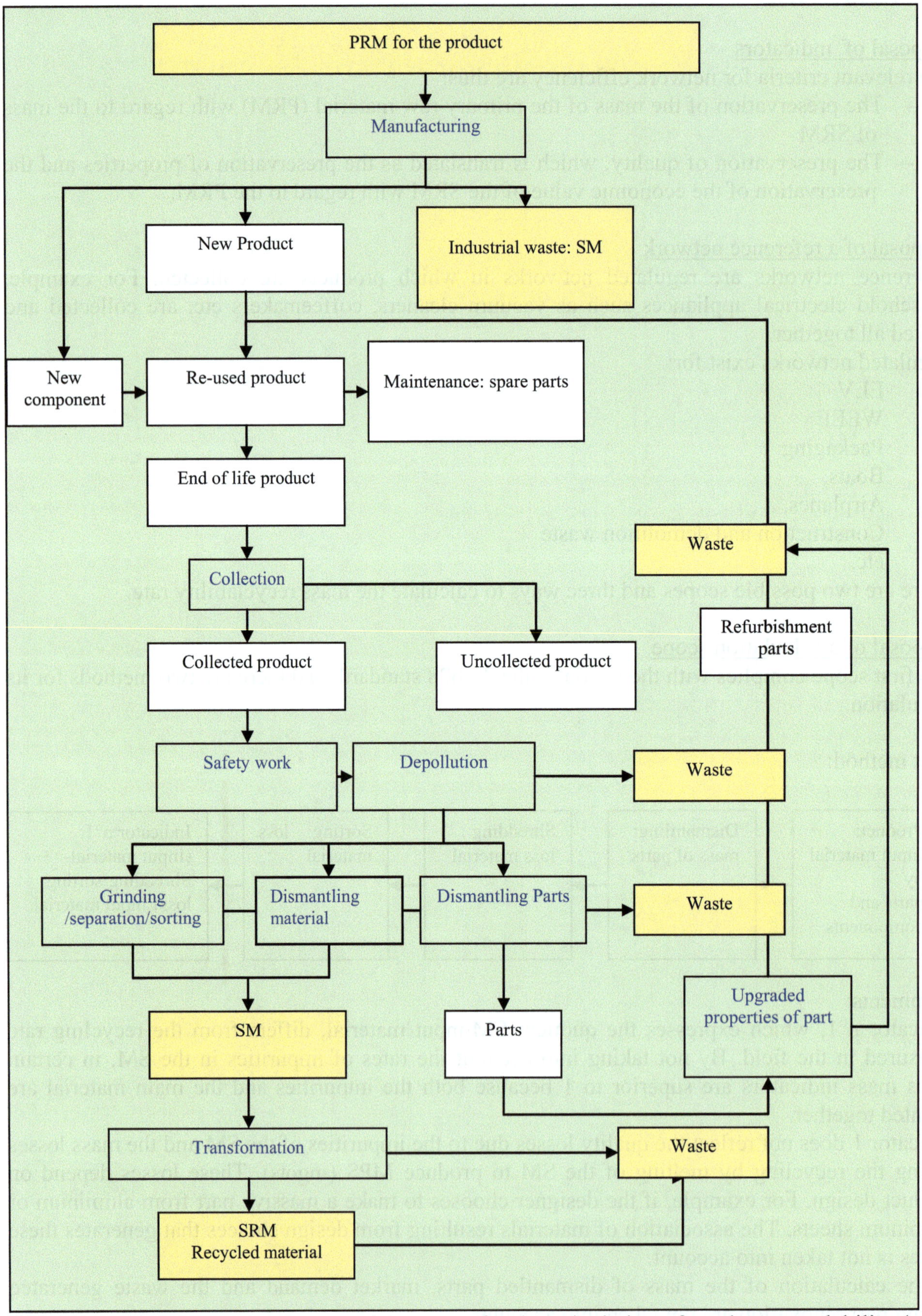

Figure 1. Definition of the scope for the calculation of product recyclability (color code: yellow: materials, white: products/components, green: transformation processing)

Second Method:
Scope 1

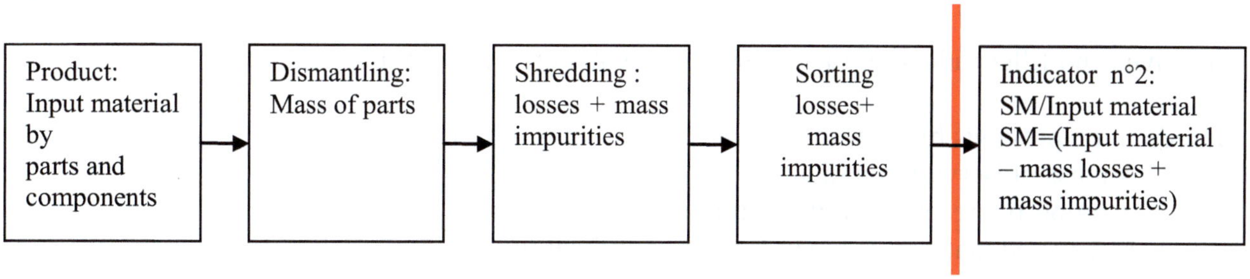

Comments:
This indicator can be linked to the measured recycling rate in the field. It is however necessary to estimate the impurity rate resulting from material associations for each type of design. This requires the use of databases built from field tests.
It also makes it possible to estimate the quality of the SM, but does not integrate the mass losses during the recycling of the SM to produce SRM.

Scope 2:

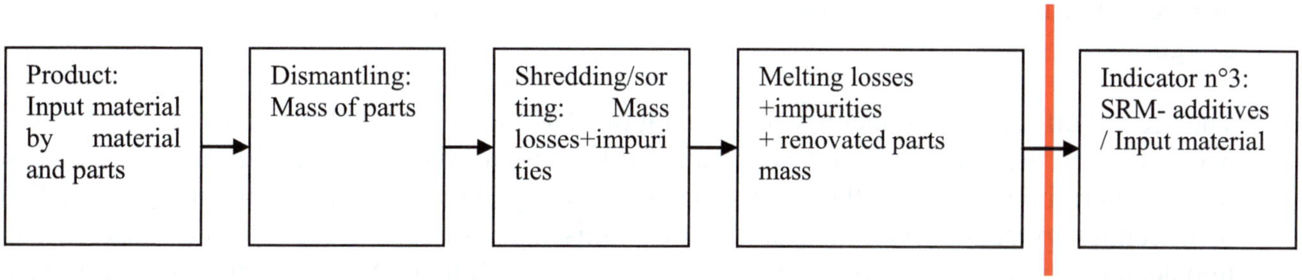

Comments:
This indicator takes into account the impurities during grinding and sorting, the SM losses and the additives added during melting to restore materials to their previous level. With regard to the previous scenario, it requires good knowledge of the main material/impurities phenomena during SM melting. This indicator is more precise for designers with product recycling objectives, since it considers the whole recycling process for various types of material associations. It is correlated with the production of secondary raw materials. Indicator n°3 can also be defined as indicated in the following diagram.

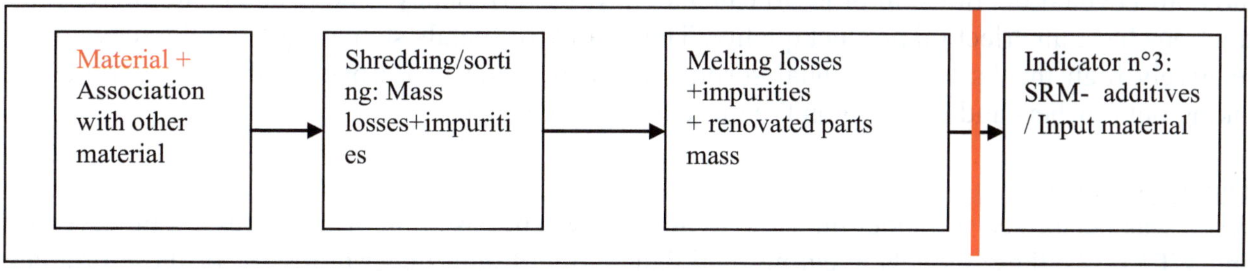

<u>Discussion of the scopes with regard to the proposed indicators</u>
The economic value preservation indicator becomes integrated into scope 1, since it takes into account the quality of the SM and thus the associations of materials in the design, and indirectly in scope 2 since the market prices integrate the value of the SRM. However, it takes into account the material losses at the grinding/sorting level.

The mechanical property preservation indicator takes into account the notion of SRM quality; it thus becomes integrated into scenario 2. It requires precise knowledge of all the treatment stages and the abundant thermodynamic data.

Additives and fillers added to the secondary raw material must not be taken into account.

Based on the previous hypotheses, we propose 3 types of indicators:

- A mass preservation indicator
- A property preservation indicator
- A recycled material value preservation indicator

Proposal of the indicators of mass, properties and economic value

State of the arts concerning indicators

The first difficulty when calculating product recyclability rates is to procure data on typical networks, i.e. which products go to which network, since a network treats several categories of products. The recyclability rate of a product by a network will actually be that of a set of product categories.

The second difficulty is to identify the geographical scope of a typical network. Indeed, it will be necessary for the various European countries to estimate the performances of their networks before determining which is the most successful. Today, every country is attempting to make its own technology prevail with the European Commission. The technical scope must also be taken into account. In this study, we propose that the scope include the SRM, i.e. that produced by the refiners, and the re-used parts after their restoration to standard.

The third difficulty is to procure data on the mass efficiency of the product treatment - what are the mass losses at the level of the grinding, the sorting and the melting of the materials into ingots, and how do we define the notion of mass efficiency?
In any transformation process, according to the laws of thermodynamics, mass losses occur [2]. To estimate them for each material composing a product, one must conduct shredding tests and analyze the residual product remaining in the recycled fractions as well as in the composition of the non-recycled fractions, in order to calculate process efficiency. The residuals in the recycled fractions have an effect on product quality. The quality of metals stemming from the recycling is relatively high thanks to the manual sorting which is made in Europe on the big particle size taking out crushers and in China on the more low particle size.

The mass indicator:
Work was performed to model the ELV treatment system based on grinding tests. For ELVs [2] and WEEE [3], these models integrate field data and highlight material losses for these products during the treatment stages. These losses depend on:

- Grinding particle size. This determines the rate of liberation of the associated materials [4] [5] and is different for ELVs and small electric devices. The more one crushes at the end, the more material losses there are.
- The reactivity of metals with respect to oxidation. Material losses are greater for example for aluminum and magnesium than steel.
- The material sorting, whose efficiency depends on processes and is not 100%.

- The melting to produce ingots, since the SM entering the refiner is not pure.

Because the material (metals, plastics, glass etc.) must be brought back to its original level, a dilution with virgin or recycled materials is necessary, depending on the outlets targeted. This is true for steel, diluted to decrease the brass impurity rate, which must be much lower than 0.2%. Wrought and cast aluminums are mixed after grinding because they cannot be sorted out with the existing techniques. The only application for this mixture is for cast aluminum, but a dilution is also necessary to reduce the iron impurity rate. The USGS publishes the average rates of recycled material inputs in primary metals.
For ELVs, plastics with densities between 0.9 and 1.1 are sorted out at the industrial level by certain recycling companies, but this is not a typical technique at the European level. The PP sorting efficiency communicated by the shredder company Galloo in 2001 was 44%. To satisfy noble outlets, the purity rate must reach 97% (with overall property losses lower than 5% [6]). Plastic sorting is not widespread in Europe today, since crusher companies' know-how concerns mainly metal recovery, and practically not at all that of plastics. Crushers did not anticipate the changes in product design, which has been incorporating increasingly less metal and more plastic for several years now. Yet due to the ELV lifespan, plastic must increasingly be taken into account today when recycling.

In standard 22 628, product recyclability is a standardized calculation, similar to an efficiency calculation for recovered materials. Due to the lack of data concerning the composition of the output of crushed and sorted fractions, the mass losses and the impurity rate are neglected in the calculations. Our bibliographical research shows that these two types of data are not insignificant. An outgoing fraction "A" can be larger in quantity than the incoming fraction due to the impurities. In theory, to calculate the mass efficiency one should take into account the rate of losses of the incomers and the impurities rate, just as in any efficiency rate calculation.
Mass efficiencies differ according to product. The mass efficiency of electronic products is lower than that of ELVs, with higher impurity rates. On the other hand, it includes significantly more precious metals. Plastics are styrenics containing toxic flame retardants, which depreciate their value if not sorted out.
For all of the reasons evoked above, one must take material associations into account when calculating recyclability.

<u>Economic indicator:</u>
The economic indicator can be a good way to express quality loss [7] [8] [9]. This indicator depends on the degree of development of recycling techniques as well as on the costs generated by recycling processes. We assume that this indicator is reflected by the monetary value of the recycled and virgin materials. The closer the value of the recycled material to the virgin material, the higher the recyclability rate. If this indicator is used by an engineering consulting firm, the designer can determine which types of economic outlets exist after recycling.
Since the quoted market prices of raw and recycled materials are volatile, the annual average prices of the secondary primary and primary raw materials as well as the trend curves for price evolutions during periods of at least 10 years is taken into account.

<u>The property indicator:</u>
The quality preservation indicator for the properties can be estimated using the exergy conservation indicator. Any irreversible process causes a loss of exergy leading to a reduction of the useful effects of the process, or has an increase in energy consumption. The data for exergy calculation is stable and fixed with regard to the economic values. This indicator can be

calculated with software. We can thus calculate the exergy losses due to mass loss, contamination with impurities, and dilution [10] [11] [12] [13].
In this context and using the available data, the recyclability of several products was calculated using these three indicators.

Conclusions
Conclusion 1:
There are two ways to calculate the material efficiency:
If we consider the input material and the SM, then the efficiencies can be greater than 1, which is incompatible with an efficiency calculation,
If we consider the incomers, the SM and their impurities, and the losses during transformation into ingots, then the efficiencies are lower than 1. [14]

Conclusion 2:
There are several types of mass losses during the product treatment cycle producing the secondary raw material [15] [16]:
- During the grinding, due to oxidation for reactive metals,
- Through the imperfect liberation of the associated materials,
- During the melting to produce ingots

Conclusion 3: The mass losses during the grinding / sorting / melting are not insignificant and vary depending on the metal.

Conclusion 4: the rate of liberation of the associated materials, as for ferrous metals, copper and aluminum, is different for ELVs (99, 89, 88%) [2] and WEEE (94, 78, 82%) [16]. It is lower for the electronic products. The losses of ferrous metals, copper etc. thus differ depending on the network.

Conclusion 5:
There is a difference between a product's mass recyclability and its mass recycling rate in the real networks [17]:
Example: Small household and electronic appliances:
- Theoretical mass recyclability rate: 82%, versus recycling rate achieved by the networks: 73%,
- Recycled metals: 73.2% versus 61%, MPS produced efficiency: 90% on average,
- Recycled plastics: 19% versus 7.2%, MPS produced efficiency: 38% on average.

Conclusion 6: the material efficiencies (SM/material/incomers) are different according to the type of equipment because the associated materials are different:
- Large household appliances, mass efficiency: ferrous: 99%, non-ferrous: 94%, plastic: 74%,
- Small electrical household appliances: mass efficiency: ferrous: 96%, non-ferrous: 98%, plastic: 38%.

Conclusion 7:
Metals and plastics are alloys. After grinding and sorting, alloys are mixed, for example aluminum, zinc, iron, copper alloys, etc., with the following consequences:[10]
- Material property losses
- Value losses: alloys have different values when they are mixed,

- Resource losses due to the dilution.

Conclusion 8:
According to the laws of thermodynamics [15], the efficiency of a transformation cannot be 100%, and the increase in mass efficiency occurs to the detriment of quality preservation. It is thus necessary to optimize the mass indicator and the quality preservation indicator (exergy) according to the economic outlets and thus to the economic value preservation indicator. In the future, it will also be necessary to take into account the irreversible contamination that accumulates in metals since they all include a fraction of recycled metals.

Conclusion 9:
According to the hypotheses taken into account to calculate a quality preservation recyclability indicator based on the economic value losses of the SM [8] [9]:

- The loss in value of recycled materials depends on the raw material value. Precious metals lose less value when they are recycled (mining/extraction costs quotient and recycling costs),
- The economic value of a recycled material depends on its quality, since recycled materials with properties close to the virgin material have values that are close but always lower,
- The loss of value depends on the use which is made of the SRM in a product.

Conclusion 10:
The application of the recyclability calculation methods from our bibliographical research shows that calculating recyclability according to standard 22 628 results in a significant underestimation and sometimes in an unrealistic image of the real recyclability.

<u>Proposal of a recyclability calculation method</u>

Three types of losses were highlighted during the treatment of end-of-life products with a recycling objective:

- loss of mass due to material losses,
- quality losses due to the association of materials, which has the effect of creating impurities after grinding (imperfect liberation) and sorting,
- economic value losses due to the material specificities and to their uses in a product; this indicator is also a quality indicator.

From the bibliographical research, three indicators and their calculation methods were tested for case studies. These indicators enable assessment of a product's recyclability:

1. Mass preservation indicator, Rm,
2. Exergy preservation indicator, Re,
3. Economic value preservation indicator, Rv.

The data used to calculate the economic value preservation recyclability indicator was updated for the year 2007. Databases for exergy preservation allow exergy calculation for any compound made from the chemical elements.
The recyclability of a product is ideal when these indicators approach 1.

The mass preservation indicator Rm is determined according to the mass efficiency ξmi (see Table 1) and is specific to every material during the various stages of the treatment of end-of-life products.

$$R\,m = \Sigma\, mi * \xi mi / \Sigma\, mi$$

- Rm: Mass preservation indicator,
- Mi: Mass of each material i (kg),
- ξmi: Mass efficiency indicator specific to a network and a material,

ξmi is the product of the mass indicators for the shredding (grinding), sorting and refining, ξmi= ξg* ξs* ξr,

- If ξ = 0, no recycling in SRM or re-use.

For product re-manufacturing:

$$R\,m = \Sigma\, \xi mi \text{ parts} * \text{Market share}$$

Table 1. Mass efficiency indicators

Accepted material	Unaccepted material	Refused material	Dismantled parts
ξm > 0 % recyclability	ξm = 0 No recyclability	ξm≥ 0 obligatory dismantling + restoration to standard, % recyclability	ξm ≤1 restoration for re-use % recyclability

The exergy preservation indicator Re, determined according to the loss during recycling, is estimated using the exergy efficiency indicator ξe.

$$Re = \Sigma\, mi * \xi e = Rm * \xi e$$

$$\xi e = e\,srm / e\,srm$$
$$e\,srm = em + ei + ed$$

- Re: Exergy preservation indicator,
- em: Loss of exergy due to the mass loss (MJ/kg),
- ei: Loss of exergy due to impurities (MJ/kg),
- ed: Loss of exergy due to the addition of virgin material for dilution in order to restore to the SM standard.

Comment: Aspect or color quality losses cannot be estimated using the exergy. In this case, they are taken into account in the ENSAM/Renault method [6].

The economic value preservation indicator Rv is determined with respect to the economic efficiency indicator ξv ", which estimates the preservation of the material's value after use and recycling.

$$\mathbf{R_v = f(Vprm, Vsm, Vsrm)}$$
$$\mathbf{= \Sigma m_i\ \xi v / \Sigma m_i}$$

Vprm: PRM value ($/kg),
Vsm: SM value (/kg),
Vsrm: SRM value (/kg),

ξv: Economic efficiency. When it is equal to 1, this means that the value of the secondary raw material is practically equal to that of the primary raw material. The values for various materials are expressed in the graph in Figure 2.

$\xi v = 1$ if $R = [0.8 ; 1]$,
$\xi v = 0$ if $R = [0 ; 0.8]$,
The value of 0.8 was chosen as a threshold.

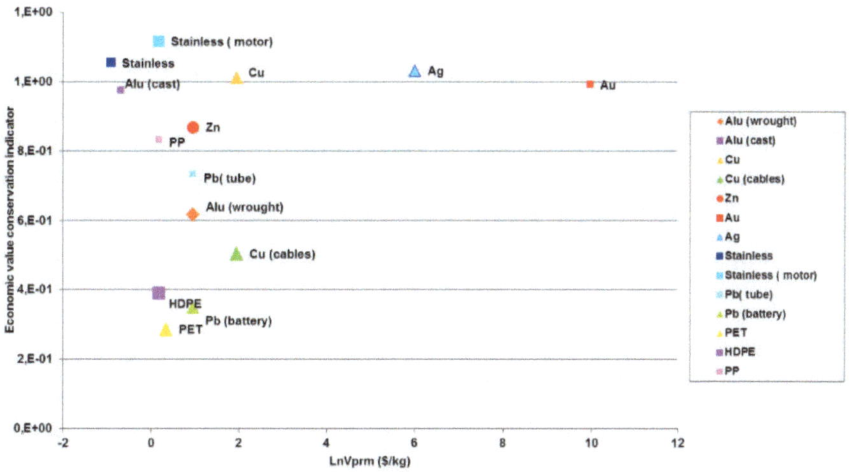

Figure 2. Economic value preservation indicator (updated in 2007) [8] [9]

Comment:
According to the laws of thermodynamics, the efficiency of a transformation cannot reach 100%, and any decrease in mass losses occurs to the detriment of material quality preservation. It is thus necessary to optimize the mass losses and the quality preservation according to the economic outlets and thus to the economic value preservation.

These three indicators (see Table 4) are defined for typical networks, for example for ELVs or WEEE. The mass indicator for the ELV network is higher than that of the WEEE, according to the studies quoted previously. However, for certain electronic products containing precious metals, the economic indicator can be higher.

Table 4 . Product recyclability defined by the preservation of mass, quality and economic value efficiency

Accepted material	Unaccepted material	Refused material	Dismantled part	Mass efficiency	Exergy efficiency	Economic value efficiency
ma_i	mi_i	mr_i	md_i	$\xi_i = \xi_g * \xi_s * \xi_r * \xi d_i$	$\xi_e = e_{srm} / e_{prm}$	$\xi_v = \Sigma m_{i\ R=(0,8;1)} / \Sigma m_i$

Case study

The bibliographical research made it possible to identify indicator calculation methods estimating the performances of end-of-life product recycling as well as the quality and value of the recycled materials produced. To calculate these various indicators, one must collect data on the mass balance assessment of the dismantling, crushing and sorting as well as the material losses up until ingot production by the network operators. This data is still unknown, but certain studies have modeled the efficiency of the systems. The databases and software used to calculate these various indicators are not always available. For these reasons, it was not possible to test all of the methods. As an illustration, mass recyclability according to the standard is compared with the mass recyclability integrating losses during the transformation process and a scope including the production of SRM.

Case study: a car body part

Hypothesis: A car body part (Figure 4) is dismantled. It has a steel core and an aluminum cover. During the grinding / sorting, the steel and the aluminum are not separated; the fraction after sorting is captured in the ferrous fraction, due to the ferrous mass. During material melting in an electric oven, the aluminum is oxidized since it is associated with the steel and thus is found in the slag, and it is therefore not recycled. Only the steel is recycled.
In the recyclability calculation, standard 22 628 does not require that the parts be treated in a representative network.

If there is at least one process allowing treatment of this part, the recyclability rate is considered to be 100%.

In the case of the car body part, the mass recyclability (see Table 3, 4) is 47%, because there are losses at the level of the separation by grinding but especially during the melting. The indicator takes into account the material associations.
The economic value preservation indicator does not take into account mass losses due to the melting but it does take into account the loss of value of the SM, in this case the aluminum in the second melting. With respect to the primary raw material, this loss is significant for the aluminum. The recyclability of the body part is 48%. The mass indicator and exergy preservation is 35%; it takes into account the losses during the steel refining.

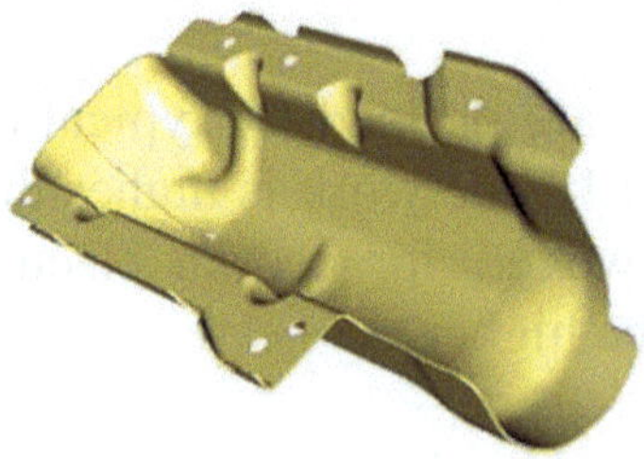

Figure 4. Case study: a car body part

Table 2. Recyclability according to standard 22 628, mass, economic value and exergy preservation indicators

Car body part: ELV channel	Mass (kg)	ξ % Standard	R Standard	ξ Mass%	R Mass	ξ v%	R Economic value	ξ exergy %	R exergy
Steel	0.17	1	0.48	0.98	0.47	1	0.48	0.74	0.35
Wrought Aluminum	0.18	1	0.52	0	0	1	0	0.8	0
Total	0.35	1	1		0.47		0.48		0.35

Table 3. Calculation of the mass efficiency

Car body part: ELV channel	ξ oxidation	ξ Separation	ξ Melting	ξ m total
Steel	1.00	0.99	0.99	0.98
Wrought Aluminum	0.91	0.88	0	0

Summary of the case study:

According to the calculation method used in the standard [1], for a product to be recyclable, a process or network must exist that allows every component or material in the product to be recycled. The scope of calculation takes into account the input materials/components and does not take into account the shredding, sorting, and refining processes leading to the production of SM and SRM. In the case of a car body part, recyclability is considered to be 100%.
If the recyclability calculation is carried out with a scope covering the MPS, the results are quite different from the standard.

This calculation is often very far from reality, as shown by our comparison of the various recyclability calculation methods tested in the case studies. There are indeed several types of losses.

Conclusion

This study shows that the recyclability calculation method according to standard 22 628 is not satisfactory for designers as concerns the scope of calculation. In order to make more relevant choices of materials and their associations to improve product recyclability, we recommend that they choose a scenario with a calculation scope covering the production of secondary raw materials.
This scenario allows more realistic theoretical product recyclability calculation and responds to the European ratification objectives. As for the quality preservation approach, a first step is to use an economic indicator to guarantee the sustainability of a recycling network. Secondly, when recycling processes can be better modeled, it will be possible to apply a quality indicator based on the notion of exergy, which we consider more reliable from an environmental standpoint.

References

1. Standard ISO 22 628. 2002-02-15. Road vehicles, Recyclability and recoverability, Calculation method.

2. M.A., Reuter et al., 2006/4. Fundamental limits for the recycling of end-of-life vehicles. Minerals Engineering, 19(5): 433-449.

3. J. Huisman, A.L.N. Stevels and I. Stobbe. 2004. Eco-efficiency considerations on the end-of-life of consumer electronic products. IEEE Transactions on Electronics Packaging Manufacturing, 27(1): 9-25.2004

4. A. van Schaik et al.. 2004. The influence of particle size reduction and liberation on the recycling rate of end-of-life vehicles. Minerals Engineering, 17(2): 331-347.

5. M.B.Castro et al., Simulation model of the comminution–liberation of recycling streams: Relationships between product design and the liberation of materials during recycling. International Journal of Mineral Processing, 2005/2/7

6. D. Froelich et al., 2007. State of the art of plastic sorting and recycling: Feedback to vehicle design. Minerals Engineering, 20(9): 902-912.

7. Robert U. Ayres. Metals recycling: economic and environmental implications, Resources, Conservation and Recycling, 1997

8. G. Villalba et al.,2004/10/1. Using the recyclability index of materials as a tool for design for disassembly. Ecological Economics, 50(3-4): 195-200.

9. G.Villalba et al., 2002/12. A proposal for quantifying the recyclability of materials. Resources, Conservation and Recycling, 37(1): 39-53.

10. S.H. Amini et al. 2007. Quantifying the quality loss and resource efficiency of recycling by means of exergy analysis. Journal of Cleaner Production, 15(10): 907-913.

11. M.B.G. Castr et al., Exergy losses during recycling and the resource efficiency of product systems Resources, Conservation and Recycling. 2007 11.

12. O., Ignatenko, A. van Schaik and M. A. Reuter, Exergy as a tool for evaluation of the resource efficiency of recycling systems. Minerals Engineering, 2008

13. O. Ignatenko, A.van Schaik and M.A. Reuter, Recycling system flexibility: the fundamental solution to achieve high energy and material recovery quotas; Journal of Cleaner Production, 2007

14. A. van Schaik, M.A. Reuter, The time-varying factors influencing the recycling rate of products. Resources, Conservation and Recycling. 2004/3

15. M.B.G. Castro et al., A thermodynamic approach to the compatibility of materials combinations for recycling. Resources, Conservation and Recycling. 2004/12 5.

16. J. Huisman, C.B. Boks and A.L.N. Stevels, Quotes for environmentally weighted recyclability (QWERTY): Concept of describing product recyclability in terms of environmental value. International Journal of Production Research. 2003

17. L. Flahaut et al, Etude pour une filière de recyclage des déchets d'équipements électriques et électroniques sur le territoire national.
http://www.dechetcom.com/comptes/jcamille/screlec%20rapport.pdf. 2004.

REWAS 2013: Enabling Materials Resource Sustainability
Edited by: Anne Kvithyld, Christina Meskers, Randolph Kirchain, Gregory Krumdick, Brajendra Mishra, Markus Reuter, Cong Wang, Mark Schlesinger, Gabrielle Gaustad, Diana Lados, and Jeffrey Spangenberger
TMS (The Minerals, Metals & Materials Society), 2013

ROCK SMELTING OF COPPER ORES WITH WASTE HEAT RECOVERY

Terry Norgate, Sharif Jahanshahi and Nawshad Haque

CSIRO Minerals Down Under National Research Flagship, Clayton South, Victoria 3168, Australia

Keywords: Smelting, Embodied energy, Greenhouse gases, Ore grade, Waste heat

Abstract

It is generally recognised that the grades of metallic ores are falling globally. This trend can be expected to increase the life cycle-based energy requirement for primary metal production due to the additional amount of material that must be handled and treated in the mining and mineral processing stages of the metal production life cycle. Rock (or whole ore) smelting has been suggested as a possible alternative processing route for low grade ores with a potentially lower energy intensity and environmental impact than traditional processing routes. In this processing route, the beneficiation stage is eliminated along with its associated energy consumption and greenhouse gas emissions, but this is partially offset by the need for more solid material to be handled and heated up to smelting temperatures. A life cycle assessment study was carried out to assess the potential energy and greenhouse gas benefits of a conceptual flowsheet of the rock smelting process, using copper ore as an example. Recovery and utilisation of waste heat in the slag (via dry slag granulation) and offgas streams from the smelting step was also included in the study, with the waste heat being utilised either for thermal applications or electricity generation.

Based on the assumptions made, the results of the study indicated that this conceptual processing route does offer the potential for significant reductions in the embodied fossil fuel energy and associated greenhouse gas emissions of copper metal production from low grade ores compared to the most likely other competing processing route, heap leaching, for these ores. This is particularly the case where waste sulphide material is available, as the energy from oxidation of the waste sulphide during smelting replaces fossil fuel energy required for smelting. However, in the absence of any waste sulphide material, the embodied energy was only less than for heap leaching when the recovered waste heat was utilised in thermal applications. Other issues, particularly economics, will also play a significant role in determining the viability of this processing route for low grade ores.

Introduction

Metallic ore grades are falling globally as higher grade reserves are exploited first and are progressively depleted. At the same time, the demand for primary metals extracted from these ores is expected to increase, despite increased levels of dematerialisation and recycling. Sustainability concerns have highlighted the need to minimise resource consumption and environmental emissions while meeting these demands. However, this is becomingly increasingly difficult as ore grades fall and deposits become more mineralogically complex. This raises the issue of developing possible alternative processing routes for low grade ores with lower energy intensities and/or environmental impacts [1].

Rather than beneficiating an ore by grinding and flotation to produce a concentrate for smelting, a possible alternative is to smelt the whole ore directly (ie. "rock smelting"). While this processing route eliminates the mineral beneficiation stage and its associated impacts (including energy consumption and greenhouse gas emissions), more solid material has to be handled and heated up to smelting temperatures (typically 1250°C for copper) requiring additional energy input (and associated greenhouse gas emissions). However, many copper mines have waste sulphide dumps containing significant amounts of iron sulphides (pyrite (FeS_2) and pyrrhotite (FeS)). Using this waste material as a fuel and flux source (the oxidation of sulphides is exothermic) in the rock smelting stage has the potential to reduce the fossil fuel-based energy requirements of the process as well as disposing of a waste material that can give rise to acid mine drainage (AMD - the outflow of acidic water from mine sites). However, this reduction in smelting energy is partially offset by additional electrical power requirements for the smelting stage due to the additional amount of input material to be treated, as well as for the acid plant to convert the additional amount of suphur dioxide in the furnace off-gas into sulphuric acid. The other advantage of the proposed processing route is a reduction in the amount of sulphide waste materials that could end up in the tailings dam and hence result in AMD.

Sustainability concerns have seen the metal production sector, like other industrial sectors, come under increased pressure to reduce its environmental footprint over the various processing stages in the supply chain from ore mining though to metal production and refining. Life cycle assessment (LCA) is a methodology that has been widely used to examine these concerns. In a previous LCA study of the rock smelting of low copper grade ores [1], a conceptual version of this processing route was considered, with no particular smelting technology specified, although some of the emerging technologies for direct ironmaking could be amenable to this type of processing. The possible use of waste sulphide material (assumed to be FeS) in the smelting stage was also included in this earlier study. The embodied energy[1] and greenhouse gas (Global Warming Potential, GWP) footprints of rock smelting of copper ore were compared with those of the following processing routes in this earlier study:

- concentration and smelting
- pressure acid leaching
- in-situ leaching
- heap leaching

As lower grade ores tend to be more mineralogically complex, it is anticipated that they will require finer grind sizes to achieve mineral liberation, and this will significantly increase the energy required for the concentration and smelting route. Pressure acid leaching was shown to have similar energy requirements to concentration and smelting at low ore grades, while in-situ leaching still has a number of development issues to be addressed. This leaves heap leaching as the most likely competing processing route to rock smelting for low grade copper ores, and these two processing routes are compared in this paper.

No recovery of the heat contained in the waste streams (slag and offgas) from the smelting stage was considered for the rock smelting case in the earlier LCA study [1]. However, CSIRO has recently developed a new integrated slag treatment process based on a two-step process involving a dry granulator and a packed-bed counter-current heat exchanger [2-4]. A follow up

[1] Fossil fuel energy accumulated over the various life cycle stages of primary metal production.

study was therefore carried out to investigate the impact of slag waste heat recovery using CSIRO's dry granulation process combined with offgas waste heat recovery on the embodied energy and greenhouse gas footprints of copper metal production by rock smelting of copper ores. The results of this latest study are presented in this paper.

Smelting Energy

Modelling of rock smelting of copper ores was carried out using CSIRO's MPE (multiphase equilibrium) model [5] in order to establish suitable smelting conditions to produce a high grade copper matte (65% Cu). FeS was assumed to be available on site from a waste sulphide dump as outlined earlier. The MPE modelling results were used to derive a series of reactions (along with reaction extents) that were incorporated into a METSIM simulation model of the rock smelting process [1]. This METSIM model was then used to examine the effect of decreasing ore grade on the smelting energy requirement and thereby on the embodied energy of the copper metal subsequently produced from these ores. For comparative purposes, the case where no waste sulphide material is available as a fuel and flux source for smelting was also considered. A simple schematic flowsheet of rock smelting of copper ore as modelled by METSIM is shown in Figure 1. The composition of the ore used for the base case was as follows:

Chalcocite (Cu_2S)	1.3% (ie. 1.0% Cu)
Pyrite (FeS_2)	33.2%
Quartz/silica (SiO_2)	65.5%

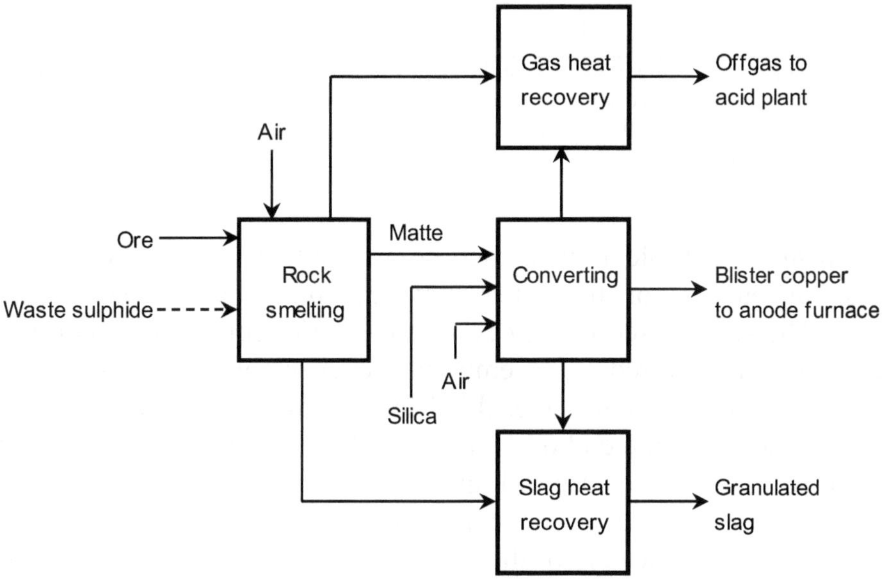

Figure 1. Schematic flowsheet of rock smelting of copper ore.

The results of the METSIM simulations showed that the energy balance around the converting stage was very small relative to the smelting stage, which is a reflection of the respective mass flows into each stage (ie. 100 t/h of ore into smelting, 1.6 t/h matte into converting). Therefore the focus was on the energy balance around the rock smelting stage and this is shown in Table I. The heat loss for the base case (waste sulphide available) was assumed to be 0.77 GJ/t ore (or

0.45 GJ/t total solid[2] (ore + waste sulphide charge) to give a net smelting energy input of zero. This heat loss value is similar to that previously estimated for conventional copper smelting [6]. The heat loss is less for the no waste sulphide case as it is assumed that the heat loss is proportional to smelting furnace surface area and for fixed cyclindrical geometry and percentage charge volume, surface area is approximately proportional to charge mass[3].

In the waste sulphide case, the external smelting energy input is zero as the heat of reaction from sulphide oxidation balances the energy required to heat the input streams (ore, waste sulphide and air) to 1250°C as well as the heat loss. In the no waste sulphide case, less energy is required to heat the input streams as there is no waste sulphide input stream, but this is partially offset by the lower amount of energy derived from sulphide oxidation, resulting in an external smelting energy requirement of 1.08 GJ/t ore. The waste sulphide case with no slag waste heat recovery (second row Table I) was the case considered in the earlier study [1].

Table I. Energy balance for rock smelting stage (1% Cu ore).

	Heating input streams[1] (GJ/t ore)	Smelting reaction heat[2] (GJ/t ore)	Heat loss (GJ/t ore)	Net smelting energy (GJ/t ore)
Waste sulphide available	5.96	-6.73	0.77	0.00[3]
No waste sulphide available	2.82	-2.19	0.45	1.08[4]

1. This term is the amount of heat required to heat the input streams in the smelting stage (ore, waste sulphide and air) from ambient temperature (20°C) to the smelting temperature (1250°C).
2. This term is the amount of heat released by the exothermic metallurgical smelting reactions.
3. 5.96 + 0.77 – 6.73 = 0.00
4. 2.82 + 0.45 – 2.19 = 1.08

The smelting stage results in Table I were incorporated into a LCA model of copper metal production that also accounted for the other material and energy inputs over the copper production life cycle. The LCA-based energy consumption (ie. cumulative energy consumption over all life cycle stages) is referred to as embodied energy (or Gross Energy Requirement, GER). Figure 2 shows the energy inputs over this life cycle for the two cases (1% Cu ore with and without waste sulphide), with the embodied fossil fuel energy indicated. The embodied fossil fuel energy for the no waste sulphide case in Figure 2 is about 2.0 GJ/t ore. This corresponds to the fossil fuel-based energy input over the copper metal production life cycle from mining through to smelting (indicated as the sum of the "other stages" and the "smelting stage" in Figure 2). It does not include the energy input derived from the exothermic sulphide oxidation reactions in the smelting furnace as this is not fossil fuel based. The smelting stage fossil fuel energy input for the no waste sulphide case in Figure 2 is 1.08 GJ/t ore which corresponds to the value shown for this case in the "Net smelting energy" column in Table I. There is no fossil fuel energy input for the smelting stage in the waste sulphide case shown in Figure 2 as smelting reaction heat provides the required energy (see Table I) with no external fossil fuel energy required. The combined amount of fossil fuel energy and sulphide oxidation energy in Figure 2 less heat losses,

---

[2] 0.77 GJ/t ore x 100 t ore/(170 t ore + waste sulphide) = 0.45 GJ/t solid charge.

[3] Alternatively, more ore can be charged into a fixed furnace volume if no waste sulphide is added, but the furnace heat loss is essentially unchanged, meaning a lower heat loss per tonne of ore charged.

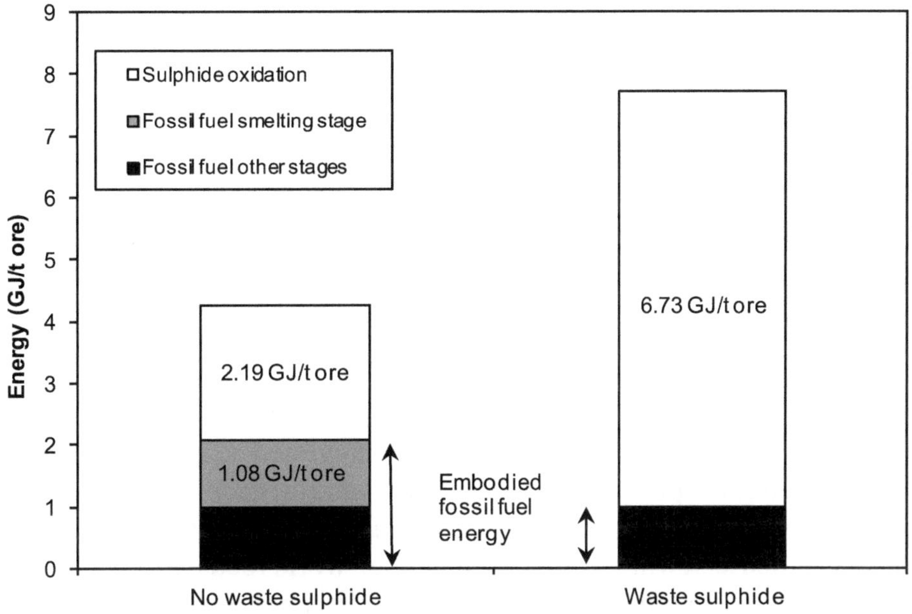

Figure 2. Energy inputs over copper production life cycle (base case, 1% Cu ore).

represents the maximum amount of input energy that could potentially be recovered as waste heat (in slag, matte and offgas) from the smelting stage in each case (viz. 5.96 GJ/t ore for the waste sulphide case and 2.82 GJ/t ore for the no waste sulphide case, corresponding to the energy to heat the input streams in Table I). In this study it was considered that in practice, waste heat is only recovered from slag and offgas as the mass flowrate of matte is small.

If a significant proportion of this energy input can be recovered from the slag and offgas streams, this figure indicates that the embodied fossil fuel energy of copper production could be reduced substantially, perhaps even to less than zero (i.e. a fossil fuel energy credit) as the sulphide oxidation energy substitutes for fossil fuel energy. The potential for such an outcome is reflected by the relative amounts of sulphide oxidation energy and smelting fossil fuel energy in Figure 2, with the sulphide oxidation energy for the waste sulphide case in Figure 2 being many times greater than that of the smelting fossil fuel energy.

Waste Heat Recovery and Utilisation

As noted above, recovery of the waste heat contained in the smelting slag and offgas has the potential to significantly improve the energy and greenhouse gas footprint of copper ore rock smelting. However, in order to do this, there must be opportunities to utilise the recovered waste heat, either on-site or off-site[4]. On-site utilisation of the waste heat could involve thermal use (direct or indirect) or it could be used to generate electricity to partially offset plant electrical power consumption. Off-site utilisation would more than likely be in the form of generated electricity unless the off-site thermal application was nearby. While preheating the smelting input streams with the recovered waste heat is possible, when waste sulphide is present the heat of reaction from sulphide oxidation provides this preheating energy, so this application would only be an effective use of the recovered waste heat in the no waste sulphide case. In the following analysis, the following two extreme scenarios are considered:

[4] Quality of the waste heat is key factor to be considered when analysing potential applications for utilising the recovered waste heat, with high grade heat usually referring to heat sources of 600°C and over.

- all recovered waste heat is utilised thermally;
- all recovered heat is used to generate electricity.

The potential amounts of slag and offgas waste heat available for recovery are based on both slag and offgas being cooled from 1250°C to 50°C, with the actual amounts recovered and utilised discussed in the following sections. While it may prove difficult to extract high grade heat (>600 °C) if the slag is cooled to 50°C, with a practical limit perhaps closer to 100°C, the former value was used to estimate the maximum potential waste heat available in the slag.

Slag

As mentioned earlier, CSIRO has developed a new integrated slag dry granulation and heat recovery process involving a spinning disc granulator and a packed-bed counter-current heat exchanger which has been tested up to semi-industrial pilot plant scale. Development of this process has been described elsewhere [2-4], with initial work focussed on the slag granulation stage, but currently focussed on the heat recovery stage. This process has the potential to recover a significant proportion of the available smelting slag waste heat in the form of hot air at temperatures up to 600°C which can then be utilised in the applications outlined above. Vatanaku et al [7] reported that 80% of total slag thermal energy was recovered in the slag dry granulation plant that operated at the Fukuyama Steel Works in Japan during the 1980's and early 1990's and it was assumed that same percentage of slag waste heat was recovered to the hot air stream for this analysis.

Offgas

A range of waste heat recovery equipment has been developed to recover the waste heat from flue gases (or offgases) from furnaces, kilns, boilers and ovens. Recuperators are one of the most widely used devices for this purpose. A recuperator is a gas-to-gas heat exchanger placed on the stack of a furnace. There are numerous designs, but all rely on tubes or plates to transfer heat from the outgoing offgas to the incoming combustion air, while keeping the two streams from mixing and cross-contaminating each other. The thermal energy efficiency of these devices is reportedly similar to that assumed for slag waste heat recovery above, ie. 80%. While it may be necessary to strip the SO_2 and fumes from the offgas before it can be utilised, which may mean lower thermal efficiency, for the purpose of the study an offgas waste heat recovery efficiency of 80% was assumed.

Other assumptions made for the purposes of the analysis were as follows:

- efficiency of hot air utilisation in thermal applications is 80%
- efficiency of electricity generation using hot air is 32%
- electricity consumption of dry slag granulation plant is 9.5 kWh_e/t slag [3] or 0.14 GJ_{th}/t ore (waste sulphide), 0.08 GJ_{th}/t ore (no waste sulphide)
- electricity consumption of offgas waste heat recovery system is 3.3 kWh_e/t slag [3] or 0.05 GJ_{th}/t ore (waste sulphide), 0.03 GJ_{th}/t ore (no waste sulphide).

Based on the above assumptions, the amount of waste heat recovered and utilised either thermally or for electricity generation from slag, offgas and both combined are shown in Table II for the waste sulphide and no waste sulphide cases. The combined slag and offgas waste heat

Table II. Recovered and utilised waste heat.

	Slag			Offgas			Slag + offgas	
	Maximum energy available (GJ/t ore)	Thermal application[1] (GJ/t ore)	Electricity[2] (GJ/t ore)	Maximum energy available (GJ/t ore)	Thermal application[3] (GJ/t ore)	Electricity[4] (GJ/t ore)	Thermal application (GJ/t ore)	Electricit (GJ/t ore
e	2.04	1.19	0.48	3.72	2.94	1.17	4.13	1.65
te e	1.36	0.81	0.32	1.36	1.06	0.43	1.87	0.75

x 0.8) - 0.14) x 0.80 = 1.19 (waste sulphide)
x 0.8) - 0.08) x 0.80 = 0.81 (no waste sulphide)
x 0.8) - 0.14) x 0.32 = 0.48 (waste sulphide)
x 0.8) - 0.08) x 0.32 = 0.32 (no waste sulphide)
0.05) x 0.80 = 2.94 (waste sulphide)
0.03) x 0.80 = 1.06 (no waste sulphide)
0.05) x 0.32 = 1.17 (waste sulphide)
0.03) x 0.32 = 0.43 (no waste sulphide)

Table IV. Embodied fossil fuel energy results for base case (1% Cu ore).

	Smelting stage energy credits				Embodied energy of other stages	Overall embodied energy of copper metal production	
	Thermal application (GJ/t ore)	Electricity generation (GJ/t ore)	Thermal application (GJ/t Cu)	Electricity generation (GJ/t Cu)	(GJ/t Cu)	Thermal application (GJ/t Cu	Electricity generation (GJ/t Cu)
Waste sulphide	-4.13	-1.65	-425.8	-170.1	100	-325.8	-70.1
No waste sulphide	-0.79	0.33	-81.4	34.0	100	18.6	134.0

recovered and utilised for the two applications from Table II is compared in Table III with the smelting energy requirements from Table I. The results in Table III indicate that with combined slag and offgas waste heat recovery there is an excess energy credit both with and without waste sulphide for thermal application of the recovered waste heat. Electricity generation gives an excess energy credit only when waste sulphide is available, although the smelting energy required is still reduced (ie. from 1.08 to 0.33 GJ/t ore) for this application with no waste sulphide available. However, these results are strongly dependent on the composition assumed for the ore as well as the thermal efficiencies.

Table III. Net smelting energy with utilisation of slag and offgas waste heat.

	Smelting energy	Smelting slag and offgas heat recovered		Net smelting energy required or excess energy credits[1]	
		Thermal application	Electricity generation	Thermal application	Electricity generation
	(GJ/t ore)	(GJ/t ore)	(GJ/t ore)	(GJ/t ore)	(GJ/t ore)
Waste sulphide	0.00	4.13	1.65	-4.13	-1.65
No waste sulphide	1.08	1.87	0.75	-0.79	0.33

1. A positive value indicates that thermal energy over and above that recovered from the slag and offgas is still required for smelting, while a negative value indicates an excess of energy, ie. an energy credit.

Embodied Energy and Greenhouse Gas Emissions

The excess energy credits or net smelting energy requirements shown in Table III were incorporated into the copper ore rock smelting LCA model developed as part of the earlier study [1], and the model was then used to examine the effect of ore grade by varying the input ore grade over the range 0.1-1%[5] for the various scenarios considered in the study. The embodied fossil fuel energy of copper metal produced via the rock smelting route for the base case ore grade (1% Cu) for the different scenarios is given in Table IV while the effect of ore grade on embodied energy is shown in Figures 3 and 4. Table IV includes the fossil fuel energy consumption for the other stages in the copper metal production life cycle (viz. 0.97 GJ/t ore – see Figure 2 – or 100 GJ/t Cu). The corresponding effect of ore grade on the GWP[6] results is shown in Figures 5 and 6. Figures 3-6 also include the corresponding results for heap leaching.

The results in Figures 4 and 6 indicate that if waste sulphide is available for use as a fuel and flux source in the rock smelting stage, copper metal production by this route with waste heat recovery, based on the assumptions made, has a lower embodied energy and greenhouse gas footprint than heap leaching. In fact the embodied energy and greenhouse gas footprints are negative (ie. a credit) when waste sulphide is available. However, if waste sulphide is not available, Figures 3 and 5 indicate that copper metal production by the rock smelting route only has a lower embodied energy and greenhouse gas footprint than heap leaching if the recovered waste heat is all used in thermal applications.

[5] In the earlier study [1], it was assumed that rock smelting of copper ores would likely be used for ore grades of 1% Cu or less.
[6] The GWP calculations are based on the assumption that any recovered waste heat is used to replace fossil fuel in the application considered with a mean greenhouse gas emission factor of 75.8 kg CO_2e/GJ (oil 79.9, coal 92.7, natural gas 54.8 kg CO_2e/GJ).

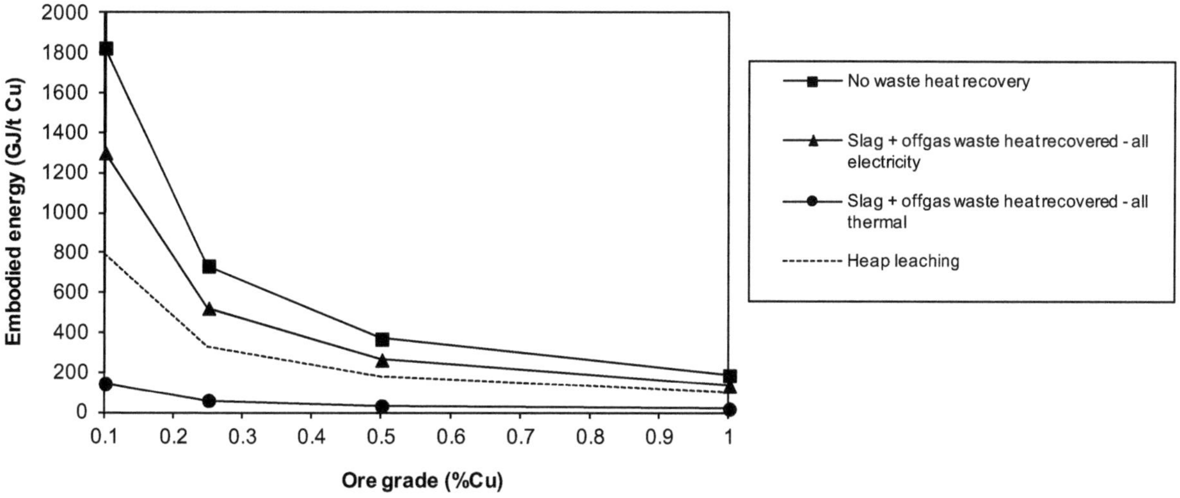

Figure 3. Embodied fossil fuel energy of copper production (no waste sulphide case).

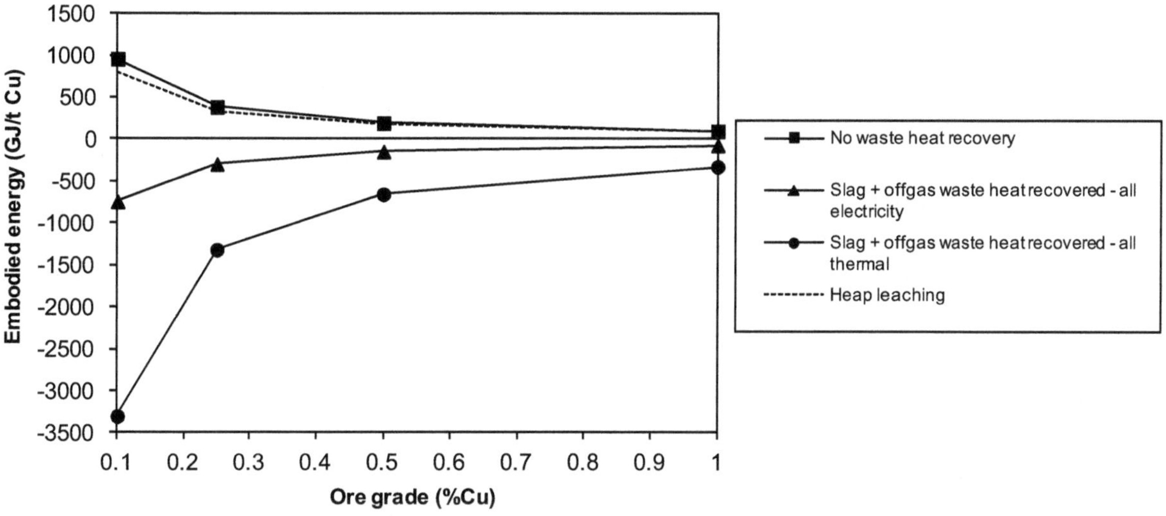

Figure 4. Embodied fossil fuel energy of copper production (waste sulphide case).

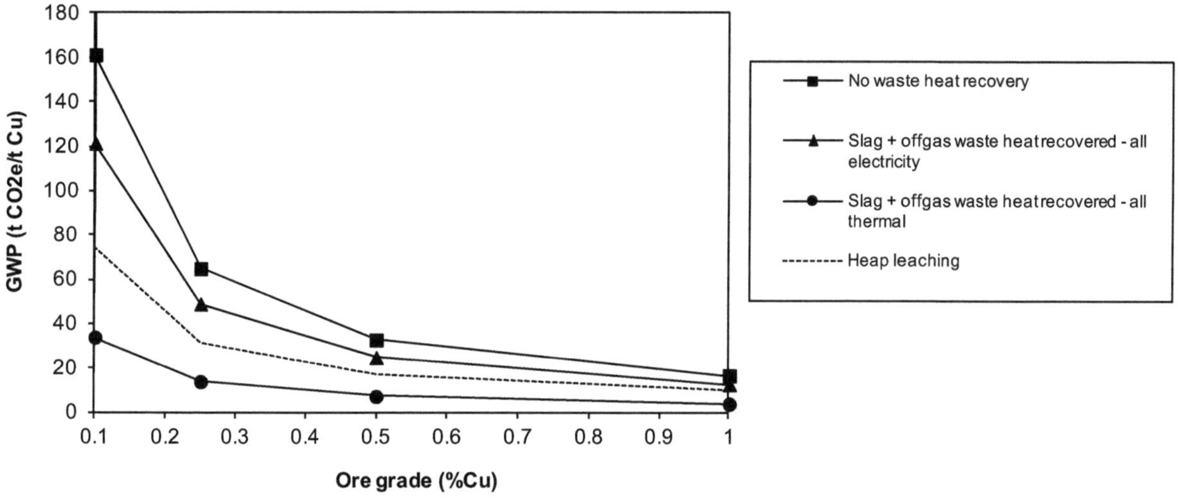

Figure 5. GWP of copper production (no waste sulphide case).

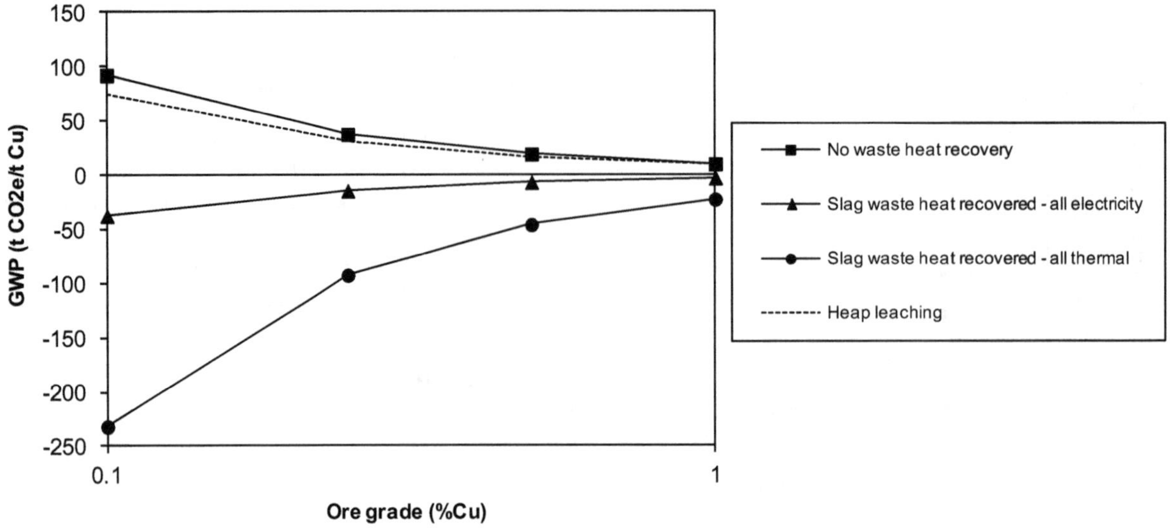

Figure 6. GWP of copper production (waste sulphide case).

Discussion

The assessment of rock smelting of copper ores combined with waste heat recovery carried out in this study is a broad analysis of this processing route, and as such, the results should be considered as indicative only. The objective of the study was to evaluate the potential sustainability benefits of this conceptual process in terms of fossil fuel energy consumption and associated greenhouse gas emissions, rather than provide definitive results. Nevertheless, it is worth noting that the results obtained in the study are affected by a number of variables/parameters, including:

- ore composition (particularly sulphide content)
- waste sulphide mineralogy (ie. sulphur content)
- the fossil fuel actually replaced with the recovered waste heat – the GWP calculations here assumed an emission factor that was the mean for oil, black coal and natural gas
- thermal efficiencies assumed for waste heat recovery technologies and for waste heat utilisation applications.

It is worth noting that the case studied here, i.e smelting with air rather than with oxygen enriched air, represents a base/worst case scenario for smelting of sulphide material with respect to energy requirement and GHG emissions. With oxygen enrichment, a considerable reduction in thermal smelting energy requirement results, and this along with other benefits such as a smaller volume of offgas to treat, will make this case much more attractive.

Some of the various issues influencing this processing route are discussed below.

<u>Thermal Efficiencies</u>

The thermal efficiencies assumed for the various waste heat recovery technologies and waste heat utilisation applications were as follows:

- slag dry granulation – 80% of slag energy is recovered to hot air

- air/gas preheating – 80% of hot air or offgas energy is recovered to air or gas being preheated
- electricity generation – 32% of hot air thermal energy is converted to electrical energy (70% boiler, 45% turbine, ie. 70 x 45/100 = 32% overall).

To illustrate the likely effect of lower thermal recovery and electrical generation efficiences on the results, the above efficiencies were reduced by half and the embodied energy calculations repeated. The results of this analysis are shown in Figures 7 and 8, corresponding to Figures 3 and 4 but with the lower efficiencies. Figure 7 indicates that with these lower efficiencies for the no waste sulphide case, the embodied energy for thermal and electricity applications are now both higher than for heap leaching. However, if waste sulphide is available, Figure 8 indicates that even with these lower efficiencies the embodied energy for both applications is still less than for heap leaching, although only the thermal application continues to have an embodied energy credit.

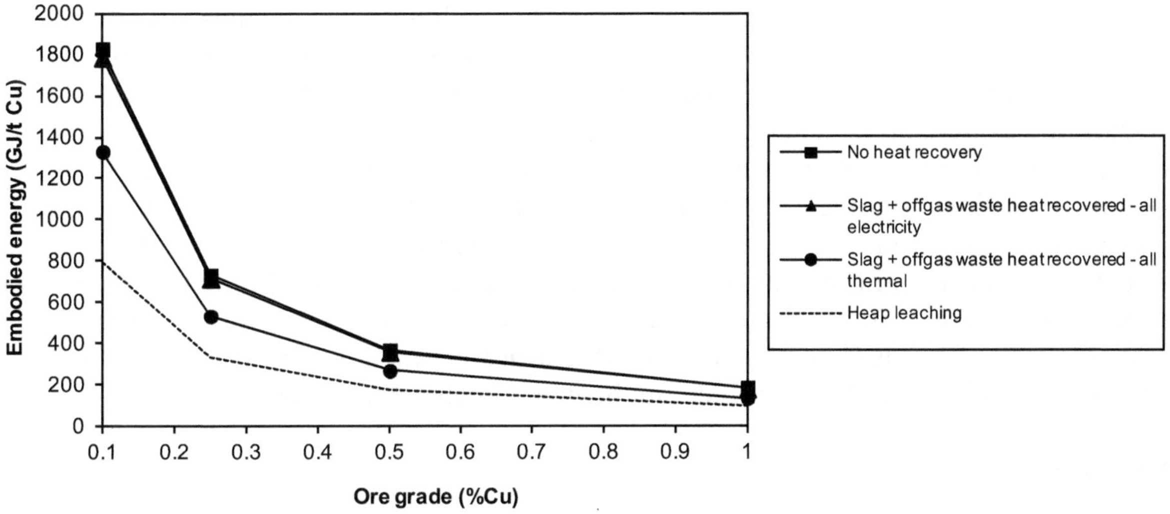

Figure 7. Embodied energy of copper production with reduced thermal efficiencies of waste heat recovery and utilisation (no waste sulphide case).

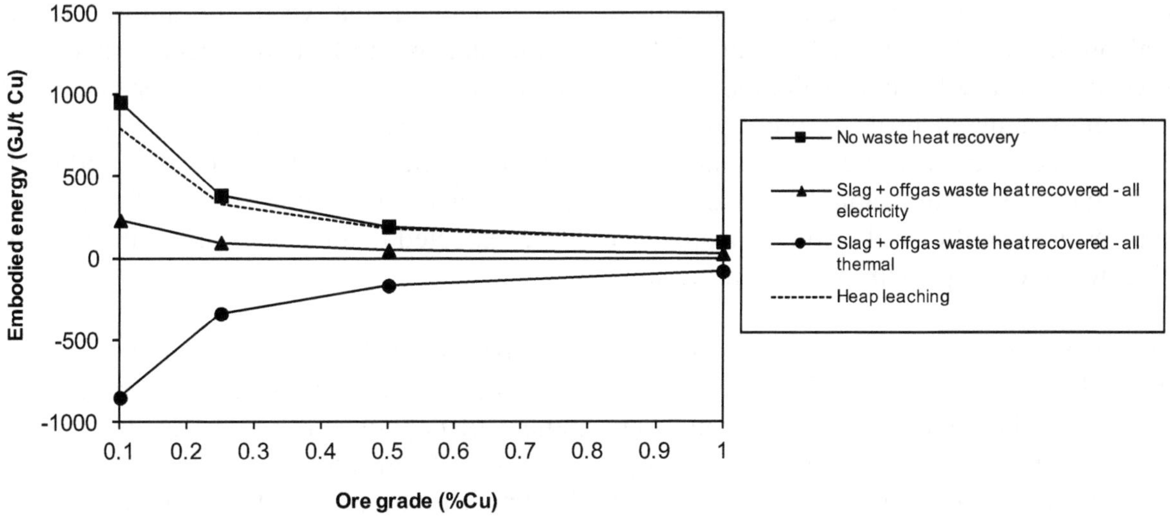

Figure 8. Embodied energy of copper production with reduced thermal efficiencies of waste heat recovery and utilisation (waste sulphide case).

Drivers For Waste Sulphide Use

It was pointed out earlier that many copper mines have waste sulphide dumps containing significant amounts of iron sulphides. Exposure of these waste sulphide dumps to air and water can give rise to acid mine drainage (AMD) which can enter waterways and result in significant (and often long-lived) environmental impacts. As illustrated in this study, this waste sulphide material can be used as an energy source for smelting, replacing fossil fuel consumption and thereby reducing the embodied fossil fuel energy and associated greenhouse gas footprints of copper production. If energy supply is the main driver for use of the waste sulphide, the most energy and resource efficient approach to its utilisation is to use an amount that is just sufficient to cover all smelting energy requirements as for the waste sulphide case in this study (see Table I). This approach utilises the sulphide oxidation energy directly in smelting while at the same time maximising the waste sulphide resource for ongoing use.

On the other hand, if AMD is the main driver for waste sulphide use, the most effective approach for its utilisation would be to use as much of this material as possible in the smelting stage, thereby removing the source of the AMD problem as quickly as possible. In this case it would be necessary to ensure that the smelting operating conditions were such that all the additional iron and sulphur in the waste sulphide are oxidised so that they report to the slag and offgas streams respectively, without significantly influencing the matte phase. It is likely that copper losses would increase as a result of the increased slag volume. This approach is not the most energy efficient approach to utilising the waste sulphide material because the thermal inefficiencies associated with waste heat recovery and utilisation mean that less of the excess sulphide oxidation energy is utilised compared to its direct use in smelting. However, the focus of this approach is to address the AMD problem in the most timely manner possible.

Other Issues

While the results of the LCA study described in this paper have shown that conceptually rock smelting with slag and offgas waste heat recovery has the potential to reduce the embodied fossil fuel energy (and greenhouse gas emissions) for copper metal production from low grade ores, there are a number of other issues that must be considered in assessing the feasibility of this processing route. One is the large amount of slag generated relative to the amount of copper metal produced which must be handled through the slag waste heat recovery system. This is shown in Figure 9 for both sulphide scenarios considered. Economics is another issue that has not been considered in this study which will have a significant impact on the viability of the process. The large volumes of ore and slag relative to the amount of copper produced will substantially affect the capital costs of the process and also the operating cost if transportation distances are significant. For this reason, it would be expected that the rock smelting plant would be located in close proximity to the mine site.

The concentration of copper oxide in the slag for the base case shown in Table I (1%Cu ore) was 0.03%. The high slag volume produced by rock smelting, particularly as the ore grade falls, could affect the copper recovery through entrapment of copper matte prills in the slag. Thus settling is required to ensure good copper recovery from the emulsified matte in slag. Furthermore, rather than aiming for a high grade copper matte of 65% Cu as mentioned earlier, lowering the target matte grade could potentially reduce copper losses to the slag. The use of a slag cleaning furnace to recover the dissolved copper oxide as well as entrapped sulphide copper could also assist in this regard.

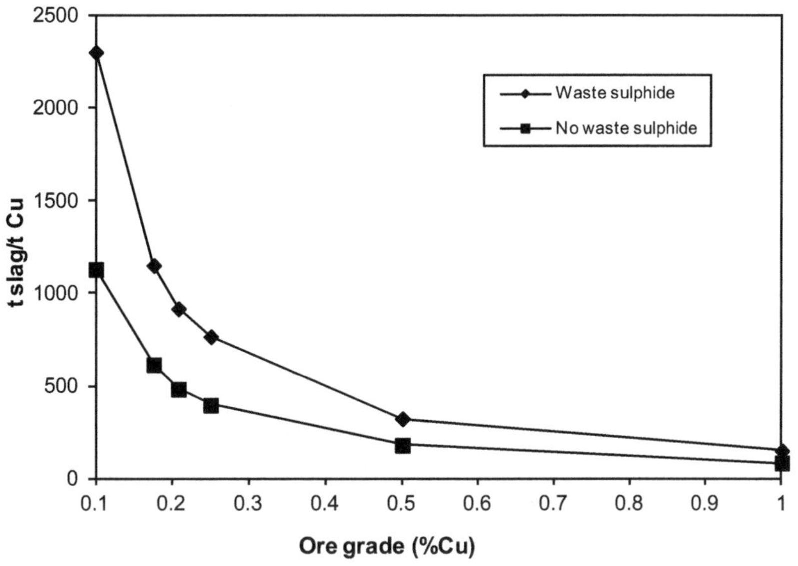

Figure 9. Slag/copper ratio as a function of ore grade.

Conclusions

Falling metallic ore grades together with an increasing focus on sustainability has raised the issue of developing possible alternative processing routes for low grade ores with lower energy intensities and/or environmental impacts than traditional processing routes. Rock (or whole) ore smelting has been suggested as one such alternative, and a study was carried out to assess the potential energy and greenhouse gas benefits of a conceptual version of the rock smelting process using copper ore as an example, including recovery of the heat contained in the waste streams (slag and offgas) from the rock smelting stage.

The results of the study indicated that this conceptual processing route does offer the potential for significant reductions in the embodied fossil fuel energy and associated greenhouse gas emissions of copper metal production from low grade ores compared to the most likely other competing processing route, heap leaching, for these ores. This is particularly the case where waste sulphide material is available, as the energy from oxidation of the waste sulphide during smelting replaces fossil fuel energy required for smelting, and with utilisation of the recovered waste heat in either thermal applications or electricity generation. Depending on whether the driver for waste sulphide use is energy supply or AMD management, different approaches to the utilisation of this material (ie. zero net smelting energy or maximum smelting usage) could be expected in practice. However, in the absence of any waste sulphide material, the embodied fossil fuel energy was only less than that for heap leaching with thermal application of the recovered waste heat.

It should be appreciated that as the study was a broad analysis of the rock smelting processing route only, the results should be considered as indicative rather than definitive at this stage. Furthermore, while realistic thermal efficiencies were assumed in the study for waste heat recovery and utilisation technologies based on data reported in the literature, the results will be significantly influenced by thermal efficiencies achieved in practice. Other issues, particularly economics, will also play a significant role in determining the viability of this processing route for low grade ores.

References

1. Norgate, T and Jahanshahi, S. Low grade ores – smelt, leach or concentrate? Minerals Engineering, 23, 2010, 65-73.

2. Xie, D, Jahanshahi, S and Norgate, T. Dry granulation to provide a sustainable option for slag treatment. Sustainable Mining Conference, Kalgoorlie, Western Australia, August 2010, 22-28.

3. Norgate, T, Xie, D and Jahanshahi, S. Technical and economic evaluation of slag dry granulation. Proceedings of AISTech Conference, May 2012a, Atlanta, USA, 35-46.

4. Norgate, T, Xie, D and Jahanshahi, S. Economic and environmental evaluation of slag dry granulation. Proceedings of 4th International Conference on Process Development in Iron and Steelmaking (SCANMET IV), June 2012b, Lulea, Sweden, 507-516.

5. Jahanshahi, S, Zhang, L, Sun, S, Langberg, D, Xie, D, and Chen, C, 2004. Modelling physico-chemical properties of melts and kinetics of coal-slag reactions, in Proceedings 7th International Conference on Molten Slags, Fluxes and Salts, 493-502.

6. Somerville, M, Frazer, E, Vecchio-Sadus, A, Bruckard, W, Jeffery, P, Thurlby, J, Norgate, T, Trang, S, Smith, L and Jahanshahi, S. Copper recovery from direct copper converting slag, CSIRO Minerals Report, DMR-598, June 1997.

7. Vatanaku, M et al. Waste heat utilisation to increase energy efficiency in the metals industry. Energy Technology 2011, The Minerals, Metals & Materials Society, 5-16.

REWAS 2013: Enabling Materials Resource Sustainability
Edited by: Anne Kvithyld, Christina Meskers, Randolph Kirchain, Gregory Krumdick, Brajendra Mishra, Markus Reuter, Cong Wang, Mark Schlesinger, Gabrielle Gaustad, Diana Lados, and Jeffrey Spangenberger
TMS (The Minerals, Metals & Materials Society), 2013

RE-PROCESSING OF MINING WASTE: AN ALTERNATIVE WAY TO SECURE METAL SUPPLIES OF EUROPEAN UNION

Anne-Gwénaelle Guézennec, Françoise Bodénan, Guillaume Bertrand, Annabelle Fuentes, Gael Bellenfant, Bruno Lemière, Patrick d'Hugues, Daniel Cassard, Maurice Save

BRGM, 3 Av. C. Guillemin, 45060 Orléans cedex 2, France

Mining wastes, critical metal, metal recovery, databases, ProMine (EU project)

Abstract

In France, a recently started project handled by the French geological survey (BRGM) is aimed at identifying interesting old mining wastes deposits at the national level and assessing the metal recovery potential of these dumps. By crossing several databases and information from BRGM archives, 95 old mining sites with sizeable tailings dumps were identified. Selection criteria used to draw up this list were chosen mainly on the basis of the "Criticality Report" compiled for the European Commission in 2010, in which 14 mineral raw materials -12 critical metals- have been explicitly named as highly critical for the industrial development of the European Union. In most of these mines which date back hundreds of years or more, only a single or at best a couple of metals were extracted with processes whose performances were considerably lower than those used today. Knowing the type of ore commodities and the processes characteristics, it has been thus possible to assess the presence of valuable elements for each tailings dump. From this list an Ag-Pb French abandoned mine has then been selected as a case study to evaluate the potential of extraction of metals still remaining in the tailings with special focus on Ag and Sb. A global site characterization methodology is proposed which can be extrapolated to other sites according to key parameters.

Introduction

Until recent times, access to mineral resources and especially to metals was not a major challenge since they were considered as abundant and easily extractable. However, with the strong economic growth in emerging countries such as China, India and Brazil, combined to the increase of the world population, the increasing demand in raw materials raises growing concern regarding their availability. Many metals, metalloids or rare earth elements which didn't have any application in the past are now critical elements for the manufacture of high added value products especially in the domain of ICT or "green technologies". For example, new low carbon technologies like wind turbines, electric cars, energy saving light bulbs, fuel cells and catalytic converters, require rare and precious metals for their production. Many of them are by-products of base metals production and their reserves are often poorly known and/or very limited. These elements are not the only ones which are facing scarcity problem. Many more elements that play a crucial role in our daily lives, including nickel, cobalt and copper, are being depleted at a remarkable rate.

Supplying and securing the mineral resources with minimum environmental footprint is a serious challenge, especially for the European Union which accounts for 25 to 30% of the world's metal consumption but produces only approximately 3% of the world's ore. European dependency on metal import is growing every year despite efforts in the development of recycling technologies and in material science. These tensions point out the need to associate the development of the recycling industry to the identification of new potential sources which could be used for the recovery of rare and valuable materials in order to close the gap in raw materials supplying. In addition to deposits of secondary post-consuming waste, old waste deposits related to past mining and metallurgical activities can also be significant reserves of valuable metals. It is important to note here, that these wastes also contain residual quantities of base and precious metals (Cu, Ni, Zn, Co, Au, Ag) which must not be neglected in today's context of resource scarcity.

In France, a recently started project handled by the BRGM (French geological survey) is aimed at identifying interesting old mining wastes as possible secondary resources deposits at the national level and assessing their potential for metal recovery. This paper presents the methodology developed in this project in order to select suitable mining sites where wastes will then be sampled and used to test new recycling processes for metal recovery.

Selection methodology

The methodology developed for the selection of potentially interesting mining wastes deposits has mainly consisted in cross-analyses of the database dedicated to mining wastes developed in the European research project ProMine and archives produced and managed by the BRGM.

Presentation of ProMine Database

ProMine is a European Union (EU) co-funded project, focused on "Innovative concepts and processes for strategic mineral supply and for new high added value mineral-based products". A major goal of the project was to develop a Pan-EU GIS data management and visualization system for natural and man-made mineral endowment. In order to reach this objective, the project produced pan-European databases of primary and secondary mineral resources, the ProMine Mineral Deposits (MD) and Anthropogenic Concentrations (AC) databases, respectively (Cassard, 2012; Cassard et al., 2012). These databases were included in a homogeneous multi-layer information system covering the whole European territory and available through the Internet (http://ptrarc.gtk.fi/ProMine/default.aspx).

The ProMine AC database stores all the information related to anthropogenic mineral concentrations in Europe (mine products and wastes, ore processing wastes and treatment wastes). Each site is described through about 35 fields distributed into four groups: (1) general information, including status, owner, location and the list of processes that have been implemented on the site; (2) information on wastes and products including the type of storage, the type of waste, the mineralogy, estimation of volume/tonnage, the type of commodity available and the grade, with calculation of potential resource, per commodity, at the site scale and the type of environmental impact; (3) the environmental aspects, with per environmental impact, the type of environmental pathways and receptors, the type of water treatment and the description of the type of restoration used; (4) general ccomments, Iconography and Bibliography.

Because it is nearly impossible to compile an exhaustive directory of mining wastes covering the whole European Union, the ProMine AC database focus on major anthropogenic concentrations and on the most interesting in terms of volume/tonnage and content (i.e. possible presence of strategic metals). It is thus an as exhaustive as possible inventory of concentrations which could be processed for the recovery of strategic/high-tech/critical commodities. The total number of records in the database is 3,412, among which 644 are located in France (Figure 1) and were analyzed for the present study.

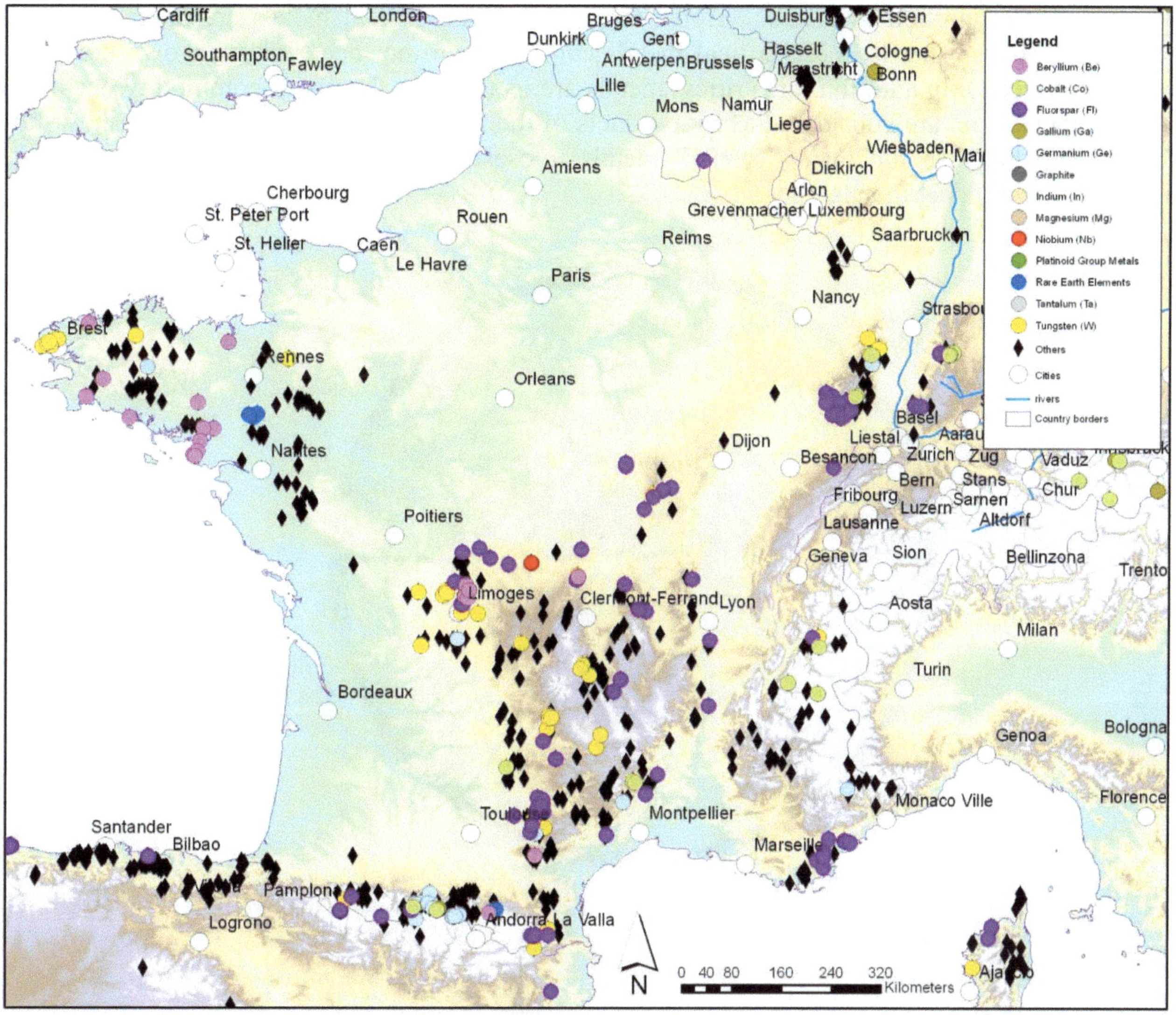

Figure 1 – Past Mining and metallurgical sites in France (with major commodities) with mining wastes, extracted from the ProMine AC database.

Selection criteria

Selection criteria used to perform the researches in the databases were chosen mainly on the basis of the well-known list of strategic metals. They are generally defined as metals that are vital to modern technology and industry, but that have sources susceptible to disruption. Strategic metals are distinct from "precious" metals, like silver and gold, because most are not as vital to either technology or industry in the same way. Likewise, they differ from "base" metals, such as copper, lead, iron and zinc, in that those metals are relatively abundant in locations around the world.

In July 2010 a working group of the European Commission under the umbrella of the Raw Materials Supply Group identified, among the list of strategic metals, 12 critical metals with significant concern for future supply in the European Union: Antimony – Sb, Beryllium – Be, Cobalt –Co, Gallium -Ga, Germanium –Ge, Indium – In, Magnesium -Mg, Niobium – Nb,

Platinum Group Metals - PGMs, Rare Earths - RE, Tantalum- Ta, Tungsten –W,. The main parameters for identifying the 12 critical metals were supply risk and economic importance. Taking into account the specificities of the French industrial context, some other metals, like silver or titanium, have been added to the European list. It is important to note that in many cases a significant proportion of primary global resources of the critical metals are hosted in poly-metallic deposits. That is, these critical metals commonly do not occur on their own but along with other metals. In addition, the critical metals are usually not the major commodity occurring in the deposits and are considered as by-products of the main carrier metals. Figure 2 shows the relationship between the main "carrier" metals (closest to the center of the circle) and the associated co-product metals.

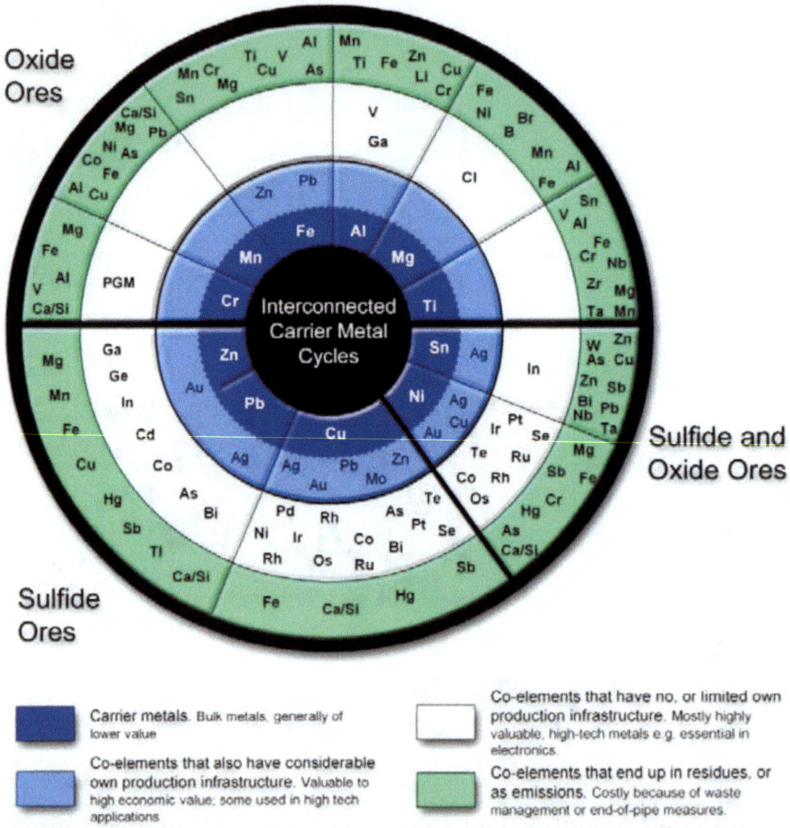

Figure 2. The 'Metals circle' shows the occurrence of minor and critical metals with the primary carrier metal ores. The co-metals surround the inner circle of carrier base metals. Poly-metallic copper, nickel, zinc and lead resources will likely be accompanied by several of the critical metals.

The querying and cross-analysis of the databases was made through the list of critical metals but also through other relevant criteria such as the volume of wastes, the operating dates, the treatment process, the environmental impacts. This work led to the selection of 95 old mining sites with sizeable tailings dumps. In most of these mines, among which some date back hundreds of years or more, only a unique or - at most - a couple of metals were extracted then processed with performances that were considerably lower than what could be achieved today. Knowing the type of ore commodities and the processes characteristics, it has been possible to assess the presence of valuable elements for each tailings dump.

Assessment of metal recovery potential: application to a case study

From the selected sites, the study focused then on an ancient Ag-Pb French mine, abandoned since the end of 19th century. The purpose was to develop a holistic concept for the recovery of trace metal remaining in the wastes that could be applied more generally to other mining wastes.

Presentation of the case study

An initial geological/mineralogical assessment of the mining wastes dumps remaining on site was carried out from available historical documents. The ore mineralization was dominantly composed of silica and sulphides: silver and bismuth-bearing galena, sphalerite, pyrite, bournonite ($PbCuSbS_3$) and chalcopyrite in a silicate gangue. The mine was operated from early 19th century for Pb and Ag production. The historic beneficiation schemes comprised successive steps of grinding and physical separation at millimetric scale. No flotation was performed; the treatment mainly consisted in gravimetric separation and manual sorting. Gravimetric separation handled at that time did not enable to treat particles lower than 100 µm, which usually lead to significant losses of silver.

The wastes were stockpiled in four dumps accounting for approximately 87,000 m^3 with different particle size aspects:

- Two dumps are made mainly of homogeneous and fine materials which are supposed to be the wastes produced by the gravimetric techniques used in this mine. The volume of those dumps is estimated respectively to 48,400 m^3 (Dump I) and 5,200 m^3 (Dump II).
- One dump is made of coarse particles probably removed from the treatment by gravity separation with shaking tables. Its volume is estimated to 13,500 m^3 (Dump III).
- The last dump is composed of more heterogeneous materials which are probably a mix of the previous ones. Its volume is estimated to 20,100m^3 (Dump IV).

Sampling and characterization methods

For a reliable estimation of the geochemical characteristics of the materials contained in the four dumps a representative sampling method is required. For this purpose a sampling map was established for each dump. On-site XRF chemical screening analyses were performed on the surface and on bulk samples of approximately 5 kg taken at regular intervals down to 5m depth. In the lab samples were dried at 40°C, homogenized and re-analyzed. On a selection of sample ICP-AES, X-ray diffraction and mmicroscopic observations were also performed to investigate metal speciation and liberation size (how far tailings should be ground for metal extraction)..

Some of the samples were sieved to produce several particle size fractions (+2 mm, 800 µm, 400 µm, 200 µm and 80 µm) which were weighed and analyzed by FPXRF.

Preliminary results

The preliminary results demonstrate that the materials of each dump present quite homogeneous metal contents. As an example, the figure 3 shows Pb, Fe, S and Ag contents obtained from FPXRF analysis on the samples coming from the dump IV. The preliminary results obtained from analysis of particle size fractions demonstrate that particles upper to 800 µm and lower to 80 µm are enriched in lead and silver. This is probably due to the type of ore beneficiation treatment used in this mine where fine and coarse particles where rejected in the tailings.

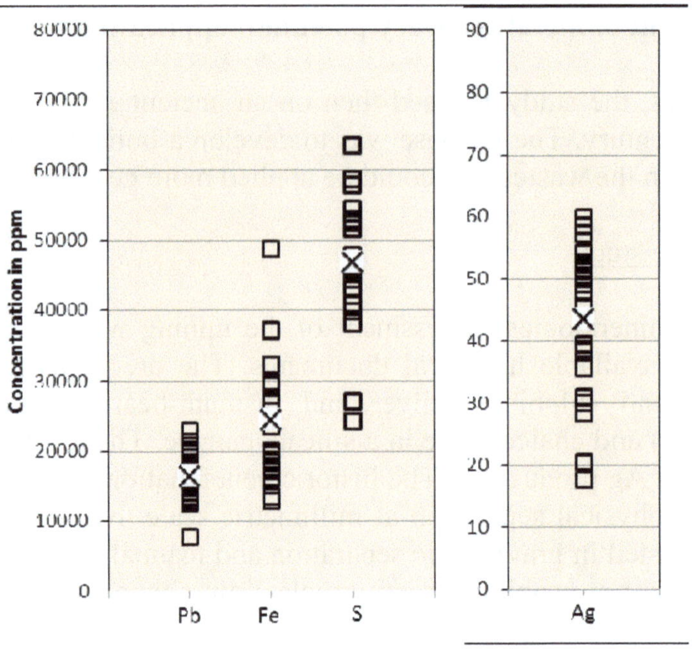

Figure 3: Results of FPXRF analysis for Pb, Fe, S, Ag, obtained from samples collected on dump IV (squares) and average contents (crosses)

From these data a treatment process flowsheet was discussed and is in progress to concentrate Pb and then Ag by sieving separation and gravimetric techniques. Metallurgical balances will then be established to affine knowledge on this site.

Conclusions and Perspectives

The study presented in this paper aims at (i) identifying on the French territory mining wastes with valuable metals recovery potential, and (ii) developing a methodology to characterize the mineralogical and chemical composition of those wastes. It is based on a multidisciplinary approach which must be led to the development of suitable and innovative processing technologies for further recovery of remaining metals.

By cross-analyses of suitable databases and archives from the French geological survey, 95 abandoned mining sites with sizeable wastes dumps were identified as potential targets for future extraction of remaining metals. One of them, an old Ag-Pb mine operated during the 19th century, has been selected in order to test an integrated approach to assess the metal contents of those wastes. Inventory of historical data related to the mine and a sampling campaign following by detailed investigations of mineralogy and geochemistry were performed. Preliminary results show homogeneous metal contents in the dumps and an enrichment of lead and silver in the fine and coarse particles. From the whole results of chemical and mineralogical analysis, an appropriate process flowsheet will then be proposed for the recovery of silver which is the target metal in this case study.

Acknowledgements: We thank P. Gentil for access to data and site and M Beaulieu for precious help during field trip.

References

Cassard D., 2012. The use of standard exchange EarthResourceML in the ProMine project. 34th International Geological Congress (IGC), 5-10 August 2012, Brisbane, Queensland, Australia, #1993

Cassard D., Bertrand G., Maldan F., Gaàl G., Juha K., Aatos S., Angel J.M., Arvanitidis N., Ballas D., Billa M., Christidis C., Dimitrova D., Eilu P., Filipe A., Grazea E., Inverno C., Kauniskangas E., Maki T., Matos J., Meliani M., Michael C., Mladenova V., Navas J., Niedbal M., Perantonis G., Pyra J., Santana H., Serafimovski T., Serrano J.J., Strengel J., Tasev G., Tornos F., Tudor G., 2012. ProMine pan-European Mineral Deposit database: a new dataset for assessing primary mineral resources in Europe. Mineral Resources Potential Maps : a Tool for Discovering Future Deposits. 12th-14th March 2012, Nancy, France

REWAS 2013: Enabling Materials Resource Sustainabi
Edited by: Anne Kvithyld, Christina Meskers, Randolph Kirchain, Gregory Krumdick, Brajendra Mish
Markus Reuter, Cong Wang, Mark Schlesinger, Gabrielle Gaustad, Diana Lados, and Jeffrey Spangenber
TMS (The Minerals, Metals & Materials Society), 20

Potential of Steelmaking Slag as New Phosphorous Resource in terms of Total Materials Requirement

Eiji Yamasue[1], Kazuyo Matsubae[2], Kenichi Nakajima[3] and Tetsuya Nagasaka[2]

[1] Kyoto University

[2] Tohoku University

[2] National Institute for Environmental Studies

Key Words: Phosphorous, steelmaking slag, total materials requirement, resource dependency

Extended Abstract

Despite of importance of phosphorous for agricultural food production and for the chemical industry, it is present in nature only as a trace element, and what is worse is that high-quality ores are drastically decreased.

It is therefore important to consider the quantity and availability of untapped phosphorous resources. One of the possible candidates would be some steel-making slags because the total amount of phosphorous in such slags in Japan is comparable to the imported amount. Such the slags consist of phosphorous rich (more than 10 mass%) and free phases, the each of which can be separated under the strong magnetic field after pulverization. The separated phosphorous-rich phase is possibly utilized as a new phosphorous resource and the residual matrix can be recycled to iron and steel-making processes as flux.

One of authors has proposed the novel evaluation method of recyclability of materials in urban mine in terms of total materials requirement (TMR), that is, the TMR for recycling (urban ore TMR, UO-TMR).

In many of industry, phosphoric acid is rather preferable to elemental phosphorous. Thus the aim of this study is to evaluate the quality of phosphoric acid derived from a steel-making slag compared with that from natural ore from the viewpoint of TMR.

The TMR of phosphoric acid produced from natural phosphorous ore (NO-TMR) and recycled from a steel-making slag (UO-TMR) using strong static magnetic field were estimated and compared. Furthermore, not only TMRs of phosphoric acid but also those of by-products such as gypsum were estimated as well.

As the results of estimation, the NO-TMR of phosphoric acid was estimated to be 6.0~11.8 kg/kg and UO-TMR was 5.6~29.4 kg/kg. The fluctuation is ascribed to utilization efficiency of by-products. These results mean that the steelmaking slag has potential as new phosphorous resource in terms of TMR.

TMR can be thought of as the indicator that can evaluate how much mining sites are developed to obtain resources. This means that the degree of self-sufficiency of phosphorous from two resources can be evaluated by analyzing the country-by-country breakdown of TMR for these materials.

ASSESSING A RECLAIMED CONCRETE UP-CYCLING SCHEME THROUGH LIFE-CYCLE ANALYSIS

Sylvain Guignot[1], Kathy Bru[1], Solène Touzé[1], Yannick Ménard[1]

[1]BRGM; 3 avenue Claude Guillemin; Orléans, 45060, France

Concrete, aggregate, electro-fragmentation, recycling, life-cycle analysis

Abstract

The present study evaluates the environmental impacts of a recycling scheme for gravels from building concretes wastes, in which the liberated aggregates are reused in structural concretes while the residual mortar fines are sent to the raw mill of a clinker kiln.

The evaluation follows a life-cycle analysis approach performed according to the ISO standard 14040, and whose scope encompasses the production of clinker through a dry kiln technology, the mining processes of the raw materials needed in the kiln, the extraction of round and crushed natural aggregates, and the crushing of concrete wastes using usual jaw crushers or pulsed-power electrical fragmentation. Insofar as possible, the inventory data are collected at the national scale of France and are recovered from the supplier of the fragmentation device, from local quarries and from an estimated mean-technology of clinker production. The choice of the impact assessment indicators is restricted to midpoints according to a problem-oriented methodology, and primarily focuses on a potential reduction in the natural resources depletion and in the CO_2 emissions. The study specifically addresses the influence of *(i)* the amount of recovered cement paste added to the kiln raw mill, and *(ii)* the distance of transportation modalities of concrete wastes to the crushing processes and of the recycled aggregates to construction sites. The results establish links between significant environmental gains and the various distances of transportations that intervene in the alternative processing of concrete wastes. These links will be probed more deeply in a future work.

Introduction

The current management policy of reclaimed concrete in France directs 99 % of recycled coarse gravels to low added-value roadway pavement applications. In 2008, 38.2 Mt of wastes were generated by C&D operations, from which 15 Mt of gravels were recovered, a figure that could reach 23 Mt provided the best available techniques for deconstruction and source separation were more widely implemented. On a technical point of view, gravels in reclaimed concretes are recovered through impact and jaw crushing, sieving stages and over-band magnetic separations of scrap metals [1]. When replacing part of the natural aggregates in concretes, recycled aggregates (RA) entail: *(i)* an increased drying shrinkage, *(ii)* decreased compressive strength, splitting tensile strength and modulus of elasticity, but *(iii)* no significant variations concerning frost resistance, chloride penetration and rate of carbonation [2, 3]. The loss in compressive strength is ascribed to the weaker adhesion of RA to the surrounding cement matrix in concrete, which is due to higher water absorption of the residual mortar covering the surface of the aggregates [4]. This absorption can vary between 4 % and 8 % of the aggregate initial mass, while only between 0.8 – 3.7 % for NA [3]. Until recently in France, such absorption capacity banned any use of recycled aggregates in concretes [5].

Over the last three decades, several avenues have been explored to address this absorption issue. The first of them consisted in prewetting the recycled aggregate to the condition termed "water saturated surface dry" before concrete batching, hence preventing the attached mortar from further drawing local water [4]. However, *(i)* from a practical point of view, such saturated conditions are difficult to be maintained on a construction site, and *(ii)* prewetting the aggregates also implies a concomitant increase in cement, to ensure correct compressive strength and drying shrinkage at 28 and 182 days [6] and maintain a good concrete consistency and workability [7]. The second approach to deal with the residual mortar lie on studies having evidenced a better reliance towards local cracks propagation in concretes made from recycled aggregates, reliance which could be due to local reductions in stiffness promoted by the adhering mortar [9]. The idea is therefore to mitigate the negative effect of RA with natural aggregates, and thus, designing mixed-aggregates concretes. However, the precise quantification of the upper limit for RA addition is still difficult by its dependency to the chemical properties of the RA, to the quality of concretes from which they stem, to the kind of cement used and to the technology of concrete production. A broad common estimate for RA replacement ratio in concrete is about 30 % [8]. Lastly, the third approach to manage cement paste on the RA was to design cleaning process. Several methods have been put forward [6], among which gravity concentration [10], screw grinding [11], and heating and rubbing of concrete [12]. Gravity concentration and screw grinding seem to only fit to coarse aggregates (size above 20-25 mm), with low recycling rates, about 20 % and 34-45 % respectively. Crushed concrete comprises 49.1 % of aggregate plus adhered mortar and 43 % of clean aggregates, the remaining part being composed of ceramic, bitumen and other materials [14]. The use of organic extractants has also been proposed by [13], but the environmental gains brought about by the reuse of the aggregates are counterbalanced by the use of organic extractants and the numerous washing steps associated.
On these considerations at the present time, a choice seems to be made between *(i)* producing cement-covered aggregates after a few basic liberation and separation processes, assess their potential use in structural concretes of various grades through standardized indicators related to their physical properties, and determine the extra amount of cement required to maintain comparable concrete properties, *(ii)* or ridding the aggregates of their mortar layers through more advanced, less environment-friendly separation processes to ensure their use in high-quality concrete.

Recently, our team extended the work of [17] and demonstrated that, by pulsing electrical discharges into concrete blocks, it was possible to separate coarse clean aggregates from a fine fraction having a high grade of a cement paste. Such a fraction could provide substitutes to natural limestone, clay and sand in the raw mill of a clinker kiln, while the aggregates could be up-cycled in high quality structural concretes. Both of these outlets would provide numerous advantages. First, using recycled aggregates could help to meet the forthcoming stringent legislation on the opening of new quarries of natural aggregates. Second, recycling aggregates could also save money by disposing of a cheaper aggregates supply (the pre-tax price of clean recycled aggregates ranges from 15 to 25 euros per m^3 in France [16]), while avoiding paying tipping fees to landfill inert wastes (amounting to 5 euros per metric ton in France). Third, substitutes to raw mill result in reduced consumption of natural calcareous resources, coupled with reduced carbon dioxide emissions stemming from less dissociation of carbonated materials [15]. Using recycled aggregates would lessen the environmental burdens of landfilling installations, while reducing the consumption of non-renewable mineral resources. Besides it would offset local shortages of aggregates, thus reducing the need for transportation of aggregates over long distances and the various greenhouse gas emissions associated. Thus, the environmental impacts of this recycling scheme remain to be quantified.

To the best of our knowledge, studies addressing this specific issue are rather scarce. In a LCA study addressing the production of low and high quality concretes, [18] underlined the significance of even the slight additional amount of cement required by the low quality aggregates (about 5 % according to [2]), which tended to increase some environmental categories linked to respiratory effect, climate change and energy use. On the contrary, incorporating the aggregates into low quality concretes entailed no increase in cement content, and hence all environmental indicators were reduced. [19] recently reported an evaluation of a recycling scheme in which the aggregates were recovered from a heating and rubbing process and were incorporated to new concretes, while mortar was used either as a soil stabilizer or as a raw material in clinker. As long as mortar was valued, the CO_2 balance became significantly negative (about -217 kg_{CO2} per ton of concrete block).

Yet, these works seems to miss the fact that using RA in structural concrete rather than in roads implies that the part of gravels diverted from road to construction will have to be made up for by virgin aggregates from quarries. The overall mass balance might be relatively unaffected, but the environmental balance may be changed owing to the differences in the cleaning steps aiming at removing mortar. While part of the reclaimed concrete gravels have to meet harsher technical specifications to be used in structural concrete, the aggregates they replace will follow the reverse pattern and will not require an advanced treatment to be included in road bases. Lastly, as underlined by [2], the transportation distances of any recycling scheme dealing with heavy materials have to be heeded for, as they can set the borders between environmentally beneficial recycling schemes and the others. Therefore, various scenarios should be built up to examine the sensitivity of the environmental impacts to the transport distances of aggregates and raw materials, as well as to the way they are transported, in different concrete recycling schemes.

The present paper reports an evaluation of the environmental impacts of this alternative management of building concretes, performed according a Life Cycle Analysis methodology. For each operation considered in the processes, inputs and outputs are collected and aggregated to determine their impact on various environmental categories. These inputs/outputs cover consumptions of minerals, water or energy, as well as waste heat, air emissions, water emissions, and solid wastes to the ground. The implementation of LCA comprises four successive steps: the definition of the goal and scope of the study, the inventory of data from the considered processes, the impact assessment of these data, and the interpretation of the results [20].

Goal And Scope

Goal

Current practices in concrete recycling consist in largely non-selective crushing and screening processes, which do not allow a straight separation between the clean aggregates and the cement paste. The resulting gravels are thus restricted to low-added value applications mostly in earthworks (excavation filling, roadbeds, or floor foundation).
Fragmentation of concrete rubbles by means of pulsed electrical discharges can enhance the aggregate separation from the cement paste in the concrete gravels [21], thereby broadening their range of use towards structural concretes. As an additional advantage, the hydrated cement paste mixed with sand (also termed Recycled Cement Paste or RCP) contains numerous chemical elements that could advantageously be reused to manufacture new clinkers.

Therefore, incorporating fragmentation in the technological handling of concrete wastes result in defining a new global processing scheme for these wastes, scheme which comprises new outlets for the clean aggregates and the RCP, as well as quantitative modifications in the raw materials production for earthworks, building concrete and clinker production.
The objective of the present study is therefore to compare these two processing schemes of concrete wastes according to their respective environmental impacts. The study is thus primarily intended for aggregates suppliers, cement and concrete manufacturers, but can also be of interest to the industry of solid waste management, to several environmental organizations, and to LCA practitioners.

System Boundaries And Functional Unit

A preliminary system would include the emissions and mineral resources and energy consumptions associated with: *(i)* the production of the raw materials needed in the clinker kiln, and *(ii)* the usual concrete crushing and the pulsed-power fragmentation processes.
For both processing schemes of concrete wastes, and whatever the functional unit considered - processing 1 kg of waste building concrete or producing 1 kg of aggregates – a multi-functionality issue occurs due to the fact that the aggregates produced by the two systems have different outlets, either for structural concretes or for road constructions. This multi-functionality is unraveled in accordance to the recommendations of the ISO 14040 standard [22] by extending the system boundaries to include subsystems fulfilling the functions of *(i)* producing natural aggregates for concrete production and *(ii)* producing crushed aggregates for road embankments. In a processing view, the final system is therefore divided into 6 sub-units (Figure 1), and the common function of the two recycling scheme included is "processing 1 kg of waste concrete".
Both inputs are concretes pre-sorted from the building rubbles which contain no scrap metal that would come from steel-reinforced structures. The system is not extended beyond aggregates production to concrete manufacturing, since the produced aggregates have the same quality, hence the same use, as natural aggregates. Thus the comparison is drawn between two extreme situations – with a 0 % or a 100 % RA reuse in structural concretes. This deviates only slightly from current practices in France, from which almost 2 Mt in 17 Mt of RA are reused without further treatments. All the processes of production are followed by the cleaning operations needed for the materials to meet the quality standards of their respective function.
Second, in line with the recommendations of [23], the capital goods are included in the present LCA since part of it is intended to compare the production and the operation of alternative equipment (pulsed power fragmentation) with the continued use of existing equipment (concrete crushing). These infrastructure data are fed into the inventory table after being averaged on the totalized "functional unit" outputs of the process on its estimated lifetime. In order to be consistent, infrastructures data are also accounted for in the inventory.
Lastly, the LCA involves several transportations of heavy materials, which results in significant emissions to air, and for that reason, transports of the raw materials, of the building concrete wastes, and of the natural and recycled gravels are within the system boundaries.

Figure 1. Boundaries of the systems considered. Up: new scheme, down: current scheme

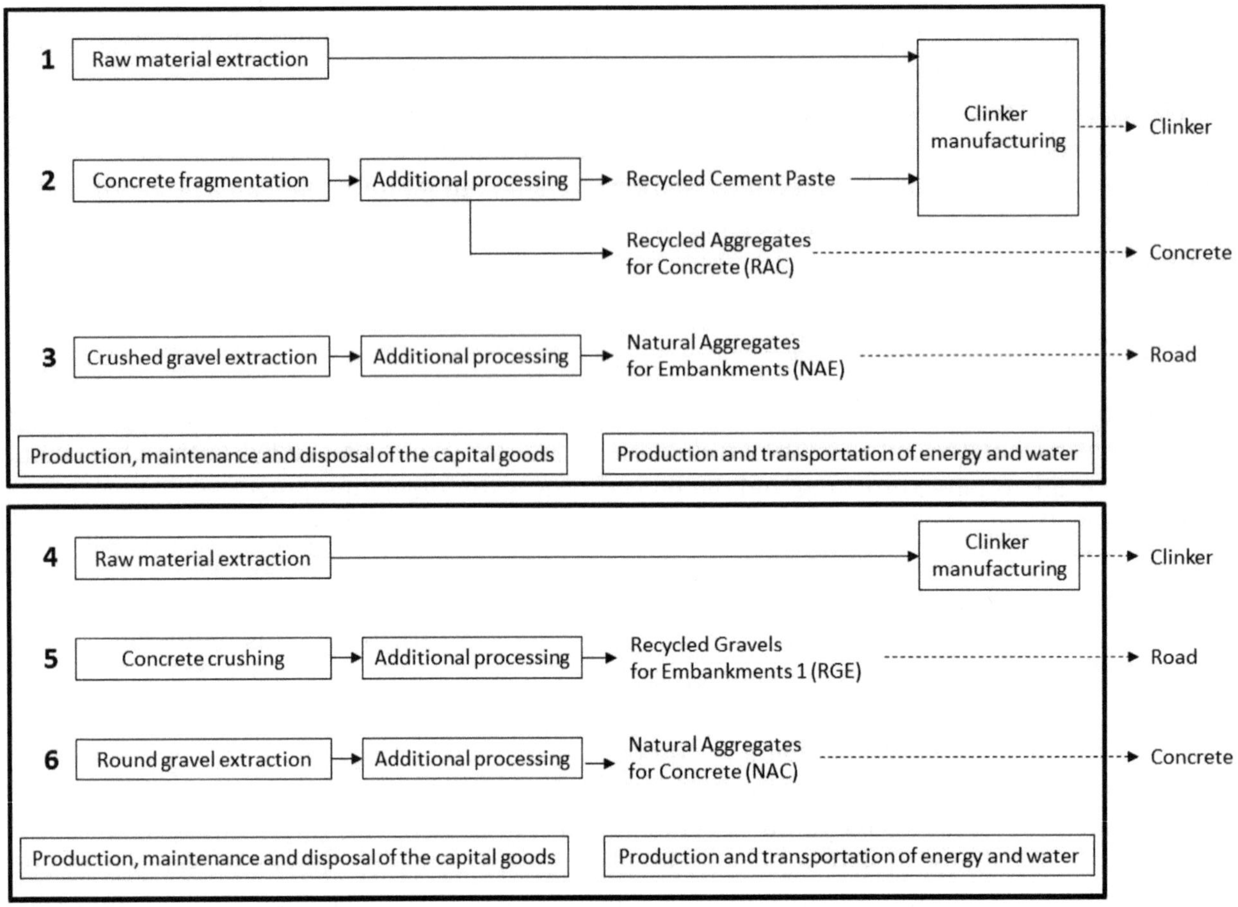

In Figure 1, it is assumed that crushed and round aggregates correspond respectively to aggregates for building concrete and road applications. The reverse case will be also implemented in the discussion part of the present LCA.

Comparison Of The Usual And The New Recycling Scheme

The relative increase in the natural aggregate consumption for embankments is accounted for by subtracting the environmental impacts associated with sub-unit (5) from that of (3). The decrease in the contribution of NA to concrete manufacturing is heeded by the subtraction of the impacts associated with (2) from the impacts of (6). The decrease in the consumption of natural resources for clinker manufacturing is obtained by subtracting the environmental impacts of (4) from (1). Let p the recovery rate (on a mass basis) for aggregates in the sub-unit (2). Given that the common functional unit for both recycling schemes is 1 kg of concrete waste, (2) produces p kg of aggregates and $(1-p)$ kg of recycled cement paste (see Table 3). Therefore, the environmental impacts calculated from [(3)-(5)] and [(2)-(6)] must be multiplied by p. Considering the recovered cement paste, calculations presented in the Inventory section enable to estimate its allowed range of content in the clinker raw feed. Let P_{RCP}^{ck} this mass proportion of recycled cement paste related to the production of 1 kg of clinker. The mass of clinker which can be formed from the remaining $(1-p)$ kg of recovered cement paste is thus $(1-p)/P_{RCP}^{ck}$. Therefore, the environmental impacts of [(1) – (4)] must be calculated for a throughput of $(1-p)/P_{RCP}^{ck}$ kg.

As a result, the environmental extra-cost (EEC) brought by the use of the alternative process of concrete fragmentation will be obtained by applying the following formula:

$$EEC = p.[(3)-(5)+(2)-(6)]+(1-p)/P_{RCP}^{ck}.[(1)-(4)] \quad (1)$$

This data treatment provides a solution for the allocation issue arising from the concomitant production of aggregates and recycled cement paste in the concrete fragmentation process. The environmental burdens are allocated on the basis of the mass repartition of the materials since the hydrated cement paste could not be economically balanced against the aggregates.
Lastly, note that this system excludes the landfilling of non-valuable materials stemming from process (5), and thus minimizes the environmental impact of the usual recycling scheme.

Impact Assessment

The environmental categories considered for the impact assessment depend on the position chosen on the impact pathway of a given substance, they may be chosen at a midpoint or at an endpoint position. As defined in [24] the mid-points indicators concern the human toxicity, the ozone layer depletion, the formation of photooxidant species, the inorganic respiratory effects, the climate change, the aquatic and terrestrial ecotoxicity and eutrophication, the acidification, the abiotic and biotic resources depletion, and the land use. These indicators are further aggregated in indicators accounting for the final environmental impacts, or endpoints, on the human health, the abiotic and biotic natural environment, the abiotic and biotic natural resources and the abiotic and biotic artificial environment. For example, sulphur dioxide causes acid rain (mid-point indicator), and damages both human health and the natural environment (end-point indicators). The farther the indicators are on the impact pathway, the greater the associated uncertainties are. In this study, priority was given to the precision of the results, and therefore the indicator point was put as close as possible to the associated emissions, following a problem-oriented, rather a damage-oriented, methodology.
The environmental impact of a given process can be global or local in scale [25], and the choice is made in the present LCA to assess global effects from local inventory data, as advised by [26]. The study aims at assessing the long-term impacts of an alternative management scheme of concrete wastes on the resource availability, the ecosystem diversity, and the human health. For these reasons, the ReCiPe MidPoint method 1.06, egalitarian version, is chosen for impact assessment [27]. The alternative scheme should enable to reduce the CO_2 emissions, the natural resources consumption, and the energy use. CO_2 affects the human health and the ecosystem integrity through its contribution to the radiative infrared forcing of the atmosphere, therefore, climate change (CC) is a relevant indicator. Less consumption of natural aggregates can be evaluated by the impact category "natural land transformation" (NLT), and conversely, the decrease in the energy use by the category "fossil fuel depletion" (FD). In addition, and owing to the fact that the present processes can be well documented regarding their emission in SOx and NOx, the terrestrial acidity (TA) is also considered, a relevant choice according to a study dealing with the environmental impact of cement production performed by [28]. By contrast, and due to a lack of data, the various indicators related to eutrophication, ecotoxicity, water depletion and particulate matter are discarded from the impact assessment.
The four selected indicators are therefore climate change (CC), natural land transformation (NLT), fossil fuel depletion (FD) and terrestrial acidity (TA).

Data Quality

Given that the present LCA is intended to be used in a decision-making process, the quality of technical data is crucial. A specificity of this work is that it involves the extraction and use of aggregates from earth and from building wastes. Considering that the physical properties of the natural aggregates depend on the geographical situation of the extraction site, and that the extraction process differs with the nature of aggregate, a special attention must be paid in using local rather than global averaged process data. Obviously, a precise description of every quarry procedure cannot be envisaged. But, documenting two extremes extraction process (respectively from an alluvial source or from a solid rock) allows taking into account this situation to some extent. Another consequence is that transportation from a natural mine or from a demolition site to the pulsed power or to the usual crushing process are to be precisely defined, and again, at a regional scale. Data from processes are collected on ongoing industrial processes as far as possible. Lastly, any missing inventory data is supplied by the EcoInvent 7.3.3 database.
The Life Cycle Inventory was compiled in the SimaPro software (Pré Consultants), which also performs the impact assessment using a user-selected method of calculus

Choice Of Variables Assessed

The present study aims at catching the influence of the main variables defining a recycling scheme of building concrete gravels based on a straight separation of aggregates and hydrated cement paste. This separation is performed by using specific equipment based on pulsed power fragmentation. Furthermore, the recycling scheme addresses heavy materials (may it be aggregates, raw materials for clinker kiln, or concrete gravels) and consequently, the transport modalities are of uttermost significance in an environmental evaluation.
On such consideration, the variables taken into account are listed in Table 1. Their specific values are specified in the following Inventory part.

*Table 1. Variables considered in the study. *Waste Concrete Treatment Unit*

Variables		**Evaluation**
Recycled cement paste (RCP) content in clinker kiln		Chemical analysis of RCP
Transportation distances	Raw materials local quarry → clinker plant	Data from statistics offices and scientific literature
	Natural aggregates quarry → road construction site quarry → building construction site	
	Concrete gravels WCTU* → road construction site	
	HCP : WCTU → clinker plant	
	Recycled aggregates WCTU → building construction site	
Mode of transportation	Hauling	
	Shipping	

Life Cycle Inventory (LCI)

Life Cycle Inventory is carried out by considering each process unit as a stand-alone system, for which the surroundings environment acts as a source of all the materials and fuels inputs to the system and as a sink for all outputs from the system. For each unit process described on Figure 1, an inventory table is established and gathers the materials and energy inputs as well as the individual emissions to the air, to natural water and to the ground.
To account for the data sensitivity to the location and the geographical scale, data are collected from local quarries operators. Furthermore, data quality was improved by contacting the pulsed-power and the jaw crusher manufacturers. Finally, most of the data are reliable with time since they come from the well-established processes of clinker manufacturing, and rock and concrete crushing.

Pulsed Power Fragmentation Of Concrete

A schematic representation of a reclaimed concrete processing is shown Figure 2, along with the alternative part of process under study in the present work, and composed of the high voltage pulse power fragmentation device marketed by Selfrag (and termed Selfrag in what follows). The materials entering the Selfrag must be below 40 mm in size, and therefore a jaw crushing must be maintained in the global line of process. According to Figure 2, the +40 mm fraction is directed towards an impact crusher to recover materials under 40 mm in size. Thus, the Selfrag can replace the impact crusher according to a scheme highlighted by dotted lines on the figure. Several cycles of Selfrag processing are considered to obtain the maximum amounts of hydrated cement paste on one hand and of clean aggregates on the other hand.

Figure 2. Schematic representation of a process for recovering aggregates from concrete. Dotted lines correspond to the electrical pulsed fragmentation device

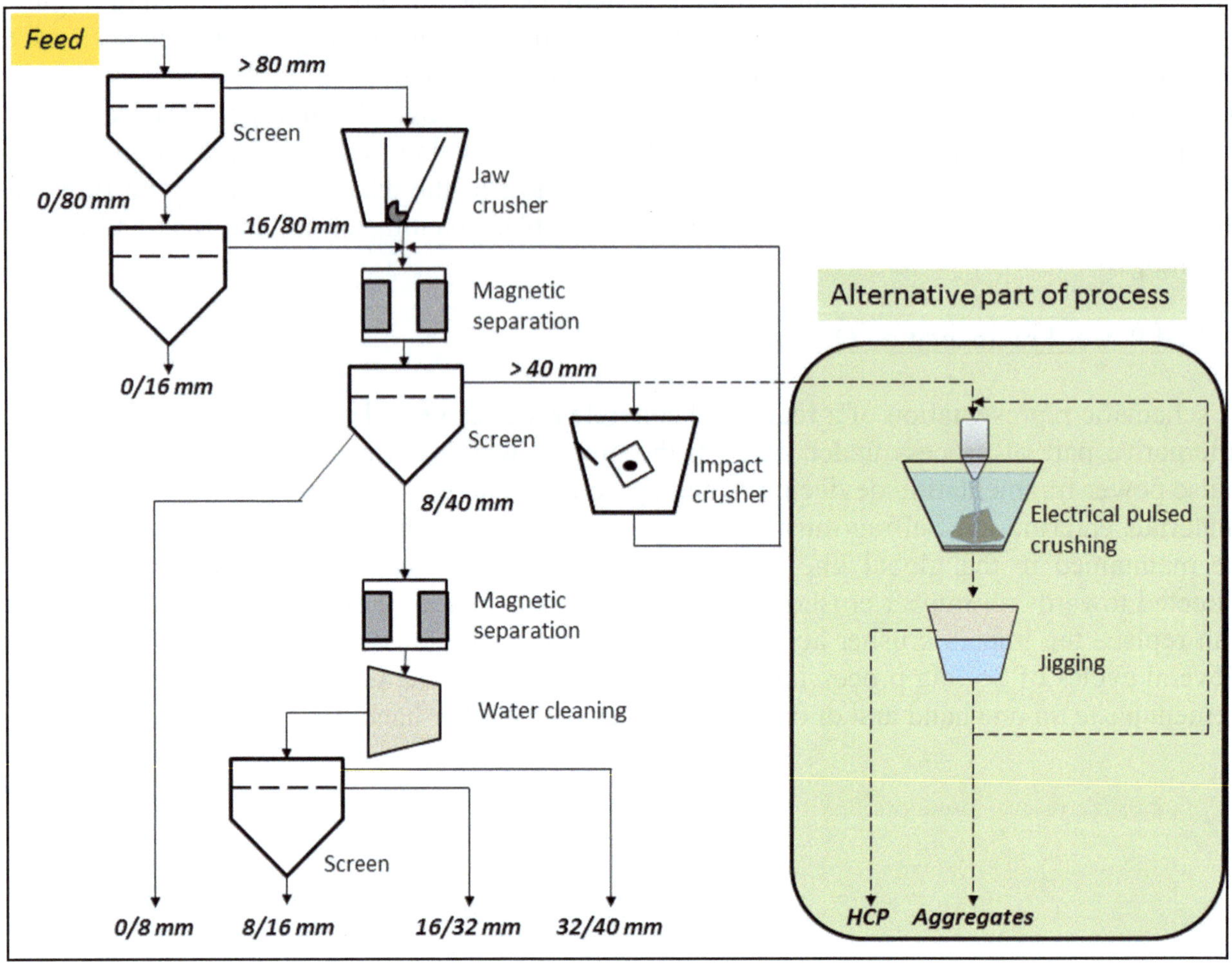

The variable parameters of the pulsed-power fragmentation are the number of pulses and the applied voltage. Experiments were carried out in our laboratory on test pieces of concrete to determine the values of these parameters to obtain the greater liberation rate of hydrated cement paste. The characteristics of the concrete specimens are summarized in were cylindrical, their length and diameter were 40 mm and 40 mm respectively, and their precise formulation is given in Table 2.

Table 2. Formulation of the concrete specimens

Shape and dimension	Cylindrical, h = 40 mm, ø = 40 mm
Cement type	CEM I 52.5 N
Mass ratio water/cement	0.6
Mass of cement	400 kg per m^3 of concrete
Mass of aggregates	1620 kg per m^3 of concrete
Water	240 kg/m^3

According to the chemical analysis, up to 32 % in weight of the initial material could be recovered in the -2 mm fraction after 15 pulses of 125 kV were applied. The process was run a second time on the +2 mm fraction. The chemical contents of CaO, SiO_2, Al_2O_3 and Fe_2O_3 in the two -2mm fractions obtained enable to determine the proportions and the chemical compositions of the -2mm and the aggregates fraction leaving the process at the steady state.

These values are summarized in Table 3, and they are used to evaluate the environmental extra-cost, see the Goal and Scope section.

Table 3. Steady-state composition of the -2 mm and the aggregate fraction after pulsed-power electrical fragmentation (operating conditions are 125 kV, 15 pulses)

	(%)wt	HCP content (%)wt	Aggregates content (%)wt
- 2 mm	47 (=1 − p)	48	52
Clean aggregates	53 (= p)	10	90

Datas concerning the lifespan of the equipment for pulsed-power fragmentation, the chemical composition of its main parts and the total electrical consumption over its lifespan were collected from the industrial manufacturer, and added to the module defining the production of Recycled Aggregates for Concrete RAC (see below in the LCI section). For reasons of confidentiality these data are not disclosed in the present paper, but remain available upon request.

Clinker Manufacturing With And Without Recycled Hydrated Cement Paste

Global Process. Clinker is supposed to be produced in a dry process kiln with multistage preheaters and precalciners. Apart from the internal plant transportations, the main CO_2 emissions come from the chemical reactions in the clinker kiln, the fuel combustion during clinker production and the electrical power consumed for raw materials and clinker grinding and milling to the desired sizes. These data fluctuates with time and from one cement plant to another, according to a seven-year monitoring of SOx, NOx, CO_2 and clinker dust emissions performed on six cement plants by [30]. These fluctuations are related to the process stability and the efficiency of the dust filters used. Consequently, only mean values for the main emissions/consumptions can be used in the inventory, and they are documented from literature data.

Estimation Of CO_2 Emissions And Minerals And Energy Consumption At The Kiln Scale. The usual feeding of the kiln is made of limestone, sand, iron ore and clay, which are the main mineral carriers of Ca, Si, Fe and Al. Part of these elements can be supplied by the hydrated cement paste recovered from the alternative concrete fragmentation. The extent of replacement ratio in the raw feed is determined using a calculation method described in a previous communication [31] which extends the pioneering work of [32]. This calculation allows singling out the potential decrease in the CO_2 emissions and in the energy consumption brought by using recycled RCP in raw material.

This methodology relates the raw mix composition (proportions of calcite, clay, iron ore and recycled HCP) with the mineralogical composition of the clinker formed (mass proportions of C3S, C2S, C3A and C4AF), through a system of four equations determined by using Bogue equations [33] and simple mass balances between the inlet and the outlet of the kiln. The CaO, SiO_2, Al_2O_3 and Fe_2O_3 contents in RCP are determined by chemical analysis, and its mineralogical composition is assumed to be described by a mix of hydrogarnets C3AH6 ($3CaO.Al_2O_3.6H_2O$) and C3FH6 ($3CaO.F_2O3.6H_2O$), CSH silicate hydrate gel C3S2H3 ($3CaO.2SiO_2.3H_2O$) and ettringite ($C3A.3CaSO_4.32H_2O$), in residual amounts for mature concretes. Four specific parameters used by clinker manufacturers and related to the practical operating of the kiln are expressed from the mineralogical composition of the clinker. They are the Lea and Parker lime saturation ratio (LaP), the silica ratio (SR), the Kühn alumino-ferric module (MK) and the sum of CaO, SiO_2, Al_2O_3 and Fe_2O_3 contents in clinker (their definition can be found in [34]). These ratios are allowed to vary within precise ranges to ensure a good

clinker workability, which therefore constrains the previous system of four equations. Using a Sequential Quadratic Programing algorithm implemented in Matlab, it is then possible to obtain the acceptable ranges for proportions of calcite, clay, iron ore and RCP in the raw feed.
Choosing a definite composition within these ranges enables to perform two calculations: *(i)* the theoretical enthalpy at 25 °C required for clinker reaction with RCP (ΔH_f^{RCP}), *(ii)* and the specific thermic energy consumed, which encompasses the previous enthalpy, the various heat losses around the equipment (conductive and radiative losses), and the heat losses leaving the process through the exhaust fumes, the clinker dusts, and the clinker itself. All these losses are assumed to be not affected by the change in the raw feed, and thus equal to a reference value - estimated to be 1.3 MJ/kg clinker [35]. Thus, by calculating the enthalpy for clinker reaction, the specific energy (E_S^{RCP}) is obtained by $E_S^{RCP} = 1.3 + \Delta H_f^{RCP}$.
Knowing the specific thermic energy and the nature of the fuel used, it is possible to calculate the amount of fossil fuel required to fulfill the heat demand of the kiln, and thereby, the amount of SO_2, NOx and CO_2 released during the combustion. The former value is added to the CO_2 arising from the limestone dissociation, and to the emissions related to the electrical power consumptions in grinding and milling sections to obtain a global value for the CO_2 emissions of the process.
Construction Of The Module In Simapro Software. The generic dry process of clinker production is chosen in the Ecoinvent database. The grounding, crushing, milling and mixing of raw materials, as well as the grinding of clinker and the collection of dusts are already allocated. The electricity used for raw meal preparation depends on the hardness of the material and it is assumed that the recycled cement paste displays the same workability as the materials it replaces. Two modules are elaborated from this generic process. The first one corresponds to a reference clinker composed of 57.9 % of C3S, 21.1 % of C2S, 12.6 % of C3A and 8.4 % of C4AF, starting from pure minerals (calcite, kaolinite, silica and goethite). The associated specific thermic energy E_S^0 is 3.14 MJ/kg of clinker, which corresponds to a mean value found in dry process kilns operated in France according to [35]. The second module corresponds to a clinker manufactured from recycled cement paste and which meets the same module values as the reference clinker. The solid combustible for both processes is a bituminous coal with a lower heating value of 35.6 MJ/kg and whose elemental composition is 88.5 % of C, 5.0 % of H, 4.5 % of O, 1.0 % of S and 1.0 % of N.
The internal plant transportations depend on the internal logistics of each cement plant (spaces allocated to the receipt of raw materials, to the process of production and to the silos containing the manufactured cement), which implies that only a mean value can be chosen. For the sake of simplicity, the Ecoinvent value is chosen. The external transportations are considered to be negligible owing to the fact that cement plants are often located near their respective quarry.
Table 4 summarizes the values of the data modified in the Ecoinvent module corresponding to clinker formation. As a preliminary step, two cases are considered, depending whether the LaP, SR and MK mineralogical ratio are chosen to correspond with the values for the reference clinker or with the highest decrease in the CO_2 emissions.

Table 4. Data modified in the clinker module from the Ecoinvent database. Data refer to 1 kg of clinker

			Reference clinker	**Clinker with RCP**	
				Case 1 : Equal modules	*Case 2 : Lowest CO_2 emissions*
Raw	Limestone	g/kg*	1217	1093	1048

materials	Clay/marl	g/kg	164.8	137	86.6
	Iron Ore	g/kg	30.7	25.3	29.0
	Sand	g/kg	149.3	(-)	(-)
	RCP	g/kg	(-)	277	340
Energy	Hard coal	g/kg	88.2	82.3	78.9
	Specific thermic energy	kJ/kg	3140	2895	2769
Gases	CO_2	g/kg	821.6	748	717
	SO_2	g/kg	1.77	1.65	1.58
Transport	Raw materials	kg.km	41[43] – 170[41]	33 - 137	30 - 127
	Fuels	kg.km	98[43] – 406[41]	79 - 326	73 - 302

The specific heat energies are determined for an output of a 1 kg milled clinker. In the assessment of the global system, they will be weighted by the theoretical clinker mass which can be produced given the extent of substitution of the natural limestone by recycled concrete in the raw material, this extent stemming directly from the respective efficiencies of the recycling technology considered (pulsed power or impact crushing) and processed on a 1 kg input of waste concrete (see the Goal and Scope section).

Crushing Processes Of Concrete And Natural Gravels

Crushing processes refer to the production of aggregates from natural sources and from concretes for building concretes and for road embankments.

LCI For RAC And RGE Processes Of Production. Regarding the crushing of concrete to produce RAC, technical data are not well documented in the literature since the process is barely implemented in the industry. On an energetic point of view, the main difference between the processes of production of RAC and RGE is the replacement of the secondary impact crusher by the high voltage pulsed power device (Figure 1). According to the technical documentation from crusher suppliers, typical jaw and impact crushers need 2 kWh and 1 kWh per metric ton of waste concrete, as opposed to 2.5 kWh/t for the Selfrag. On this basis, the total energetic costs for a RGE and a RAC process of production are respectively 3 and 4.5 kWh/t.

Regarding the various emissions and fluid consumptions of the RGE process, Table 5 summarizes inventory data for coarse recycled aggregates obtained from demolished reinforced concrete structure and crushed in a typical mobile plant [36]. These data complement the values proposed by the Ecoinvent 7.3.3 database for the module "Gravel crushed at mine", since both processes of production involve heavy crushing steps, which liberate dusts and consume fluids.

Table 5. LCI data for the recovery of recycled aggregates for embankments[36]

	Consumption/emission	Unit	Value
	Electricity	kWh/t	(-)
	Fuel	MJ/t	18.2
GHG	CO_2	g/t	1695
	CH_4	g/t	1.59
	N_2O	g/t	0.67
Toxic gases	NO_x	g/t	19.2
	SO_x	g/t	6.7
	CO	g/t	4.27
Organics	Non-methane VOC	g/t	0.48

	POP	g/t	$2.49\ 10^{-3}$
Particles	> 30 µm	g/t	41.2

The module corresponding to the RAC process is completed with the data supplied by the Selfrag Company, see above the definition of the pulsed-power fragmentation of concrete (Figure 2).

LCI For Natural Crushed Aggregates For Concrete (NAC) And Round Aggregates For Embankments (NAE). As underlined by [37], three main factors contribute to the environmental load of the processes of natural aggregates recovery: their type, accessibility and regional availability. Accessibility and regional availability impact on the transport of the recovered aggregates to their place of consumption, and have no influence on the extraction process (issues related to transport are dealt with in a following section). By contrast, the impacts are quite different according to the kind of retrieved aggregates, round or crushed. Round aggregates result from the external erosion and the exposure to weather conditions. Except for their excavation, optional transport to a processing plant, washing and grading to meet the requirements of the national standards, neither complex nor energy-costly processes are needed. This is contrasted by the blasting, transportation, crushing/milling, sieving and washing operations performed on crushed aggregates. It is assumed those crushed and round aggregates are respectively dedicated to building concretes and road applications.

Precise field data concerning these two schemes of treatment remain scarce; however, a thorough study carried out by [38] earned our attention and will be the main source of information for the present inventory. The study focuses on three sites of aggregate extraction in France, differing either by the physical characteristic of their respective aggregates or by their level of production. For the sake of simplicity, only data concerning two sites with the same production rate but a different kind of aggregates are used. Both inventories include the production of fixed assets and maintenance vehicles.

Round aggregates are assumed to be excavated from a pit by means of a wheel loader, transported to the processing unit by a conveyor belt and further processed to produce four size fractions, 0/4, 4/6.3, 6.3/10 and 10/14. In 2005 125,000 tons of aggregates have been produced with 13 % of extracted materials being lost during the washing steps or being removed owing to their poor quality. 80 % of the production is used in concrete batching plants or for the production of roof tiles.

Crushed aggregates are supposed to be blasted from rock mass and transported to a crushing and milling unit which falls into three successive parts, composed of a jaw and cone crushers and sieving units. The physical properties of the recovered aggregates are suitable with a use in asphalt concrete. In 2005 180,000 tons of aggregates have been produced with 24 % of extracted materials being non-compliant with a road use according to the French standard NF EN 12620.

The values reported on Table 6 refer to a metric ton of materials produced during the year 2005 in France. Blasting is excluded from this inventory.

*Table 6. LCI data for the extraction of round and crushed aggregates. *includes fuel burned in building machine and diesel needed for internal transportations. **includes loading and unloading operations*

			Round aggregates NAE	**Crushed aggregates NAC**
	Consumption/emission	**Unit**	**Value**	**Value**

	Raw material	t/t	1.13	1.24
	Electricity	kWh/t	5.82	3.94
	Fuel	MJ/t	31.0*	20.4*
GHG [38]	CO_2	g/t	2390	1552
	CH_4	g/t	0.48	0.380
	N_2O	g/t	3.39	2.66
Toxic gases [38]	NO_x	g/t	53.02	67.61
	SO_2	g/t	4.05	2.66
	CO	g/t	34.38	29.51
Organics [38]	Non-methane VOC	g/t	14.42	7.82
	POP	g/t	$2.49\ 10^{-3}$	$1.61\ 10^{-3}$
Heavy metals [38]	Cd	g/t	$1.94\ 10^{-2}$	$4.83\ 10^{-6}$
	Cu	g/t	$7.51\ 10^{-6}$	$8.24\ 10^{-4}$
	Cr	g/t	$1.28\ 10^{-3}$	$2.43\ 10^{-5}$
	Ni	g/t	$3.75\ 10^{-5}$	$3.40\ 10^{-5}$
	Se	g/t	$5.26\ 10^{-5}$	$4.83\ 10^{-6}$
	Zn	g/t	$7.51\ 10^{-6}$	$4.85\ 10^{-4}$
Particles	> 30 μm	g/t	30.9** [36]	1239** [39]

Lastly, in each of these processes the water consumption is missing. The choice of discarding it in an inventory of aggregates extraction may seem irrelevant, particularly since the consumption strongly varies between the two extreme processes. However, choosing one the sparse data available in the literature is equally questionable since the distinction between the water withdrawn from the environment then released on one hand, and the water polluted after washing out clays from the aggregates (spent water) is often not apparent. The volume of spent water also depends on the recirculation system, which is installation-specific. As an example water consumption needed to produce 1 metric ton of aggregates in the UK ranged between 37 and 401 t/t in 2009 [40]. Thus, giving a precise estimation for the water consumption in aggregates production lines seems unrealistic.

Regarding the specific case studied in the present work, the alternative scenario developed for the management of concrete wastes needs extra water not only for the electrical pulsed-power crushing, but also for the extra natural aggregates consumption, meaning that the various water consumptions at every stage of the global process do not balance each other. Neglecting this additional water consumption results in underestimating the depletion in abiotic resources.

Transports

Distances. Table 7 summarizes all the transports heeded in the present study. The values are found in These data represent input variables impacting the environmental balance, and their initial value are found in reports form national statistics offices and from the relevant scientific literature and by a broad estimation considering the cement plant network at a country scale.

*Table 7. Transports considered in the present study. *see Figure 2. **Waste Concrete Treatment Unit.*

Module*	Name	Material	Starting point	Destination	Values
(1) (4)	D_1	Raw Materials	Quarry	Cement plant	41[44] - 170[41] km
(3) NAE	D_2	Natural aggregates	Quarry	Road	Road : 20[38] - 100[41] km
					Boat : 100 - 150 km
(6) NAC	D_3	Natural aggregates	Quarry	Building	Road : 40[41] - 59[43] km
					Boat : 100 - 150 km
(2) RAC (5) RGE	D_4	Concrete gravels	Demolition site	WCTU**	20[41] - 100 km
(2) RCP	D_5	RCP	WCTU	Cement plant	150 - 200[46] km
(5) RGE	D_6	Concrete gravels	WCTU	Road	10 - 30[45] km
(2) RAC	D_7	Recycled aggregates	WCTU	Building	20 - 100 km

Raw materials

The transport distances of raw materials (first line) show a wide range of variations, mirroring two extreme situations for supply of the quarries.

Concrete gravels from demolition sites (RAC and RGE)

The waste concrete treatment unit (WCTU) handles concrete waste through usual crushing or pulsed power/microwave fragmentation: transport distances of concrete gravel are assumed to be equal whatever the treatment applied. The value can be different from zero if WCTU is not located at the demolition site, *i.e.* if this is not a mobile unit. According to [41], the distance between road demolition sites to landfilling installations is about 20 km. This value will be assumed to hold for the distance to the non-mobile WCTU.

Aggregates from quarry to road site (NAE and NAC)

INSEE (the French National Statistics Office) gives an estimate for the mean transportation distance of raw materials in 2003 of 35 km [42]. This distance has been inching up over the last decades, an evolution mainly accounted for by several closings of quarries on environmental grounds and not compensated by the matching openings. For virgin round gravels, [38] states that the distance between the extraction and the production site is less than 2 km, and that the graded aggregates are not sent over distances above 20 km to road construction/resurfacing sites. This figure falls short of the distances - between 75 to 100 km - considered by [41]. This discrepancy illustrates the need to precise the geographical scope heeded, as [38] refers to one specific quarry while [41] provide an averaged value. However it is important to stretch that none of this figure is closer to the truth than the other, since both suffers from a lack of representativeness respectively caused by a too small and a too large scale considered. Consequently and for the sake of completeness, both values will be taken into account in the simulation. Regarding natural aggregates supply for structural concretes, [43] gives a precise value of 59 km, valid in the administrative area surrounding Paris.

Aggregates from various origins to road site (RGE)

Recycled aggregates travel 30 km on average to reach the site of road construction, according to the French Union of Carriers and Building Products Industries in a recent publication [45]. No other reliable figure was found in literature concerning local data.

Aggregates from demolition sites to building construction centers (RAC)

The corresponding distances of the aggregates whereabouts remain without a precise value, and therefore will constitute the main variables of the study. Distances considered will be kept above 20 km.

Recycled Cement Paste
A detailed mapping of the cement plants in France is available in [46], and enables a broad estimation of the mean cement plant distance in the area surrounding Paris, around 150 km. Various distances, not smaller than this value, will be tested in the study.
Mode Of Transportation. Materials are mainly transported by roads with usual 38 t lorries, whereas alternative transports such as inland waterway transport could be used. Road transportation represents 95 % of total weight transported, which is explained by the fact that this mode of transportation fully comply with the constraints facing the building sector, which are *(i)* short transportation distances, *(ii)* a seasonal production cycle, *(iii)* temporary nature of construction site, *(iv)* small loads transported, and *(v)* reliable delivery schedules. For such small transportation distances, handling operations in unloading areas are far too significant for rail or waterway transports to compete. In addition railway stations or inland port be located near the quarry or the concrete batching plant are required, and this requirement matter-of-factly increases as natural aggregates are more and more sought in far-away quarries. Still, a closer look at regional specificities in France indicates that in Paris region, waterway is comparable to road in the importation of aggregates (5.3 Mt as against 8.5 Mt in 2002). The Administrative Region which mostly contributes to this importation is the neighbor Haute-Normandie Region (3 Mt), its main inland port being at around 150 km from Paris along the Seine river.
Given these considerations, LCA simulations will also include the case of a waterway transportation of aggregates of 150 km. This value will represent the upper bound considered, as the French professional Union in the field gave the estimation of 100 km for a waterway transportation averaged at the national scale.

General Infrastructures

The emissions associated with the electrical consumption are modified in every sub-process of the global system to take into account the specificity of the French production mix of electricity, which strongly differs from the default Swiss one proposed in the Ecoinvent database. The share of electricity produced by nuclear plants is more important in France, and is made up for in the Swiss system by more hydrothermal and conventional thermic source [47].

Impact Assessment (IA) and Discussion

The system definition leads to various scenarios depending on the fragmentation technology used, the recycling rate of the cement paste, and the distance of transportation of the various materials involved. In what follows, two recycling rates of the RCP obtained by pulsed electro-fragmentation have been considered, along with two distances of transportation for the concrete gravels to the WCTU and the aggregates leaving the WCTU to building construction (D_4 and D_7), see Table 7. The other transportation distances are not modified and are respectively $D_1 = 41$ km for raw materials, $D_2 = 20$ and $D_3 = 40$ km for natural aggregates, $D_5 = 150$ km for RCP, and $D_6 = 10$ km for concrete gravels (see Table 7).
This set of parameters allows evaluating the sensitivity of the results to the location of the WCTU. Values of D_4 and $D_7 = 0$ would correspond to a WCTU close to the demolition/construction sites, a condition which is not always met in practice owing to the minimal space required for the WCTU operations.

Table 8. Variables studied in the present paper.

Transport		RCP rate	Case
Recycled aggregates D_7 (RAC)	Concrete gravels D_4 (RAC/RGE)	P^{ck}_{RCP}	
20 km	20 km	0.340	1
		0.277	2
	100 km	0.340	3
		0.277	4
100 km	20 km	0.340	5
		0.277	6
	100 km	0.340	7
		0.277	8

Results for the environmental extra-cost brought by the alternative processing of reclaimed concrete are summarized Table 9.

Table 9. Balance of the environmental impacts between the current and the alternative scenario of processing 1 kg of reclaimed concrete

Case	Climate change *kg CO_2 eq*	Terrestrial acidification *kg SO_2 eq*	Natural land transformation *m^2*	Fossil depletion *kg oil eq*
1	-1.04	$-1.09.10^{-2}$	$-6.37.10^{-6}$	$-6.74.10^{-3}$
2	-1.02	$-1.08.10^{-2}$	$-6.25.10^{-6}$	$-3.52.10^{-3}$
3	-1.03	$-1.08.10^{-2}$	$-2.91.10^{-6}$	$-3.20.10^{-3}$
4	-1.01	$-1.07.10^{-2}$	$-2.79.10^{-6}$	$-1.32.10^{-5}$
5	-1.05	$-1.09.10^{-2}$	$-8.72.10^{-6}$	$-9.14.10^{-3}$
6	-1.03	$-1.09.10^{-2}$	$-1.04.10^{-5}$	$-7.79.10^{-3}$
7	-1.04	$-1.08.10^{-2}$	$-5.52.10^{-5}$	$-5.60.10^{-3}$
8	-1.02	$-1.08.10^{-2}$	$-6.97.10^{-6}$	$-4.25.10^{-3}$

All the impacts categories score to negative values, meaning that the proposed alternative processing allows reducing the environmental footprint of concrete wastes. The decrease is impressive concerning the avoided emissions in CO2, about 1 kg saved per kg of concrete wastes handled. The terrestrial acidification shows a constant value, apparently not affected by the various distances of transportation considered. Regarding the natural land transformation and the fossil depletion categories, it can be noted that the gain brought by the alternative processing scheme diminishes when D_4 is increased (except for case 5) or when D_7 is lowered. An important point is that the gain remains positive, whilst RAC is sent over 100 km to building construction sites.

To better understand the global results, the respective costs associated with the production processes *(i)* of clinker, *(ii)* of aggregates/gravels for embankments, and *(iii)* of aggregates for structural concretes are reported on Figure 3. These costs are expressed as differences between the impacts of the various modules presented in Figure 1: (3) – (5) gives the environmental cost to produce aggregates for embankments in the alternative scenario, (2) – (6) gives the costs for aggregates for structural concretes and (1) – (4) the cost for clinker manufacturing. The subtractions are weighted by the mass flows indicated (see Figure 1), to determine their respective significance in the global system.

The maximum allowed recycled cement paste in the clinker kiln ($P^{ck}_{RCP} = 0.340$).

Figure 3. Costs (in term of environmental impacts) of different processes in the global system described on Figure 1. The numbers on the horizontal axis correspond to the case described in Table 9. The content of RCP in the clinker kiln corresponds to the lowest value for CO2 emission ($P^{ck}_{RCP} = 0.34$).

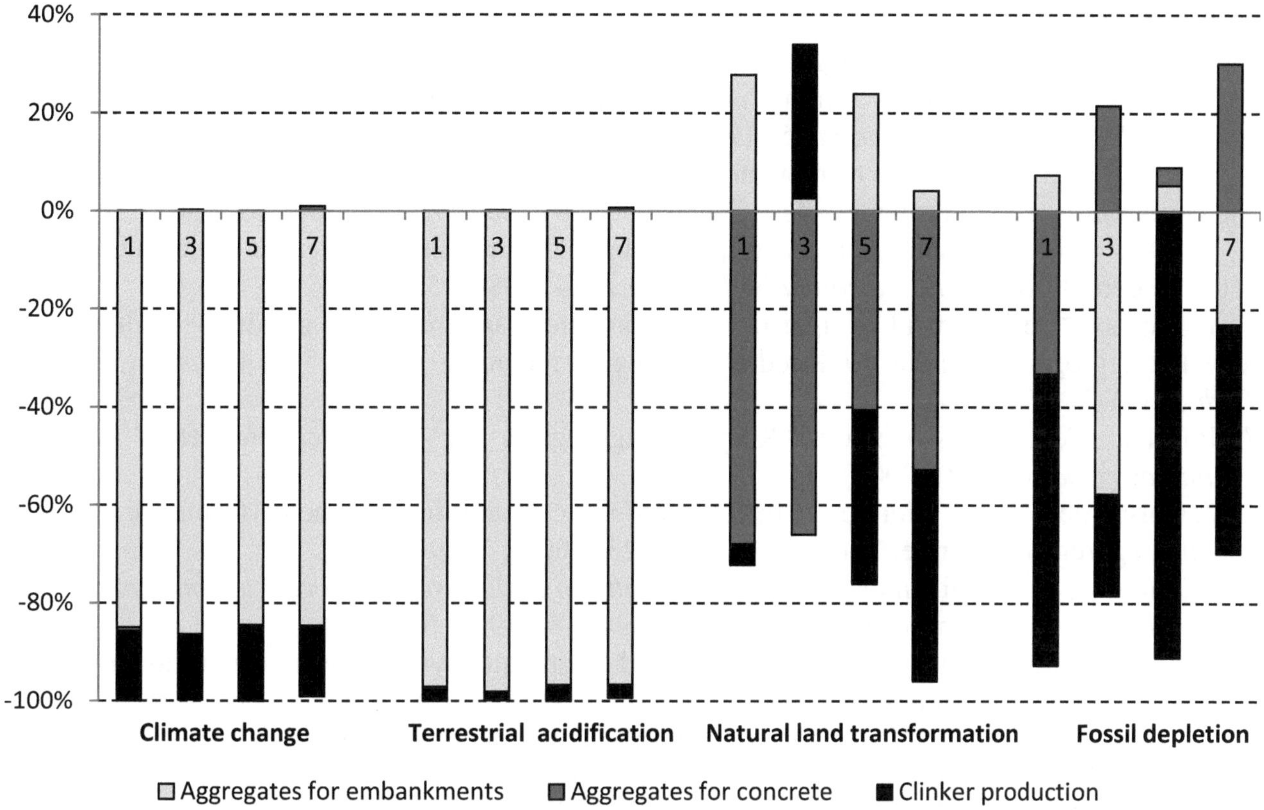

It can be seen from this figure that the production of aggregates for embankments is the main contributor to the decrease in climate change and terrestrial acidification. This is explained by the fact that crushing concrete is more energy-intensive than crushing solid rocks to recover aggregates, and therefore more fuel is consumed, resulting in higher (CO_2 and SOx) emissions.

The gain in natural land transformation is mainly accounted for by the production of aggregates for concrete, which is in agreement with the replacement of natural aggregates by aggregates from concrete. Similarly, the positive cost to produce aggregates for embankments (case 1) comes from the fact that gravels initially obtained from concrete wastes and now diverted to building concretes have to be found in natural quarries. This cost seems to decrease for case 3 and 7, which corresponds to D_4 = 100 km. This is ascribed to the fact that, in ReCiPe Midpoint, an increase in the distance of transportation indirectly results in a higher occupation of natural land by road. Regarding fossil depletion, an important gain seems to accompany the alternative process. This is partly due to the feeding of the clinker kiln with less carbonated materials, requiring less amount of combustible to obtain the free lime. The production of aggregates for concrete depletes more fossil fuel in cases 3, 5 and 7, which is explained by the distances of transportation of gravels to the WCTU or of RAC to building sites. The difference (3) – (5) reflects the same trend by becoming positive when the distance of transportation of concrete wastes to the crusher is 100 km.

These preliminary results establish links between potential environmental gains in an alternative processing of concrete wastes and the various distances of transportations that intervene. Future work will therefore be conducted to investigate these links.

References

1. D. Chisholm, "Best practice guide for the use of RA in new concretes" (CCANZ technical report TR14, ISSN 1171-4204, 2011).
2. S. Marinkovic et al., "Comparative environmental assessment of natural and recycled aggregate concrete," *Waste Management*, 30 (2010), 2255 – 2264
3. T.C. Hansen and H. Narud, "Strength of recycled concrete made from crushed concrete coarse aggregate," *Concrete International*, 5 (1983), 79 – 83.
4. C.S. Poon et al., "Influence of moisture states of natural and recycled aggregates on the slump and compressive strength of concrete," *Cem. Concrete Research*, 34 (2004), 31 – 36.
5. M. Bauchard, "Utilisation en technique routière de granulats provenant du concassage de béton de démolition," *Bull. liaison Labo P. et Ch.*, 134 (1984), 53 – 57.
6. Y. Dosho, "Development of a sustainable concrete waste recycling system – application of recycled aggregate concrete produced by aggregate replacing method," *J. Advanced Concrete Tech.*, 5 (1) (2007), 27 - 42
7. Portland Cement Association, "Recycled aggregates for reinforced concrete?" *Concrete Technology Today*, 23(2) (2002).
8. V. Corinaldesi and G. Moriconi, "Influence of mineral additions on the performance of 100 % recycled aggregate concrete," *Con. Build. Mat.*, 23 (2009), 2869-2876.
9. M. Kikuchi, T. Mukai, and H. Koizumi, *Demolition and reuse of concrete and masonry: reuse of demolition waste* (London: Chapman and Hall, 1988), 595 – 604.
10. M. Abe, "Study on production of recycled aggregate by gravity concentration method," (1997). *JCA Proceedings of Cement and Concrete*, 51, (1997), 482 – 487.
11. Y. Sakasume et al., "Development for production technique on high quality recyclable aggregate. Part 4. Method of screw grinding" (Paper presented at the Annual Meeting Architectural Institute of Japan, 2002), A-1, 1027 – 1028.
12. Y. Kuroda and H. Hashida, "A closed-loop concrete on a construction site" (Paper presented at the International symposium on sustainable development of cement, concrete and concrete structures, Toronto, Canada, 2005), 667 – 683.
13. M. Etxeberria et al., "Influence of amount of recycled coarse aggregate and production process on properties of recycled aggregate concrete," *Cem. Concr. Res.*, 37 (2007), 735 – 742.
14. M. Tsujino et al., "Application of conventionally recycled coarse aggregate to concrete structure by surface modification treatment," *J. Advanced Concrete Tech.*, 5 (2007), 13 – 25.
15. J.S. Damtofts et al., "Sustainable development and climate change initiatives," *Cem Concr Res.*, 38 (2008), 115 – 127.
16. Clamens (2012) http://www.groupeclamens.org/recyclage-beton les_tarifs_du_recyclage. html? qs= tarif.
17. S. Maeda et al., "Research on concrete aggregate collection technology by pulsed power discharge" (Paper presented at the 34th Conference on Our World in Concrete & Structures, Singapore, 2009 August 16th-18th).
18. A. Braunschweig, S. Kytzia and S. Bischof, "Recycled concrete: environmentally beneficial over virgin concrete?" (Paper presented at LCM 2011 – Towards Life Cycle Sustainability Management, Berlin, 2011).
19. Y. Ishikawa, "Issues in LCA in the concrete industry," *JLCA Newsletter*, 7 (2009), 13 – 16.
20. SETAC, *Guidelines for life-cycle assessment: a code of practice* (Society of Environmental Toxicology and Chemistry Press, Pensacola, FL, 1993).
21. N. Lippiat and F.Bourgeois, "Investigation of microwave-assisted concrete recycling using single-particle testing," *Min. Eng.*, 21 (2012), 71 – 81.

22. ISO (International Organization for Standardization) 14040. Environmental Management (1997). Life Cycle Assessment. Part I. Principles and Framework, Geneva, CH.
23. A. Tillman, T. Ekvall, H. Baumann, "Choice of system boundaries in life cycle assessment," *J. Clean. Prod.*, 2 (1994), 21 – 29.
24. H. Udo de Haes et al., *Life-cycle impact assessment: striving towards best practice* (Pensacola, FL: SETAC Books, 2002)
25. D.N. Huntziger and T.D. Eatmon, "A life-cycle assessment of Portland cement manufacturing: comparing the traditional process with alternative technologies," *J. Clean Prod.*, 17 (2009), 668 -675.
26. J.W. Owens, "Life –Cycle Assessment – Constraints on moving from Inventory to Impact Assessment," *J. Ind. Ecol.*, 1 (1997), 37 – 49.
27. M. Goedkoop et al., "ReCiPe 2008 – a life cycle impact assessment method which comprises harmonized category indicators at the midpoint and the endpoint level. First edition", (http://www.leidenuniv.nl/cml/ssp/publications/recipe_characterisation.pdf).
28. C. Chen et al., "Environmental impact of cement production: detail of the different processes and cement plant variability evaluation," *J. Clean Prod.*, 18 (2010), 478 – 485.
29. S. Solomon et al., *Climate change 2007. The physical science basics. Contribution of Working Group I to the Fourth Assessment Report of the Intergovernmental Panel on Climate Change* (Cambridge: Cambridge University Press, 2007).
30. B. Von Bahr et al., "Experiences of environmental performance evaluation in the cement industry. Data quality of environmental performance indicators as a limiting factor for benchmarking and rating," *J. Clean Prod.*, 11 (2003), 713 – 725.
31. S. Guignot et al., "Development of an innovative technique for demolition concrete up-cycling: a response to mitigate GHG's emissions and natural resources depletion" (Paper presented at the XXVI International Mineral Processing Congress – IMPC 2012, New Delhi, India, 2012).
32. C.W. Moore, "Chemical control of Portland cement clinker," *Am. Ceram. Bull.*, 61 (1982), 511 – 515.
33. R.H. Bogue, "Calculation of the compounds in Portland cement," *Industrial and Engineering Chemistry*, 1(4) (1929), 192 – 197.
34. H.F.W. Taylor, *Cement chemistry* (ICE Publishing, 1997).
35. G. Baudet, "Production de CO2 dans l'élaboration des ciments Portland. Bilans et possibilités de réduction des émissions" (Report 410, BRGM/French Geological Survey, 2004).
36. S.B. Marinkovic and I.S. Ignjatovic, *Innovative materials and techniques in concrete construction (ACES Workshop). Chapter 7. Recycled Aggregate Concrete for structural use. An overview of Technologies, Properties and Applications* (Michael N. Fardis Editor, Springer Science, 2010).
37. P. Van den Heede and N. De Belie, "Environmental impact and life cycle assessment (LCA) of traditional and 'green' concretes: Literature review and theoretical calculations," *Cement Concrete Comp.*, 34 (2012), 431 – 442.
38. T. Martaud, "Evaluation environnementale de la production de granulats naturels en exploitation de carrières" (PhD Thesis, Orleans University, 2008).
39. U.M. Mroueh et al., *Life cycle assessment of road construction* (Finnra Reports 17/2000, Finnish National Road Administration, 1999).
40. Aggregate Industries, (Sustainability report, 2009). From: http://www.aggregate.com/sustainability/. Accessed 11/05/12.
41. S. Sayagh, "Approche multicritères de l'utilisation de matériaux alternatifs dans les chaussées" (Ph.D Thesis, ENPC, 2007).

42. INSEE - Service de Statistiques Nationales d'Entreprises (French National Statistics Office). La protection de l'environnement. Le transport des matériaux de construction. Accessed 11/05/12.
43. G. Habert et al., "Development of a depletion indicator for natural resources used in concrete," *Resources, conservation and recycling*, 54 (2010), 364 – 376.
44. ATILH - Association Technique des Industries des Liants Hydrauliques (2002). Module d'informations environnementales des ciments. From : http://www.infociments.fr/developpement-durable/construction-durable/icv-ciments
45. UNICEM "Etudes d'impact environnemental. Plus de complexité," *Unicem Magazine*, 760 (2012).
46. P. Lebret, "Inventaire des carrières d'approvisionnement des cimenteries en France métropolitaine" (Report BRGM/RP-56986-FR, BRGM/French Geological Survey, 2009).
47. Eurostat (2011). Electricity production and supply statistics. From: http://epp.eurostat.ec.europa.eu/statistics_explained/index.php/Electricity_production_and_supply_statistics. Accessed on 11/05/12.

REWAS 2013

Enabling Materials Resource Sustainability

Battery Recycling

Session Chairs

Gregory Krumdick
John Sullivan

Modeling of synergistic effect of Cyanex 302 and D2EHPA on separation of nickel and cadmium from sulfate leach liquors of spent Ni–Cd batteries

Ehsan Vahidi[1], Ataollah Babakhani[2], Fereshteh Rashchi[2], Alireza Zakeri[3]

[1]Department of Civil and Environmental Engineering, University of South Florida, Tampa, FL 33620, USA
[2]School of Metallurgy and Materials Engineering, University College of Engineering, University of Tehran, Tehran, 11155, Iran
[3]School of Materials and Metallurgical Engineering, Iran University of Science and Technology, Tehran, 16846, Iran

Keywords: Spent Batteries, Solvent Extraction, D2EHPA, Cyanex 302, Nickel, Cadmium

Abstract

A model was developed to predict the synergistic effect of Cyanex 302 and D2EHPA on co-extraction and separation of nickel and cadmium from sulfate leach liquors of spent Ni–Cd batteries with the aim of increasing separation efficiency. The stoichiometric coefficient of cadmium was determined for sole D2EHPA and three different D2EHPA to Cyanex 302 ratios by applying the slope analysis method. The experimental data of cadmium extraction in pH range of 0.5–3, temperature of 25, 40 and 60 °C and various proportions of organic solvents (D2EHPA and Cyanex 302) were used to propose correlations of distribution coefficient of cadmium by multiple linear regression. The equation was found via multiple linear regression for the estimation of distribution coefficient of cadmium and the result showed that the proposed correlation is in good agreement with the experimental values. The extraction equilibrium constants, enthalpy change, and entropy change of the extraction reaction were also determined.

Introduction

Nickel–cadmium batteries are used in many portable electronic devices, military and defense applications, but due to the presence of toxic cadmium, they are classified as hazardous waste [1]. Recycling of spent batteries is important, both from an environmental and economic point of view [2, 3]. Several countries in Europe, Asia and South America have already explored the possibilities of establishing labeling, collection and recycling programs for Ni–Cd batteries, while others are in the organization stage [4, 5]. Solvent extraction has been widely used in separation, purification, and recovery of metal ions from aquatic solutions. This technology is characterized by lower energy consumption, higher metal selectivity, and higher purity of products compared to other processes and therefore, it has also been proposed for waste treatment and recycling processes such as separation of metals from spent primary and secondary batteries. Solvent extraction of Cd^{2+} from leach solutions has been studied by Cyanex 301 [6], D2EHPA [7, 8], Cyanex 302 [9], Cyanex 923 [10], Cyanex 272, Tops 99 and PC88A [11]. Almela and Elizalde [9] analyzed the suitability of Cyanex 302 for extraction of Cd (II) compared with other similar organophosphorus extractants, such as Cyanex 272 and they found Cyanex 302 to be effective in extracting cadmium in very acidic conditions. Cadmium extraction with D2EHPA involves co-extraction of both Ni and Co [7], which causes issues in the separation process efficiency. Moreover, application of phosphorus based extractants for the solvent extraction studies of Ni such as TOPS 99 [12], D2EHPA [13], and Cyanex 272 [14,15] has been investigated.

In the present research, a novel synergistic extraction system using Cyanex 302 and D2EHPA as extractants has been investigated. In addition, the extraction mechanism and thermodynamic parameters of cadmium were studied in detail as a comparison of the single D2EHPA with the mixture of Cyanex302 and D2EHPA.

Experiments

Materials

Cadmium sulfate (97% pure) and nickel sulfate (98% pure) salts were obtained from Merck, Germany. The organic solvents used in this work were industrial-grade di-(2-ethylhexyl) phosphoric acid (D2EHPA), 97% pure, from Bayer, Germany, and bis-(2,4,4-trimethylpentyl) monothiophosphinic acid (Cyanex 302), 98% pure, from Fluka, Canada. The extractants were used without further purification. Kerosene, from Tehran Oil Refinery Co., was used as diluent. Sulfuric acid (98% pure) and sodium hydroxide 0.1 M, from Merck, Germany, were used as pH modifiers.

Experimental procedure

Stock solutions of nickel and cadmium (2 g/L each) were prepared by dissolving sufficient amount of their sulfate salts in distilled water. Each extraction test was performed by mixing 200 mL of the aqueous phase solution with 200 mL of the organic phase containing 20% v/v of the extractant (either sole D2EHPA or the mixture of D2EHPA and Cyanex 302) and 80% v/v kerosene. This aqueous to organic phase ratio (A/O) of 1:1 was used in all tests. During mixing, temperature of the system was controlled by a thermostatic bath. Experiments were carried out at 25, 40 and 60 °C. A PY-11 pH meter Sartorius made in Germany was used to monitor pH during the experiments. The mixture was agitated very well by a mechanical agitator for 10 minutes after when pH was stable. pH was fixed by adding either sulfuric acid or sodium hydroxide solution. The mixture was then transferred to a separatory funnel, equilibrated and allowed to disengage. After phase separation, the aqueous phase was analyzed for nickel and cadmium directly after suitable dilutions by AAS. The concentration of metal ions in the organic phase was calculated from the difference between concentrations in the aqueous phase before and after extraction by mass balance.

Results and discussions

Influence of Cyanex 302 as Synergist

Fig. 1 shows the percentage of cadmium and nickel ions extracted as a function of pH at different D2EHPA to Cyanex 302 ratios at 25 °C, which illustrates significant synergistic shifts of cadmium and nickel isotherms due to the addition of Cyanex 302 to D2EHPA. As it can be seen from the figure, increasing the ratio of Cyanex 302 to D2EHPA shifts the cadmium extraction curve more to the left and that of the nickel extraction curve to the right. In other words, addition of Cyanex 302 to D2EHPA significantly improves the separation of cadmium over nickel.

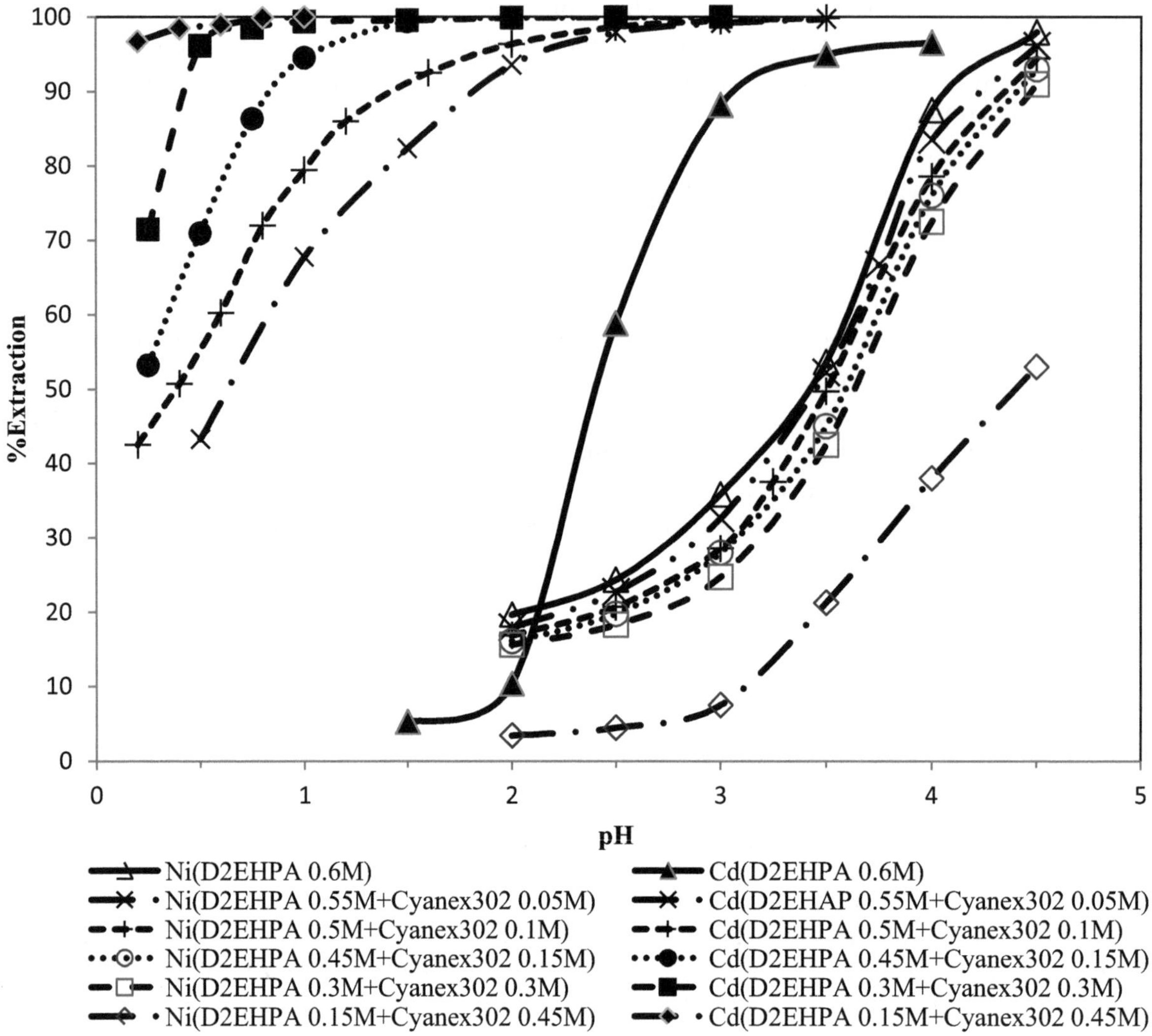

Figure 1- Effect of additions of Cyanex 302 to D2EHPA on the extraction of Cd (II) and Ni (II) at 25 °C and A/O ratio of 1:1.

Extraction mechanism of cadmium by D2EHPA

In order to determine the formation of metal–organic complex during extraction of cadmium, a dimer state of D2EHPA has been considered as reported by various authors [16, 17, 18, 19]. Thus, the extraction mechanism of the metal ion (in this case cadmium) with D2EHPA in kerosene may be expressed as follows:

$$M_{aq}^{n+} + (n+p)/2\overline{(H_2A_2)_{org}} \xleftrightarrow{K_{ex}} \overline{(MA_n(HA)_p)_{org}} + nH_{aq}^{+} \tag{1}$$

where (H_2A_2) is extractant in dimeric form, M is metal, n is valence of the metal or metal complex ion and p number of molecules of extractant engaged in reaction. The equilibrium constant of the extraction reaction K_{ex}, can be given as a function of molar concentration, provided that the ionic strength of the aqueous solution is constant.

$$\left[k_{ex} = [MA_n(HA)_p]_{org}[H^+]_{aq}^{n} / [M^{n+}]_{aq}[H_2A_2]_{org}^{(n+p)/2}\right] \tag{2}$$

The distribution coefficient, D, is defined as the ratio of metal concentration in organic phase to the metal concentration in aqueous phase at reaction equilibrium, and is substituted into the above equation.

$$D = K_{ex}[H_2A_2]_{org}^{(n+p)/2} / [H^+]_{aq}^n \tag{3}$$

By taking the logarithm of the above equation:

$$\log D = \log K_{ex} + (n+p)/2 \log[H_2A_2]_{org} - n\log[H]^+ \tag{4}$$

$$\log D = \log K_{ex} + (n+p)/2 \log[H_2A_2]_{org} + npH \tag{5}$$

Plot of log D vs. equilibrium pH: As shown in Fig. 2, for cadmium, the plots of logD vs. pHe gave straight lines with a slope of about 2.0 for sole D2EHPA (0.5 M), suggesting the totally released protons included two molecule of H^+ during extraction.

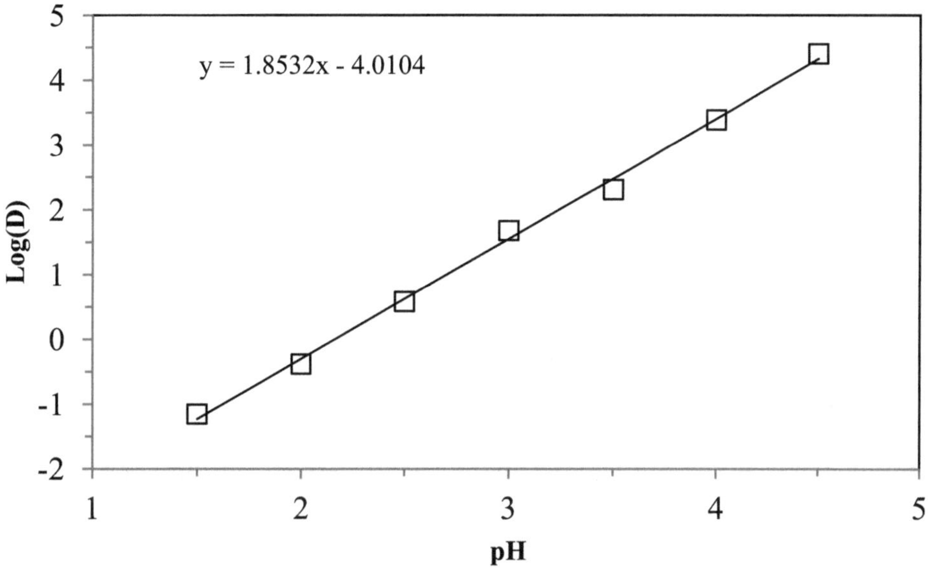

Figure 2- Effect of pHe on the extraction of cadmium with D2EHPA at concentrations of 0.5M extractant.

Plot of log D vs. log $[H_2A_2]_{org}$: In this investigation concentration of D2EHPA was varied in the range 0.15–0.6 (M) in kerosene. Fig. 3, shows a plot between log D vs. log $[H_2A_2]_{org}$ at equilibrium pH = 3, which corresponds that the extraction of cadmium was linearly related with extractant concentration. The slope value {(n+p)/2} was determined to be ~1.5 thus assuming the value of 2 for n, the p value becomes 1.

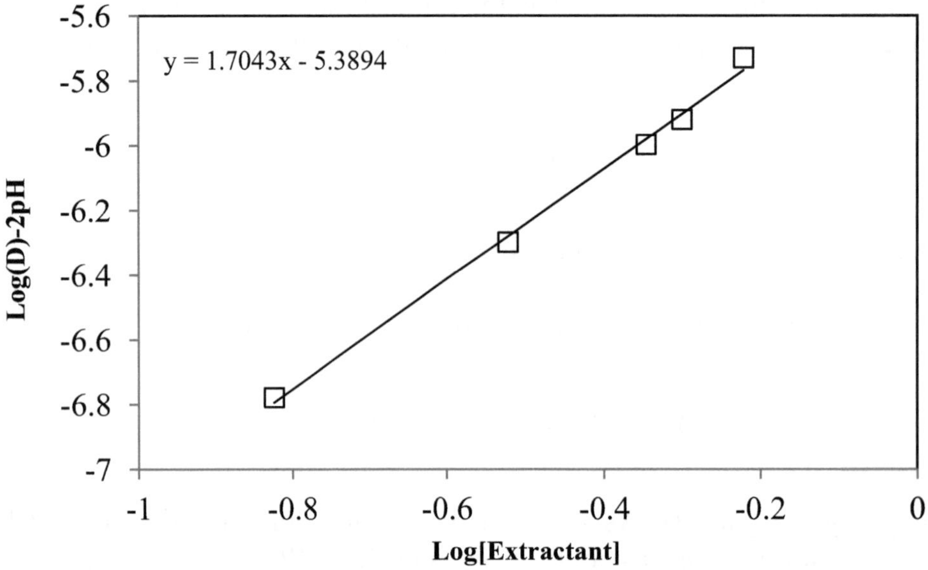

Figure 3-Effect of D2EHPA concentration on the extraction of cadmium, pHe = 3.

The extraction mechanism for cadmium extraction by sole D2EHPA may be represented as:

$$Cd^{2+}_{aq} + 3/2\overline{(H_2A_2)_{org}} \xleftrightarrow{K_{ex}} \overline{(CdA_2(HA))}_{org} + 2H^{+}_{aq} \qquad (6)$$

It indicates that cadmium is solvated with 1.5 molecule of dimeric D2EHPA with formation of $\overline{(CdA_2(HA))}_{org}$ complex in the organic phase. This approves the findings reported by Kumar et al [20].

Extraction mechanism of cadmium by mixture of D2EHPA and cyanex 302

Plot of log D vs. equilibrium pH: Similarly, in order to determine the extracted complexes in the mixed-extractant systems, a series of experiments were carried out and the conventional slope analysis method was used. As shown in Fig. 4, the plots of $\log D_m$ versus pH_e at fixed concentrations of Cyanex 302 (0.1M) and D2EHPA (0.5M) give a straight line with a slope of about 1.0, which suggests that there is one released protons in the extraction reaction equation.

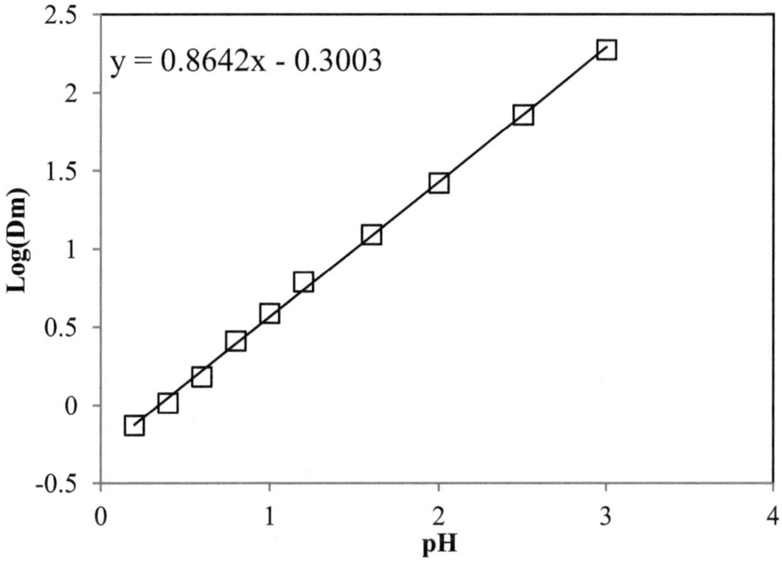

Figure 4- Effect of pHe on the extraction of cadmium by mixtures of D2EHPA and Cyanex 302.

<u>Plot of log D vs. log $[H_2A_2]_{org}$:</u> Similarly, at fixed aqueous acidity (3) and concentration of the cyanex 302 (0.1M), the plots of $logD_m$–pH_e are linear with slopes of about 1.0 for various concentrations of D2EHPA (Fig. 5), which indicates that one molecule of D2EHPA is involved into the extraction reaction.

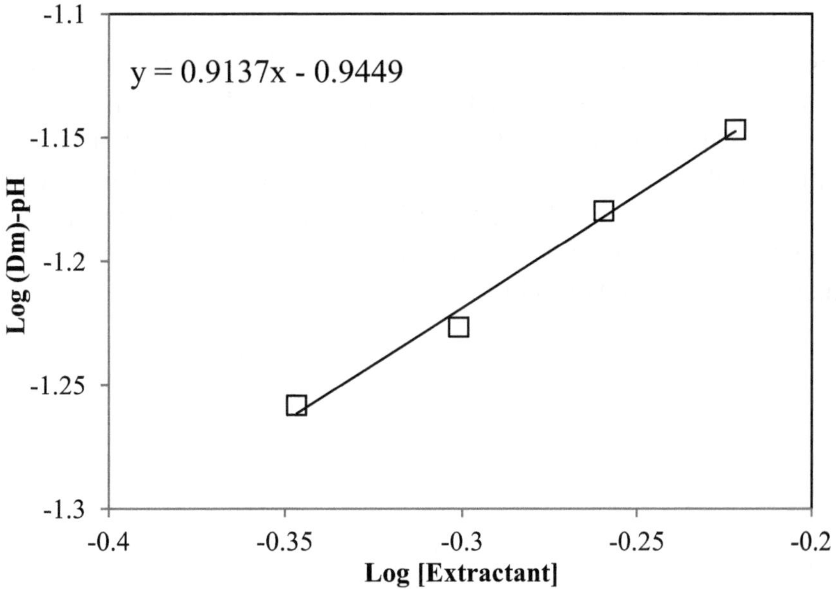

Figure 5-Effect of D2EHPA concentration on the extraction of cadmium, pHe = 3.

The extraction mechanism for cadmium extraction may be represented as below, provided when extracted compounds are not associated with each other.

$$Cd^{+}_{aq} + \overline{(H_2A_2)_{org}} \xleftrightarrow{K_{ex}} \overline{(CdA\ (HA)\ )}_{org} + H^{+}_{aq} \quad (7)$$

Distribution coefficient

A general multiple linear equation was used for finding the equations of distribution coefficient for cadmium as a function of temperature (T) and pH independent variable. The coefficients of the equations were calculated via multiple linear regressions. At fixed concentrations of extractants (D2EHPA=0.5 M, Cyanex 302=0.1 M) the insignificant terms were eliminated from the equations with not much change in the precision of the correlation by applying the ANOVA technique. The simplified equation of distribution coefficient for cadmium is as follows:

$$Log(D_{Cd}) = -20.65 + 0.038 \times T + 7.32 \times pH - 1.104 \times pH^2 \quad (8)$$

The results of the model are tabulated in Table 1. The model F-value of 16 implies that the model is significant. There is only a 0.1% chance that a "Model F-Value" this large could occur due to noise. Values of p-value (Prob>F) for this model is less than 0.0500 indicating that the model term are considered to be statistically significant. This is desirable as it demonstrates that the terms in the model have a significant effect on the response.

Table 1- Analysis of variance table

Source	Sum of squares	Mean Square	df	p-value	F- value	
Model	10.42	3.47	3	0.0010	16	significant
A-T	3.67	3.67	1	0.0034	16.90	
B-pH	5.84	5.84	1	0.0008	26.90	
B^2	0.91	0.91	1	0.0743	4.21	
Residual	1.74	0.22	8			

Fig. 6, Compares the experimental data and the model predicted values of extraction of cadmium. It can be seen in this figure that the predicted values from the model are in good agreement with the experimental data. At low pHs, the proposed correlation has excellent ability to predict distribution coefficient with an appropriate accuracy while discrepancies at higher pHs can be attributed to the sensitivity of the distribution coefficient to pH in this region. At high pH region (high extraction percentages), distribution coefficient considerably increases while extraction percentage increases only slightly.

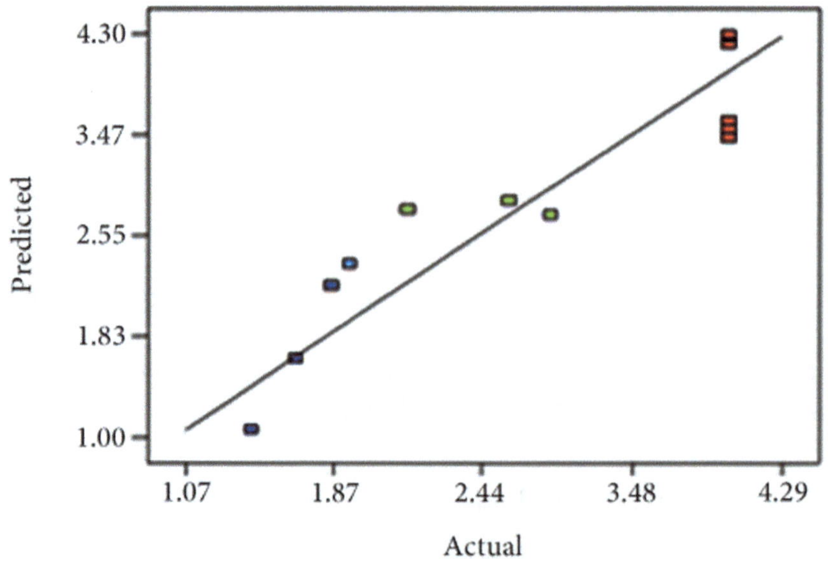

Figure 6- Comparison of experimental data of distribution coefficient with calculated results employing modular approach for cadmium.

Thermodynamic study

Experimental temperature plays an important role in a liquid-liquid extraction process. At fixed concentrations of pH (2) and extractants (D2EHPA=0.5 M, cyanex 302=0.1 M) the influence of temperature has been studied in the present work. If the relationship between the distribution ratio and the temperature is obtained, the change of enthalpy of the reaction, ΔH, can be calculated according to the following:

$$\frac{\Delta D}{\Delta \frac{1}{T}} = \frac{-\Delta H}{2.303R} \quad (8)$$

The change of Gibbs free energy (ΔG) and the change of entropy (ΔS) can thus be obtained as follows:

$$\Delta G = -RT \ln K \quad (9)$$

$$\Delta G = \Delta H - T\Delta S \Rightarrow \Delta S = \frac{\Delta H - \Delta G}{T} \quad (10)$$

Where ΔG and ΔS are determined when T is 298 K.

Plots of log D versus [1000/T (K)] are shown in Fig. 7 and the K values, calculated from Eq. (5). The ΔH, ΔG, and ΔS values were calculated as 29.19 kJ mol^{-1}, 1.59 kJ mol^{-1}, and 92.62 J mol^{-1} K^{-1}, respectively. It can be seen that the sign of ΔH is positive, indicating that synergistic extraction processes is endothermically driven. The ΔS value is positive, implying that the synergistic extraction may occur easily, which is in accordance with the theory of increasing entropy from the view of statistics.

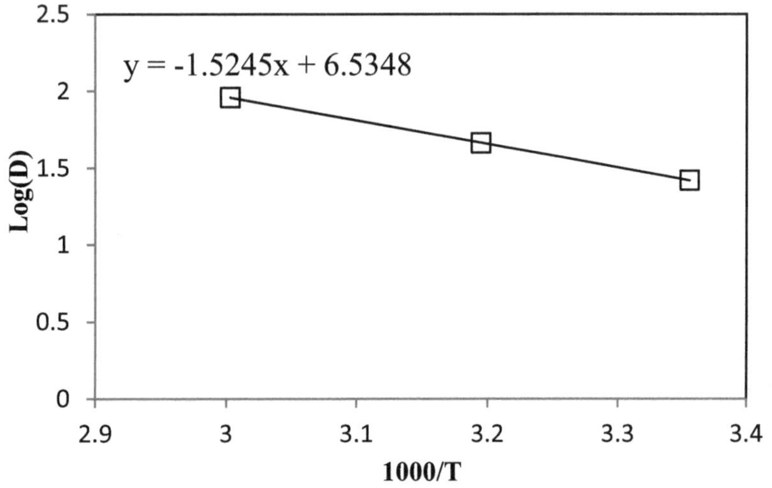

Figure 7- Relationship between distribution coefficient and temperature, pHe = 2.

Conclusion

The following conclusions are drawn from solvent extraction studies of cadmium and nickel from sulfate solution:

1- Adding Cyanex 302 to D2EHPA caused a synergistic effect and shifted the extraction curve of cadmium and nickel to the left and right, respectively. As a result, the separation factor and $\Delta pH_{0.5}$ value increased.

2- The extraction mechanism and behavior cadmium by single D2EHPA are different from that of the synergistic extraction by mixtures of D2EHPA and Cyanex302. The extraction studies showed the formation of $\overline{(CdA_2(HA))}_{org}$ species in the organic feed with sole D2EHPA and $\overline{(CdA(HA))}_{org}$ with the mixture of D2EHPA and Cyanex 302.

3- An equation was found via multiple linear regression for estimation of distribution coefficients of cadmium. The distribution coefficients of cadmium calculated based on these correlations is in good agreement with the experimental values.

4- The thermodynamic parameters, ΔH, ΔG, and ΔS, have been calculated, indicating that synergistic extraction reaction of cadmium is endothermically driven.

References

1. R.A. Sharpek, "Local government household battery collection programs: costs and benefits," *Resources, Conservation and Recycling,* 15 (1995), 1 – 19.

2. C.J. Rydh, and M. Karlström, "Life cycle inventory of recycling portable nickel–cadmium batteries," *Resources, Conservation and Recycling,* 34 (2002), 289–309.

3. C.J. Rydh, and B. Svärd, "Impact on global metal flows arising from the use of portable rechargeable batteries," *Science of the Total Environment,* 302 (2003), 167–186.

4. A.M. Bernardes, D.C.R. Espinosa, and J.A.S. Tenório, "Recycling of batteries: a review of current processes and technologies," *Journal of Power Sources,* 130 (2004), 291–298.

5. J. David, "Nickel–cadmium battery recycling evolution in Europe," *Journal of Power Sources,* 57 (1995), 71–73.

6. B. Ramachandra Reddy, D.N. Priya, and K.H. Park, "Separation and recovery of cadmium(II), cobalt(II) and nickel(II) from sulphate leach liquors of spent Ni–Cd batteries using phosphorus based extractants," *Separation and Purification Technology,* 50 (2006), 161–166.

7. C.A. Nogueira, and F. Delmas, "New flowsheet for the recovery of cadmium, cobalt and nickel from spent Ni–Cd batteries by solvent extraction," *Hydromettalurgy,* 52 (3) (1999), 267–287.

8. V. Kumar, M. Kumar, M.K. Jha, J. Jeong, and J. Lee, "Solvent extraction of cadmium from sulfate solution with di-(2-ethylhexyl) phosphoric acid diluted in kerosene," *Hydrometallurgy,* 96 (2009), 230–234.

9. A. Almela, and M.P. Elizalde, "Solvent extraction of cadmium (II) from acidic media by Cyanex 302," *Hydrometallurgy,* 37 (1995), 47–52.

10. B. Gupta, B. Akash Deep, and P. Malik, "Extraction and recovery of cadmium using Cyanex 923," *Hydrometallurgy,* 61 (2001), 65–71.

11. B. Ramachandra Reddy, D. Neela Priya, and J. Rajesh Kumar, "Solvent extraction of cadmium (II) from sulphate solutions using TOPS 99, PC 88A, Cyanex 272 and their mixtures," *Hydrometallurgy,* 74 (2004), 277–283.

12. B. Ramachandra Reddy, S. Venkateswara Rao, P. Kyung Ho, Solvent extraction separation and recovery of cobalt and nickel from sulphate medium using mixtures of TOPS 99 and TIBPS extractants, Miner. Eng. 22 (2009) 500–505.

13. G. Owusu, "Selective extraction of Zn and Cd from Zn–Cd–Co–Ni sulfate solution using di-2-ethylhexyl phosphoric acid extractant," *Hydrometallurgy,* 47 (1998), 205–215.
14. B. Ramachandra Reddy, D. Neela Priya, S. Venkateswara Rao, and P. Radhika, "Solvent extraction and separation of Cd(II), Ni(II) and Co(II) from chloride leach liquors of spent Ni–Cd batteries using commercial organo-phosphorus extractants," *Hydrometallurgy,* 77 (2005), 253–261.
15. B. Ramachandra Reddy, and D.N. Priya, "Chloride leaching and solvent extraction of cadmium, cobalt and nickel from spent nickel–cadmium, batteries using Cyanex 923 and 272," *Journal of Power Sources,* 161 (2006), 1428–1434.
16. A. Mellah, and D. Benachour, "The solvent extraction of zinc and cadmium from phosphoric acid solution by di-2-ethylhexyl phosphoric acid in kerosene diluents," *Chemical Engineering and Processing,* 45 (2006), 684–690.
17. E.K. Alamdari, D. Moradkhani, D. Darvish, M. Askari, and D. Behnian, "Synergistic effect of MEHPA on co-extraction of zinc and cadmium with DEHPA," *Minerals Engineering,* 17 (2004), 89–92.
18. J.A. Golding, C.D. Barclay, "Equilibrium characteristics for the extraction of cobalt and nickel into di(2-ethylhexyl)phosphoric acid," *The Canadian Journal of Chemical Engineering,* 66 (1988), 970–979.
19. E. Vahidi, F. Rashchi, K. Pashayi, "Effect of Additives on Kinetics of Liquid-Liquid Extraction in a ZnSO /D2EHPA/Kerosene System," *Canadian Metallurgical Quarterly,* 49 (2010), 235-240.
20. V. Kumar, M. Kumar, M. K. Jha, J. Jeong, J.-ch. Lee, "Solvent extraction of cadmium from sulfate solution with di-(2-ethylhexyl) phosphoric acid diluted in kerosene," *Hydrometallurgy,* 96 (2009), 230-234.

REWAS 2013: Enabling Materials Resource Sustainabi
Edited by: Anne Kvithyld, Christina Meskers, Randolph Kirchain, Gregory Krumdick, Brajendra Mish
Markus Reuter, Cong Wang, Mark Schlesinger, Gabrielle Gaustad, Diana Lados, and Jeffrey Spangenber
TMS (The Minerals, Metals & Materials Society), 20

RECYCLING OF EXHAUST BATTERIES IN LEAD-FOAM ELECTRODES

Girolamo Costanza, Maria Elisa Tata

University of Rome Tor Vergata
Via del Politecnico, 1 – 00133 Rome (Italy)

Keywords: lead-foam electrodes, Pb recycling, exhaust batteries.

Abstract

Lead and lead-alloy foams have been investigated in this research. In particular low-cost techniques for the direct production of lead-based electrodes have been analyzed and discussed in this work. The relevance of the main process parameters (powder compacting pressure, granulometry, base metal composition, sintering temperature and time) have been focused and the effect on foam morphology has been discussed too. In particular "Sintering and Dissolution Process" (SDP) and "Replication Process" (RP) have been employed and suitable modified. Both spherical urea and NaCl have been adopted in the SDP method. In the replication process it has been evidenced that the viscosity of the melt is fundamental. Furthermore the research examines lead recovery and recycling of exhaust batteries into foam-based electrodes. A novel method for the direct conversion of Pb scrap into lead foam is discussed too.

Introduction

Porous materials, interesting for many industrial applications, can be manufactured by many processing routes [1]. One of the most employed method has been developed at the Fraunhofer Institute of Bremen and is based on the Powder Metallurgy (PM) [2]. According to this method metal powders of the base metal, blowing agent and the addition of ceramic particles (for modifying melt viscosity and stabilizing metal foam) are required. In a previous work [3] a lot of parameters affecting foaming process and the effect of the powder mixing composition have been analyzed. Despite up to now the scientific community has focused the attention mainly on Al foams, a lot of other metals can be foamed. In particular this work deals with lead and its alloy foams. As explained in some papers [4-6], lead foams can be used as light-weight electrodes in lead-acid batteries. In particular in [4] Irretier and Banhart attempted to produce lead foams analyzing various lead powders (twelve) from many manufacturers. Negative drawbacks of the method set-up are: high powders cost, long mixing time (30 min.), hot pressing for the precursor production (250 °C and 110 MPa for 20 min.), hot extrusion (275 °C after a conditioning time of 2 hours) and finally foaming with a furnace temperature in the range 350-550 °C. So this solution is difficult to be applied as alternative to the traditional method (grid + paste) usually employed for the production of standard electrodes. Complex and expensive also the method by Dai et al. suggested in [5-6]: lead foams negative grids are produced by electrodeposition of lead on a copper substrate.

Automotive batteries recycling for the recovery of the secondary lead is under severe regulations in highly developed countries. The main task is the production, utilization, collection and recycling of most of the chemical power sources [7]. The electrochemical power source most widespread is lead-acid battery representing 65% of all primary and secondary cells. 100% of junk batteries, theoretically, can be collected and recycled [8]. This practice is already introduced in many countries all over the world. Despite this effort, the complete recovery of a battery is difficult because the scrap is quite inhomogeneous, in consistency and in phase chemistry, with many different wastes: lead oxide, soft lead dross, wet paste etc. [7]. Lead recycling process employed by the recycling industry is the pyrometallurgical one in a rotary furnace [7], consisting in four stages: 1) grinding of the battery to separate plastic, electrolyte and lead plates; 2) lead reduction in a rotary furnace; 3) separation of metallic lead from slag; 4) refining of recycled lead. In the period 1999-2006 an exploratory study of lead recovery in lead-acid batteries lifecycle in USA was conducted with the following concluding remarks: 1) lead recovery and recycling was stable in the examined period; 2) lead consumption was increased at a annual rate of 2,25%; 3) slag resulting from recycling technology process inefficiencies must be minimized.

In this work a new process has been developed that can solve the last one drawback. The novel proposed method allows the direct production of foam-based lead electrodes from waste material. In literature very little informations have been found about lead foams in general [4] and for electrodes in lead acid batteries applications in particular [6]. Furthermore the experimental solution proposed is fundamental for the improvement of lead-acid battery efficiency and reduction, at the same time, of the battery weight. Rational operation includes maximum active-material utilization, high scrap recycling rate, direct conversion of lead scrap into electrodes in an economical way. Finally an experimental battery has been prepared with lead-foam electrodes, both for negative and positive plate. It is believed that positive plate determines the performance of a lead-acid battery [5]. However, because power batteries always work under a partial state of charge, the negative plate has a great influence on the battery's performance, due to the lower charged ability caused by a serious sulfation on the negative plate [5]. Results achieved in this work by the authors have been developed employing sintering and dissolution process (SDP) [9] and replication process (RP) [10]. Both methods have been analyzed and modified in order to manufacture low cost lead foams. The main task is the definition of a production method which could be competitive with the standard one usually employed by batteries manufactuter. The recycling of lead electrodes from waste batteries is discussed too and an experimental battery has been built with lead-foam electrodes, at the moment with great porosity size, both for negative and positive plates.

Material and methods

Two experimental techniques have been employed in this work: SDP (sintering and dissolution process) [9] and RP (replication process) [10]. In particular on four main goals the attention of the investigators has been focused: homogeneous porosity distribution inside the metal, easy filler removal, ability to produce manifold geometries and finally acceptable mechanical strength. The SDP technique has been adopted and modified starting from both commercial lead powder (200 mesh) and scrap derived from the negative electrode of junk batteries. Urea and NaCl have been employed as filler agent. This choise was driven by the easy dissolution process (water at 70 °C is enough) and also by the relative low cost. Pb powders and filler have been mixed together in order to obtain a uniform distribution of both components inside the mix. After mixing uniaxial cold compaction has been performed, applying from 5 to 12 tons on a cross section of diameter 15 mm. The successive step of this process was sintering. According to the many experiments performed an optimal range has been identified (200-250 °C). Finally such a precursor was washed in hot water (70 °C) in order to remove the filler (urea or NaCl).

Replication process has been adopted with the main goal of employing bulk Pb and its alloys instead of expensive powders. The base process described in [10] has been suitably modified according to the different base material and experimental conditions. Lead and its alloys melting occurs inside a small furnace heated from outside. The crucible, made of copper, steel or aluminum, has been filled with a mix of coarse lead particulate and NaCl particles, heated above the melting point and finally a small pressure has been applied by a suitable piston. Melt infiltration was improved thanks to this expedient. For the production of two lead electrodes employed in the set-up of the first experimental battery a rectangular cross-section crucible (13 mm x 27 mm) has been adopted. Foam density was analyzed too.

SDP method: experimental results

Powders recycled by grinding paste from a waste battery have been mixed with NaCl. The precursor (diameter 15 mm) has been manufactured by cold uniaxial pressing (5 tons) of the mix in a mould. After many experiments the optimal sintering time and temperature have been determined in 250 °C for 30 minutes. After sintering NaCl was removed with hot water (70 °C) in 30 minutes. Many samples have been manufactured and a sample image is shown in Fig. 1.

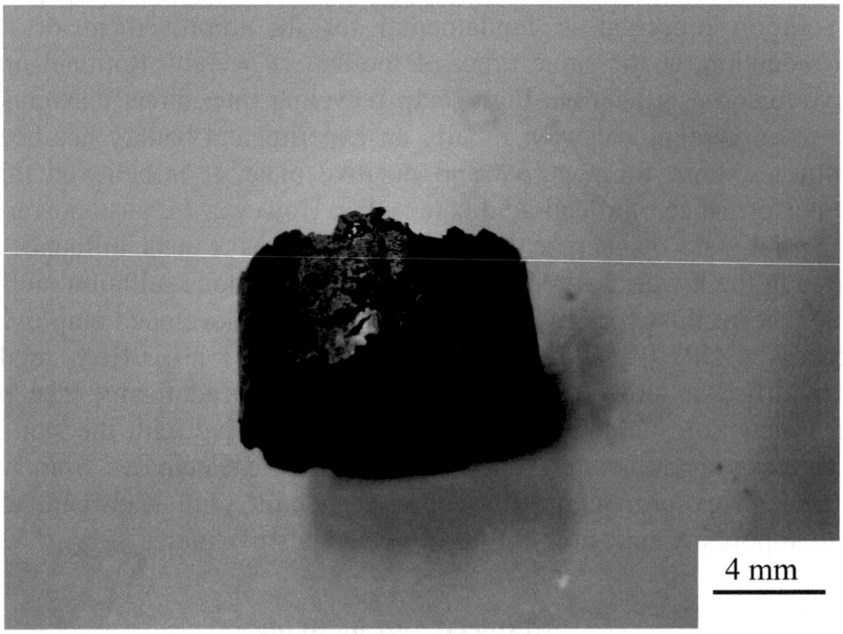

Fig. 1 Lead foam produced with SDP method from waste paste of a battery.

In sueccessive experiments the filler change, employing urea instead of NaCl, didn't show satisfactory results. Despite the precursor, produced with the same experimental procedure previously described, appeared uniform and compact, while washing in hot water it broke up. The strategy change was to produce foam from pure lead powder.

In the attempt to manufacture lead from Pb powder many experiments have been carried out. The two fundamental steps (washing and sintering) of the SDP method have been alternatively performed on the precursor. If washed and then sintered, precursor appears too fragile and easy to be broken. If dissolution follows sintering the evaporated urea deposits on the surface of the foam and it is quite difficult to be removed despite a long permanence in hot water. In general the powder method is not the best way to manufacture Pb-foam for the reason that powders are too expensive, pollutant and it is more difficult to control the final porosity.

RP method: experimental results

In the replication process Pb melting is fundamental. In this work NaCl is adopted as spacer and successively removed in hot water after lead solidification. Nevertheless viscosity of pure lead doesn't allow small porosity size (about 1-2 mm). An example is shown in Fig. 2. As a consequence some changes have been made to the process. The viscosity must be reduced, for example employing Pb alloy (Pb-Sn 60-40, Fig. 3) instead of pure lead. An alternative solution could be pressure application on the melt through a piston in order to promote the infiltration of the melt between adiacent filler particles.

Fig. 2. Lead foam produced by RP process employing coarse NaCl as filler.

Fig. 3 Pb-Sn (60-40) produced by RP process employing coarse NaCl as filler.

Good properties in terms of melt viscosity have been found in Pb-Sn-Sb alloy employed for the production of grids in lead-acid batteries. For this reason we decide to use this material as base metal for foam production. The mixture (50-50 vol. %) was prepared with NaCl (average diameter 1 mm) and small lead alloys fragments (2-3 mm) derived from grids of junk batteries. Flame applied on the external surface of the crucible causes alloy melting. A small pressure is applied through a piston after melting on the top of the liquid metal for optimal infiltration. After solidification NaCl has been removed in hot water. Circular and rectangular cross-section samples (Fig. 4) have been manufactured in our experiments. In particular the rectangular ones (density 3 g/cm^3) have been employed as electrodes for the assembly of a test battery. Also if pores morphology is not optimal for size, shape and distribution, we decided to test the performance of the new electrodes.

Figure 4. Rectangular cross-section lead alloy foams manufactured starting from grids of junk batteries.

Experimental battery

Light electrodes manufacturing for high-efficient lead-acid batteries could be an important challenge, for starters as for traction accumulators in automotive applications. Electrodeposition of Pb on copper foam substrate [5-6] is a low-speed process, difficult to be carried out in an economical way in the industrial production. The process suggested in this work, despite further studies and experiments are required, is extremely easy and economical. Waste grids from junk batteries are directly converted into new electrodes. The specific surface area is an important parameter for lead foam material. The higher the specific surface area, the higher the utilization efficiency of the active material. Further experiments are underway in order to reduce as more as possible porosity size.

After electrodes manufacturing, a small scale test has been performed: a 2V lead-acid cell employing commercial sulfuric acid (31%) was charged and discharged (Fig. 5). After 4 hours of charge (first time) the battery has shown a voltage of 2.2 V. After charging, a small lamp (3 W) has been turned on for 30 minutes. At the moment it is just a qualitative analysis of the potentiality of the new production method. Further tests will be performed in the future to quantify the charge-discharge performance of the battery. At the same time foam-electrodes manufacturing must be improved in order to reduce porosity size, sintering time and temperature, compaction condition and filler removal.

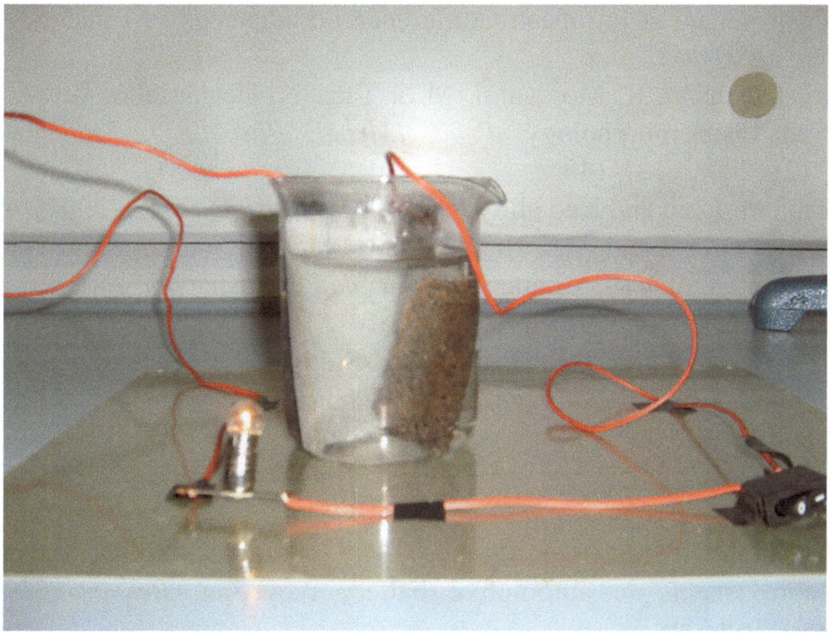

Figure 5. The 2V lead-acid cell with electrode foam charged and discharged.

Discussion and conclusions

High-quality lead-foam electrodes can be produced selecting carefully process parameters. Fundamental is an appropriate selection of particles employed, alloy composition, compaction conditions, sintering temperature and time. The most remarkable results of this research are reported in the following:

1) Lead and lead alloy foams can be produced modifying the sintering and dissolution process or the replication process well known for Al alloys;
2) Uniform foams with density of 3 g/cm^3 have been manufactured. Lower density can be reached;
3) Foam lead-plates have been produced by replication process starting from waste grids of junk batteries;
4) An experimental battery with foam lead-plates has been assembled and tested with a charge/discharge cycle;
5) Porosity size is not optimal for the improvement of battery efficiency and weight reduction, so further efforts must be focused on the production of small-pore lead-alloy foams;

The main results of the preliminary investigation are encouraging. Lead-foam porosity must be reduced at least 2 degrees of magnitude. Furthermore the effect of lead oxides and sulfate on the battery efficiency must be evaluated too.

References

1. J. Banhart, Manufacture, characterization and application of cellular metals and metal foams, *Progress in materials science*, 46 (2001), 559-632

2. J. Banhart, J. Baumeister, M. Weber, Powder metallurgical technology for the production of metal foam, Paper presented at European Conference on advanced PM materials, Birmingham, UK, 23-25 October 1995.

3. G. Costanza, G. Gusmano, R. Montanari, M.E. Tata, N. Ucciardello, Effect of powder mix composition on Al foam morphology, *Proc. IMechE Part L: J. Materials: Design and Applications*, 222 (2) (2008), 131-140.

4. A. Irretier, J. Banhart, Lead and lead alloy foams, *Acta Materialia*, 53 (2005), 4903-3917.

5. C.S. Dai, B. Zhang, D.L. Wang, T.F. Hi, X.G. Hu, Preparation and performance of lead foam grid for negative electrode of VRLA battery, *Materials Chemistry and Physics*, 99 (2006), 431-436.

6. C.S. Dai, B. Zhang, D.L. Wang, T.F. Hi, X.G. Hu, Study of influence of lead foam as negative electrode current collector material on VRLA battery charge performance, *Journal of Alloys and Compounds*, 422 (2006), 332-337.

7. J. Kéri, J. Precskò, Development and use of a new system for environmentally clean recycling of lead battery scrap. *Journal of power sources*, 53 (1995), 297-302.

8. M.A. Kreusch, M.J.J.S. Ponte, H.A. Ponte, N.M.S. Kaminari, C.E.B. Marino, V. Mymrin, Technological improvements in automotive battery recycling. *Resources conservation & recycling*, 52 (2007), 368-380.

9.Y.Y. Zhao, D.X. Sun, A novel sintering-dissolution process for manufacturing Al foams, *Scripta Materialia*, 44 (2001), 105-110.

10. C. San Marchi, A. Mortensen, *Handbook of cellular materials*, (Weinheim, 2002), 43.

REWAS 2013: Enabling Materials Resource Sustainability
Edited by: Anne Kvithyld, Christina Meskers, Randolph Kirchain, Gregory Krumdick, Brajendra Mishra, Markus Reuter, Cong Wang, Mark Schlesinger, Gabrielle Gaustad, Diana Lados, and Jeffrey Spangenberger
TMS (The Minerals, Metals & Materials Society), 2013

TECHNICAL STATUS AND PROGRESS OF LEAD RECYCLING OF BATTERY

Wei-feng Li[1,2], Li-hua JIANG[3], Jing Zhan[*1], Chuan-fu Zhang[1]

[1]School of Metallurgical Science and Engineering, Central South University; Central South University , Changsha 41003, China
[2]Henan Yuguang Gold & Lead Co., Ltd., Jiyuan 454650, China;
[3]Jiyuan Vocanionala and technical College, Jiyuan 454650, China

*Corresponding Author: zhanjing2001@hotmail.com

Keywords: Waste lead acid battery, recycling, lead paste reproductive process, redox bath melting

Abstract

The characteristics of various components in waste lead acid battery are analyzed in this paper. The present status and the study progress situation in industry production and research field of recycling of waste lead acid battery and lead paste used broken-separation technology are introduced. The comparison of advantages and disadvantages in different industry processes is carried. The advantages of redox bath smelting of lead concentrate and lead paste are analyzed. The method of redox bath smelting will be a low-carbon, environmentally friendly and efficient processes of secondary lead production and can be intensive to desulfurize for high temperature pool.

Introduction

Since the 90s of last century, China has become the fastest increasing country in the lead production and consumption. China's refined lead production was 4.2 million tons in 2010. With the development of transport, mobile communications and other industries, the lead demand will continue to increase. It can be predicted that China's lead demand in 2015 will reach 6 million tons. The consumption of lead-acid battery accounts for about 85 percent of the total lead consumption in the lead consumption structure. The yields of China's lead battery occupy about 1/3 of the world production. Each year the scrapped battery has been up to over 1.5 million tons. However, the overexploitation of raw ore resources, resources exploration lag and shortage of lead ore make the increasingly cumulative amount of waste battery has been a tremendous, renewable and secondary resource.

Though regenerated lead industry in China has been made remarkable progress and initially formed an independent industry, there exists a wide gap compared to highly efficient and mechanized treatment process in foreign. Recycling of waste battery in China is still in a state of disorder. Due to the lower technical level and backward equipment level, resources have not been used rationally, recovery rate of metal is low, energy consumption is high and environmental pollution is severe. A wealth of waste sulfuric acid is arbitrarily poured out. Also, waste lead that is one of the most dangerous solid waste is often abandoned everywhere by small workshops owing to improper handling or custody, which not only seriously pollutes soil and water, but also often directly harms human health. Therefore, it will have a significant and profound impact on secondary lead industry if resources in the lead-acid batteries are effectively used, new technology of clean metallurgy is developed, and efficient utilization of secondary resources in the life cycle of lead is achieved.

Analysis of composition and characteristic of waste lead-acid battery

Composition of waste lead-acid battery

A complete waste lead-acid battery is usually composed of four types of substances, namely electrolyte (H_2SO_4 solution), organic matter (polypropylene, polyethylene, PVC, formic, etc), metallic lead (grid, connections) and lead paste. Some general scrapped lead-acid batteries were broken up, and then we got the composition as follows: grid 24%-30%, lead paste 30-40%, diaphragm 2-3%, plastic shell 5-7.5%, bakelite shell 15-20%, H_2SO_4 solution 11-30%. The content of lead in the waste lead-acid battery is approximately 60%, in which lead connection section accounts for 13.8%, grid accounts for 38.2%, filler accounts for 48%. The average concentration of lead in these lead materials is about 87%. Grid metals contain about 3% Sb and a small amount of other metals.

Analysis of materials in the waste lead-acid battery

As the use of bakelite is restricted, the shell and diaphragm of battery is mostly plastic organic matter, such as polypropylene, PVC, and its density is generally 1144g/cm3. Most organic compounds are easily returned to be recycled after completely separated from lead. However, a little diaphragm material is difficult to be isolated from lead, which usually can be treated separately during smelting at high temperature, avoiding dioxin gases during combustion at low temperature.

Waste lead-acid battery involves electrolyte which density is 1124-1128 g/cm3 and contains 30-36% H_2SO_4. At present, the electrolyte is basically in situ disintegrated and drained anywhere. Every year about 50000 tons acid electrolyte is dumped into the earth, in which soluble lead is up to 7g/L ~ 10g/L, causing the average level of lead in soil reached 1g/kg ~ 50g/kg which is 200 times of the general soil. Therefore, receiving water bodies and groundwater are severely polluted. Moreover, once the soil is contaminated by heavy metal, it is difficult to remove its impact. Hence, the recycling of waste electrolyte which contains lager amounts of suspended solids is also the key of the effective cleaning process of waste lead-acid battery.

Waste lead-acid battery mainly contains two lead materials, namely grid and lead paste which is concerned with the processing methods, technology and capacity of battery. Originally grid mainly used Pb_2Sb alloy as material, and with the expansion of less and free maintenance battery, Pb_2Ca and low antimony alloy are gradually replacing traditional Pb_2Sb alloy, which cause the content of antimony in grid will be reduced. Lead paste actually is the mixture of $PbSO_4$, PbO_2 and PbO which are formed after pasting, formation and using during battery manufacture. Its constitution depends on cycle numbers and life-span of waste battery, causing the different composition and shapes of grid and lead paste. Table 1 shows the typical phase composition of grid and lead paste.

Table 1 The typical phase composition of grid and lead paste (wt%)

name	Total Pb	Metallic Lead	PbO	PbO_2	$PbSO_4$	Sb	Density (g/cm^3)
grid	92-95	92-95	microscale	--	microscale	3-6	9.4
Lead paste	67-76	-	10-15	15-20	25-30	-0.5	3.3

The treatment of waste lead-acid battery generally consists of two stages: broken separation and efficient regeneration of waste material. Then we can separate grid, lead paste and organic waste plastic. Therefore, the comprehensive utilization of various components of waste lead battery and smelting process of lead materials have typical characteristics of the recycling economy.

Status and trends of broken separation technology in battery

Broken separation of waste lead-acid battery is developed to use special saws and machinery knife from broken down by ax and artificial separation. In the 80-90s of last century, modern enterprises in the world developed a full mechanical system of broken separation based on "wet sieving method". The representative is CX system of Italian Engieec Company. The broken separation process can be divided into two methods, that is, artificial separation and mechanical separation, in which mechanical separation process is classified into heavy medium separation and hydraulic separation according to the difference of separation medium. Both methods have a high mechanization and automation.

Artificial broken separation

Currently broken separation of waste lead-acid battery is still dominated by artificial separation in China. Namely, dismantle the case of waste battery, then take out lead sheet and lead dust. During disassemble process residual electrolyte drains into the soil and lead dust fills the air, settling on the ground. Therefore, there exist the following problems: manual work, high energy consumption, serious pollution and low recovery. Artificial separation can only effectively isolate plastic, but can't separate grid and lead paste, which determines the efficient regeneration of waste material can only use simple mixed smelting to recover lead.

Mechanical crushing--heavy medium separation

The method is the products of wet sieving with three different size faction (fine, medium coarse and coarse) after breaking waste battery using a hammer crusher. Parts of the fine material which are lead paste are smelted after dried and some as the weighting agent are used for heavy medium separation of medium coarse and coarse materials. The organic components and hard lead in the medium coarse and coarse materials are transited into different heavy medium hydrocyclone to separate out various products and pile alone. Although the technology can achieve large-scale mechanization, automatically separate waste battery, and has high efficiency, low basic construction investment and high lead antimony recovery, there exist some disadvantages, such as lower degree of automation and processing capacity, not complete separation of products, high mutual contain rate and low yield of lead grid.

Mechanical crushing--hydraulic separation

Currently the foreign mostly uses mechanical crushing--hydraulic separation to dispose waste battery. Waste battery with shell is upgraded to hammer crusher hopper by belt conveyor. During upgraded process the battery shell is broken down by hydraulic punching machine, and then the acid electrolyte is spilled to the container. Waste battery is shattered to smaller than 20mm by crusher with a hook-like structure of gravity hammer before discharged. Broken material is continuously sent to hydraulic grade box by a horizontal spiral conveyor. Adjusting the pressure of hydraulic grade box, broken material is graded according to density differences. Big density of lead grid sinks in the bottom of grade box, and then be carried away by a spiral conveyor. Small density of lead paste and organic compound flows into the fixed sieve with water. The screenings is lead paste of fine size which can be unloaded by stepping cream-removed machine. Organic compound on the screen flows into another hydraulic grade box which separates plastic and rubber, then is discharged by respective spiral conveyor. The equipment material in the system is acid-resistant stainless steel which has a long life. The use of wet broken process can avoid dust. The recovery of lead is over 99%, and the remaining is lost in the organic compound. The mutual contain rate of separated products is less than 0.5%.

Trends of broken separation technology

With the further aggravation of environmental situation, artificial broken separation which is commonly used by the small domestic business has a tendency of elimination. On the contrary, the environmental friendly, high quality mechanical crushing--hydraulic separation CX system is favored by many world famous producers of secondary lead. Environmental protection agencies in many countries have approved the construction of CX facilities, which make mechanical crushing-hydraulic separation method will be the mainstream of waste lead-acid battery disposal.

Status and trends of efficient regeneration technology of waste

The recycling technologies of waste battery are complex and various because of complicated lead paste and difficult recovery technologies. In the practical production the processing technologies of lead paste may be broadly classified in three methods, namely, simple mixed reduction, transformed desulfurization reduction and redox of lead paste with the lead concentrate. Research and development of new processes for treatment of lead paste are mainly wet solid electrolysis and preparation of lead salts with wet method. At present, the disposal of lead paste has been focusing on two topics, namely, how to remove sulfur and how to reduce lead paste to metallic lead.

Production technology of lead paste regeneration

Simple mixed reduction smelting

The method is a traditional way of lead paste recycling, which mainly uses traditional blast furnace, shaft furnace, rotary and reverberatory furnace to recover lead. The traditional fire process has high energy consumption, serious pollution and low recycling rate etc.

Transformed desulfurization reduction smelting

The method firstly makes lead paste desulfurization by alkaline wet process before smelting. So that we can not only reduce the smelting temperature, but also to some extent decrease the generation of lead steam and dust, and alleviate environmental pollution. There is no need to consume other flux, consequently reducing slag yield, lowering lead loss and improving the metal recovery. Besides, organic waste can replace some fuel, which facilitates reducing energy consumption of the smelting process. However, sulfur can only be transformed into useless glauber's salt. Based on extensive research, broken-separation-desulfurization-pyroetallurgy treatment process was proposed in china, which mainly includes crushing sorting, desulfurization of lead paste, short furnace smelting and refining. The process can eliminate the pollution of lead vapor and SO_2, improve the recovery of lead to 95% and reduce energy consumption.

Redox smelting of lead paste with lead concentrate

The method treats lead paste containing S 4% of waste battery together with lead concentrate. The chemical reaction heat of lead concentrate can be fully used to achieve high-temperature smelting of lead paste and lead concentrate. SO_2 obtained from lead paste is merged with SO_2 generated from concentrate smelting, then sent for production of sulfuric acid, no single treatment. We can not only efficiently recover the lead in the lead paste, but also recycle the sulfur.

Conventional sintering blast furnace lead smelting process use low temperature oxidation roasting process, and the lead sulfate cannot be effectively removed, so the conventional sintering technique is not suitable for treatment of lead paste from waste lead acid battery. There are some problems in new technology of the direct lead smelting process, Such as high sulfur residual in the middle materials, poor separation of product and smelting slag or matte. Oxygen bottom blowing lead melting process can achieve the interaction reaction of lead sulfate and lead sulfide materials and achieve high-temperature pool desulfurization, to avoid the dispersion of sulfur in lead paste, and the sulfur in high lead slag residue can be reduced to below 0.5%, reducing environmental pollution. Shuikoushan Mining Bureau and Yunnan Xiangyun Feilong companies use oxygen bottom blowing - blast furnace smelting reduction process to deal with lead paste of lead-acid batteries and lead concentrates, achieving good results. Yuguang Gold & Lead Company began the research that recovering lead from lead-acid battery in 2004. A production line model dealing with 360 thousand tons of waste lead-acid battery each year was established by 2009 used the technology of New secondary lead smelting process of automatic separation of waste lead-acid batteries - bottom-blowing melting invented by Yuguang Gold & Lead Company and Central South University. In addition, Sweden's Boliden's Kaldor smelting method and Australian Ausmelt and Isasmelt methods have been used to recycle secondary lead resource .

Research of new technology for lead paste recycling

To eliminate fume and dust derived from smelting and refining, research and development of new technology for lead paste recycling mainly focused on wet process, solid-phase electrolysis and electrowinning process.

(1)The electrorefining and electrodeposition process of U.S. Rolla Research Center is the physical dissociation of waste lead battery. Waste acid is regenerated with activated carbon column and the plastic shell is recycled. The grid is casted into anode for conventional electro-refining. Lead plaster is dissolved by silicon fluoride acid after transformed subsequent electrodeposition with Ti plated PbO for anode. The purity of both electrolytic metallic lead is up to 99%.

(2) Italy "GS" Process is the physical dissociation of waste lead battery. Separated plastic is sold, and the grid and lead plaster is electrodeposited with the graphite for anode after dissolved by fluoroboric acid.

(3)Solid-phase electrolysis was developed by Institute of Process Engineering, Chinese Academy. The process first separates lead-acid battery into plastic, diaphragm, grid and lead mud. Plastic is directly sold. Diaphragm is harmless incinerated. Grid is made into six-lead alloy ingot which is used in the production of new lead battery after being low-temperature melted and deployed component. Lead mud is coated on the cathodic board for electrolysis after treated. $PbSO_4$, PbO_2 and PbO are reduced to lead, and then melted, casted into ingot which is supplied to battery manufacturing plant. The power consumption for the production of one ton of lead is 600kWh, the recovery of lead is up to 95%, and the purity of electric lead is over 99.99%. The method is a clean production process of lead recovery.

(4) Pre-desulfurization-electrodeposition process was independently researched and developed by Shenyang Academy of Environmental Sciences. Characteristics of the process are pre-desulfurization treatment for the lead mud and regeneration of the desulfurization solution. And then acid leaching of desulfurization material obtains lead-rich electrolyte. Lead-poor electrolyte

is returned to leaching process. Its main feature is no waste gas and slag, and the lead recovery is up to 95-97%.
(5) Hunan Institute of Rare Earth Materials develops lead salt products. The lead recovery is over 96%. The hydrometallurgical technology and methods can separate and recover lead in lead-acid battery and prepare yellow lead. Red lead produced by pyrometallurgy meets market demand. The technology is reliable, low investment, no secondary pollution and significantly economic and social benefits.

Advantages and disadvantages of lead paste recycling processes

Although the wet treatment of lead paste in waste battery can effectively prevent air pollution and increase the recovery rate and have a better environmental benefit, the process is too long, the equipment investment is large, and it can't be connected with the existing pyrometallurgical recycling production line. Besides, wet technology is still in research and development stage, which can't meet the actual conditions. Therefore it does not possess the conditions of development and promotion.

The pyrometallurgy process of lead paste has a short flow sheet, less investment in equipment, large capacity of treatment and suitable for all types of plant. Simple mixed reduction smelting has a serious pollution, high energy consumption and great environmental pressure, so it will be gradually phased out. Transformed desulfurization reduction smelting will also lose competitiveness along with increasing environmental pressure and the rising cost of conversion of desulfurization. With the gradual replacement of sintering oxidative desulfurization by bath smelting desulfurization, it is possible that oxidative desulfurization for high temperature pool deal with lead sulfate material. The redox of lead paste bath smelting with lead concentrate will be the most effective method of dealing with lead paste produced from waste batteries.

Trends of efficient regeneration technology of waste

Research and development of secondary lead smelting technology has been developed towards large-scale, environmental protection and intensification. Apart from all wet solid electrolysis for smelting method of waste lead-acid battery belonging to technical research, the primary lead smelting enterprises enter the field of secondary lead. The treatment of lead paste mixed with lead concentrate is also effective for production development. The direct lead smelting methods of bath smelting for primary lead mainly include Jifusaite, Ausmelt, ISA, QSL and domestic oxygen bottom-blowing-blast furnace reduction The treatment of lead paste with these methods not only recover the lead, but also effectively recover the sulfur.

Conclusion and suggestion

(1)The artificial broken separation has a tendency of elimination, the mechanical crushing-heavy medium separation which can't completely separate products and has a high mutual contain rate is also limited and the environmental friendly, high quality mechanical crushing-hydraulic separation will be the mainstream of waste lead-acid battery disposal.
(2) Conventional sintering blast furnace lead smelting process use low temperature roasting process, causing the sulfur in the lead sulfate can not be effectively removed, thus the traditional sintering process is not suitable for treatment of waste lead acid batteries. Some new technology of direct lead smelting process have problems such as difficult to control the process, high sulfur residual existing in the middle materials, poor separation of rude product and smelting slag or matte and so on, causing it can not effectively deal with the lead sulfate material.
(3) Oxygen bottom blowing process can cause the interaction reaction between lead sulphate and

lead sulphide material to achieve high-temperature pool intensive desulfurization, and residual sulfur of high-lead slag can be reduced to 0.5%. The residual sulfur of high lead slag can be reduced to below 0.5%. Further equipped with direct reduction of liquid high lead slag technology with obvious energy saving ,, the redox of lead paste bath smelting with lead concentrate can not only make full use of mineral potential, realize self-heating melting and recycle lead, but also recycle the sulfur in the lead paste containing 7% sulfur, protect the environment and achieve short process, large-scale, intensive production. It will be a low-carbon, environmentally friendly and efficient process on secondary lead.

Reference

[1] Li Wei-feng, Zhang Xiao-guo, Guo Xue-yi, Zhang Chuan-fu. Status and progress of lead smelting technology in china, China Nonferrous Metallurgy,2010,(2):29-33

[2] Guo Xue-yi, Tian Qing-hua. Theory and methods of non-ferrous metal resources cycle[M]. Changsha: Press of Central South University,2007

[3] Bai Ding (Translation). Bell's policy Niushen secondary lead metal production technology[J]. World Nonferrous Metals, 2003,(10):68-69

[4] Guo Cui-xiang, Zhao You-Cai.Review of Hydro-metallurgical Processes for Recovering Lead from Scrap Lead-acid Batteries. Journal of Dongguan university of technology(Natural Science Edition).2006,13(1):81-86

[5] Chen Xi, Research on the new technology of secondary lead abroad. Resource Recycling ,2009,(1):32-33

REWAS 2013

Enabling Materials Resource Sustainability

Enabling Sustainability through the Physics of Metals & Materials Processing

Session Chairs

Gabriella Tranell
Markus Reuter

REWAS 2013: Enabling Materials Resource Sustainability
Edited by: Anne Kvithyld, Christina Meskers, Randolph Kirchain, Gregory Krumdick, Brajendra Mishra, Markus Reuter, Cong Wang, Mark Schlesinger, Gabrielle Gaustad, Diana Lados, and Jeffrey Spangenberger
TMS (The Minerals, Metals & Materials Society), 2013

CYANIDE AND COPPER RECOVERY FROM BARREN SOLUTION OF THE MERRILL CROWE PROCESS

José R. Parga[1], Jesús L. Valenzuela[2] and J.A. Díaz[1]

[1]Department of Metallurgy and Materials Science Institute Technology of Saltillo
V. Carranza 2400, Saltillo Coahuila, CP. 25000, México.
[2]Department of Chemical Engineering and Metallurgy, University of Sonora, Hermosillo, Sonora, México

Keywords: Cyanide recovery, Copper recovery, Precipitation copper

Abstract

This paper is a brief overview of the role of inducing the nucleated precipitation of copper and cyanide in a flashtube serpentine reactor, using sodium sulfide as the precipitate and sulfuric acid as pH control. The results showed that pH had a great effect on copper cyanide removal efficiency and the optimum pH was about 3 to 3.5. At this pH value copper cyanide removal efficiency could be achieved above 97 and 99 %, when influent copper concentration ions were 650 and 900 ppm respectively. In this process the cyanide associated with the copper, zinc, iron cyanide complexes are released as HCN gas under strong acidic conditions, allowing it to be recycled back to the cyanidation process as free cyanide.

Introduction

The greatest amounts of cyanide-containing wastes are produced by precious metals milling operations, electroplating industry and coal processing or coking effluents processes. Because of high toxicity and to comply with Federal and State regulations, the treatment of wastewater with cyanide complex is required before safe discharge of cyanide wastes.

In mining operations, cyanidation is the predominant method by which gold and silver are recovered from their ores. In practice, the dissolution of gold and silver in aqueous cyanide solution is typically carried out with 0.03-0.3% NaCN, at pH greater than 10, and aeration to keep the pulp or solution saturated with oxygen (>7 ppm). The overall reaction for the dissolution of gold and silver in dilute, aerated, alkaline cyanide solutions may be expressed by the classic Elsner equation:

$$4\,Au + 8\,CN^- + O_2 + 2\,H_2O = 4\,[Au(CN)_2^-] + 4\,OH^- \quad (1)$$

Which has the following mechanism:

$$Au + 2\,CN^- = Au(CN)_2^- + e^- \quad (2)$$

$$O_2 + 2\,H_2O + 2\,e^- = 2\,OH^- + H_2O_2 \quad (3)$$

$$H_2O_2 + 2\,e^- = 2\,OH^- \qquad (4)$$

In these mechanisms the cyanide ion is the complexing agent or ligand and oxygen is the oxidant [1]. The reaction with silver is similar. As the cyanide concentration increases, the rate controlling step for the reaction moves from cyanide diffusion (anodic) to oxygen diffusion (cathodic). It was also been established that trace levels of impurities (Cu, As, Zn, Pb and Fe ions) are critically important to the leaching and recovery process, a finding which has rationalized why many gold and silver cyanidation studies have produced poor leaching and low quality Dore. The association of gold and copper mineralization in commercially viable ore is a common occurrence. The concentration of cyanide used in practice to dissolve gold in ores is typically much higher than the stoichiometric amount required, owing to the solubility of other minerals. Free cyanide produces complexes with several metallic species, especially transition metals, which show a broad variation in both stability and solubility [3-6] :

$$M^{2+} + {}_{y}CN^- \rightarrow M\,(CN)_y^{(x-y)} \qquad (5)$$

Many common copper minerals are soluble in the dilute cyanide solutions typical of leach conditions found in gold cyanidation processes. Minerals such as azurite, malaquite, are fast leaching and soluble in dilute cyanide solutions. Enargite and chalcopyrite leach more slowly but are sufficient soluble to cause excessive cyanide loss and contamination of leach solutions with arsenic [4]. This has a cost impact on the cyanidation process, which may be manifested in poor gold and silver extraction and recovery, high cyanide consumption and high precipitate-management and bullion-refining costs.

Precipitation Of Copper After The Merrill–Crowe Process

Most copper minerals react readily with cyanide and the presence of cyanide-soluble copper affects gold and silver recovery from the cyanide solutions. In the Merrill-Crowe process, the copper is precipitated along with gold and silver, resulting in a higher consumption of zinc dust, fluxes in the smelting of the precipitate and decrease the live of the crucible. This research has focused either on the removal of copper before smelting the gold and silver precipitate or the prevention and/or minimization of the impact of copper in the barren solution after the filter press in the Merrill-Crowe process. Therefore, the increment of copper in the barren solution poses serious metallurgical problems in the cyanidation circuit and it is necessary to include a process to strip the copper prior the leaching the silver and gold. Failure to do so will result in lower dissolution of precious metals and production of high-copper-silver/gold bullion. The treatment of high-copper-silver/gold leach solutions, either before or after precious metals recovery, have focused on precipitation of copper as chalcocite (Cu_2S) and cyanide recovery.

Precipitation of Copper and Cyanide Recovery

There has been growing interest for the recovery of both copper and cyanide from silver and gold barren solutions, because in our case we suffer high cyanide consumption costs as a result of the presence of base metals (principally copper) that react with and consume cyanide during the silver-gold cyanidation leach. The cost of recovering and recycling this cyanide from the barren leach solution will often be lower than the cost of purchasing new cyanide. It has been almost a century since the Mills-Crowe process for cyanide regeneration was developed by the Mining Company Beneficiadora de Pachuca in México (England Patent No. 241669, 3.9.24) [5] and until today no significant changes to the process have been made. The simplest process for cyanide recycling involves acidifying the barren clarified solution (pH between 2 and 5). During acidification, free cyanide and relatively weakly complexed cyanide (Ag, Cu, Zn, Fe) are converted into HCN gas, which is then volatilized by passing a stream of air bubbles through the solution. The air/HCN gas stream is scrubbed in a caustic solution in a second tower reactor to convert the HCN back into free cyanide ions for recycling [6]. In this process copper and silver are not recovered for resale. This has prompted interest to also recover copper by selective metal sulphide precipitation. The copper sulphide precipitate is then recovered by conventional clarification and filtration to produce a filter cake (45 to 60% Cu) which can be shipped to a copper smelter.

Precipitation of Copper. Precipitation is employed for the removal of heavy metals from wastewater. Among the chemical precipitation methods, precipitation of metal hydroxides is most conventional, but it suffers from shortcoming, such as high solubilities. Sulphide precipitation of metals is a viable alternative process for copper recovery from the barren cyanide solutions because of the possible high degree of metal removal over a broad pH range, however hydrogen sulfide is odorous and highly toxic [7] .

Cyanide Recovery After cyanidation, the key advantage of a sulphide precipitation process is the ability to operate in the barren solution to recover first the copper/silver and then the acidic conditions in the solution result in rapid release of free cyanide (HCN_{gas}) that is easily recoverable by volatilization at lowered pH value.
If the cyanide is present in the barren solution after precipitation from the Merrill Crowe process as free cyanide (pKa= 9.4), it is possible to convert 99% of the cyanide into HCN gas by lowering the pH of the solution to about 6:

$$CN^{-} + H^{-} \rightarrow HCN_{(gas)} \quad (6)$$

If, on the other hand, the cyanide is present as a metal-cyanide complex, the pH must be reduced to more acidic values to break down the complex and produce HCN gas. For example, the copper cyanide complex does not break down completely, even in strong acid solution, unless there is an oxidant present in the solution. In the absence of oxidant, the copper tricyanide species (which is the most stable copper complex under normal cyanidation conditions: log B_3 = 28) decomposes to form a CuCN precipitate, plus 2 moles of HCN gas, at pH values less than 3. Hence, 33% of potentially recoverable cyanide is lost to the precipitate:

$$Cu(CN)^{2-}{}_{3} + 2\,H^{+} \rightarrow CuCN(s) + 2\,HCN_{(gas)} \quad (7)$$

Also, with the addition of sulphide ions (Na_2S) to the acidified cyanide solution results in the precipitation of cuprous sulphide (chalcocite), which is favored because of its extremely low solubility (Ksp = 2.3 x 10^{-48}) (2). The following reaction takes place:

$$2\ Cu(CN)^{2-}{}_3 + 2\ H_2SO_4 + H_2S_{(gas)} \rightarrow Cu_2S(s) + 6\ HCN_{(gas)} + 2\ SO_4^{2-} \qquad (8)$$

In addition to precipitating copper, sulfide addition also results in the near-complete precipitation of silver, as shown in the following reaction.

$$2\ Ag\ (CN)^-{}_2 + H_2S(gas) \rightarrow Ag_2S(s) + 4\ HCN_{(gas)} \qquad (9)$$

Based upon the reactions [6-9], acid conditions may cause the dissociation of the complexes, due to the formation of some copper precipitate and subsequent liberation of HCN by volatilization. Also accord with these reactions, up to 99 % of copper could be recovered and HCN gas is stripping from the barren solutions and then adsorbed in an alkali solution of NaOH. The simplified chemistry of the process is presented in the following reaction:

$$HCN_{(gas)} + NaOH \rightarrow NaCN + H_2O \qquad (10)$$

The precipitate is a saleable copper product in its own right or can be blended with the arsenopirite flotation concentrate made in the flotation sulphide plant.

Materials And Methods

Copper precipitation and cyanide regeneration experiments were performed to determine the effect of different process conditions on the solids of copper/silver sulphide produced by sulphide precipitation. Precipitation experiments were carried out in a 1 liter round-bottomed reaction vessel with ports for an overhead stirrer, a gas sparger and a pH electrode. The pH meter is VWR 8005 Scientific and stirring motor with a glass impeller driven BDC 1850 CAFRAMO and cone size sedimentator (1000 ml). The barren solutions used containing copper, silver, zinc and iron ions of varying concentration. The pH of the barren solution was adjusted to the required level with sulfuric acid and then a mixture of Na_2S/water was added. In all precipitation experiments samples of the liquor and solid were taken at known times and then, solutions and solids from the process were separated by filtration through cellulose filter paper. The sludge from the precipitation step was dried either in an oven or under vacuum at room temperature. Analysis of copper, silver, zinc, iron, and arsenic were performed by ICP/Atomic Emission Spectrometry and free cyanide content was determined directly via titration, whereas the total cyanide was measured by means of titration after distillation. At the end of the experiment, HCN volatilization reached efficiencies above 95% and the capture of cyanide gas by NaOH (1 M) solution was almost 95%.

Results And Discussion

The experiments were carried out at different pH under atmospheric pressure in glass reactor. In all the tests the barren initial solution from the Merrill-Crowe plant was from the same batch (0.1 Ag ppm, 184 Zn ppm, 636 Cu ppm, 4 Fe ppm). The following conditions were also fixed: temperature 25 ^{0}C; stirring speed 200 rpm; 0.5 to 2 g/l Na_2S and reaction time 90 seconds. The experimental results of the copper, silver, zinc and iron precipitation at different pH are presented in Table 1.

Table 1 Results of copper, silver, zinc and iron sulphide precipitates.

	Ag	Zn	Cu	Fe	Na_2S (g)	pH
Feed Barren Solution(ppm)	0.1	184	650	4	0	10.95
Solution 4 (ppm)	0	8	0	0	1.0	5.0
Precipitate 4, (%)	119	13.4	51.27	1.0	-	-
Solution 5 (ppm)	0	32	0	0	1.0	4.5
Precipitate 5, (%)	138	9.92	56.34	0.9	-	-
Solution 6 (ppm)	0	54	0	0	1..0	4
Precipitate 6, (%)	118	1.49	62.68	1.1	-	-
Solution 7 (ppm)	0	40	0	0	1.0	3.0
Precipitate 7, (%)	129	9.53	62.24	1.1	-	-
Solution 8 (ppm)	0	134	0	0	1.0	2.5
Precipitate 8, (%)	106	11.24	60.5	0.9	-	-

The results showed that pH had a great effect on copper cyanide removal efficiency and the optimum pH was about 3 to 4.0. At this pH value copper cyanide removal efficiency could be achieved above 99 %, when influent copper concentration ions were 636 ppm. Some black precipitates were observed in the solution samples of the pH 2 to 6 experiments; which suggested that there were copper, silver, arsenic, zinc and iron as sulphide. The production of this sulphide is confirmed by X-ray diffraction, as shown in Figure 1.

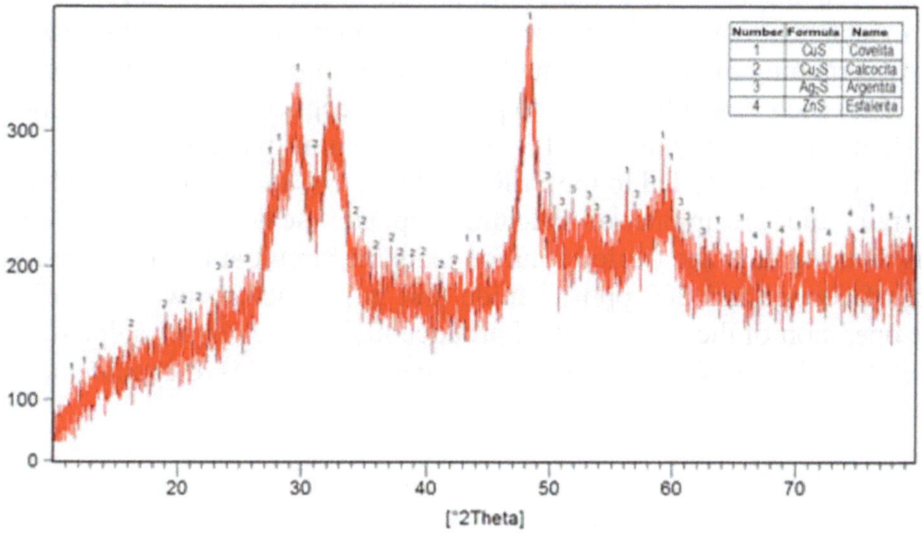

Figure 1. X-ray diffraction spectrum obtained from the sulphide precipitate at pH = 3.

The measured sample, which was collected from experiment pH = 3 (see Table 2), gives rise to peaks corresponding to covellite, calcocite, argentite and esfalerite . The size, EDAX and morphology of the solids in experiment pH are also shown in Figure 2 by SEM and EDAX analysis.

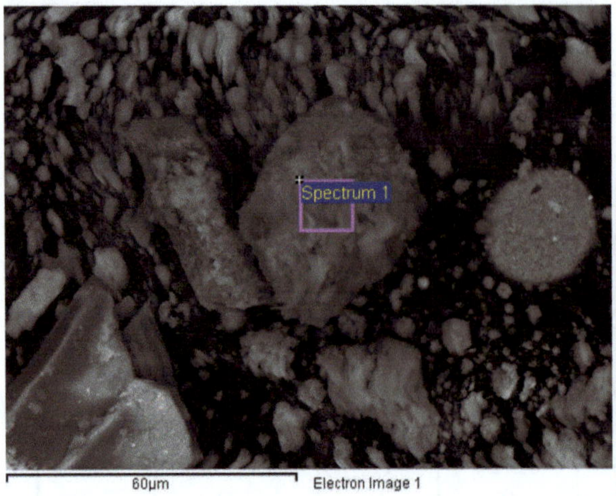

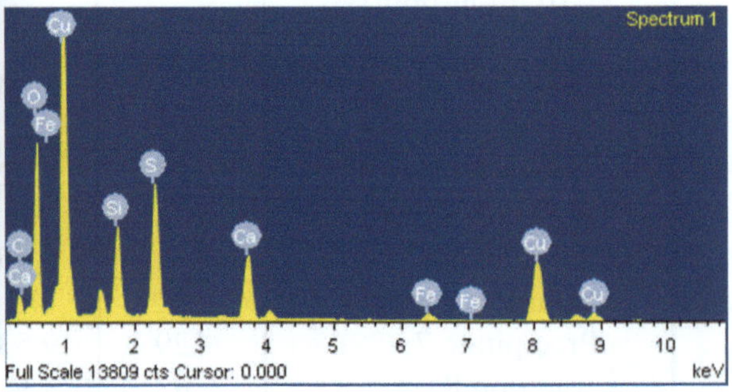

Figure 2. SEM image and chemical composition of solid product as determined by EDX, which shows the presence of copper, sulphur and iron in the sulphide particle.

Industrial Application

In base on the experimental evidence obtained with the sulphide precipitation study for copper and cyanide removal from the barren solution after the Merrill-Crowe process. This process has been installed on a mine site at full scale. The SERPENTINE (flushtube) system is a viable technology for the recovery of copper, silver and subsequent recovery of HCN gas by scrubbing in NaOH. A simplified process flow diagram in which uses sodium sulphide (Na_2S) such as sulphide ions, to precipitate copper and silver and convert cyanide to HCN gas, under acid conditions (pH 3 to 4) is shown in figure 3.

Barren solution is currently fed to the SERPENTIN at 10 to 15 liters/second at a pH 11. At this flow rate, the precipitate of calcium sulphate (scale) would not occur. In five continuous working days the treated solution exits the circuit at a pH of 4, carrying about 0 to 10 ppm of copper and 200 ppm cyanide and is pumped to two neutralizing (pH= 7) tanks. The operation of the SERPENTIN produce high grade copper sulphide precipitate in the range of 40 to 55% of Cu with 130 gr/ton Ag and recoveries of cyanide of 80%.

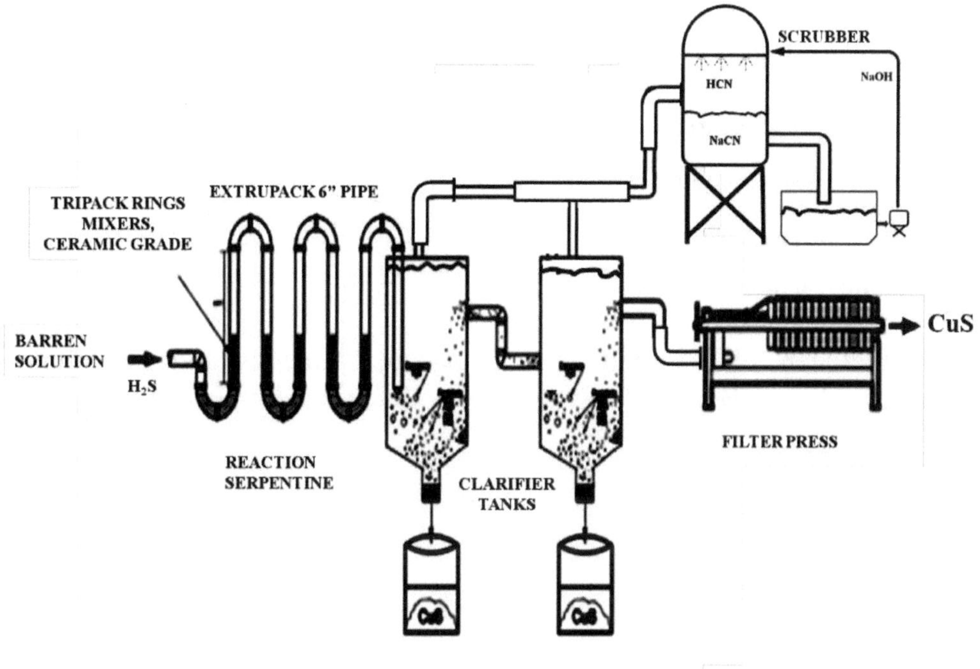

Figure 3. Simplified diagram showing the SERPENTINE process for copper, silver recovery and cyanide recycle.

Conclusions

The advantages of the SERPENTIN is high precipitation rate of copper and silver (99%), compact treatment facility, relatively low operation cost, the precipitate is a saleable copper/silver product, and the main advantages is that produce high grade copper sulphide precipitate in the range of 40 to 55% of Cu with 130 gr/ton Ag and recoveries of cyanide of 80%.

Acknowledgements

This study was supported by Minera William in Mexico, CONACYT and to Dirección General de Educación Superior Tecnológica (DGEST) from Mexico.

References

[1] J.R. Parga, J.L. Valenzuela and F. Cepeda, "Pressure Cyanide Leaching for Precious Metals Recovery", *Journal of Metals*, 10, (2007), 43-47.

[2] F. Habasi, "Kinetics and Mechanism of Gold and Silver Dissolution in Cyanide Solution", Bulletin No. 59, Bureau of Mines and Geology, State of Montana, April, 1967.

[3] J.R. Parga, et al., "Cyanide Detoxification of Mining Wastewaters with TiO_2 Nanoparticles and its Recovery by Electrocoagulation", *Chemical Engineering and Technology*, 32, (2009), 1901-1908.

[4] C.A. Fleming, "Cyanide Recovery". *Developments in Mineral Processing 15*, ed. M.D. Adams, Elsevier B.V., (2005) 29, 703-727.

[5] C.W. Lawr, "Cyanide Regeneration as Practiced by Compañia Beneficiadora de Pachuca", Mexico, Technical Publication AIME, No. 208, 06 (1929), 1-37.

[6] J.R. Parga and D.L. Cocke, "Enhance Cyanide Recovery by Using Air-Sparged Hydrocyclone", *Chemical Engineering and Technology*, 26 (2003), 1-5.

[7] Z.M. Shareefdeen, W. Ahmed and A. Aidan, "Kinetics and Modeling of H_2S Removal in a Novel Biofilter", *Advances in Chemical Engineering and Science*, 1 (2011), 72-76.

REWAS 2013: Enabling Materials Resource Sustainability
Edited by: Anne Kvithyld, Christina Meskers, Randolph Kirchain, Gregory Krumdick, Brajendra Mishra, Markus Reuter, Cong Wang, Mark Schlesinger, Gabrielle Gaustad, Diana Lados, and Jeffrey Spangenberger
TMS (The Minerals, Metals & Materials Society), 2013

NORTHERN REGIONS OF RUSSIA AS ALTERNATIVE SOURCES OF PURE WATER FOR SUSTAINABLE DEVELOPMENT: CHALLENGES AND SOLUTIONS.

V.A. Tsukerman[1], A.V. Gudkov[1], S.V. Ivanov[1]

[1] Kola Science Centre, Russian Academy of Sciences
Apatity, Murmansk region, Russia

Keywords: Northern regions, Water resources, Pollution.

Abstract

The paper discusses problems associated with the existing crisis of water scarcity in the modern conditions of the global water use. Available alternative sources of fresh water may be underground and surface waters of the North and the Arctic. Investigated the current situation and condition of fresh water resources in the technological and industrial development of the North and Arctic. The necessity of developing and using green technologies and measures to prevent pollution of surface and ground water from industrial sectors of the Northern regions is shown. Studied modern technologies and techniques for monitoring groundwater and determination of their age in order to avoid and prevent the effects of environmental contaminants. The ways of use of innovative production technologies of fresh and clean water of north Russia for sustainable development, and delivery of water in the needy regions of the world are investigated.

I

The current total water consumption in the world makes 10% of the world's renewable fresh water per year, of which deadweight loss is 50%. World water resources are distributed unevenly across the Earth, they are not unlimited, and have become a major limiting factor for sustainable economic development in many regions.

In most countries the need for fresh water increases to meet the needs of the world population, urbanization, industrial development, irrigation for the production of food, etc. This situation is undoubtedly worse with the growth of population, pollution of surface and ground water and the threat of climate change.

Russia takes a leading position in the world. With most of the water is located in the northern regions. Therefore, one of the main alternative fresh water sources in the world may be waters from the northern and arctic regions. There are huge reserves of fresh water, concentrated in the arctic: glaciers, surface water (rivers and lakes) and groundwater sources. The exact number of these sources is difficult to quantify, because for example, in only one of the Murmansk regions there are more than 25,000 freshwater lakes and rivers. There is no alternative for the preservation of these sources for the future sustainable development of mankind. [1]

The relatively small population of the North should have a positive impact on the reservation of the water resources. However, in northern territories the main areas of mining are focused (extraction and processing of minerals). Herewith an increase in recycling is provided. Northern territories provide 25% of gross domestic product, 18% of electricity, 25% of the forest products and provides 50-60% of foreign exchange earnings of the Russian Federation. Supplies more than 90% natural gas, 75% oil, 80% gold, 90% copper and nickel, almost all diamonds, cobalt, platinum group metals, apatite concentrate. However, there are problems associated with the contamination of fresh water. Therefore, in areas where active mining and processing enterprises are located now the maximum permissible concentration in surface water sources is exceeded. [2]

II

The rivers of Murmansk region belong to the basins of the White and Barents Seas. The main watershed, which stretches close to the east-west direction, to the west of the ridge is the border with Finland on several mountain tundra (Puytsi, Vine, Wolf, Lovozero), in the east - on the elevated part Keyvskoy ridge. The area of river basins of the northern slope watershed, facing the Barents Sea, is 64 400 km^2 and the area of the southern slope (White Sea) is 80 500 km^2. From here to the north and south the main rivers of Kola Peninsula are flowing. Taking a central position, the watershed determines the short length of the flowing river from him, the nature of fault of their longitudinal profiles and small catchment area. Exception to the length is river Ponoy whose length is 426 km. In the east-west direction only rivers Yokanga and Ponoy are flowing. The valleys of the rivers Tuloma and Kovda have extended to the north-east.

The characteristic feature of the drainage system structure of the Murmansk region is the large number of small rivers. Thus, 95% of the rivers are streams of less than 10 km, and the length of these ads up to 63% of the total length of all the rivers. River network density of whole territory is 0.46 km/km^2. Most of the rivers flowing from lakes and flows through them. On their way they crossed a number of lakes, form drops, rapids and waterfalls. Such rivers are properly called lake-river systems and they collect water from large areas and are characterized by high water levels. The main rivers of Murmansk region are Tuloma, Kola, Voronya, Umba, Niva, Kovda and others.

Factors affecting the water objects by changing the surface of river basins have a particularly noticeable impact on the ecological condition of the small rivers. One of the main characteristics of small rivers is a close link between runoff formation and landscape basin. This makes an extraordinary vulnerability of rivers with intensive development of the watershed. Deforestation and the draining of wetlands in their catchment areas, construction of large livestock farms and poultry farms without concomitant environmental activities and discharge of waste water into the rivers without proper treatment quickly lead to a breach of environmental conditions and accelerated aging of small rivers. Without a reasonable regulation of increasing water load on small rivers it becomes increasingly difficult to manage the rational use and protection of large areas of large rivers.

The lakes are located on the territory relatively uniformly. Their total number is 6%, in the basins of the north coast - 6 - 11%, in the basins of the White Sea - 3 - 8%, the most number of lakes in the basin of river Varzina - 21%. The main number of lakes - 99% - applies to small lakes with surface area of less than 1 km^2. On the average there is one lake per 1 km^2 on the whole territory. The high water content and the favorable composition of the longitudinal profile of most lake-river systems of the territory allowed using hydro energy resources to produce the electricity needed for the rapid industrial development of the Murmansk Economic Region. Natural mode of most of the largest lakes and rivers is regulated by waterworks HPP.

Large reservoirs provide perennial regulating of water runoff: in the Barents Sea basin runoff is regulated to 52% of the catchment area, the White Sea - from 32%, which is 41% of the territory of the Murmansk region. The territory in the Murmansk region has a large number of cities and large enterprises. According to the data for the periods from 1998 to present to detect contamination in the region constant monitoring of surface water is performed. Committee on Natural Resources and the Environment on the basis of observation of the quality of surface

waters, which was held on 30 rivers, 8 lakes and 4 reservoirs notes and extremely high levels of pollution in the water bodies of Apatity, Polar, Kirovsk, Kola, Monchegorsk, Murmansk, Nickel, and Korzunovo Luostari. This in turn begins to affect negatively on the local population. Since many of the cities get their water from surface water bodies, rivers and lakes. [3]
Various timely innovative programs aimed for greening industries will help preserve the available water resources, and clean contaminated, prevent future shortages of clean water in the northern regions. Main environmental operation associated with water bodies is to perform monitoring. In the Murmansk region studies conducted only in areas of potential risk: 30 rivers, 8 lakes and 430 rivers, eight lakes and 4 reservoirs. Compared with the total number of ponds in the region applied studies are insufficient because of the large amount of missing out observations of water sources. Observations are carried out only in places of active water use and do not cover water sources which can be used in the future to provide fresh water for the population.

III

The Murmansk region is characterized by a high degree of concentration of production and processing enterprises belonging to mining and metallurgical complex. In Monchegorsk, Zapolarniy, Nickel included in the "Severonikel" and "Pechenga", included in the MMC "Norilsk Nickel", in Kandalaksha - Kandalaksha aluminum plant. Steelmakers are JSC "Olkon" (Olenegorsk), JSC "Kovdor" (Kovdor). In Kirovsk and Apatity largest Russian enterprise for the production of apatite and nepheline concentrate for the production of mineral fertilizers are working. The share of the above companies represents about 70% of all emissions.
In the area of the negative impact of "Apatit" the towns of Kirovsk and Apatity as well as Big Vudjavr lake and White river are located. The main pollutants in these water bodies are nitrogen compounds, organic and suspended solids, phosphates, petroleum products, which are integral components of urban domestic and industrial waste water processing plants apatite (ANOF) and mines of "Apatit". The extraction and enrichment of apatite ore natural waters contaminated with fluorides - specific pollutants from mining, mining and industrial wastewaters of main workshops of "Apatit".

Big Vudyavr Lake.

According to the available data in the Big Vudyavr lake "Apatit" annually dumped more than 54 million cubic meters of mine waters containing about one thousand tons of sulfates, chlorides 300 tons, more than 115 tons of nitrogen and 200 tons of particulate matter, 75 tons of fluoride, 85 tons of organic substances, 10 tons of phosphates, more than 4 tons of oil products. Concentration of specific substances exceed maximum permissible concentration (MPC) - in 91% of the samples - the content of fluoride, more than 30% - phosphate and nitrite nitrogen, 25% - organic substances. Maximum concentrations of pollutants exceed the MPC: nitrites - 5 times, fluoride - 3 times, organic matter and phosphates - reached MPC and observed: fluoride and phosphate - the autumn-winter low water, and organic matter and nitrite nitrogen - in July.
The annual average concentration of fluoride greater than 1.5 times the MPC, and the concentration of the remaining indicators were below the MPC.
The concentration of copper and phenols were within background concentrations and exceeded MPC by 3 times. Suspended solids ranged from 0 to 9 mg/dm3 and were higher than last year.

White River

The White river flows out of Lake Big Vudyavr, receiving household and stormwaters from Kirovsk and Apatity, filtration and waste waters from the tailings apatite processing plant "Apatit" (ANOF-2) and discharges of small businesses on its way.
More than 80% of the samples exceeding the permissible concentration were observed in the fluoride content, organic matter and copper, 50-60% - nitrite nitrogen, zinc, mercury, phenols, in every third sample - phosphates in some samples of water - ammonia nitrogen, petroleum and manganese.
Exceeding of MPC for pollutants in the river almost observed throughout the year, but the highest concentrations were recorded, mainly in autumn and winter low flows under adverse hydrological conditions when dilution of waste water was the smallest. Thus, the maximum content of fluoride (3.8 MPC) was observed in December, phosphate (2.5 MPC) - In November, organic matter (2 MPC) - In January, manganese, and zinc (2 MPC) - In February, suspended solids (16-19 mg / dm3) - from December to February, nitrite nitrogen (41 MPC) - in July. During the year the concentration of nitrite nitrogen exceeded MPC by 4 times, copper - 2, polluted fluoride - 1.9, organic matter - 1.5 times. Compared to previous years (25%), water quality in the White River has not changed, as evidenced by the complexity factor of 22%. [3]

Lake Imandra / Apatity

Lake Imandra from which the inhabitants receive drinking water in Apatity is referred to the water risk group, although such monitoring on the quality produced water in the lake is not carried out by the State Inspection and held only by the technical control of the enterprise, as well as control by the organization of water utility and is not systematic.
On Lake Imandra observations were in the coastal area, from May to October in an area of businesses and communities. Annual average copper content was 4 - 6 MPC at all cross-sections of the lake, except for Monchegorsk, where the average concentration was 11 times higher than the MPC. Increased levels of molybdenum - 4 MPC in an average year is noted in Apatity. The message is clear - people of Apatity need a transition to alternative sources of water supply in particular to groundwater.

IV

Potential operational value of groundwater resources in the Murmansk region is 2557 m^3/day, what is significantly higher than values of total water withdrawal of groundwater for 2012 - 411.4 m^3/day.
Underground drinking waters are associated with different crystalline rocks, and rocks to the Quaternary. 98.3% of approved groundwater accounts for aquifer system of Quaternary deposits, 1.2% - in the crystalline rock aquifer system, and 0.5% - in the mixed type (aquifer system of quaternary sediments and crystalline rocks). Main water intake is performed from complex of crystalline rocks and is 84%.
As of early 2012 the Murmansk region has 31 explored deposits that passed state examination with the reserves, approved by the State Reserves Committee (SRC) and the Territorial Reserves Committee, with amount of 379.03 thousand m^3/day. Of these, 27 fields for drinking water supply, 3 deposits for industrial water supply and 1 mineral waters deposit. [3]
There is a project to transfer residents of Apatity on underground water sources, from water intake "Malaya belaya", but due to lack of funding it is put on hold. In Apatity-Kirovsk district residents of Kirovsk receive water supply from groundwater sources. From all the areas of water users, focused on the underground water supply, there is the highest percentage of water use - 80%. [4]

Recently, the water intake of Kirovsk, have been observed excess nitrogen and aluminum. This is due to the fact that ground water is the medium of migration and transfer agent substances in underground geosphere and play an important role in the evolution of the Earth's crust. Therefore it is needed to study more careful underground water sources, which can also be exposed by different contaminants, and water pollution in water intakes can occur decades later, and grow even after the withdrawal from pollution sources.

V

Analysis of the current state of resources and quality of groundwater as a result of human impact and rapid climate change is particularly relevant for the conservation of these waters for the future of humanity. Conduction of research on the dating of ground water sources will help to determine what is time reserve before the waste water will go straight to the consumer and this research will take arrangements to clean them.

Measurement Method And Outcomes

Traditional methods for solving problems of filtration and mass transport in subsurface hydrosphere are laborious, expensive and in many cases do not allow to extrapolate data obtained in a short period of relatively small areas on long time periods and hydrogeological structures. Analysis of current research shows that almost all the innovative approaches, largely free of these disadvantages are based on isotopic methods including use of radioactive isotopes allowing to define the most important parameter of all processes - time.

Measurement method The use of "radioactive" clock to determine the age of water is complicated by the fact that there is only one radioactive isotope, part of the water molecule – tritium - and that the separate "portion" of water with dissolved components is generally not a "closed" system, and when dating it is necessary to consider the process of mixing of waters of different ages and sources. Despite these inherent difficulties for dating hydrogeological processes a number of techniques are suggested. One of the most common of these is the tritium dating of groundwater.
In Russia, there is only one laboratory which can date water in this way and it is located at the Geological Institute KSC RAS. Similar observations have been started there since 2010 and the waters from intake "Centralny" were examined, whose water is used by the residents of Kirovsk.
3H - 3He dating of natural waters was first proposed by Tolstikhin and Kamensky (1969) and is widely used in study of open water and groundwater in Europe and the U.S. Age of water is defined as the time past from the termination of the exchange of gases dissolved in the water with the atmosphere gases until the moment of sampling.
In Russia this method can be called innovative. All equipment for testing and experimentation has been created from scratch with the exception of the mass spectrometer which identified the tritium content.

Outcomes Water intake "Central" age was determined which is on average 22 years. It is water that gets into the ground in the early 90's, are now actively enters the house residents in Kirovsk. With the trend growth of contaminants in surface water from the 90's and ending of present time, we can assume further deterioration of groundwater quality. [5]
The results of the studies of wells Little White River Area, dating to early 90s, showed no excess of MPC in water intake. Further analyzes were not carried out.

According to sanitary-microbiological data and regulatory documents part of the surface water catchment area of the Khibiny massif is characterized as polluted. Pollution is particularly high in areas receipt domestic wastewaters of municipal treatment plants, in areas of housing estates, bakery waste, that is in place when more organic.
The only water bodies that are clean on the areas of mining operations of "Apatit" are Small Vudyavr lake, Vudyavryok river, and upstream of Yuksporryok river. All other surface water reservoirs are subject to constant contamination.

Treatment Measures

The main measure of treatment is reducing the amount of waste water. The present level of technological development cannot completely abandon the discharge. There is not a company that would fully supported by the 100% recycling. [2]
Another innovative solution to help improve water quality is to centralize wastewaters. Each company has its own slop lake or reservoir which receives wastewaters. Combine them into one big will help to reduce maintenance costs. Previously the problem of creating such a system was not appropriate in this project for several reasons:

1) The high cost of treatment facilities. 100 small treatment plants cost higher than one large. Winning gave no lying of sewerage system but it did not compensate for this problem.
2) A large number of local treatment facilities have discharge to more water bodies if the sources of wastewater were higher in upstream than intakes designed for water and it could have a negative impact on public health.
3) The sum of the designated areas with their health and safety zones for each treatment plant in total would be greater than the area for a one major treatment plant.

Development of innovative technologies allows approaching to the problem for wastewater treatment in new ways. It is become possible to use a decentralized method of cleaning with minimal use of chemical cleaners. This requires that in the complex treatment plant the biological treatment unit should take the main load.
Purification systems with highly productive aerotank can satisfy these conditions. In aerotank it is need to maintain a special environment for bacteria and their high concentration. [1] Such treatment closed type reactors meeting these requirements invented and used abroad. Implementation of these treatment technologies in the Murmansk region will significantly reduce waste water discharge thus greatly reducing the load on the open waters.
However, currently used treatment methods based on any type of filtering are imperfect. The main risk is that it is impossible to fully escape from the contaminant. In water treatment there are two products in output: pure water, concentrated and separated dirty wastes. There is always a risk that the storage of waste can be destroyed. The presence of large doses of pollutants and other contaminants can enter the water and bring much more harm to the environment than in low concentrated form. Therefore the implementation of decentralized wastewater treatment plants requires detailed study.

Conclusion

A considerable part of monitored water shows that waters in the area of "Apatit" are contaminated and treatment technologies used in this area are not perfect. Identification of substances as contaminants in other areas is not held. Therefore an increase of monitoring net of natural water quality and the transition to a new way of decentralized wastewater treatment can reduce emissions of various substances to the drinking water pools. Attention also should be paid to the improvement and development of dating groundwater technology. These technologies will allow knowing the age of the water and in the case of a man-made disaster or increased emissions of enterprises will give an opportunity to predict and forecast change in the quality of groundwater for decades.

References

1) Arctic. NORTH RIVER. Russian North. Basic Sciences: Proceedings of the Fifth Northern Social and Environmental Congress, [Ed. Ed. VA Chereshnev] Committee of the State Duma of the Federal Assembly of the Russian Federation for Science and High Technology, Ural Branch of the Russian Academy of Sciences. - Moscow: Advertising and Publishing Complex "Gallery", 2010. - 222 p. - ISBN 978-5-9911-0040-3

2) North: Arctic vector of socio-ecological research / Collective authors Ed. Ed. VN Lazhentsev. - Syktyvkar, 2008. - 408 p. (Scientific Council for Regional Development, Komi Scientific Center, Ural Branch of RAS)
136 - 148

3) http://www.gov-murman.ru/envcond/

4) IN Tolstikhin and Kamensky IL (1969) On the possibility of determining the age of groundwater tritium - helium-3 method. Geochemistry 8, 1027-1029.

5) Tolstikhin IN, Kamensky IL, I. Tokarev, Hannibal M. Skiba, VI, Gudkov AV, Melekhova GS (2010) 3H - 3He isotope tracer evolution of groundwater (aquifers slopes Khibiny, Kola Peninsula). XIX Symposium on geochemistry name A.P.Vinogradova, Vernadsky Institute, Moscow, 16-18 November (Abstracts). Moscow, Publishing House of the "Watercolor", 370-372.

REWAS 2013: Enabling Materials Resource Sustainabil
Edited by: Anne Kvithyld, Christina Meskers, Randolph Kirchain, Gregory Krumdick, Brajendra Mish
Markus Reuter, Cong Wang, Mark Schlesinger, Gabrielle Gaustad, Diana Lados, and Jeffrey Spangenber
TMS (The Minerals, Metals & Materials Society), 20

Selective extraction of vanadium from the APV-precipitated waste water

Cui Li, Hong-Yi Li*, Chun-Bin Tu, Tao Zhang, Hai-Xing Fang, Bing Xie

College of Materials Science and Engineering, Chongqing University, Chongqing 400044, China

Key words: Waste water; Selective extraction; Ion exchange; Vanadium

Abstract

In the process of precipitating ammonium polyvanadate (APV) to produce vanadium pentoxide in Pan-steel in China, rest waste water usually contains about 24~333mg/L V(V), 2~100g/L Cr(VI) ,20~500mg/L Si(IV) and 20~100g/L Na_2SO_4. In order to recover valuable and also toxic metal ions contained in the waste water, effective extraction method of using anion exchange resin was realized to extract Vanadium selectively, leading to effective separation between vanadium and chromium. To ensure vanadium was absorbed by the resin, V(V) and Cr(VI) were reduced to V(IV) and Cr(III) by $NaHSO_3$, respectively, and then V(IV) was oxidized by H_2O_2 to V(V) anions. Effects of temperature, solution pH, concentration of ions and absorbing time on vanadium absorption rate were investigated. Chromium was precipitated from rest solution while vanadium was eluted from resin by NaOH solution and then precipitated. Results showed that vanadium recovery of 73% could be obtained in optimized condition. The resin could be regenerated by 3% hydrochloric acid, which indicated the recyclability of the resin and thus low cost of this established method.

Introduction

Vanadium is widely used in chemical industry for its chemical activity and in metallurgical industry for its capability of improving the intensity and toughness of alloy [1-2]. With abundant Vanadium-titanium magnetites in Panzhihua area in China, the vanadium slag produced by BF-BOP process is the main raw materials for V_2O_5 production in Pan-steel [3]. In the final stage of vanadium recovery process, vanadium is extracted by being precipitated as APV, which is then separated from the upper clear solution, and the rest waste water is generated. High content of vanadium and chromium in the water is a waste of natural resources and also pollution to environment due to their toxicity. In order to reduce the environmental contamination, $Na_2S_2O_5$ was usually added into the waste water to reduce V(V) to V(IV) and Cr(VI) to Cr(III), followed by alkalization and precipitation [4]. However, in this method Vanadium Hydroxide and Chromium Hydroxide were produced simultaneously but difficult to recover the valuable elements of Vanadium and Chromium separately for utilization. Therefore, method to recover vanadium and chromium separately is in demand. Ion exchange resins are effective media to separate and collect metal ions [5-6]. Recent research on vanadium extraction by ion exchange resins mainly focused on investigation of effects on the vanadium adsorption[7-8]. However, the waste water contains relatively high content of chromium, ammonium and silica compared to vanadium, which caused pretreatment was essential before extraction and enrichment of vanadium.

Thus, a method ultilizing anion exchange resin to recover and separate vanadium and chromium has been established in this paper. The pretreatment procedure was conducted in our method and its negative influence on vanadium adsorption was inhibited. The established method consisted of steps of deammoniation, desilication, reduction, oxidation, adsorption, elution and precipitation. The static adsorption showed that the adsorption effects were mainly related to reaction temperature, solution pH, concentration of several salts and absorbing time. By this method, the toxicity of waste water was eliminated; vanadium was extraction and concentrated, which increases vanadium recovery rate, and chromium was also recovered at the same time. The used resin was recyclable and results in a low cost of this established method. This work was supported by National Natural Science Foundation of China No. 51090382, Fundamental Research Funds for the Central Universities of China No. CDJRC10130010 and No. CDJXS12132238 and Sharing Fund of Chongqing University's Large -Scale Equipment No. 2012061506.

Materials and Experimental

The chemicals used were all of analytical regents. Porous weak-base anion-exchange resin named DEX-V was studied with static state method. Experimental sample was waste water rested from the precipitation of APV in Pan-steel.

Pretreatments

Deammoniation and Desilication

In the process of deammoniation, solution pH was kept at about 11.5. Heat the solution at temperature above 90℃ in water bath for 5~6 hours to facilitate the escape of NH_3.aluminum sulfate in Al/Si molar ratio of 0.95~1 was added into solution immediately to promote the formation of sulfuric Al-Si complex in solution at pH 9.0 around, followed by heating at about 90℃ for more than 3 hours. The process of desilication was completed after precipitating and filtering.

Reduction and Oxidation

The solution pH was maintained at 1.6 to 1.9. The solution was placed into water bath at 70°C for 3 hours with $NaHSO_3$ addition in stoichiometric ratio. H_2O_2 was put into the treated solution at pH around 3.0, with stirred at 30 rounds per second for about 4 hours. The mixed solution was heated at 100 °C until no gas bubbles escaped in order to oxidize V completely and eliminate redundant H_2O_2.

Adsorption

Optimal conditions including solution pH, ratio of liquid/solid and temperature were studied first. Solution with vanadium concentration of 4.667mmol/L was mixed with fresh water in ratio of 1:2; wet resin was then added to the mixed solution in liquid/solid ratio of 20 and stirred at 2rps. The fresh water was substituted by solution of sodium sulfate, chromic chloride and sodium chloride to investigate influences of the three salts on vanadium adsorption respectively.

Pretreated waste water was mixed with anion exchange resin in liquid/solid ratio of 50 around to enrich vanadium in optimized adsorption conditions.

Elution

After separating the resin from solution, vanadium was eluted from resin to solution by 2% NaOH as eluent. NaOH was mixed with loaded resin for 25 min, in liquid/solid ratio of 4 for 5 times. The elution yield of Vanadium reached as high as 95%. The resin could be recycled by 3% hydrochloric acid.

Precipitation and Roasting

Since the pH was at 8.8 and excessive amounts of ammonium chloride was added, vanadium was precipitated in the form of ammonium metavanadate, which was then treated by filtering, drying and roasting at 560℃ for 30 minutes[8]. The vanadium pentoxide was thus obtained.

pH of solution with vanadium removed was maintained at around 8.8, aiming at extracting chromium by precipitation as chromic hydroxide, as shown in reaction (1.6), the precipitate was roasted at 1150℃ for 3 hours , chromium products was obtained as Cr_2O_3.

Results and Discussion

Pretreatments

Because of high concentration of ammonium ions in solution, vanadium could be easily precipitated in the form of ammonium metavanadate, which would be removed together with desilication products and lead to loss of vanadium. To avoid this, it is necessary to remove ammonium before silicon.

After removing ammonium and silicon, the vanadium loss was only about 9%. Compositions of deammoniated and desilicated waste water and initial waste water were determined by ICP-AES as shown in Table 1.

Table 1 Chemical composition of waste water

Sample	V mg/L	Cr g/L	P mg/L	Si mg/L	Ca mg/L	Al mg/L
Initial waste water	91.58	1.18	2.28	411.24	134.92	2.14
Pretreated waste water	82.81	1.17	0.93	3.99	117.02	11.42

In order to separated vanadium from chromium efficiently, $NaHSO_3$ was added to reduce V(V) and Cr(VI) to V(IV) and Cr(III), respectively, and then V(IV) was oxidized by H_2O_2 to V(V) while Cr(III) remained unchanged. Reduction reactions were represented as equations (1.1) and (1.2). Vanadium existed as VO_4^{4-} at pH 7.0~8.5 as shown in reaction (1.3). Solution pH was adjusted to 4.0~5.5 3 hours later, additions of H_2O_2 enabled reaction (1.4). Reaction (1.5) proceeded at 95℃ and $VO_2(O_2)_2^{3-}$ ions changed into $H_2V_{10}O_{28}^{4-}$ due to the decomposition of H_2O_2.

$$Cr_2O_7^{2-} + 3HSO_3^- + 5H^+ \rightarrow 2Cr^{3+} + 3SO_4^{2-} + 4H_2O(1.6 \le pH \le 1.9) \quad \textbf{(1.1)}$$

$$2VO_2^+ + HSO_3^- + H^+ \rightarrow 2VO^{2+} + SO_4^{2-} + H_2O(1.6 \le pH \le 1.9) \quad \textbf{(1.2)}$$

$$VO^{2+} + 6OH^- \rightarrow VO_4^{4-} + 3H_2O(7.0 < pH) \quad \textbf{(1.3)}$$

$$2VO_4^{4-} + 5H_2O_2 + 2H^+ \rightarrow 2VO_2(O_2)_2^{3-} + 6H_2O(4.0 < pH \le 5.5) \quad \textbf{(1.4)}$$

$$26H^+ + 10VO_2(O_2)_2^{3-} \xrightarrow{\Delta} H_2V_{10}O_{28}^{4-} + 10O_2\uparrow + 12H_2O \quad \textbf{(1.5)}$$

$$Cr^{3+} + 3OH^- + nH_2O \rightarrow Cr(OH)_3 \cdot nH_2O\downarrow(pH = 8.8) \quad \textbf{(1.6)}$$

Adsorption

The pretreated water was mixed with anion exchange resin. Vanadium was absorbed to the resin in the form of $H_2V_{10}O_{28}^{4-}$ while cations still remained in the solution (as shown in equation (1.7).

$$H_2V_{10}O_{28}{}^{4-}{}_{(aq)} + 4R - Cl_{(resin)} \rightarrow R_4 - H_2V_{10}O_{28\,(resin)} + 4Cl^-{}_{(aq)}\,(pH = 3.0) \quad \textbf{(1.7)}$$

Solution pH

Attention should be paid to the solution pH during adsorption. Vanadium has tendency to exist as VO_2^+ in the solution when the pH value was lower than 2.0, which was disadvantageous to the exchange reaction due to the fact that only anions' adsorption onto the resin was allowed. Besides, impurity ions could absorb onto the resin more easily at high pH. Resin has wide working pH range of pH 1.0-7.0. Therefore, solution pH of 2.0, 3.0, 4.0 and 5.0 were compared in this work. In Figure 1(a), solution with vanadium concentration of 4.667mmol/L was mixed with fresh water in ratio of 1:2. Resin was then added to the solution in liquid/solid ratio of 20 at various pHs respectively. Results showed that vanadium enrichment capacity of resin at pH 2.0 was inferior to that at higher pH when absorbing at 25℃ for 280 minutes. It could be illustrated by reaction (1.8), in which vanadium existed as VO_2^+ instead of $H_2V_{10}O_{28}{}^{4-}$ in the solution. Compared to pH at 4.0 and 5.0, in solution at pH 3.0 similar exchange adsorption ability of vanadium with weaker physical adsorption of chromium appeared due to of the fact that chromium existed as Cr^{3+} at pH 3.0, which was consequently the optimized solution pH during adsorption.

$$H_2V_{10}O_{28}{}^{4-} + 14H^+ \rightarrow 10VO_2^+ + 8H_2O(pH < 2.0) \quad \textbf{(1.8)}$$

Temperature

Figure 1(b) presented the effect of temperature on vanadium absorption rate in conditions as described in Solution pH. As shown, the absorption rate of vanadium increased slightly with increased temperature (at pH 3.0). It indicated that the exchange process was endothermic. Temperature was fixed at 25℃ in following work due to the little impact of temperature on the adsorption rate of vanadium.

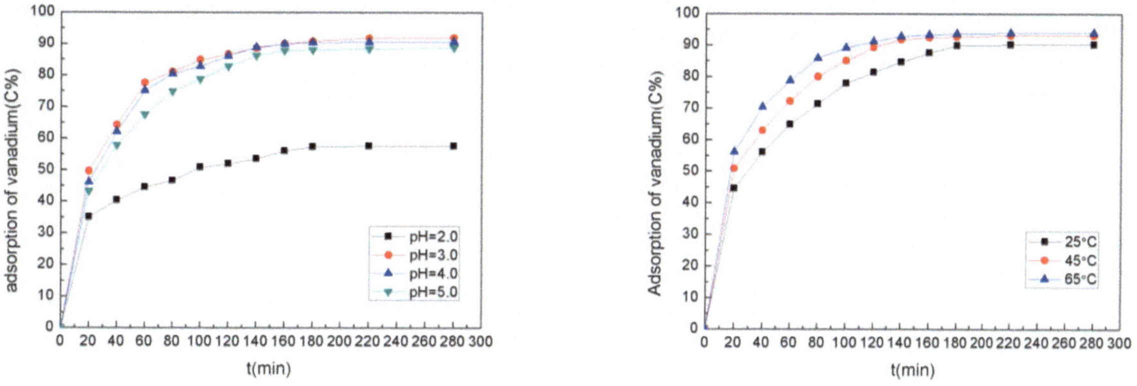

(a) (b)

Figure1. Influences of (a) solution pH value (b) absorbing temperature on vanadium adsorption rate.

Liquid/solid Ratio

Solution containing vanadium was mixed with resin at pH 3.0 in liquid/solid ratio of

15~75with other conditions as described in Solution pH. As shown in Figure 2(a), it demonstrated that the adsorption capability of vanadium was gradually improved as absorbing time increasing until the resin reached saturation (fully loaded). With increased volume of resin, effective sites, where the exchange reaction happened, on the surface of resin increased accordingly. To save the experimental cost and ensure the adsorption efficiency of vanadium, liquid/solid ratio was optimized to be 20. When extracting vanadium from the pretreated waste water, liquid/solid ratio was 50 according to the different concentration of vanadium between the waste water and the solution prepared to investigate optimized conditions.

Compared to 20 minute later, higher exchange speed was obtained at the first 20minutes, that's mainly due to the great concentration gradient existed between internal and external of resin, which was propitious to the external diffuse of vanadium ions in dynamical analysis. As time increasing, the concentration gradient of vanadium between solution and loaded resin decreased, which restrained external diffuse of vanadium, leading to the reduction of exchange speed. Besides, the reaction interfaces advanced from surface on resin towards its center and thus internal diffusion of vanadium was gradually become a restricting factor. Figure 2(a) also showed that the resin could reach saturation when kept stirring at 2rps for about 280min and the optimum exchange time was chosen to be 280min.

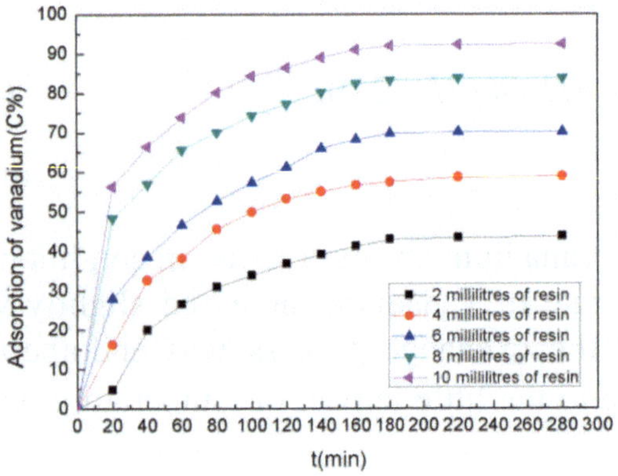

(a)

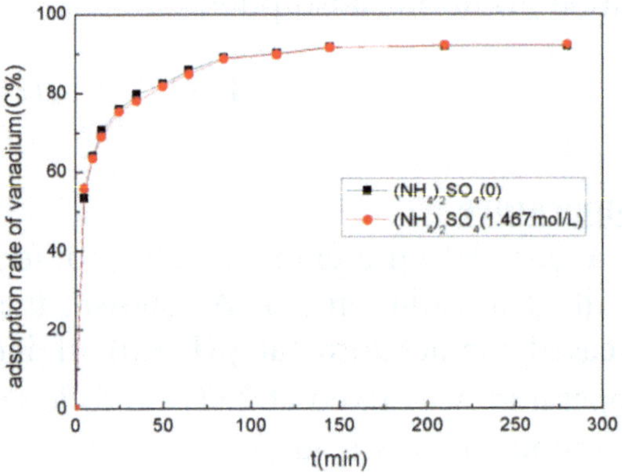

(b)

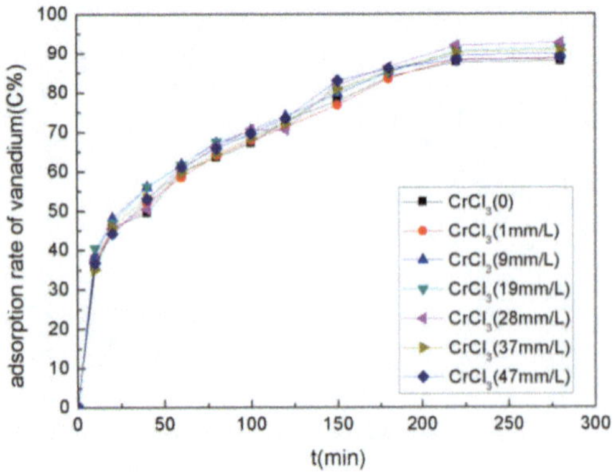

(c)

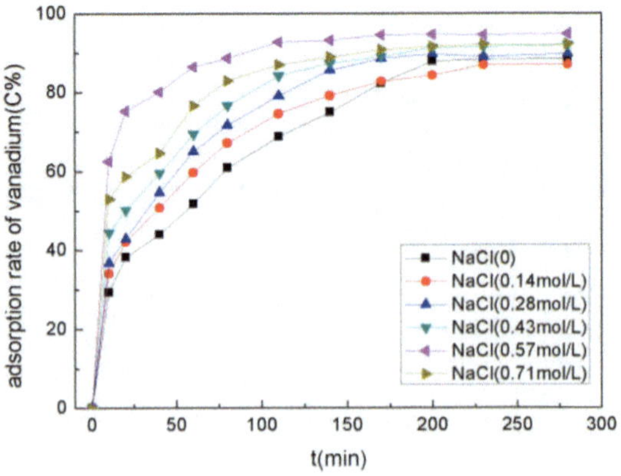

(d)

Figure2. Influences of (a) Volume of Resin (b) Concentration of ammonium sulfate (c) Concentration of chromic chloride (d) Concentration of sodium chloride on vanadium absorption rate.

Concentration of Impurity Ions

Since the initial waste water contained highly concentration of sodium ions, chromium ions, chlorine ions and sulfate ions and the two later ions were cumulatively increased in process of pH adjusting, attentions were paid to the impacts of the main impurity ions as mentioned above. Influences on vanadium adsorption were investigated by mixing resin with several salts, including ammonium sulfate, sodium chromate and sodium chloride. Other conditions were kept the same as described in Liquid/solid Ratio. As shown in Figure 2(b)-2(c), $(NH_4)_2SO_4$ and $CrCl_3$ had little effect on absorption of vanadium, respectively.

When NaCl was added, it had slightly positive effects on the vanadium adsorption. As shown in Figure 2(d), sodium chloride promoted the adsorption and the optimal addition concentration was 0.57mol/L, which was comprehensive reflection of the state of resin and the inhibition of chlorine ions on vanadium exchange reaction. With NaCl addition, it was observed that the loaded resin settled down gradually onto the bottom of container. It suggest that density of the solution was increased with NaCl addition and the resin can suspend in the solution, which lead to increased contact area between resin and ions and thus increased exchange efficiency. However, the greater amounts of NaCl added, the greater concentration gradient of chlorine ions between the solution and resin declined, which made chlorine ions leave the resin surface more difficultly. The available adsorption sites thus decreased and this was disadvantageous to vanadium adsorption. When 0.71mol/L NaCl was added, vanadium adsorption rate was lower than that with addition of 0.57mol/L. High concentration of chlorine ions restrained the exchange reaction. Besides, the resin floated to the surface of solution with 0.71mol/L NaCl added, leading to an insufficiency contact with vanadium ions in solution. The concentration of chlorine ions in initial waste water was within the range investigated. Above all, the results showed that concentration of impurity ions did not have obvious undesirable effects on vanadium adsorption.

Conclusions

In this paper, V_2O_5 was extracted from the APV-precipitated waste water by mixing ion exchange resin with pretreated waste water in optimum condition as bellows. Solution with vanadium removed was neutralized to recover Cr_2O_3 effectively. The toxicity of final waste water was decreased significantly to facilitate environmental conservation.

(1) Silicon and ammonium were removed effectively from the APV-precipitated waste water in pretreated process.

(2) The optimum conditions for adsorbing vanadium from pretreated waste water by resin were at pH 3.0 in liquid/solid ratio of 50 for 280 minutes with 2rps at 25°C, the negative influences of sodium ions, chromium ions, chlorine ions and sulfate ions on vanadium adsorption by resin nearly could be ignored.

(3) Vanadium and chromium products with purity of 96% and 93% were achieved with recoveries 72% and 95%, respectively.

References

[1] R. R. Moskalyk and A.M. Aflantazi, "Processing of vanadium: a review," *Minerals Engineering*, 16 (2003), 793-805.

[2] Ji Yunbo, et al, "Research and development of vanadium extraction technology," *Metallic Ore Dressing Abroad*, 5(2007), 10-15.

[3] D. X. Huang, Vanadium Extraction and Steelmaking (Beijing, BJ: Metallurgical Industry Press, 2000), 11-15.

[4] Fu Zibi, et al, "Increasing abilities of handling V^{5+} and Cr^{6+} containing in wastewater after vanadium precipitation," *Ferro-Alloys*, 6(2008), 40-43.

[5] Liu Anhua, et al, "Technology of vanadium extraction from v-hearing solid wastes and its prospect," *Metal Mine,* 10(2003), 61-64.
[6] Zeng Li, et al, "Study of separation of vanadium from ammonium molybdate solution by ion exchange," *Rare Metals and Cemented Carbides*, 2(34)(2006), 1-5.
[7]Chen Y, et al, "Investigations on the extract molybdenum and vanadium from ammonia leaching residue of spent catalyst," *Int. J. Miner. Process,* 1(2006), 42-48.
[8] Song Fu, et al, "Extraction of vanadium from sodium tungstate by ion exchange," *Rare Metals and Cemented Carbides,* 3(34)(2006), 5-8.

REWAS 2013: Enabling Materials Resource Sustainability
Edited by: Anne Kvithyld, Christina Meskers, Randolph Kirchain, Gregory Krumdick, Brajendra Mishra, Markus Reuter, Cong Wang, Mark Schlesinger, Gabrielle Gaustad, Diana Lados, and Jeffrey Spangenberger
TMS (The Minerals, Metals & Materials Society), 2013

Pt-doped TiO_2 nanoparticles for photocatalytic degradation of phenols in wastewater

M. A. Barakat[1,2*], R. I. Al-Hutailah[1], E. Qayyum[3], and J.N. Kuhn[3]

[1]Department of Environmental Sciences, King Abdulaziz University, Saudi Arabia
[2]Central Metallurgical R & D Institute, Helwan 11421, Cairo, Egypt
[3]Department of Chemical & Biomedical Eng., University of South Florida, Tampa, Fl, USA

Key words: Wastewater, phenolic pollutants, photocatalytic degradation, Pt/TiO_2, nanoparticles.

Abstract

Pt-doped TiO_2 nanoparticles catalysts were synthesized and evaluated for UV photocatalytic degradation of phenol and 2-chlorophenol (2-CP) in synthetic wastewater solutions. The catalysts were synthesized by immobilizing colloidal Pt nanoparticles onto titanium dioxide (rutile TiO_2). Several analytical tools, such as standard BET isotherms, X-ray diffraction (XRD), transmission electron microscope (TEM), were used to investigate the specific surface area, structure, and size distribution of the catalysts and its components. The catalytic activity was measured in a batch photoreactor containing solutions of phenol and 2-CP independently, with UV irradiation of 450 W. UV-visible spectrophotometer was used for analyzing the concentration of phenols in solution at different time intervals during the photodegradation experiment. Parameters affecting the photocatalytic process such as concentration of the catalyst, solution pH, and phenols concentration have been investigated. Results obtained revealed that Pt/TiO_2 showed a higher activity for UV- photocatalytic degradation of both phenol and 2-CP pollutants in solution (as compared to the rutile TiO_2). . The degradation efficiency values were 87.7 and 100% for both of phenol and 2-Cp, respectively, under optimized conditions (0.5 g/L catalyst with a pollutant concentration of 50 mg/L after irradiation time of 180 minutes).

1. Introduction

During the recent decades, photocatalytic applications using semiconductors have been received much attention to solve certain environmental problems [1-4]. Photocatalysis is a promising technique for the treatment of contaminated waters and ground waters, which has been widely studied in recent years due to its ability to oxidize organic molecules completely without

---

accumulation of hazardous byproducts. Among many candidates for photocatalysts, TiO_2 is one of the most suitable semiconductor material for industrial applications; recognized for its high efficiency, low cost, high physical and chemical stability, widespread availability, and non-corrosive property [5,6]. Upon the absorption of light energy that is equal to or greater than band gap energy, the electrons in the valence band of the semiconductors such as titanium dioxide (TiO_2) can be excited to those in the conduction band, leaving a positive hole (h^+) in the valence band [7,8]. The positive hole is a strong oxidant, which can oxidize a compound directly or react with the electron donors in the environment such as water or hydroxide ions to form hydroxyl radicals (OH^-) that are also potent oxidants. The photocatalytic activity of TiO_2 is due to its wide band gap and long lifetime of photo-generated holes and electrons compared to other semi conductors but high degree of recombination of photo-generated electrons and holes is a major limiting factor in controlling its photocatalytic efficiency and impedes the practical application of these techniques in the degradation of contaminants in water and air. Thus, a major challenge in heterogeneous photocatalysis is the need to increase charge separation efficiency of the photocatalyst [9]. Coupled semiconductor photocatalysts exhibit a very high photocatalytic activity for both gas and liquid phase reactions by increasing the charge separation and extending photo-excitation energy range. Recently, many researchers had shown a lot of interest in coupling two semiconductor particles with different band-gap widths. Research groups have carried out photocatalytic activity experiments using various coupled semiconductor particle systems such as TiO_2–CdS [10], TiO_2–WO_3 [11], TiO_2– SnO_2 [12], TiO_2–MoO_3 [13], TiO_2–Fe_2O_3 [14]. Research on the photocatalytic activity of TiO_2–ZnO coupled oxides was also carried out [14-15]. Surface modification by doping with metal ions and organic polymers has been proven to be an efficient route in improving the photocatalytic activity of TiO_2. Study revealed that Ag dopant accelerates the transformation of anatase TiO_2 to its rutile form, and relatively low Ag concentration (2–6%), results in increase in specific surface areas of the TiO_2 powders. The presence of Ag in crystalline TiO_2 was found to strongly enhance the photocatalytic activity of TiO_2 in various degradation processes [16-18].

Phenol and chlorophenolic compounds are of wide use in the production of wood preservers, pesticides and biocides, and constitute an important class of recalcitrant pollutants [19]. They are also found in the wastewater of petrochemical industries, plastics industry, and can be found in pulp, and insulation materials [20-22]. Wastewater emanating from oil refineries is often contaminated with aliphatic and aromatic petroleum hydrocarbons and organo-chlorinated compounds [23-25] and the current conventional processes for treatment of such effluents have caused environmental concerns and resulted in accumulation of hazardous sludge [26-28]. These compounds cause serious environmental problems, due to their high toxicity, recalcitrance, bioaccumulation, strong odor emission, persistence in the environment and suspected carcinogenity, and mutagenity [29,30]. Phenols concentration in a groundwater sample from a contaminated aquifer was reported as high as 25–55 mg.L-1 [31]. In wastewater, the concentration can reach higher than 200 mg.L-1 [32]. These findings illustrate the seriousness of the phenolic pollutants and the importance of finding an effective method for treating such hazardous wastes. The objective of this research work is to evaluate Pt/TiO_2 nanoparticles sample for the UV- photocatalytic degradation of some phenolic pollutants (phenol and 2-CP) in synthetic wastewater.

2. Experimental

2.1 Materials

Titanium dioxide (rutile TiO_2) from Alfa-Aesar was used as photocatalyst support. Chloroplatinic acid ($H_2PtCl_6 \cdot 6H_2O$, 99.9% pure; Sigma-Aldrich) was used as a precursor for Pt. Standard grades of phenol and 2-CP solutions (Merck) were utilized as pollutants in synthetic wastewater for the photodegradation experiments. Compressed gases, used for BET analysis, were UHP from Airgas. All other chemicals used in this work were of reagent-grade quality. The light source was a water-cooled 450W high-pressure mercury lamp (Hanovia 608A36, ACE Glass, NJ, USA), the spectral irradiance for the UV lamp (260 W/m^2) ranges from 228 to 420 nm at a distance of 1m from the light source, according to information provided by the manufacturer.

2.2 Synthesis and Characterization

2.2.1 Nanoparticle Synthesis and Characterization

Pt nanoparticles capped by polyvinlypyrrolidone (PVP) were prepared by colloidal routes established in the literature [33-37]. First, chloroplatinic acid ($H_2PtCl_6 \cdot 6H_2O$) was dissolved into DI water to achieve a solution of 20 mL and 6.0 mM. This solution was then diluted with 180 mL of methanol. Polyvinlypyrrolidone (PVP, molecular weight of 40,000, 133 mg) was added to this mixture. The combined solution was heated to 110°C and allowed to reflux for 3 hr to achieve the metal nanoparticles. After cooling, the suspension was triple washed (cycles of hexane and ethanol with intermediate centrifugation) to remove excess PVP. The cleaned particles were re-dispersed in an ethanol suspension. The size of particles was measured by transmission electron microscopy (TEM). TEM images were acquired at acceleration voltages of 60 kV using an FEI Morgagni 268D microscope. For each synthesis, the average particle size and a particle size histogram was measured by counting the particle size of 100 particles.

2.2.2 Supported Catalyst Synthesis and Characterization

Samples of Pt/TiO_2 catalysts were synthesized by immobilizing Pt nanoparticles onto titanium dioxide (rutile TiO_2 from Alfa-Aesar). The particles were supported by mild sonication for 3 hours at a metal : titanium dioxide mass ratio of 0.3% : 99.7%. Following sonication, the composite materials were dried with mild heating on a hot plate (T ~ 60 °C) and then in a drying over (T ~ 90 °C). Standard BET analyses were performed to measure the specific surface area from the nitrogen adsorption isotherm using an Autosorb IQ (Quantachrome Instruments). Catalysts were degassed at 200°C prior to analysis at a temperature of 77 K. X-ray diffraction (XRD) was performed using a Philips PANalytical X-Pert Pro x-ray diffractometer with a Cu Kα x-ray source.

2.3 Catalytic activity

The activity of the synthesized catalyst samples, Pt/TiO_2 nanoparticles, was evaluated by UV- photodegradation experiments. A bench-scale system consists of a cylindrical pyrex-glass cell with 1L capacity was used as a photocatalytic reactor. A 450-Watts mercury lamp was placed in a 5 cm diameter quartz tube with one end tightly sealed by a Teflon stopper. The lamp and the tube were then immersed in the photoreactor cell with a light path of 3.5 cm. The whole reactor was cooled with a water-cooled jacket on its outside and the temperature was kept at

22°C. Compressed air was purged into the solution by bubbling compressed air from the bottom to maintain an aerobic condition. Magnetic stirrer was also used to keep the solution chemically uniform. The pH values of the solutions were adjusted by adding NaOH (1 M) and HCl (1 M) using an Orion Model 801A pH meter. Solutions were radiated by UV light for specific time intervals. The experiments were carried out for 180 minutes of UV irradiation (after keeping in dark for 30 minutes in each run to reach equilibrium state). A 20-mL sample was drawn every 30 minutes and filtered for analysis through 0.45 μm syringe filters (Gelman Acrodic syringe filter with 25 mm diameter, and Nylon membrane of 0.45 μm pore size, PN4436T, Pall Gelman Laboratory, Ann Arbor, Michigan). Samples were analyzed for the residual concentrations of either phenol or 2-CP by UV-visible spectrophotometer (UV mini 1240 - Shimadzu) at 271 nm and by high performance liquid chromatograph (HPLC).

3. Results and Discussions

3.1. Structure of Pt nanoparticles

During the syntheses of the Pt nanoparticles, the precursor solution turned black indicating the formation of metal nanoparticles. A representative TEM image of the synthesized and washed Pt nanoparticles is presented in Figure 1. From this image, an average particle size and a distribution was determined to be 2.84 ± 0.61 by measuring the size of 100 particles and statistically analyzing the resulting histograms. This particle size and deviation agreed with the literature results [34, 37] for this synthesis approach.

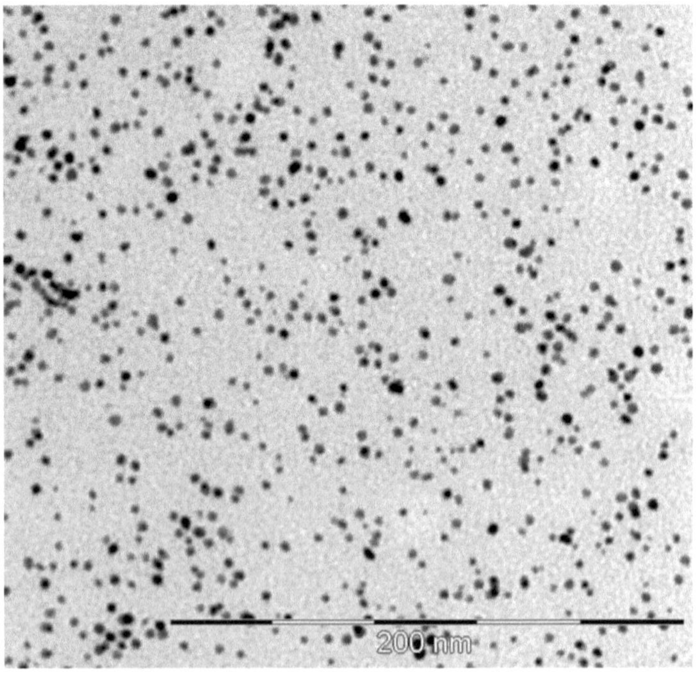

Figure 1: TEM imaging of synthesized and washed Pt nanoparticles

3.2. Characterization of Pt/ TiO_2 nanoparticles

Following incorporation of the synthesized and washed Pt nanoparticles onto TiO_2, the resulting structure was characterized by XRD and nitrogen physisorption. XRD (Figure 2) showed the rutile structure as indicated by the Miller indices. Due to the low loading and the small particle size of the Pt nanoparticles, the Pt phase is not observed by this technique. Based on nitrogen physisorption data (Table 1), the doping of Pt onto TiO_2 did not impact the porosity nor the surface area. That is, similar results were obtained for the titania substrate as for the Pt/titania sample.

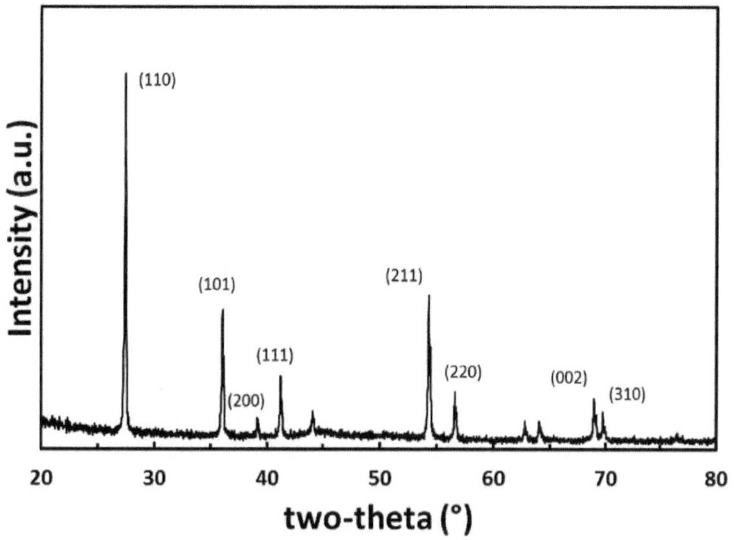

Figure 2: XRD pattern for Pt/ TiO_2 nanoparticles catalyst

Table 1: Porosimetry of Pt/TiO_2 nanoparticles as determined by nitrogen physisorption.

BET Surface Area (m^2/g)	Pore Diameter (nm)	Pore Volume (cm^3/g)
2.7	2.8	8.1×10^{-3}

3.3. Photocatalytic activity

All photo experimental runs were performed in dark with the catalyst samples for 30 minutes before UV illumination to reach adsorption equilibrium state. The pH of the aqueous solution is a key factor for photocatalytic reaction and can affect the adsorption of pollutants on the photocatalyst surface. In a previous work [38], the photocatalytic degradation of both phenol and 2-CP solutions over the TiO_2 rutile sample was evaluated (for comparison with the Pt/TiO_2 samples). The efficiency of photo degradation was very small with phenol while it was a comparatively significant with 2-CP for all pH values. However, the degradation of both phenol and 2-CP in solutions was enhanced in the acid medium rather than neutral or alkali medium, the optimum pH value was 3. The maximum degradation efficiency values for both phenol and 2-CP were 34.4 and 66.5 %, respectively, after 180 minutes at solution pH 3.

Figures 3(a) and 3(b) show the effect of pH with time on the photocatalytic degradation of both phenol and 2-CP solutions, respectively, over Pt/TiO_2 with UV light. It is clear that the efficiency of both phenol and 2-CP degradation was inversely proportional to the pH values, optimum pH value was 3. However, the efficiency values and rate of photo degradation of phenol were lower than that of 2-CP for all pH values. The degradation efficiency of phenol was gradually increased with time up to 180 minutes for a wide range of pH values. The maximum degradation efficiency values of phenol were 87.7, 38.14, 23.78 and 10.4 % with solution pH values of 3, 5, 7, and 9, respectively. On another hand, the degradation efficiency of 2-CP was increased rapidly from the beginning. The optimum pH value was 3 with 2-CP complete degradation efficiency after 180 minutes. The efficiency values of 2-CP degradation were 93.8, 81.6, and 65.5 with solution pH of 5, 7, and 9, respectively. The decrease in the degradation rate with increase in pH can be attributed to the fact that TiO_2 is amphoteric in aqueous solution. The point of zero charge (pHpzc) of TiO_2 is 6.8, thus below this value the TiO_2 surface is positively charged and above it is negatively charged. At higher pH, phenol exists as negatively charged phenolate species. Low degradation rate at higher pH is attributed to the fact that when the concentration of OH^- is higher in the solution, it prevents the penetration of UV light to reach the catalyst surface [39]. Moreover, high pH favors the formation of carbonate ions which are effective scavengers of OH^- ions and can reduce the degradation rate as confirmed from literature (Akbal and Onar [40], Naeem and Feng [41]).

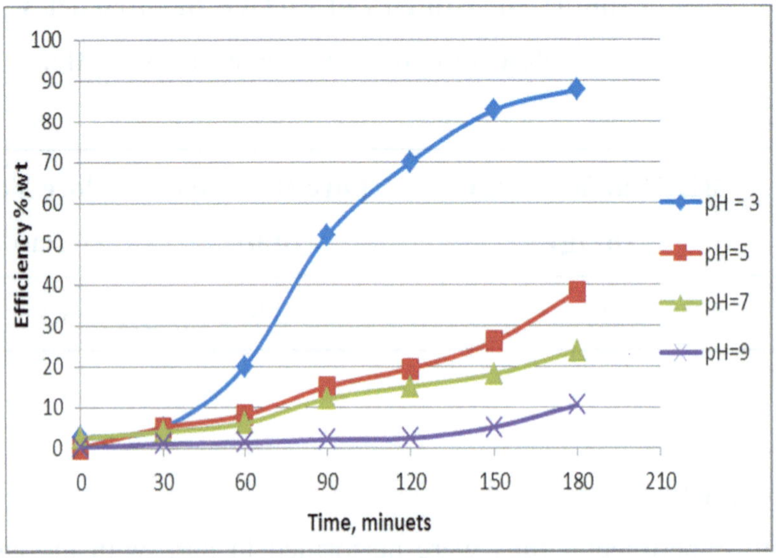

Figure 3(a). Effect of pH with time on the photocatalytic degradation of phenol over Pt/ TiO_2. *phenol concentration = 50 ppm, TiO_2 dose = 1 g/L.*

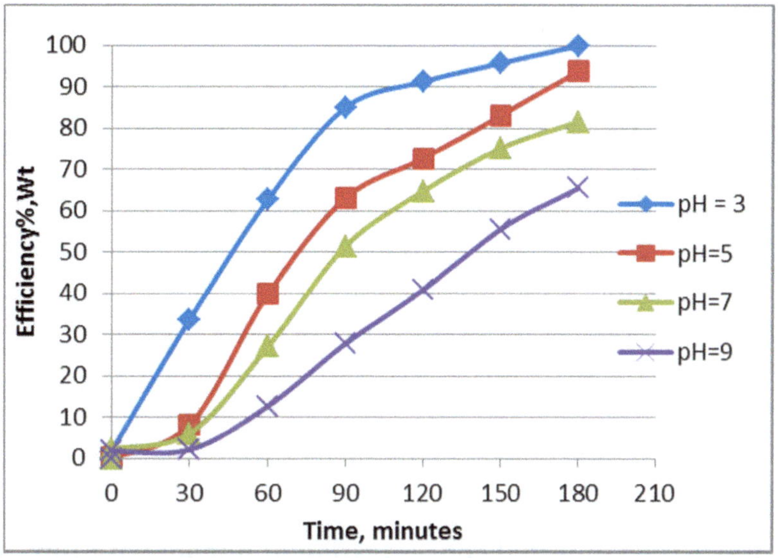

Figure 3(b). Effect of pH with time photocatalytic degradation of 2-CP over Pt/ TiO_2. *2-CP concentration = 50 ppm, TiO2 dose = 1 g/L.*

Figure 4(a) shows the effect of initial phenol concentration with time on its photocatalytic degradation over Pt/TiO_2 in the range of 50-200 ppm. As mentioned above, the degradation efficiency of phenol was gradually increased with time up to 180 minutes for the studied range of phenol concentration. On another hand, the degradation efficiency was inversely proportional to the initial phenol concentration. The optimum phenol concentration value was 50 ppm with 87.7 % degradation after 180 minutes, while the efficiency was drastically decreased to 71.4 and 7.5 % by increase the phenol concentration to 100 and 200 ppm, respectively. Figure 4(b) shows the effect of 2-CP initial concentration with time on its photocatalytic degradation over Pt/ TiO_2 with UV light. It can be seen that the 2-CP degradation rate and efficiency were higher than that of phenol for a wide range 2-CP concentration values. The optimum 2-CP concentration value was 50 ppm with a complete degradation after 180 minutes, and the efficiency was reached 84.2 and 55.8 % with initial 2-CP concentration of 100 and 200 ppm, respectively, within 180 minutes. The decrease in the degradation rate of phenols (phenol and 2-CP) at higher concentration is attributed to the fact that light absorbed by the phenolate species is more than that of TiO_2. Thus light absorbed by the phenols is not effective to carry out the degradation. Further, the equilibrium adsorption of phenols on the catalyst surface active site increases and more and more molecules of phenol get adsorbed on the surface of the catalyst. Therefore, competitive adsorption of OH^- on the same site decreases and consequently the amount of $^{\bullet}OH$ and $O_2^{\bullet-}$ on the surface of catalyst decreases [41].

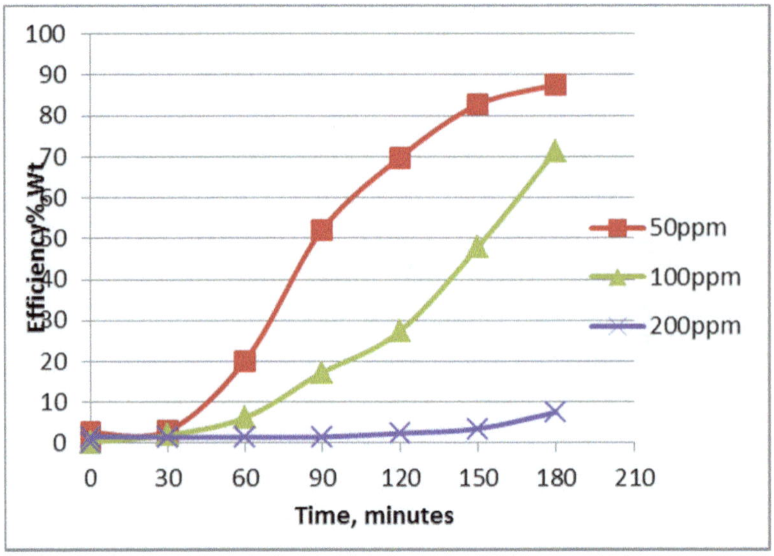

Figure 4(a). Effect of initial concentration of phenol with time on the photocatalytic degradation of phenol over Pt/ TiO_2. *pH= 3, TiO_2 dose= 1 g/L.*

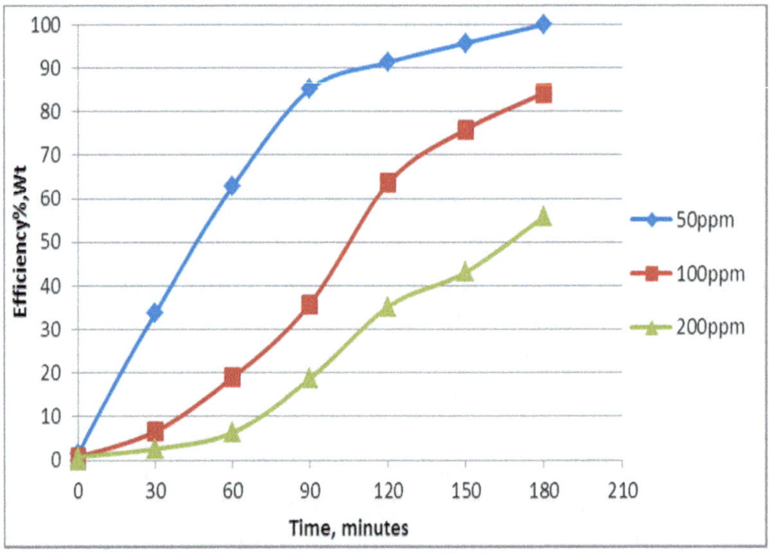

Figure 4(b). Effect of initial concentration of 2-CP with time on its photocatalytic degradation over Pt/ TiO_2. *pH= 3, TiO_2 dose= 1 g/L.*

The effect of the Pt/TiO_2 catalyst concentration on photocatalytic degradation of phenol and 2-CP has been investigated {Figures 5(a) and 5(b)}. Different Pt/ TiO_2 concentrations ranging from 0.5 to 2g/L were used. In Figure 5(a), it is clear that a high degradation rate of phenol was achieved after 180 minutes with catalyst dose of 0.5 or 1 g/L with degradation efficiency of 87.7 and 86.7%, respectively. However, increasing the catalyst dose to 2 g/L resulted in decreasing in the efficiency value to 66.1%. In Figure 5(b), the degradation rate of 2-CP increased rapidly at early time for all dosages of Pt/ TiO_2. A complete degradation of phenol was achieved after 180 minutes with catalyst dosages of 0.5 or 1 g/L. However, increasing the catalyst dose to 2 g/L resulted in decreasing in the catalytic efficiency to 93.9%. The inverse effect of the catalyst concentration on the degradation efficiency can be attributed to the fact that the increase in the number of TiO_2 particles will increase the number of photons absorbed and consequently the number of the phenol molecules adsorbed. At high concentrations, decrease in the photodegradation may be due to the aggregation of free TiO_2 particles that results in a decrease in the number of surface active sites [41-42]. Further, the excessive opacity and screening effect of excess TiO_2 act as shield as the increased number of particles in solution, and consequently hinder the light penetration. Therefore, there is a loss of available surface area for light-harvesting and as a result reduction of the catalytic activity [43-44].

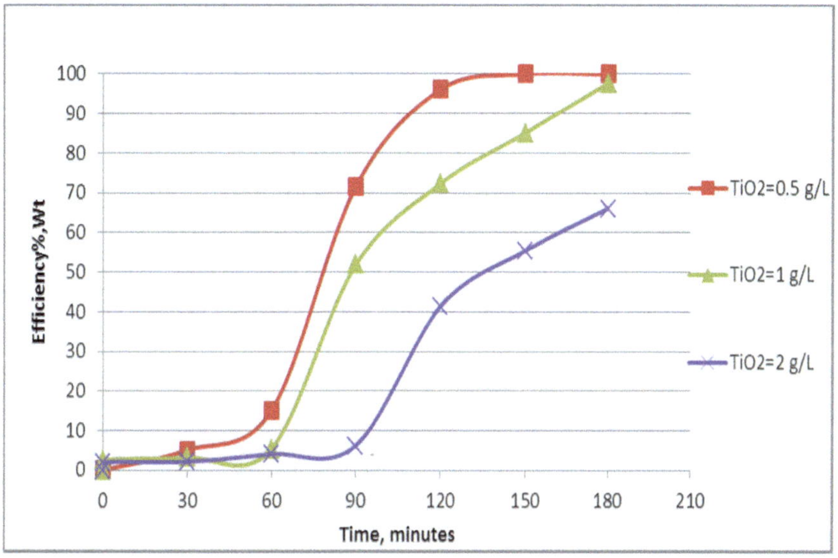

Figure 5(a). Effect of TiO_2 concentration with time on the photocatalytic degradation of phenol over Pt/ TiO_2. *pH= 3, phenol concentration = 50 ppm.*

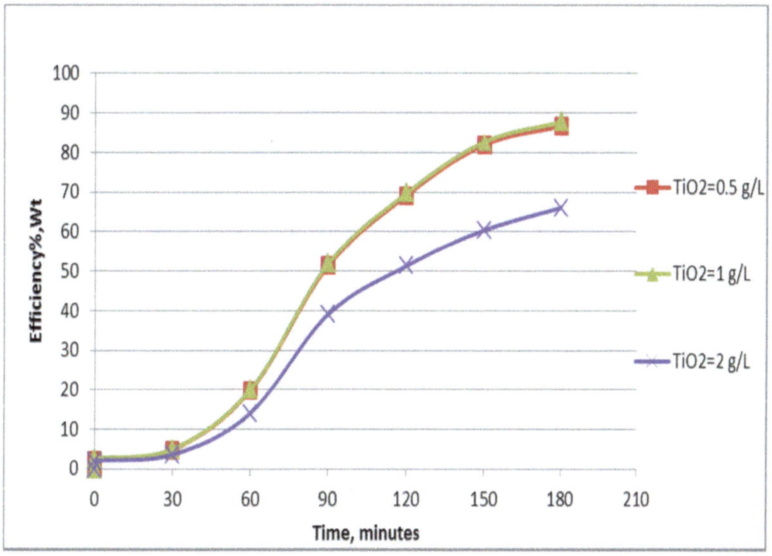

Figure 5(b). Effect of TiO_2 concentration with time on the photocatalytic degradation of 2-CP in presence of Pt/ TiO_2. *pH= 3, 2-CP concentration = 50 ppm.*

Figures 6 summarizes the efficiency of the photodegradation processes of phenol and 2-CP over both Pt/TiO_2 and TiO_2 rutile supports. The photodegradation processes were optimized by using 0.5 g/L catalyst with pollutant concentration of 50 mg/L within irradiation time of 180 minutes at different solution pH values for all the samples. The highest degradation efficiency for both phenol and 2-CP were achieved at solution pH 3 with the two catalysts samples. A further increase in the pH values resulted in a gradual decrease in the degradation rate. However, a very significant increase in the degradation efficiency of phenol and 2-CP was achieved with the Pt/TiO_2 catalyst compared to the TiO_2 rutile (from 34.4 to 87.7 for phenol, and from 94.5 to 100 % for 2-CP), at solution pH 3 after 180 minutes. This phenomena can be explained as follows; One of the most promising methods to increase the photocatalytic activity of TiO_2 catalyst is surface modification, this can be achieved by metal doping into catalyst. Dopants, such as Pt, can act as good electrons scavenger from the conduction band of the semiconductor. Thus, they hinder the recombination of photogenerated electrons and holes through increasing the charge separation [45-46]. On another hand, Noble metal incorporation into TiO_2 dielectric provides an absorption feature due to the surface Plasmon resonance (SPR) occurring over the visible range of the spectrum. In particular, Ag, Pt, and Au metals are the most popular materials due to the strong SPR character [47].

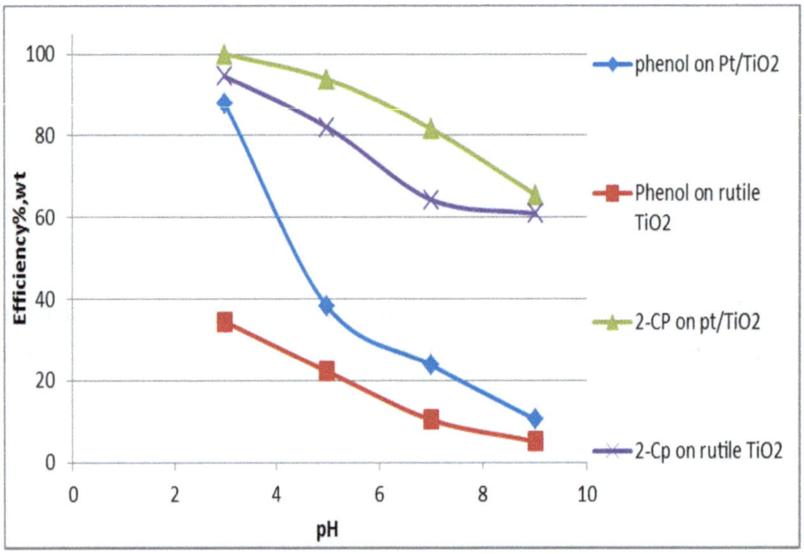

Figure 6. Efficiency of the photodegradation of both phenol and 2-CP over Pt /TiO_2 and TiO_2 rutile supports.

Phenol (2-CP) concentration = 50 ppm, catalyst dose = 1 g/L, *time = 180 minutes.*

Conclusions

Pt/TiO_2 nanoparticles catalysts were synthesized by immobilizing Pt nanoparticles onto titanium dioxide (rutile TiO_2). The average particle size of the catalyst was 2.84 ± 0.6 as measured through TEM analysis. The catalytic activity of the samples was evaluated by UV photocatalytic degradation of phenol and 2-chlorophenol (2-CP) in synthetic wastewater solution. Pt/ TiO_2 showed a higher activity for UV- photocatalytic degradation of both phenol and 2-CP pollutants in solution (as compared to the TiO_2 rutile). The photodegradation processes were optimized by using 0.5 g/L catalyst with pollutant concentration of 50 mg/L and solution pH value of 3 after irradiation time of 180 minutes. The degradation efficiency values were 87.7 and 100% for both of phenol and 2-Cp, respectively.

Acknowledgements

The authors gratefully acknowledge King Abdulaziz University (KAU, Saudi Arabia) for funding this work through the cooperation agreement with University of South Florida, USA (KAU-USF).

References

1. A. Fujishima, T.N. Rao, D.A. Tryk, Titanium dioxide photocatalysis. *J. Photochem. Photobiol. C: Photochem.* **2000** Rev. 1, 1e21.
2. M. A. Barakat, Y. T. Chen, C. P. Huang, Removal of toxic cyanide and Cu (II) ions from water by illuminated TiO_2 catalyst *J. Applied catalysis B: Environmental* **2004**, 53,13-20.
3. U.I. Gaya, A.H. Abdullah, Heterogeneous photocatalytic degradation of organic contaminants over titanium dioxide: a review of fundamentals, progress and problems. *J. Photochem. Photobiol. C: Photochem.* **2008** Rev. 9, 1e12.
4. S. Malato, Ez, P. Ferna´ndez-Iba´n˜, M.I. Maldonado, J. Blanco, W. Gernjak, Decontamination and disinfection of water by solar photocatalysis: recent overview and trends. *Catal. Today* **2009,**147, 1e59.
5. A.D. Paola, G. Cufalo, M. Addamo, M. Bellardita, R. Compostrini, M. Ischia, R. Ceccato, Palmisano. L. Photocatalytic activity of nanocrystalline TiO_2 (brookite, rutile and brookite-based) powders prepared by thermo-hydrolysis of $TiCl_4$ in aqueous chloride solutions. *Colloids and Surfaces A: Physicochemical and Engineering Aspects* **2008,** 317, 366-376.
6. M.A. Barakat, Adsorption behavior of copper and cyanide ions at TiO_2 – solution interface. *J. Colloid and Interface Science* **2005**, 291,345- 352.
7. J.M. Hermann, Heterogenous photocatalysis: fundamentals and applications to the removal of various types of aquas pollutants. *Catal.Today* **1999,**53: 115-129.
8. C. Hu, Y.Z. Wang, H.X. Tang, Destruction of phenol aqueous solution by photocatalysis or direct photolysis. *Chemosphere* **2000** 41**:** 1205–1209.
9. I. E. Braun, M. S. Pelizzetti, (ed.), Photochemical Conversion and Storage of Solar Energy, Kluwer, Dordrecht. **1991.**
10. N. Serpone, P. Maruthamuthu, P. Pichat, E. Pelizzetti, H. Hidaka, *J. Photochem. Photobiol* **1995,** A. 85: 247-255.
11. I. Shiyanovskaya, M. Hepel, Isotopic effects in cation-injected electrochromic films. *J. Electrochem. Soc.* **1998,** 145: 1023-1028.
12. K.Y. Song, M.K. Park, Y.T. Kwon, H.W. Lee, W.J. Chung, W.I. Lee, Preparation of Transparent particulate MoO_3/TiO_2 and WO_3/TiO_2 film and their photocatalytic properties. *Chem. Mater.* **2001,** 13: 2349-2355.

13. B. Pal, T. Hata, K. Goto, G. Nogami, Photocatalytic degradation of o-cresol sensitized by iron–titania binary photocatalysts. *J. Mol. Catal. A: Chem.* **2001,**169: 147-155.
14. I. Bedja, P.V. Kamat, Capped Semiconductor Colloids. Synthesis and Photoelectrochemical Behavior of TiO_2 Capped SnO_2 Nanocrystallites. *J. Phys. Chem.* **1995.** 99: 9182–9188.
15. A. A. Aal, M.A. Barakat, R.M. Mohamed Electrophoreted Zn–TiO_2–ZnO nanocomposite coating films for photocatalytic degradation of 2-chlorophenol. *Applied Surface Science* **2008,** 254: 4577–4583.
16. Y. Cao, H. Tan, T. Shi, T. Tang, J. Li, Preparation of Ag-doped TiO_2 nanoparticles for photocatalytic degradation of acetamiprid in water. *Journal of Chemical Technology and Biotechnology* **2008,** 83:546–552
17. M.M. Behnajady, N. Modirshahla, B. Rad, Enhancement of photocatalytic activity of TiO_2 nanoparticles by silver doping: photodeposition versus liquid impregnation methods. *Global NEST Journal* **2008,** 10: 1-7.
18. A.M. Barakat, M,A. Kanjwal, S.S. Al-Deyab, I.S. Chronakis, H.Y. Kim, Influences of Silver-Doping on the Crystal Structure, Morphology and Photocatalytic Activity of TiO_2 Nanofibers. *Materials Sciences and Applications* **2011,** 2: 1188-1193.
19. Q. Xiangchun, S. Hanchang, W. Jainlong, Q. Yi, Biodegradation of 2,4-dichlorophenol in sequencing batch reactors augmented with immobilized mixed culture. *Chemosphere* **2003,** 50: 1069-1074.
20. K.D. Raung, Theory and practice for the removal of phenols in wastewater. *Industrial Pollution Prevention and Control* **1984,** 3 (3): 88-103.
21. M. R. Heidi, M. W. Chien, Y. Tao, K. Jun-Kyoung, D. M. William, Catalytic hyrodechlorination of chlorophenols in aqueous solution under mild conditions. **2004,** *Applied Catalysis A: General*, 271: 137– 143.
22. D. Liu, R.J. Maguire, G. Pacepavicius, B.J. Dutka, Biodegradation of recalcitrant chlorophenols by cometabolism. *Environmental Toxicology and Water Quality* **1991,** 6: 85-95.
23. D.H. Han, S.Y. Cha, H.Y. Yang, Improvement of oxidative decomposition of aqueous phenol by microwave irradiation in UV/H_2O_2 process and kinetic study, *Water Research* **2004.**,38, 2782.
24. A. Alinsafi, F. Evenou, E.M. Abdulkarim, M.N. P. Zahraa, A. Benhammou, A. Yaacoubi, A. Nejmeddine, Treatment of textile industry wastewater by supported photo catalysis. **2007,** *Dyes and Pigments*, 74, 439-445.
25. R.H. Mills, D. W. Davies, *Chem. Soc. Rev.* **1993,** 22, 417–425.
26. E. Bessa, G.L. Sant'Anna, M. Dezotti, Photocatalytic/H_2O_2 treatment of oil field produced waters. *Appl. Catal. B: Environ.* **2001,** 29, 125–134.
27. E.R.L. Tiburtius, P. Peralta-Zamora, A. Emmel, Treatment of gasolinecontaminated waters by advanced oxidation processes. *J. Hazard. Mater.* **2005,** 126, 86–90.
28. L.-H. Cho, Y.-G. Kim, J.-K. Yang, N.-H. Lee, S.-M. Lee, Solar-chemical treatment of groundwater contaminated with petroleum at gas station sites: ex situ remediation using solar/TiO_2 photocatalysis and solar photo-Fenton. *J. Environ. Sci. Health A* **2006,** 41, 457–473.
29. S. Contreras, M. Rodriguez, F. Al Momani, C. Sans, S. Esplugas, Contribution of the ozonation pretreatment to the biodegradation of aqueous solutions of 2,4 dichlorophenol. *Water Research* **2003,** 37: 3164-3171.

30. Ch. Wang, Ch. M. Lee, Ch. J. Lu, M. Sh. Chuang, Ch. Z. Huang, Biodegradation of 2,4,6 - trichlorophenol in the presence of primary substrate by immobilized pure culture bacteria. *Chemosphere* **2000,** 41: 1873-1879.
31. R.U. Edgehill, R.F. Finn, Isolation, characterization and growth kinetics of bacteria metabolizing pentachlorophenol. *Eur. J. Appl. Microbiol. Biotechnol* **1982,** 16:179-184.
32. M.P. Ormad, J.L. Ovelleiro, J. Kiwi, Photocatalytic degradation of concentrated solutions of 2,4-dichlorophenol using low energy light: identification of intermediates. *Applied Catalysis B:Environmental* **2001.**, 32 (3): 157-166.
33. Y. Wang, J. Ren, K. Deng, L. Gui, Y. Tang, Preparation of Tractable Platinum, Rhodium, and Ruthenium Nanoclusters with Small Particle Size in Organic Media. *Chem. Mater.* **2000**, 12, 1622-1627.
34. T. Teranishi, M. Hosoe, T. Tanaka, M. Miyake, Size Control of Monodispersed Pt Nanoparticles and Their 2D Organization by Electrophoretic Deposition. *J. Phys. Chem. B* **1999**, 103, 3818-3827.
35. R. M. Rioux, H. Song, J. D. Hoefelmeyer, P. Yang, G. A. Somorjai, High-Surface-Area Catalyst Design: Synthesis, Characterization, and Reaction Studies of Platinum Nanoparticles in Mesoporous SBA-15 Silica. *Journal of Physical Chemistry B* **2005**, 109, 2192-2202.
36. H. Song, R. M. Rioux, J. D. Hoefelmeyer, R. Komor, K. Niesz, M. Grass, P. Yang, G. A. Somorjai, Hydrothermal growth of Mesporous SBA-15 Silica in the Presence of PVP-stabilized Pt Nanoparticles: Synthesis, Characterization, and Catalytic Properties. *Journal of American Chemical Society* **2006**, 128, 3027-3037.
37. J. N. Kuhn, W. Huang, C.-K. Tsung, Y. Zhang, G. A. Somorjai, Structure Sensitivity of Carbon−Nitrogen Ring Opening: Impact of Platinum Particle Size from below 1 to 5 nm upon Pyrrole Hydrogenation Product Selectivity over Monodisperse Platinum Nanoparticles Loaded onto Mesoporous Silica *Journal of American Chemical Society* **2008**, 130, 14026-14027.
38. M. A. Barakat, R. I., Al-Hutailah, M. H. Hashim, E. Qayyum, J.N. Kuhn, Titania-Supported Silver-based Bimetallic Nanoparticles as Photocatalysts, Environmental Science and Pollution Research (accepted, under publication).
39. M. Qamar, M. Muneer, D. Bahnemann, Heterogeneous photocatalysed degradation of two selected pesticide derivatives, triclopyr and daminozid in aqueous suspensions of titanium dioxide. *Journal of Environmental Management* **2006.**, 80: 99–106.
40. F. Akbal, A. N. Onar, Photocatalytic degradation of phenol. *Environmental Monitoring and Assessment* **2003,** 83: 295–302.
41. K. Naeem, O. Feng, Parameters effect on heterogeneous photocatalysed degradation of phenol in aqueous dispersion of TiO_2. *Journal of Environmental Sciences* **2009,** 21, 527–533.
42. S. F. Chen, Y. Z. Liu, Study on the photocatalytic degradation of glyphosate by TiO_2 photocatalyst. *Chemosphere* **2007**, 67(5), 1010–1017.
43. M.A. Barakat, H. Schaeffer, G. Hayes, S.I. Shah, Photocatalytic degradation of 2-chlorophenol by Co-doped TiO_2 nanoparticles. *J. Applied catalysis B: Environmental* **2004,** 57, 23-30.
44. M. A. Barakat, J. M. Tseng, C. P. Huang, Hydrogen peroxide-assisted photocatalytic oxidation of phenolic compounds. *J. Applied catalysis B: Environmental* **2005,** 59, 99-104.
45. H. Wang, Z. Wu, Y. Liu, W. Wang, Influences of various Pt dopants over surface platinized TiO_2 on the photocatalytic oxidation of nitric oxide. *Chemosphere* **2008,** 74, 773–778.

46. Barakat, N.A.M.; Kanjwal, M.A.; Al-Deyab, S.S.; Chronakis, I.S.; Kim, H.Y. Influences of silver-doping on the crystal structure, morphology and photocatalytic activity of TiO_2 nanofibers, *Materials Sciences and Applications* **2011**, 2, 1188-1193.
47. N.A.M. Barakat, K. D. Woo, M. A. Kanjwal, K. E. Choi, M. S. Khil, H.Y. Kim, Surface plasmon resonances, optical properties and electrical conductivity thermal hystersis of silver nanofibers produced by electrospin-ning Ttechnique. *Langmuir* **2008,** 24,11982-11987.

REWAS 2013

Enabling Materials Resource Sustainability

Enabling Sustainability through Education and Consumer Awareness

Session Chairs

Cong Wang
Brajendra Mishra

REWAS 2013: Enabling Materials Resource Sustainabil
Edited by: Anne Kvithyld, Christina Meskers, Randolph Kirchain, Gregory Krumdick, Brajendra Mish
Markus Reuter, Cong Wang, Mark Schlesinger, Gabrielle Gaustad, Diana Lados, and Jeffrey Spangenber
TMS (The Minerals, Metals & Materials Society), 20

THE SUSTAINABLE INORGANIC MATERIALS MANAGEMENT (SIM^2) CONSORTIUM AT KU LEUVEN

P.T. Jones[1*], T. Van Gerven[2*], K. Van Acker[1*], Y. Pontikes[1], Ö. Cizer[3], K. Binnemans[4*], B. Blanpain[1*]

[1]Department of Metallurgy and Materials Engineering, KU Leuven
Kasteelpark Arenberg 44 box 2450, 3001 Heverlee, Belgium
[2]Department of Chemical Engineering, KU Leuven
Willem de Croylaan 46 - box 2423, 3001 Heverlee, Belgium
[3]Department of Civil Engineering, KU Leuven
Kasteelpark Arenberg 40 - box 2448, 3001 Heverlee, Belgium
[4]Department of Chemistry, KU Leuven
Celestijnenlaan 200F - box 2404, 3001 Heverlee, Belgium

*All authors are members of the steering group of the SIM^2 consortium at KU Leuven
http://set.kuleuven.be/mrc/sim2

Keywords: Resource management, recycling, recovery, industry-university cooperation, rare earths

Abstract

The transition towards resource efficient, closed-loop economies is an urgent necessity as our world is facing resource shortages and unprecedented environmental challenges. Research groups in Metallurgy, Geology, Chemical Engineering, Building Materials, Chemistry, Economics, Process Psychology and Law at KU Leuven (Belgium) have recently connected their expertise in the area of recovery and recycling of inorganic materials into a structural collaboration, the 'Sustainable Inorganic Materials Management' (SIM^2) program. This has led to a growth and an intensification of research projects as well as local and international research cooperation, propelling SIM^2 to a flagship topic for the involved research groups and for the KU Leuven as a whole.

In our presentation we will highlight the vision, the organization in collaborative programs as well as the research topics of the program at KU Leuven. In addition we will elucidate the role of SIM^2 in co-organizing a potential urban mining node for an EIT-KIC on critical raw materials.

Introduction

The transition towards resource efficient, low-carbon closed-loop economies is an urgent necessity as our world is facing unstable but rising fossil energy prices, resource shortages and unprecedented environmental challenges. In the EU in particular the prospect of insufficient and often critical raw materials (metals and minerals) will hamper further economic development [1].

For example, the attempts to increase the share of wind energy, solar power and electric transportation will face a shortage of critical metals for the production of batteries, photovoltaic panels and strong magnets. New technologies are required to better separate metals and minerals from leaner ores and secondary sources [2]. The advanced treatment of secondary raw materials is part of the Benelux region's industrial strength in recycling and contributes to the vision of closed loop processes. Companies and policy makers are starting to incite knowledge centers to increase their efforts in this field as they recognize the urgency of the transition to Sustainable Materials Management (SMM). KU Leuven and its researchers are acting to take the lead in this development and to create scientific insights to support technological and policy innovations for the transition to SMM.

Context of the interdisciplinary collaboration initiatives

KU Leuven research groups have been investigating the recovery and recycling of inorganic materials, more specifically metals and minerals, for some time. Recently there have been several successful developments that have increased the focus on inorganic resource recovery and recycling, both by a more structural collaboration between the different research groups already working in this domain and by the additional interaction with research groups previously less active in the domain. These developments have propelled *Sustainable Inorganic Materials Management* (SIM^2) to a flagship topic for the involved research groups (Metallurgy, Geology, Chemical Engineering, Building Materials, Chemistry, Economics, Process Psychology, and Law) and for the KU Leuven as a whole. The philosophy of the SIM^2@KULeuven Consortium has been to work on various types of projects, ranging from fundamental science, to precompetitive projects to fully competitive, industry-driven projects. The various KU Leuven Consortia – under the umbrella of the KU Leuven Materials Research Centre – have developed a diversified research portfolio with a total budget of more than 18 M€.

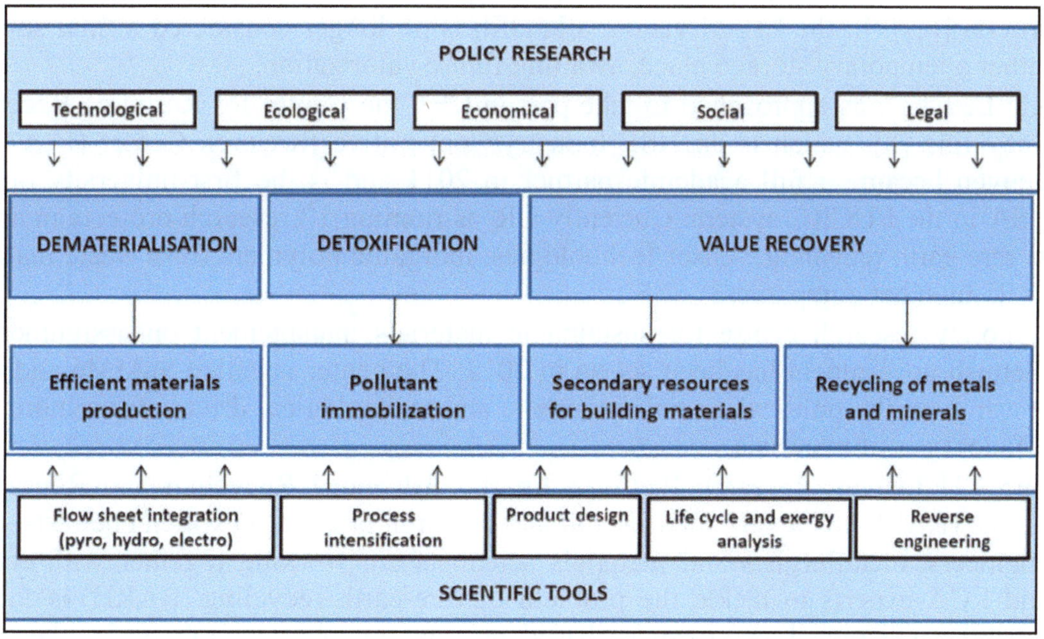

Figure 1: Research Areas of the Sustainable Inorganic Materials Management (SIM^2) Consortium at KU Leuven.

The research approach operates in two directions: fundamental science and precompetitive projects generate new ideas for breakthroughs in applied research, whereas the applied research generates new fundamental science questions. The research domains covered by SIM²@KULeuven are depicted in Figure 1, based on Fiksel's classification of SMM [3].
Among the recent developments are:

- The Center for High Temperature Processes and Sustainable Materials Management (CHTP) supported by its member companies is a flywheel for large (competitive) research projects in the areas of vessel integrity for high temperature processes, slag valorization and metal quality. CHTP was also the driving force behind the successful International Slag Valorisation Symposia in 2009 and 2011. A third Symposium [4] will be organized in March 2013.
- The KU Leuven Industrial Research Fund Knowledge Platform on Sustainable Materialization of Residues from Thermal Processes into Products [5] (SMaRT-Pro², since 2009) brings together academics from 5 research fields (chemical-engineering-and-technology, metallurgical-and-materials-engineering, building-materials-and-technology, geology-and-applied-mineralogy, economy-psychology-law) and 3 research institutions (KU Leuven, HUB and KHBO). HUB and KHBO are University Colleges in Flanders focusing on developing close-to-market solutions for industry. Working closely with industry, government and civil society, the generic goal is to strengthen knowledge on valorization of inorganic industrial by-products and provide a formal platform that can enhance the closing of industrial material cycles in Flanders and abroad.
- In 2010 the Flemish Funding Institute for Applied Science (IWT) granted a 6 M€ research project Enhanced Landfill Mining (ELFM) O&O project to Group Machiels and an academic consortium coordinated by KU Leuven (CHTP and SMaRT-Pro²). Enhanced Landfill Mining ELFM refers to the safe conditioning, excavation and integrated valorization of landfilled waste streams as both materials and energy, using innovative transformation technologies and respecting the most stringent social and ecological criteria [6,7]. In the ELFM vision, a landfill is no longer considered a final solution but rather a temporary storage place, awaiting future valorization.
- KU Leuven was approached to take part in CR³, the Center for Resource Recovery and Recycling [8], which is an NSF Industry/Cooperative Research Center (I/UCRC). KU Leuven became a full academic partner in 2011 and is the first university outside the USA in the I/UCRC system. Currently CR³ is running 12 research projects in areas such as rare earth recycling, sensor technologies, inorganic polymers from waste materials for its 25 member companies.
- A policy research centre for sustainable materials management on assignment of the Flemish government has been set up in 2012. The center is called SuMMa and conducts research on the political, social, economic and technological drivers for a more resource efficient use of materials.
- The KU Leuven Research Platform for the Advanced Recycling and Reuse of Rare Earths (RARE³) is a very interdisciplinary research project, in which chemists, chemical engineers, metallurgists and materials scientists are working together with economists and LCA experts to tackle the problem of rare-earth recycling. RARE³ is focused on breakthrough recycling processes based on non-aqueous technology for the two main applications of rare earths: permanent magnets and lamp phosphors. A user committee of 27 companies reflects the industrial relevance of this project. More information can be found on the RARE3 website [9].

- The cluster of Secondary Resources for Building Materials (SReBMat) started in 2011 as a joint effort between members of the Department of Metallurgy and Materials Engineering and the Department of Civil Engineering, both at KU Leuven. The drive behind this cluster is to perform creative, relevant, innovative and interdisciplinary work. The research focus is in four areas, namely ceramics (heavy clay ceramics as well as glass-ceramics), inorganic polymers (Al,Si-rich geopolymers as well as Fe-rich compositions), cementitious binders (for sustainable OPC production as well as on novel cementitious binders) and carbonation (for granular and non structural materials) [10].

Research scope of SIM²@KU Leuven

The scope of the technical areas is to develop intensified separation technologies in integrated flow sheets for the recovery of resource elements and minerals from solid and liquid media. The new techniques will be applicable to the treatment of ores and secondary materials such as electronic scrap and residues from industrial activities, thereby incorporating (enhanced) landfill mining, urban mining and direct recycling (see Figure 2). The following disciplines are at the core of the scope, all related to inorganic materials: physico-chemistry, electrochemistry, inorganic chemistry, hydrometallurgy and pyrometallurgy. Technologies developed in these disciplines can take advantage of process intensification and flow sheet integration, as well as from detailed material characterization.

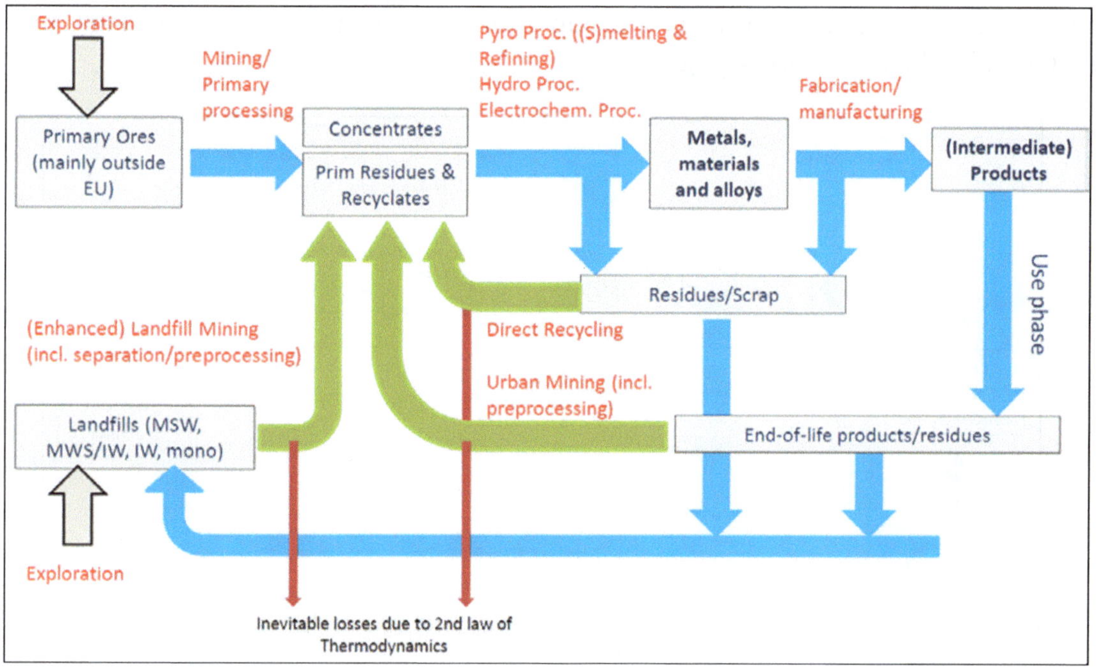

Figure 2: An integrated flow sheet concept used by SIM²@KULeuven.

Based on the expertise in KU Leuven and the current issues in industry, the following research lines in particular are currently explored:

- Immobilization of inorganic pollutants (solidification/stabilization, carbonation, sorption)
- Recycling of residues as construction materials (cement, aggregate, concrete blocks, alkali-activated or carbonated products).

- New sources for metal recovery and recycling, with on the short term focus on fayalitic slags, residues from Zn and Fe production, Li-ion batteries;
- Critical materials for clean energy technology, with focus on rare-earth elements (REEs);
- Direct reduction for the production of elements, with focus on electrowinning of metals;
- Membrane processes for the recycling of valuable minerals, with focus on phosphates;
- Process intensification of separation technology, with focus on ultrasound fields, electrical fields, magnetic fields, centrifugal fields.

Recycling of rare earths

Recycling of rare earths from a resource perspective is crucial and even more so in the European context, as Europe is largely deficient in natural rare-earth resources. An additional advantage of recycled rare earths is that the secondary raw materials do not contain radioactive thorium and uranium impurities, in contrast to most rare-earth ores, bastnäsite and monazite in particular. Due to the design of high tech products in which rare earths are currently used, such as NiMH batteries, magnets in loudspeakers, micro-phones and hard disk drives, the rare-earth elements in these devices are intimately associated with other valuable metals such as the platinum group metals and nickel. Recycling of rare-earth elements from complex, multi-material End-of-Life products will therefore need to be considered as part of complex recycling flow sheets including, comminution, physical sorting, pyrometallurgical, hydrometallurgical and electrometallurgical operations. More efficient and environmentally more benign separation methods are required, not only to separate the rare earths as a group from other metals in ores or slags, but also to separate the individual rare earths (Fig. 3)..

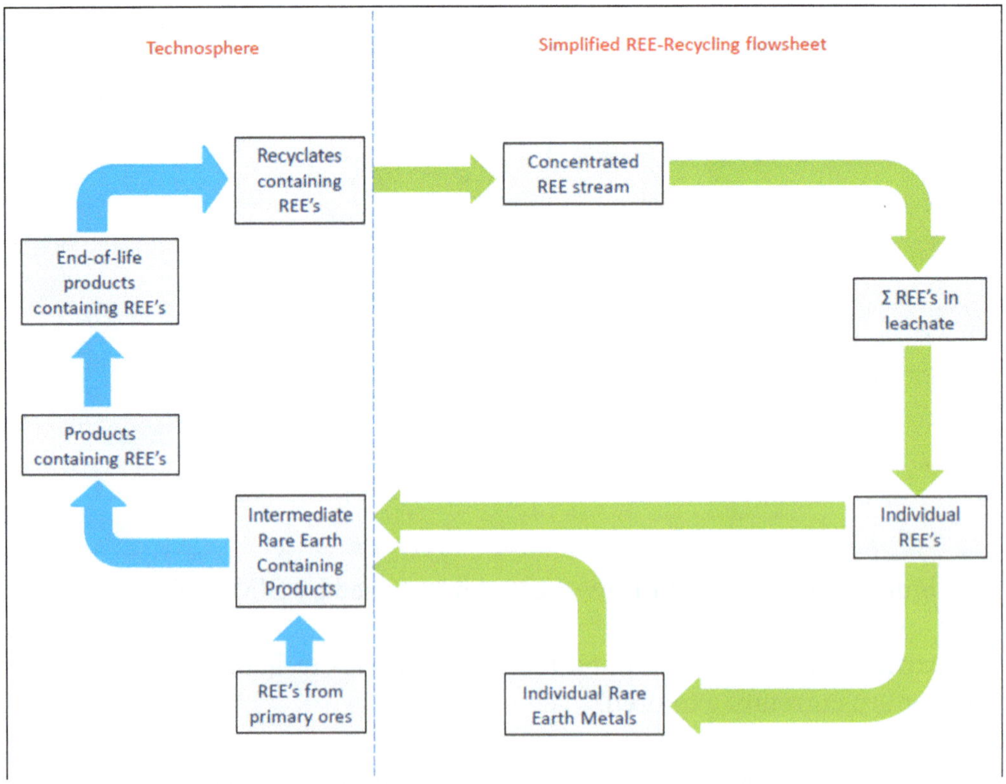

Figure 3: Closing the loop for rare earths.

An interdisciplinary approach is the key to develop the breakthroughs to enhance the recovery rate of rare earths and to contribute to the transition towards resource efficient, low-carbon, closed-loop economies. This approach is in line with the Europe 2020 Resource Efficient Europe initiative, the *Raw Materials Initiative* (RMI) of the European Union and its Communication on Commodity Markets and Raw Materials. From 2011 onwards, the SIM²@KULeuven Consortium targets a wide portfolio of fundamental and more applied projects on the recovery of rare-earth elements and has obtained 2.8 M€ research funding in 2012 through KU Leuven and EU FP7 projects.

'Technospheric mining of critical metals': a crucial theme within the future EIT KIC 'raw materials'?

To make the transition toward resource efficient, low-carbon circular economies, the EC has proposed action on the following, equally important areas [11]:

(1) Trade and investment policy;
(2) Technospheric mining of critical metals & commodity recycling;
(3) Sustainable primary mining (incl. deep-sea mining);
(4) Substitution of critical metals by less critical metals and substitutes for commodities.

These different pillars for a comprehensive raw materials policy (Fig. 4) for the EU can form the basic building blocks for the Knowledge and Innovation Community (KIC) on "Raw Materials", for which a call is expected from the European Institute of Innovation and Technology (EIT) in the near future.

The SIM^2 consortium of KU Leuven and its partners Umicore and UGent are currently constructing a node on technospheric mining of critical metals. Umicore is a world-leading WEEE recycling company with cutting-edge expertise in critical metal recovery (e.g. PGM and REE). UGent has a strong track record in biotechnology to recover critical metals from low concentration streams (Flemish and EU projects with several successful spin-offs). A distinction is made between critical metals and/or materials (e.g. rare earths) and commodity metals and/or materials, as the challenges for these resources are different. Within this node the following core activities are required:

1. Zero-waste recycling of solid/aqueous residual waste streams containing critical metals in low concentrations (e.g. REEs in mine water, bauxite residues or phosphogypsum);
2. Recycling of pre-consumer manufacturing scrap (e.g. REE-manufacturing swarf);
3. Urban mining of post-consumer End-of-Life products (e.g. recovery of REEs from electronic waste and lamp phosphors);
4. Landfill mining of historic (and future) industrial and urban waste streams containing critical metals.

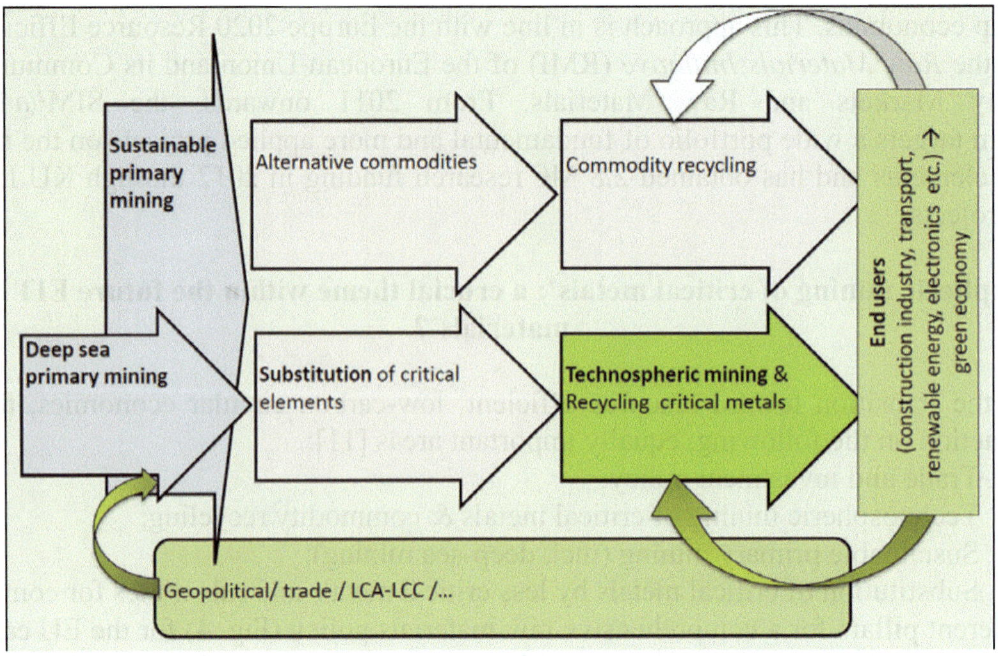

Figure 4: A comprehensive raw materials strategy including technospheric urban mining of critical metals

The advanced treatment of secondary raw materials containing critical metals is part of the Benelux strength in recycling. Umicore, KU Leuven and UGent will expand the technospheric mining node with unique partners in Flanders, France, the Netherlands, Germany and UK. Innovation in recycling critical materials is considered not only at the technological level, but also on the system level, including new business models, policy, logistics, and education. The following disciplines will be brought together in the node: pyrometallurgy, hydrometallurgy, electrometallurgy, geometallurgy, biometallurgy, enhanced landfill mining, preprocessing, characterization, design for recycling, materials life cycle assessment, policy instrumentation and business development. Collaboration with complementary nodes on mining & deep sea mining, substitutes for critical raw materials, substitutes for commodities in the automotive, aerospace and construction sector and commodity recycling is sought for.

Conclusions

Our urgent need to find new technological and policy approaches to help our society to make the transition to sustainable materials management has instigated KU Leuven research groups active in relevant technological and policy areas to form the Sustainable Inorganic Materials Management (SIM^2) consortium. The consortium is the umbrella as well as a catalyst for a number of centers and initiatives with a strong international as well as university-industry cooperative component.

References

1. European Commission, "Critical raw materials for the EU", June 2010. http://ec.europa.eu/enterprise/policies/raw-materials/files/docs/report-b_en.pdf
2. UNEP, "Recycling rates of metals: a Status Report", 2011. http://www.unep.org/resourcepanel/Publications/Recyclingratesofmetals/tabid/56073/Default.aspx

3. Fiksel, J., "A framework for Sustainable Materials Management", *Journal of the Minerals, Metals and Materials Society*, 2006, 58 (7), 15-22
4. http://www.slag-valorisation-symposium.eu
5. http://smartpro2.eu/
6. P.T. Jones et al., "Enhanced Landfill Mining in view of multiple resource recovery: a critical review", *Journal of Cleaner Production*, 2012 [http://dx.doi.org/10.1016/j.jclepro.2012.05.021];
7. P.T. Jones, et al., "Closing Material Loops: The Enhanced Landfill Mining Concept", *Journal of the Minerals, Metals and Materials Society*, 2012, **64** (7), 743-744;
8. http://www.wpi.edu/academics/Research/CR3/
9. http://www.kuleuven.rare3.eu/
10. http://www.mtm.kuleuven.be/Onderzoek/srebmat/
11. http://ec.europa.eu/enterprise/policies/raw-materials/files/docs/communication_en.pdf

REWAS 2013: Enabling Materials Resource Sustainabil
Edited by: Anne Kvithyld, Christina Meskers, Randolph Kirchain, Gregory Krumdick, Brajendra Mish
Markus Reuter, Cong Wang, Mark Schlesinger, Gabrielle Gaustad, Diana Lados, and Jeffrey Spangenberg
TMS (The Minerals, Metals & Materials Society), 20

RESOURCE EFFICIENT METAL AND MATERIAL RECYCLING

Markus A. Reuter
Outotec Oyj, Riihitontuntie 7 D, Espoo, Finland (markus.reuter@outotec.com)

Antoinette van Schaik
MARAS, The Hague, The Netherlands (a.vanschaik@marasustainability.nl)

Keywords: Design for Recycling, Design for Sustainability, Resource Efficiency

Abstract

Metals enable sustainability through their use and their recyclability. However, various factors can affect the Resource Efficiency of Metal Processing and Recycling. Some typical factors that enable Resource Efficiency include and arranged under the drivers of sustainability: Environment (Maximize Resource Efficiency - Energy, Recyclates, Materials, Water, Sludges, Emissions, Land); Economic Feasibility (BAT & Recycling Systems Simulation / Digitalization, Product vis-à-vis Material Centric Recycling); and Social - Licence to Operate (Legislation, consumer, policy, theft, manual labour.). In order to realize this primary production has to be linked systemically with typical actors in the recycling chain such as Original Equipment Manufacturers (OEMs), Recyclers & Collection, Physical separation specialists as well as process metallurgical operations that produce high value metals, compounds and products that recycle back to products. This is best done with deep knowledge of multi-physics, technology, product & system design, process control, market, life cycle management, policy, to name a few. The combination of these will be discussed as Design for Sustainability (DfS) and Design for Recycling (DfR) applications.

Introduction

Metals, their compounds and alloys have unique properties that enable sustainability in innovative modern infrastructures and through modern products as depicted by Figure 1.

Through mindful product design and high (end-of-life) collection rates, the metals and their compounds in sustainability enabling and other products can be recovered well, thus recoveries and therefore recycling of metals can be high. However, limitations on the recycling rate can among others be imposed by the (functionality driven) linkages and combinations of metals and materials in products.

Figure 2 shows various factors that can affect the resource efficiency of metal processing and recycling. The interaction therefore of primary and secondary recovery of metals drives not only the sustainable recovery of elements from minerals but also provides the recycling loop that recovers metals from complex products and therefore enables the maximum recovery of all elements from designer minerals. It is self-evident that "classical" minerals processing and metallurgy play a key role in maximizing resource efficiency and ensuring that metals are true enablers of sustainability.

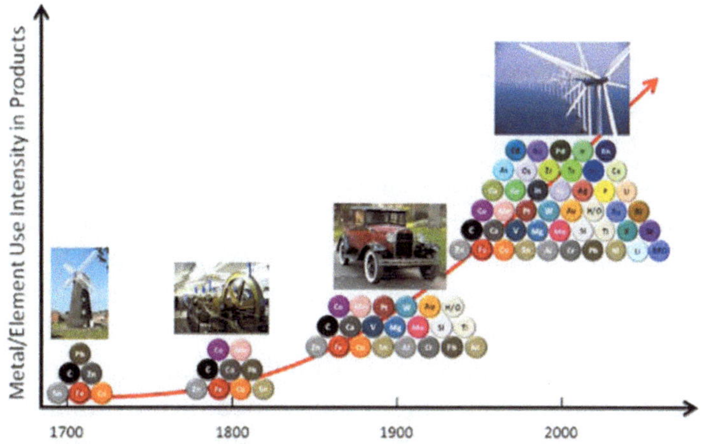

Figure 1. The intensity of use of materials in modern products requiring a "minerals" processing perspective and therefore a Product Centric view to maximize recovery of materials and metals from End-of-Life products [1].

The use of available minerals processing and process metallurgical theoretical depth to describe the system shown in Figures 2 & 3 is required to understand the resource efficiency of the complete system. A fundamental description of the system also shows what theory and methods still have to be developed to innovate the primary and recycling fields further. It is evident from Figures 2 &3 that the use of the rigorous theory developed in the classical minerals and metallurgical processing industry over the years and more recently adapted for recycling are very useful to quantify the various losses shown in Figure 3. "Classical" minerals processing and process metallurgy therefore both have a significant role to play in a modern resource constrained society and should do so vocally and actively.

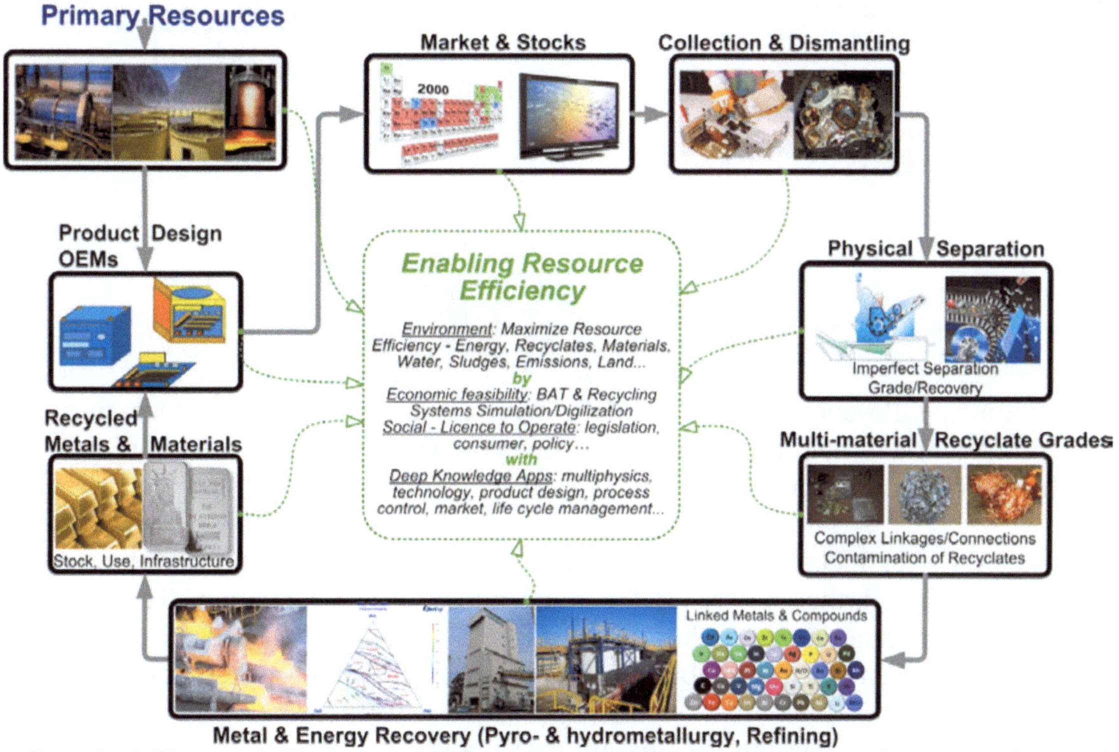

Figure 2. Sustainability and Resource Efficiency of a "closed-loop" society may be achieved by minimizing the losses from the various process steps by harmonizing the various actors in the loop including policy, economics, Life-Cycle-Management among others to minimize the use of resources in their widest sense.

The recycling and waste processing industry has much to gain to implement and adapt techniques and thinking of our industry rather than following the traditional bulk flow approaches of a material centric mindset of waste management and derived legislation, which are often coloured by this thinking [1-6].

Therefore, it would be evident that a mineral centric approach or in other words a product centric view [6] is required to maximize resource efficiency rather than a simpler metal and material centric view. It is this depth that lies at the heart of the Design for Recycling tools that will be discussed in the next section. Furthermore it is the application of this depth that will enhance the "closure" of the loop as it will permit a much deeper understanding between all actors than is the case presently.

This depth will help to better understand, sample, quantify products and recycling on element/compound/alloy level, and simulate the performance of recycling systems, also in relation to product design. This minerals processing and process metallurgical theoretical depth, knowledge and know-how will assist to optimize the system as this theoretical transparency and rigour can be linked more precisely to economics, help formulate and underpin policy and therefore as a tool assist in the maximization of metal recovery and hence recycling rates.

This rigour in the recycling field will also help to educate young talent in a more relevant manner of use to industry but at the same time also to help to increase the general level of sophistication in the field.

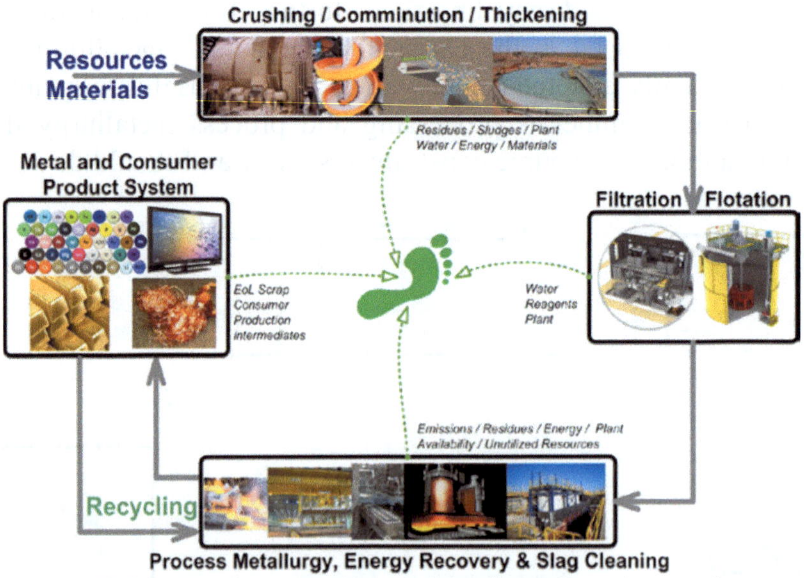

Figure 3. Harmonizing Product-Centric recycling with the mineral centric processing (i.e. primary processing of minerals) will help drive resource efficiency. Design for Resource Efficiency covers the detail of the figure minimizing the footprint.

In summary, not only do metals enable sustainability through numerous innovative products, their potential good recycling further decreases the footprint of their application. The next section will briefly discuss with reference to [2-6] how Design for Sustainability based on the abovementioned depth can be used to evaluate the recyclability of products as well increase resource efficiency.

Design for Sustainability and Recycling

The examples in this section show and discuss briefly how systems depicted by Figure 2 & 3 can be optimized by also referring to environmental footprints as well as thermodynamic indicators such as

Exergy. Two examples will be briefly given to illustrate Design for Sustainability and Recycling. This basis will help to minimize the residue and waste creation as shown in Figure 3.

Design of Sustainable Metals Processing Systems

Figure 4 gives an idea of a rigorous combination of tools that are already in use but will have to be advanced in future to embrace the bigger picture of minerals processing that includes treatment of end-of-life products. The combination of tools, in this case HSC Sim [www.outotec.com] and GaBi [www.pe-international.com] in Figure 4 permits among others the following:

- Benchmarking tool to assist/drive industry to Best Available Technique (BAT) & Resource Efficiency;
- Populating the databases of environmental software with mass, energy balance and thermodynamic consistent Life Cycle Inventories of complete BAT process systems, inclusive of for example the compounds of all relevant streams;
- Development of KPI's based on BAT and not only on global average Life Cycle Inventory (LCI) data;
- Driving Sustainability & CleanTech discussions on techno-economics basis; and
- Harmonization of language of engineers, universities' different disciplines, policy, environmentalists, etc. based on economic feasibility, rigorous technology and physics.

For example, Figure 4 shows a flowsheet of a BAT copper smelting plant including flash smelting, converting, sulphuric acid plant and electro-refining. HSC Sim functionality permits the creation of a plan in GaBi, shown in Figure 4 to include all process steps from concentrate to anode copper. Depending on which sections of plants are exported to GaBi, obviously larger scopes are included in the GaBi plan. As Outotec is a leader in the providing of copper extraction technology as well as sulphuric acid plants, BAT baselines can be created to compare all technologies on an equal basis. The possible outputs of this software are shown in Figure 4.

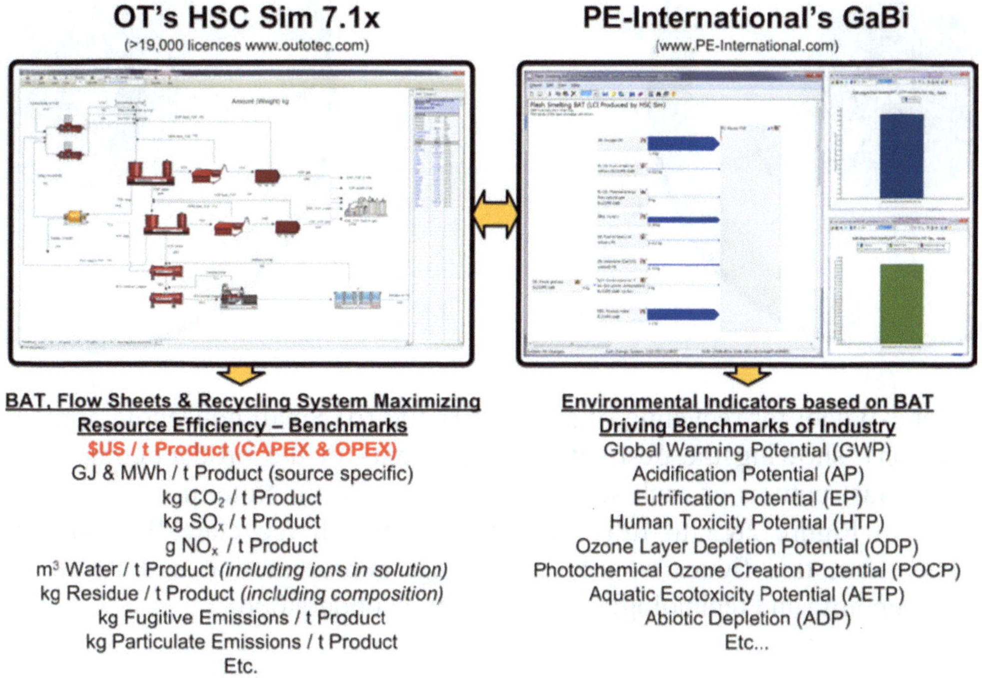

Figure 4. Design for Resource Efficiency: Linking rigorous simulation software for complete systems such as in Figures 2 & 3, complete processing plants etc. to environmental impact software.

Design for Recycling of WEEE Products

The functionality and design of (Waste) Electric and Electronic Equipment ((W)EEE), involving 50+ elements being applied in different compounds and phases, complicate recycling due to their ever more complex structures producing partially un-liberated, low grade and complex WEEE recyclates. This complex interlinked structure of materials, compounds and recyclate flows demands that DfR needs to be addressed from a product centric perspective as illustrated in Figures 2&3. Understanding the link between product design, particulate and interconnected nature of recyclates and industrial carrier metal processing infrastructures at a physics based depth is at the core of DfR [2].

In the recycling system, products and materials are usually broken or shredded in order to liberate different design-connected materials to allow for subsequent sorting of metals and materials into various recyclate fractions (ferrous, aluminium, copper, plastics, etc) and optimal recovery of materials in final treatment processes such as metallurgical processing. Due to (functional) design considerations where materials are attached to remain together or are applied and combined in very small quantities in complex chemical compounds (e.g. in electronic components such as Printed Wire Boards, capacitors, etc.) materials are likely to stay partially connected during shredding, ending up in mixtures of metal/material streams rather than as pure metal/material recyclate flows. WEEE recyclates can enter any of the different metallurgical processing infrastructures, which exist for the production of the most common commodity (carrier) metals used in society, e.g. iron, aluminium, lead and copper. The Metal-Wheel in Figure 5 illustrates for the example of a PWB (printed wire board) and an example of a Small Household Appliance (including battery) what happens to different metals entering either the Iron (Fe), Aluminium (Al), Copper (Cu), Zinc (Zn) and Lead (Pb) processing routes [3-4] – where each pie of the wheel is understood to be a complete infrastructure.

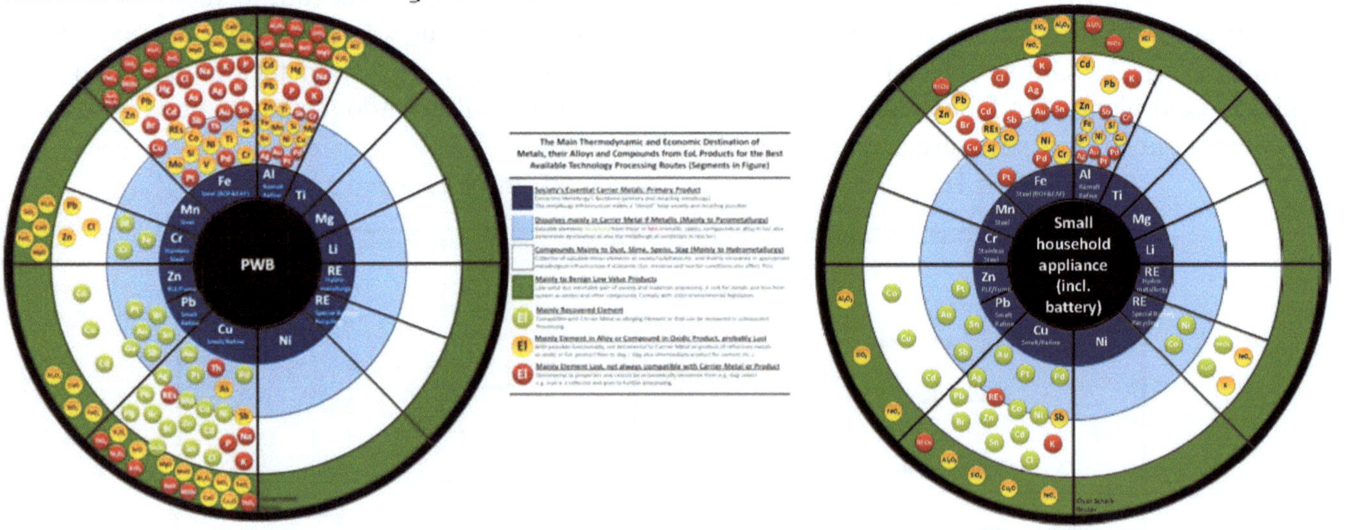

(a) (b)

Figure 5. The Metal-Wheel for recycling of a (a) PWB and (b) Small household appliance with battery (an example product), showing the destination of elements in the product/component in the carrier metal processing infrastructures of each segment (recyclate flow). Note, this is an average picture, but requires the basis of Figure 2&3 to fully determine to which recyclate flow and processing route elements go. Therefore, this figure is only an average representation and to be used only if the thermodynamic, economic and technological complexity is understood.

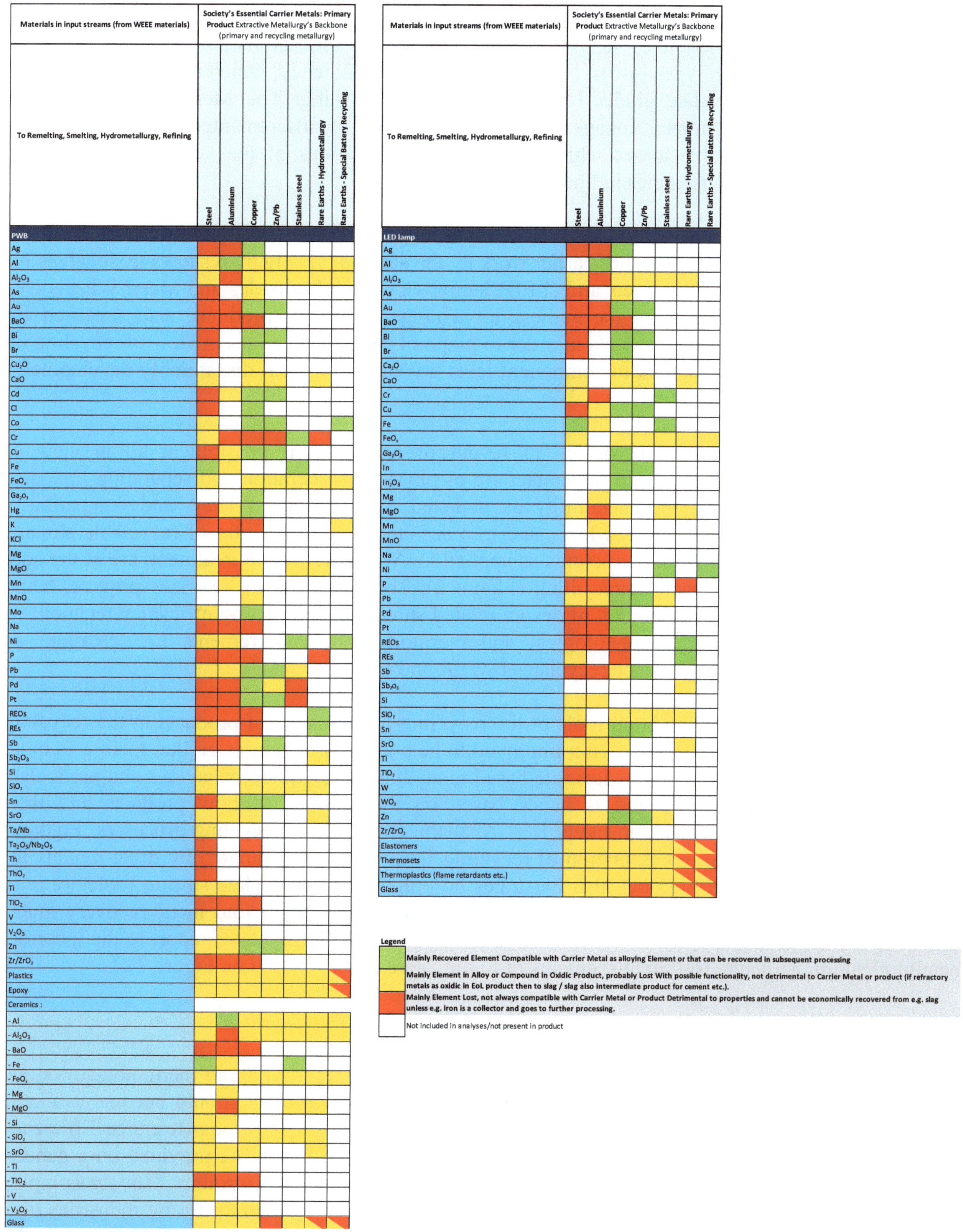

Figure 6. Design/material (in)compatibility matrices for a PWB and (LED) lamps, showing the (in)compatibility of elements in the product/component with the various carrier metal processing routes (Figure 5). Aggregated product data originating from different sources is presented. Product compositional data can also differ for different designs.

The destination of the metal and metal mixtures to one of the different carrier metal routes is determined by the combination of design choices and design characteristics (material selection, combination and connection types [5-6]), liberation and sorting efficiency, which create changes in the physical properties of recyclate grades (quality) going for recycling. The Metal-Wheel succinctly expresses what design-determined linkages mean for processing of different metals during recycling and also shows the inevitable losses which represent the limitations of the system due to design requirements and related deficiencies in liberation and sorting.

Design compatibility tables have been developed for a range of WEEE products on the basis of the Metal Wheels and on the background of earlier work by the authors [4]. Figure 6 shows these tables for PWB's from complex WEEE products and (LED) lamps. Compatibility tables and Metal Wheels (see Figure 5) have been developed by the authors for a range of WEEE products, also specified for different components within the products in order to pinpoint design challenges in view of recycling on detailed level and allow for the link to component level driven (automated) disassembly developments [7].

Innovating and optimizing product design in favour of recycling (if possible at all in view of functionality reasons) requires a deep understanding of physics of liberation and separation linked to thermodynamics, technology and metallurgy as depicted in the Metal Wheels and compatibility matrices. This recycling perspective, which take into account the composition of recyclable streams and (metallurgical) processing infrastructures as applied for WEEE recyclate treatment, are at the core of the Product Centric approach of recycling (Figure 2). The compatibility matrices provide, being linked to recyclate quality as a function of design related shredding and sorting efficiency technology, industry relevant insights into technological and thermodynamic requirements for 'creation' of economic valuable particles and hence recyclates by DfR and sorting.

Computer based-modelling/simulation tools of the recycling performance of products in the recycling system have been developed and are being improved to help guide product design that facilitates more recycling ([2], [5], [7]). It includes expert rule based modelling of liberation behaviour and particulate composition of recyclate flows and is hence based on the realities of how products and their constituents break-up and separate in likely Best Available Technology recycling processes. Figure 7 gives a simplified graphical representation of recycling flow sheets as simulated by these models for the WEEE recycling chain. Figure 7 also illustrates how the various recyclate sorting routes (e.g. to produce ferrous, aluminium, copper, etc recyclates) are connected to the various metal recovery routes of the Metal Wheel, and also shows the inevitable losses of metals if recyclates report to the incorrect metallurgical infrastructure (different segments of Figure 5 are superimposed on Figure 7).

<u>Product composition data for Design for Recycling</u>

The Metal Wheel and compatibility matrices in Figures 5 and 6 and underlying thermodynamics and technology dictate the level of depth to which product compositional data should become available from manufacturers and OEM's in order to make a sound Design for Recycling/Resource Efficiency assessment for complex multi-material products. Not only should this data be provided on a general physical material level – in line with a simple material centric approach, product data for complex multi-material designs should also be generated and made available by manufacturing industries on a mineral and therefore product centric basis. This implies that the mass composition of a product should become known on both a physical as well as chemical/compound level of detail for the entire product as well as for the different components. The latter is of importance to pinpoint Design for Recycling issues on a component level, rather than for the entire product. This will allow for much more detailed and industry relevant Design for Recycling options and hence include the "mineralogy" of the urban mine, analogous to the minerals of geological based mines. The same applies for data collection by sampling and analyses of the recyclate flows from sorting plants.

Recycling efficiency and recyclate quality are a function of design-connected materials. Therefore information on the 'layout' (i.e. construction, connections, etc) of the design is vital not only to capture and innovate material application in products, however, also to improve design (if possible!) from a material connection and combination perspective. The detail of the Metal Wheels and Design compatibility tables are hence subject to product data availability and rigour. The Product Centric approach in Figures 2&3 hence provides a theoretical basis for sampling, analyses and product design data generation.

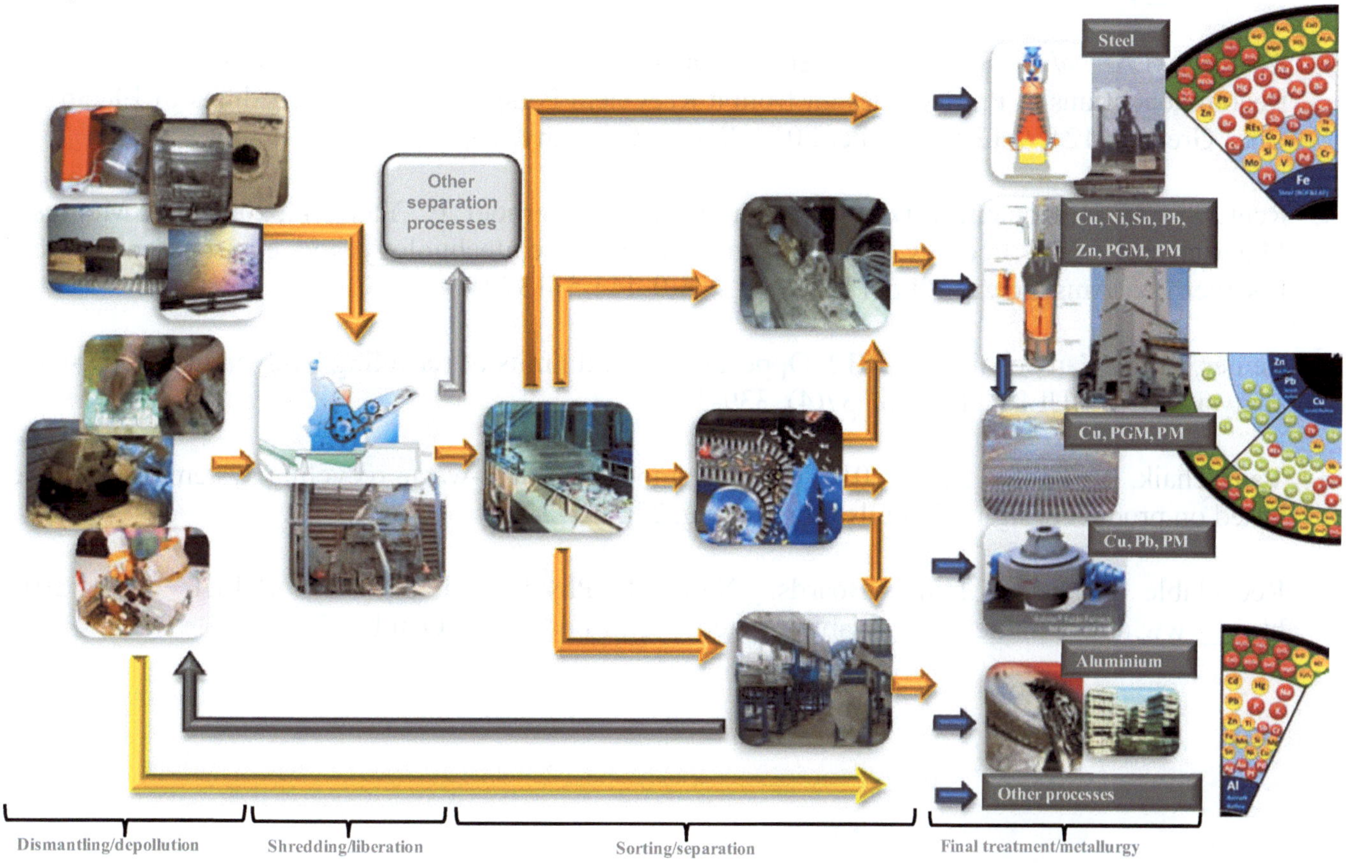

Figure 7. Simplified WEEE recycling flow sheet: An interconnected network of processes, and recyclate and material flows. Also shown is the connection to the carrier metal processing infrastructure i.e. relevant sections of the Metal Wheel (Figure 5).

Summary

Enabling Resource Efficiency and Sustainability in complex metal production and processing systems demands that a Mineral or Product Centric approach is adopted for recycling systems on the basis of available minerals processing and process metallurgical theoretical depth to ensure maximum recovery of elements, metals and compounds. Rigorous simulation tools for complete product systems and processing plants combined with environmental impact software allow for benchmarking BAT and drive Resource Efficiency. Mineral/Product Centric depth based recycling simulation tools, which capture a keen understanding of physics of sorting and design-determined breakage behaviour linked to thermodynamic possibilities and limits of metallurgical processing infrastructures drives innovation in

Design in favour of recycling and suggests physics based opportunities and technological limits of the system.

References

1. UNEP, 2013. Metal Recycling - Opportunities, Limits, Infrastructure. Reuter, M., Hudson, C., Hagelüken, C., Heiskanen, K., Meskers, C., Van Schaik, A., et al. In preparation.

2. Van Schaik, A., Reuter, M.A., 2012. Shredding, sorting and recovery of metals from WEEE: linking design to resource efficiency, In: Waste electrical and electronic equipment (WEEE) handbook (Edited by V Goodship, University of Warwick, UK and A Stevels, Delft University of Technology, The Netherlands), 163-211.

3. Reuter, M.A., Van Schaik, A., 2012. Opportunities and Limits of WEEE Recycling – Recommendations to Product Design from a Recyclers Perspective. In: Proceedings of Electronics Goes Green 2012+, 9-12 September 2012, Berlin, Germany, 10p.

4. Reuter, M.A., Heiskanen, K, Boin, U., Van Schaik, A., Verhoef, E., Yang, Y., 2005. The Metrics of Material and Metal Ecology, Harmonizing the resource, technology and environmental cycles Elsevier BV, Amsterdam, 706p. (ISBN: 13 978-0-444-51137-9).

5. Reuter, M.A., Van Schaik, A., 2012. Opportunities and limits of recycling: A dynamic-model-based analysis, MRS BULLETIN, Vol. 37(4), 339-347.

6. Van Schaik, A., Reuter, M.A., 2010. Dynamic modelling of E-waste recycling system performance based on product design. Minerals Engineering, 23, 192-210.

7. Recyclable Printed Circuit Boards. National Physical Laboratory (NPL), UK. 2012. http://www.npl.co.uk/news/recyclable-printed-circuit-boards (25 October 2012).

REWAS 2013

Enabling Materials Resource Sustainability

Enabling Sustainability through Recycling & End-of-Pipe Solutions II

Session Chairs

Jeffrey Spangenberger

Harald Oosterhof

REWAS 2013: Enabling Materials Resource Sustainabil
Edited by: Anne Kvithyld, Christina Meskers, Randolph Kirchain, Gregory Krumdick, Brajendra Mish
Markus Reuter, Cong Wang, Mark Schlesinger, Gabrielle Gaustad, Diana Lados, and Jeffrey Spangenberg
TMS (The Minerals, Metals & Materials Society), 20

Metal recovery by bioleaching of sulfidic mining wastes – Application to a European case study

A.G. Guézennec, J. Jacob, C. Joulian, S. Dupraz, Y. Menard, P. d'Hugues

BRGM, 3 av. C. Guillemin, BP.36009, 45060, Orléans, France

Bioleaching, stirred-tank reactor, recovery, sulfide ore, copper, cobalt, gold

Extended Abstract

The non-energy extractive industry (NEEI) of the EU-25 generated a direct turnover of about €40 billion, and provided employment to about 250000 people in 16629 companies in 2004. The use of primary raw materials in the production of other branches of EU industry means they have a central role in guaranteeing industrial and economic sustainability. Nevertheless current demand exceeds production, and so the EU is heavily dependent on minerals and metals imports. In this context of securing access to metals, turning mining wastes into new resources of currently unexploited valuable metals is an important challenge. The mining wastes can contain base and precious metals, but also metalloids and rare earth elements that are nowadays considered as highly critical for the industrial development of the European Union. Nevertheless, the development of alternative routes to conventional processing is still required in order to decrease the cost associated to the treatment of these unconventional resources which are more complex in composition and with lower grades.

In this study, we focused on wastes generated by processing of sulfidic ores which constitute a great part of the mining activities in Europe. These wastes (tailings) which often contain secondary valuable metals pose a threat to environment since they are sources of acid mine drainages that contain elevated contents of potentially toxic metals. Acid drainages are due to the leaching of sulfides by bacteria naturally present on mining sites. The purpose of this study was to intensify this natural phenomenon by using adapted bacterial consortia in order to recover the metals and to neutralize the wastes.

The wastes chosen for this study come from a European copper mine whose mineral of economic interest in the ore body is chalcopyrite ($CuFeS_2$). At site, the ore is grinded and valuable chalcopyrite is then separated from pyrite by flotation. Copper contained in the chalcopyrite is recovered by smelting whereas pyrite is discharged in tailings where the material used in this study was sampled. It is mainly composed of pyrite (60%) and contains cobalt (0.06 %) and gold (0.95 g/t). Bioleaching experiments were performed in 2L stirred reactors and in pilot tanks (20L) with the "BRGM-KCC" bacterial consortia whose predominant organisms are affiliated to the genera *Leptospirillum, Acidithiobacillus* and *Sulfobacillus* (Bryan et al. 2011). High dissolved metal rates were obtained (cobalt dissolution rate higher than 90%), that confirms bioleaching as an efficient process for this tailing material. Experimental results were used to design a global treatment process scheme and to perform a preliminary economic evaluation, showing good profitability of a hypothetical industrial project.

Acknowledgements: The research leading to these results has received funding from the European Community's Seventh Framework Program ([FP7/2007-2013] [FP7/2007-2011]) under grant agreement n° 228559. This publication reflects only the author's view, exempting the Community from any liability.

References

Bryan C.G., Joulian C., Spolaore P., El Achbouni H., Challan-Belval S., Morin D., d'Hugues P. (2011) - The efficiency of indigenous and designed consortia in bioleaching stirred tank reactors. Minerals Engineering, In press.

REWAS 2013: Enabling Materials Resource Sustainabil
Edited by: Anne Kvithyld, Christina Meskers, Randolph Kirchain, Gregory Krumdick, Brajendra Mish
Markus Reuter, Cong Wang, Mark Schlesinger, Gabrielle Gaustad, Diana Lados, and Jeffrey Spangenberg
TMS (The Minerals, Metals & Materials Society), 20

RECOVERY OF PLATINUM FROM DILUTE CHLORIDE MEDIA USING BIOSORBENTS

B. Zeytuncu, M.H. Morcali, O.Yucel

Istanbul Technical University, Faculty of Chemistry and Metallurgy, Maslak, 34469, Istanbul, Turkey

Keywords: Uptake, biosorbent, Isotherms, Platinum.

Abstract

Pistachio nut shells and Rice husk, a biomass residue, were investigated as adsorbents for the platinum uptake from synthetically prepared dilute chloroplatinic acid solutions. The effects of the different uptake parameters on platinum uptake (%) were studied in detail on a batch sorption. Before the pistachio nut shell material was activated, platinum uptake (%) was poor compared with rice husk. However, after the pistachio nut shell material was activated at 1000°C under an argon atmosphere, the platinum uptake (%) increased two-fold. The pistachio nut shell (inactivated and activated) and rice husk were characterized by Attenuated Total Reflection-Fourier transform infrared spectroscopy (ATR-FTIR).

The uptake equilibrium data were best fitted with the Langmuir isotherm model. The maximum uptake capacities, Q_{max}, at 25°C were found to be 38.31 and 42.02 mg/g for the activated pistachio nut shell and rice husk, respectively.

1. Introduction

The uptake methods are widely applied in environmental treatment applications around the world. Liquid–solid uptake systems are based on the ability of certain solids to preferentially concentrate specific substances from solutions onto their surfaces. This ability can be used for the removal of pollutants, such as metal ions and organic compounds, from wastewater [1-2]. Over the past decade, significant research effort has been directed to find low cost, high capacity adsorbents for the removal of metal ions. A wide range of adsorbents have been developed and tested, including several activated carbons [3-4]. In Turkey, agricultural by-products and waste materials that are available in large quantities may have the potential to be used as low-cost adsorbents. The conversion of agricultural waste materials into activated carbon would add considerable economic value, helps to reduce the cost of waste disposal, and most importantly provides a potentially inexpensive alternative to the existing commercial activated carbons. In addition, a number of low cost agricultural wastes, such as rice husk, pistachio nut shells, tree fern, peat coal and chitosan, have been used for the removal of a range of metal ions.

There are many studies in the literature on the preparation of activated carbons from agricultural wastes, such as sunflower seed hull, peanut hull, wheat bran, coir pith, banana pith, date pits, cotton stalks, palm tree, barley husk, palm kernel shell, rice husk, pinewood and soy hull [5-6].

Recently, platinum uptake from aqueous solutions via resins or waste materials has received considerable attention because platinum is appreciably present in electronic parts and plating materials due to its high resistance to oxidation [7].

In this study, pistachio nut shells and rice husk, an agricultural by-product, were used to adsorb platinum from dilute platinum chloride solutions. The objective of this study was to describe an effective uptake process using either pistachio nut shells or rice husk and to compare the two biosorbents by describing the optimal conditions and parameters for uptake platinum, including the uptake isotherms of platinum ions. Sorbent dosage, contact time and temperature parameters were studied to investigate their effect on the uptake percentage of platinum.

2. Materials and Methods

The rice husk (~250 μm thick) was obtained from the Gokbayrak Company, Turkey. The pistachio nut shells were obtained from the Gazi Fistikcilik Company, Turkey. The platinum-containing solutions for the experiments were prepared from hexachloroplatinic acid

(H_2PtCl_6) standard solution (Merck, Germany). Distilled water (TKA Smart Pure 2) was used for the wet chemical analyses.

The activated pistachio nut shell (~250 μm thick) was used for all experiments. The dry rice husks and pistachio nut shells were crushed using a crushing mill. The resulting recovered product was washed several times with distilled water to eliminate water soluble impurities and then dried in an oven at 105°C. 500 g of the small pieces from pistachio nut shell was placed in a horizontal furnace in an inert atmosphere at 1000°C for 3 hours. It was then removed from the furnace and cooled in a desiccator. After cooling, the rice husk and the activated pistachio nut shell were subsequently subjected to a homogenization treatment using a three-dimensional shaker for 1 h.

The identification of some characteristic functional groups was performed using ATR-FTIR. Figure 1, 2a and 2b displays the IR spectra analysis carried out on rice husk and pistachio nut shell and activated pistachio nut shell, respectively.

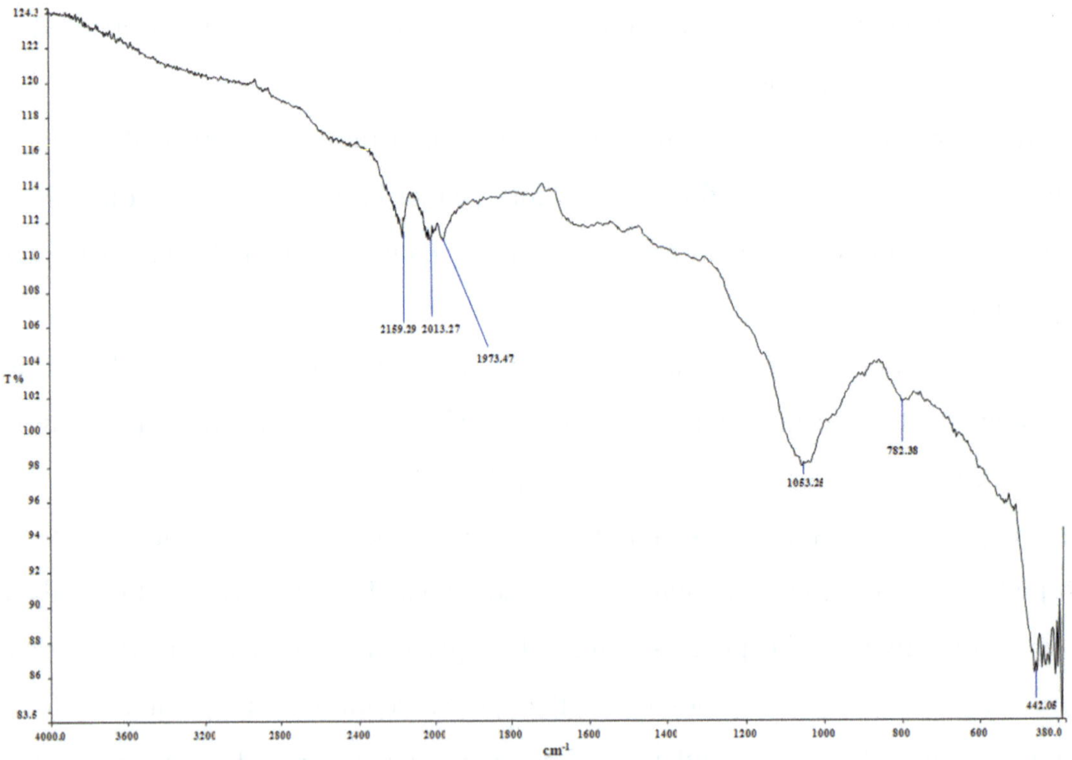

Figure1. ATR-FTIR spectra of Rice husk.

Figure 1 displays the FTIR analysis carried out on rice husk. As can be seen from the figure, IR spectra of rice husk in the region 1200–1000 cm^{-1} were considered to result from superposition of vibrations of the C-OH bond and Si-O bond in the siloxane (Si-O-Si) groups. The intense band at 1053 cm^{-1} corresponds to the stretching vibrations of silicon–oxygen tetrahedrons as (SiO_4). The high intensity of this peak was probably due to superposition of the stretching vibrations of the C-OH bond in the interval 1200–1000 cm^{-1} and the stretching vibrations of the Si-O bond. The absorbance peak at 442 cm^{-1} was due to the bending vibration of siloxane bonds [8].

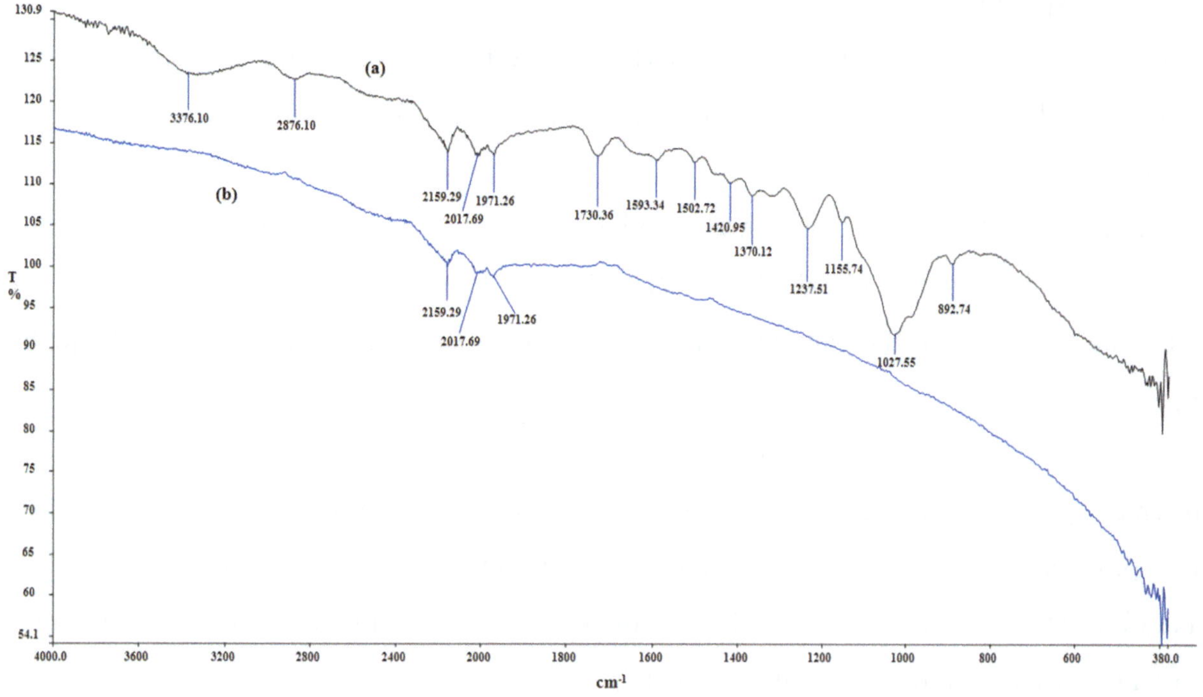

Figure2. ATR-FTIR spectra of **a)** Pistachio nut shell and **b)** Activated pistachio nut shell.

As can be seen from the Figure 2a, hydroxyl groups (O-H) is usually in the range of 3200–3650 cm^{-1} for alcohols and phenols, aliphatic groups (C-H) located at around 2876-1370 cm^{-1}. The band appearing at 1730 cm^{-1} was attributed to the carbonyl (C=O) groups. Carboxylate groups (band at 1332 cm^{-1}), esters- ethers -phenol groups (bands at 1237, 1155 cm^{-1}), alcohol groups (band at 1027 cm^{-1}) and the band at 897cm^{-1} assigned to benzene derivative vibrations are located[21]. It is evident from Figure 2b, IR spectra of activated pistachio nut shell in the region 2100-1900 cm^{-1} was considered to result from superposition of vibrations of the alkyne groups (C≡C) bond. Finally all oxygen groups like ether and alcohol and in the other groups were destroyed during the activation, resulting in activation of the carbon incorporated in the structure [7].

This study was conducted using a batch system by varying one parameter at a time. For each experiment, 5 cc of a platinum solution was brought into contact with the sorbent (Rice husk, pistachio nut shell, or the activated pistachio nut shell) in a Falcon tube to avoid exposure to air. The Falcon tubes were shaken in a temperature-controlled water bath at a manually adjusted shaking rate of 100 rpm. The initial platinum ion concentrations were 100 mg/L, 150 mg/L, 200 mg/L and 250 mg/L.

The first experimental series examined the effect of varying the sorbent dosage. The second experimental series investigated the effect of varying the contact time from 15 to 120 min. The third experimental series explored the influence of temperature, which ranged from 25°C to 45°C. Solid/liquid separation was performed following each run. For the ICP-OES analysis, each filtered solution was introduced into the instrument following an appropriate dilution.

The percentage of platinum uptake was calculated using the following equation [9]:

$$Platinum\ Uptake,\ \% = [(C_o - C_t)/C_o]*100 \quad (1)$$

The uptake capacity of the platinum ion was calculated using the following general equation [9]:

$$q_e = [(C_o - C_t)*V]/m \quad (2)$$

in which q_e is the amount of metal ions adsorbed at equilibrium per unit weight of sorbent (mg/g), C_o and C_t (mg/L) are the platinum ion concentrations present in the solution before and after uptake, respectively, V is the volume of the solution (in L), and m is the amount of sorbent (in g) used in the uptake experiment.

3. Results and Discussion

3.1. Effect of sorbent dosage on platinum uptake (%)

The dosages of the rice husk, pistachio nut shell and activated pistachio nut shell were varied, ranging from 2 to 20 mg/cm^3 of 100 ppm platinum solution in the first experimental series. Figure 3 presents the platinum uptake (%) with increasing sorbent dosages.

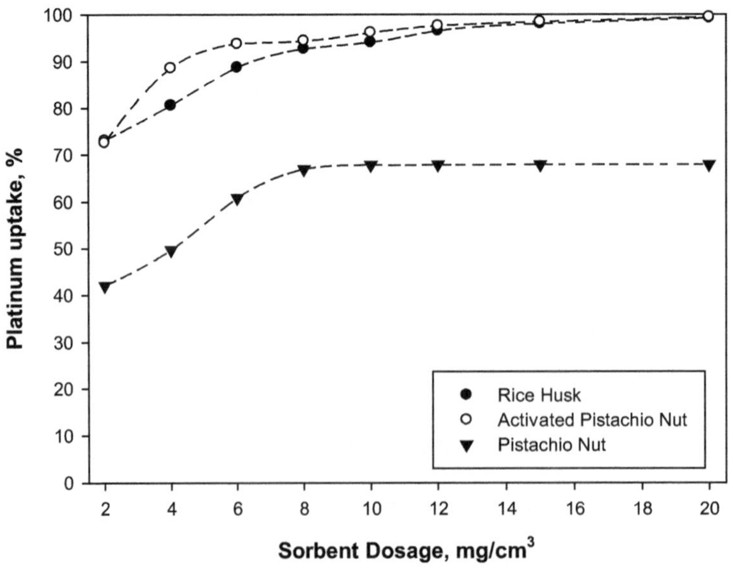

Figure 3. The effect of increasing sorbent dosage on platinum uptake (%) (60 min, 25°C, 100 rpm, pH = 1.5, and 5 cc sol.).

The platinum uptake (%) is found to increase with increasing sorbent dosage because an increasing amount of surface area is available, which exposed more active sites for binding metal ions [7]. For a 60 min contact time with the platinum-containing solution, the rice husk (4 mg/cm^3) exhibited a platinum uptake (%) of 80%; with 8 mg/cm^3 of the rice husk, 95% of the platinum was adsorbed. Similarly, for a 60 min contact time, 95% of the platinum was adsorbed using 8 mg/cm^3 of the activated pistachio nut shell.

3.2. Effect of time on the platinum uptake (%)

The effect of contact time on platinum uptake (%) was studied in the range of 15 to 120 min in this experimental series. Figure 4 presents the platinum uptake (%) as a function of the contact time.

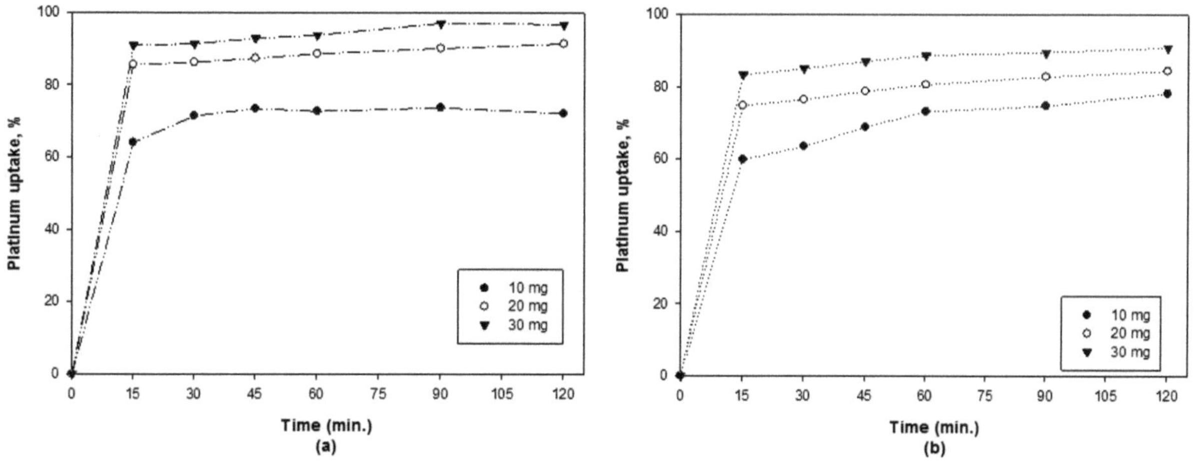

Figure 4. Platinum uptake (%) as a function of contact time by the **a)** activated pistachio nut shells and by **b)** Rice Husk (25°C, 100 rpm, pH = 1.5, and 5 cc sol.).

Figure 4 demonstrates that increasing the contact time has a positive effect on the platinum uptake (%) (i.e., the platinum uptake (%) increases with increasing time).

Figure 4a and 4b show that the uptake of platinum reached equilibrium after 15 min, i.e., there is no significant increase in uptake percentage after 15 min. Initially, this rate was higher because all the uptake sites on the rice husk and activated pistachio nut shell were vacant and the concentration was high, but after 15 min, all the uptake sites were filled with platinum ions, resulting in unchanged uptake percentages.

3.3. Effect of temperature on platinum uptake (%)

In this experimental series, the effect of temperature on platinum uptake (%) was studied in the range of 25°C to 45°C. Figure 5 presents the platinum uptake percentage as a function of temperature.

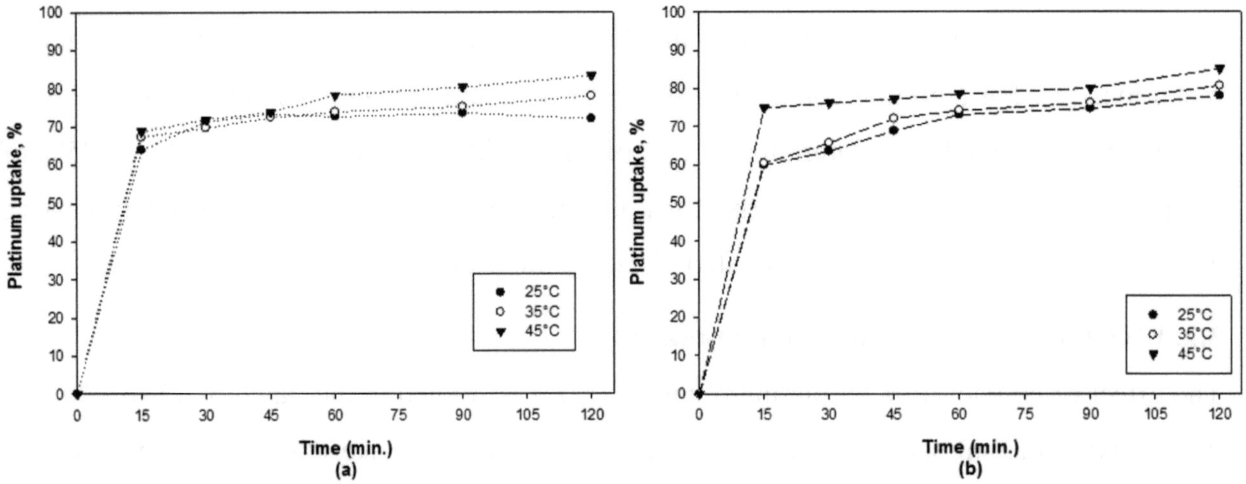

Figure 5. Platinum uptake (%) as a function of temperature by **a)** the activated pistachio nut shell; **b)** Rice husk (10 mg, 100 ppm, 100 rpm, pH = 1.5, and 5 cc sol.).

Figure 5a and 5b show that temperature has little influence on the uptake percentage for the activated pistachio nut shell. After 120 min, only an approximately 10% increment was obtained when the temperature increased from 25°C to 45°C. The platinum uptakes versus time curves at different temperatures are smooth and continuous, gradually leading to saturation, which indicates monolayer coverage of metal ions on the surface of the adsorbent [10].

3.4. Investigation of Adsorption Isotherms

The adsorption isotherms of platinum ions on these sorbents were studied at temperature, 25°C by varying the initial concentrations of the solutions from 100 ppm to 250 ppm while keeping all other parameters constant.

The equilibrium data obtained were analyzed with respect to the Langmuir and Freundlich isotherms.

3.4.1. Freundlich Isotherms

The data obtained for the uptake of platinum ions onto sorbents at an equilibrium concentration, C_o, ranging from 100 to 250 ppm were fitted to the Freundlich equation. The following linearized form of the Freundlich equation was used [11-12]:

$$\log q_e = \log K_f + 1/n \log C_e \tag{3}$$

in which, q_e is the amount of metal ions adsorbed at equilibrium per unit weight of sorbent (mg/g), C_e is the equilibrium ion concentration present in the solution after uptake, K_f is the empirical Freundlich constant or the capacity factor (in mg/g or mol/L), and 1/n is the Freundlich isotherm constant. The constants K_f and n are empirical constants that are characteristic of the system and depend on nature of the sorbent, the nature of the sorbate, the temperature, and the pressure.

The plots of log q_e versus log C_e for platinum ions uptake onto the activated pistachio nut shells and the rice husk yield straight lines with positive slopes, given by 1/n, and intercepts at $\log K_f$, as shown in Figure 6.

3.4.2. Langmuir Isotherms

The following linearized form of the Langmuir equation was used to analyze the uptake data for the uptake of platinum on the activated pistachio nut shells and the rice husk, respectively. The Langmuir equation is provided below [12-13] :

$$C_e/q_e = 1/(Q_{max}*K_L) + (1/Q_{max})*C_e \quad (8)$$

where q_e is the amount of metal ions adsorbed at equilibrium per unit weight of sorbent (mg/g), C_e is the equilibrium concentration of the sorbate in solution following uptake, Q_{max} is the maximum uptake capacity (mg/g) (which is generally called the monolayer capacity), and K_L is the Langmuir equilibrium constant (L/mg). Figure 6 shows the Langmuir uptake isotherm plot of C_e/q_e versus C_e.

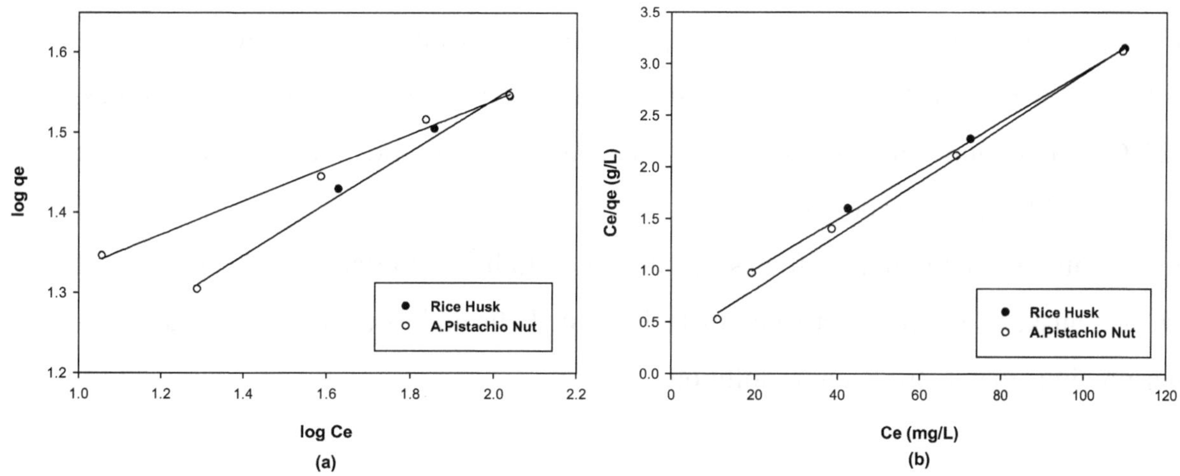

Figure. 6. **a)** The Freundlich and **b)** the Langmuir isotherm of Pt adsorbed onto the activated pistachio nut shell and rice husk (25°C, 20 mg sorbent, 60 min, 100 rpm, pH = 1.5, and 5 cc sol. at a concentration of 100–250 ppm).

The uptake isotherms of both activated pistachio nut shell and rice husk under different conditions were calculated and given in Table I. The values of each model and the correlation coefficient, R^2, were calculated from these plots. The linearity of these plots indicates the applicability of the two models. The correlation coefficients, R^2, showed that the Langmuir isotherm ($R^2 \geq 0.99$) fits better than the Freundlich isotherm ($R^2 < 0.99$). This result indicates that the uptake process of platinum ion onto the surfaces of rice husk and activated pistachio nut shell is a monolayer uptake process.

Table I. Freundlich and Langmuir constants for the uptake of platinum at 25°C temperature.

	Freundlich isotherm constants	
	Rice husk	Activated pistachio nut shells
K_f (mg/g)*(L/mg)$^{1/n}$	7.80	13.23
n	3.08	4.78
R^2	0.9916	0.9820
	Langmuir isotherm constants	
	Rice husk	Activated pistachio nut shells
K_L(L/mg)	0.09	0.04
Q_{max}(mg/g)	38.31	42.02
R^2	0.9959	0.9989

4. Conclusion

This study demonstrated that the biomass waste material, rice husk and activated pistachio nut shell, can be effective for the uptake of platinum ions from aqueous solutions. Equilibrium uptake data were well fitted by the Langmuir model. The uptake maximum capacities, Q_{max}, at 25°C of platinum ions onto activated pistachio nut shell and rice husk were found to be 38.31 and 42.02 mg/g, respectively.

The activated pistachio nut shell was demonstrated to have similar uptake capacity compared with rice husk. In addition, activated pistachio nut shells and rice husk may be used to uptake other precious metals from acidic solutions.

References

1. A. Netzer and D.E. Hughes, "Adsorption of copper, lead and cobalt by activated carbon," *Water Research,* 18 (1984), 927–933.

2. W.S.W. Ngah, C.S. Endud and R. Mayanar, "Removal of copper(II) ions from aqueous solution onto chitosan and cross-linked chitosan beads," *Reactive and Functional Polymers,* 50 (2002), 181–190.

3. K.C. Cheung and T.H. Venkitachalam, "Improving phosphate removal of sand infiltration system using alkaline fly ash," *Chemosphere,* 41 (2000), 243–249.

4. M. Rao, A.V. Parwate and A.G Bhole, "Removal of Cr^{6+} and Ni^{2+} from aqueous solution using bagasse and fly ash," *Waste Management*, 22 (2002), 821–830.

5. L.C. Romero, A. Bonomo and E.E. Gonzo, "Acid-activated carbons from peanut shells: synthesis, characterization and uptake of organic compounds from aqueous solutions," *Adsorption Science and Technology,* 21 (2001), 617-626.

6. A. Mokhtar et al., "Equilibrium and Kinetics studies for the adsorption of direct and acid dyes from aqueous solution by soy meal hull," *Journal of Hazardous Materials,* 135 (2006), 171-179.

7. S. Aktas and M.H. Morcali, "Platinum recovery from dilute platinum solutions using activated carbon," *Transactions of Nonferrous Metals Society of China*, 21 (2011a), 2554-2558.

8. S. Aktas and M.H. Morcali, "Gold uptake from dilute chloride solutions by a Lewatit TP 214 and activated rice husk," *International Journal of Mineral* Processing, 101 (2011b), 63-70.

9. Tóth J., *Adsorption: Theory, Modeling and Analysis* (New York, Marcel Dekker Inc, 2001), 573-630.

10. G. Hussain and M.A. Khan, "Adsorption of Gold (III) from Aqueous Solutions on Bagasse Ash," *Journal of the Chemical Society of Pakistan*, 33 (2011), 317-323.

11. H.M.F. Freundlich, "Over the adsorption in solution," *Journal of Physical Chemistry,* 57 (1906), 385-407.

12. S.Y. Liu et al., "Adsorption intrinsic kinetics and isotherms of lead ions on steel slag," *Journal of Hazardous Materials*, 173 (2010), 558-562.

13. I. Langmuir, "The adsorption of gases on plane surfaces of glass, mica and platinum," *Journal of the American Chemical Society*, 40 (1918), 1361–1403.

BIOEXTRACTION OF COPPER FROM PRINTED CIRCUIT BOARDS: INFLUENCE OF INITIAL CONCENTRATION OF FERROUS IRON

Ph.D. Luciana Harue Yamane, Ph.D. Denise Crocce Romano Espinosa, Ph.D. Jorge Alberto Soares Tenório.

Department of Metallurgical and Materials Engineering, University of São Paulo
Av. Prof. Mello Moraes, 2463. São Paulo, SP. 05508-030. Brazil.

Keywords: bioextraction, ferrous iron, copper, printed circuit boards, *Acidithiobacillus ferrooxidans*-LR.

Abstract

Printed circuit boards are found in all electric and electronic equipment and are particularly problematic to recycle because of the heterogeneous mix of organic material, metals, and fiberglass. Additionally, printed circuit boards can be considered a secondary source of copper and bacterial leaching can be applied to copper recovery. This study investigated the influence of initial concentration of ferrous iron on bacterial leaching to recover copper from printed circuit boards using Acidithiobacillus ferrooxidans-LR. Printed circuit boards from computers were comminuted using a hammer mill. The powder obtained was magnetically separated and the non magnetic material used in this study. A shake flask study was carried out on the non magnetic material using a rotary shaker at 30°C, 170 rpm and different initial concentrations of ferrous iron (gL^{-1}): 6.75; 13.57 and 16.97. Abiotic controls were also run in parallel. The monitored parameters were pH, Eh, ferrous iron concentration and copper extraction (spectroscopy of atomic absorption). The results showed that using initial concentration of ferrous iron of 6.75gL^{-1} were extracted 99% of copper by bacterial leaching.

Introduction

Bioleaching is based on microorganism´s capacity to produce Fe^{+3} ion, which is a powerful oxidant [1]. Oxygen availability is fundamental in the bioleaching since the bacterium *A. ferrooxidans* is aerobic and consumes O_2 in the biological oxidation of ferrous iron, as showed in Equation 1[2]:

$$2Fe^{+2} + 1/2O_2 + 2H^+ \rightarrow 2Fe^{+3} + H_2O \qquad \text{(Eq. 1)}$$

Ferric iron oxidizes metals, allowing metal´s dissolution in the acid medium (pH <2.0), e.g. copper (Equation 2):

$$Cu + 2Fe^{+3} \rightarrow Cu^{+2} + 2Fe^{+2} \qquad \text{(Eq. 2)}$$

Ferrous iron is regenerated to ferric iron through Fe^{+2}/Fe^{+3} cycle promoted by bacterial activity. Comparing with conventional recycling processes like hydrometallurgical and pyrometallurgical, bioleaching has advantages, such as mild reaction, low energy consumption, works at room temperature and normal atmospheric pressure, environmentally friendly and suitable for low-grade mine tailing and residues [3,4].

Nevertheless, bioleaching is dependent of physical, chemical and biological factors, including composition of waste or ore, pulp density, temperature, pH, ionic composition, Fe^{+3} concentration, oxidation–reduction potential, particle size, concentrations of dissolved oxygen and carbon dioxide, composition of the medium, adaptation process, bacterial strain and cell concentration[3,5,6].

Effect of each parameter on bacterial leaching is not completely elucidated. Determination of optimal conditions is essential to enhance the metals bioextractions from printed circuit boards.

Printed circuit boards are found in all electrical and electronic equipment being generally composed of polymers, ceramics and metals. Because of the amount that has been accumulated, it has become a prominent environmental problem, but also can be a source of valuable materials, such as metals [3,7].

Studies performed with/without addition of Fe^{+2} in the bioleaching using shake flasks [8,9] reported that ferric ion generated by *A. ferrooxidans* activity oxidizes metals through indirect bacterial mechanism [10].

The primary purpose of this article is to investigate the influence of initial concentration of ferrous iron on bioleaching to recover metals from non-magnetic fraction of printed circuit boards of obsolete computers using *Acidithiobacillus ferrooxidans*-LR bacteria.

Materials and methods

Characterization of printed circuit boards

Characterization of printed circuit boards from computers used in this study is reported in previous paper [11]. Samples from non-magnetic fraction of printed circuit boards from obsolete computers were obtained by comminution and magnetic separation. Figure 1 presents a flowchart of characterization steps.

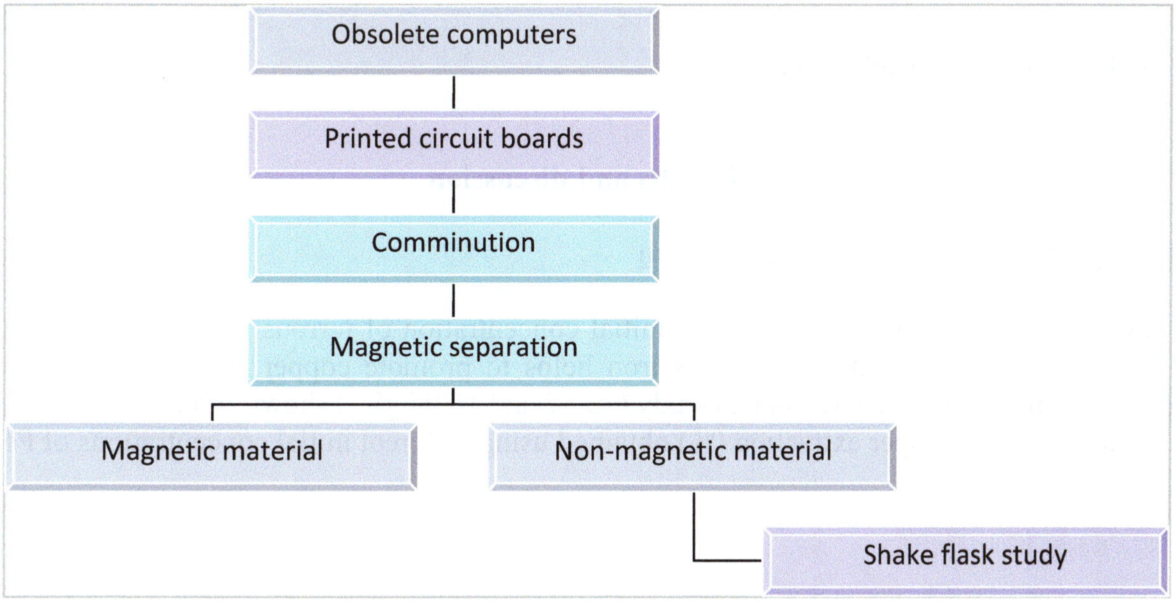

Figure 1: Flowchart of printed circuit board characterization

The following chemical analysis of the non-magnetic fraction by atomic absorption spectrometry (AAS) was obtained: 28.1% Cu, 7.8% Sn, 4.9% Pb, 4.5% Al, 3.9% Zn, 0.4% Fe, 0.2% Ni, 0.1% Ag and 0.1% Au. Metal composition was determined following analytical procedure by Park and Fray [(12)].

Effect of initial concentration of ferrous iron

A shake-flask study was carried out using rotary shaker at 30°C, 170 rpm, 15g L^{-1} pulp density and different initial concentration of ferrous iron (g L^{-1}): 6.75; 13.57 and 16.97. Abiotics controls were also studied in parallel.

Pre-adapted bacteria *Acidithiobacillus ferrooxidans* strain LR isolated by Garcia Jr. [(13)] from uranium mine effluents were inoculated (10%v/v) in Erlenmeyer flasks of 250mL capacity (previously autoclaved) with 200mL T&K medium [(14)].

Composition of T&K medium is the following: 0.625g L^{-1} $(NH_4)_2SO_4$, 0.625g L^{-1} $MgSO_4.7H_2O$, 0.625g L^{-1} K_2HPO_4 and 166.5g L^{-1} $FeSO_4.7H_2O$, which is the bacteria energy source.

The water evaporation was replenished at each sampling with sterile water acidic (pH 1.8) and the medium pH was adjusted with H_2SO_4 5M to 1.8-2.0.

Parameters evaluated

The parameters evaluated were: pH, E_h *vs* EPH, ferrous iron concentration, copper and iron total concentration by inductively coupled plasma optical emission spectrometry (ICP/OES).

The determination of Fe^{+2} concentrations was performed by titration with potassium dichromate ($K_2Cr_2O_7$).

Copper extraction was calculated by Equation 3:

$$\% \text{ Copper extraction} = [(\text{Initial CC} - \text{Leach CC}) / \text{Initial CC}] \times 100 \qquad \text{(Eq. 3)}$$

Being, CC: copper concentration.

Results and discussion

Effect of initial concentration of ferrous iron

Copper extraction using 6.75g L^{-1} of initial concentration of ferrous iron was 99%. Choi et al.[(15)] concluded that addition of ferrous iron helps to promote copper dissolution, but in the higher concentrations evaluated in this study (13.57 and 16.97 gL^{-1}) shown inhibitory effect.

Figure 2 shows copper extraction (%) obtained using different initial concentrations of Fe^{+2}.

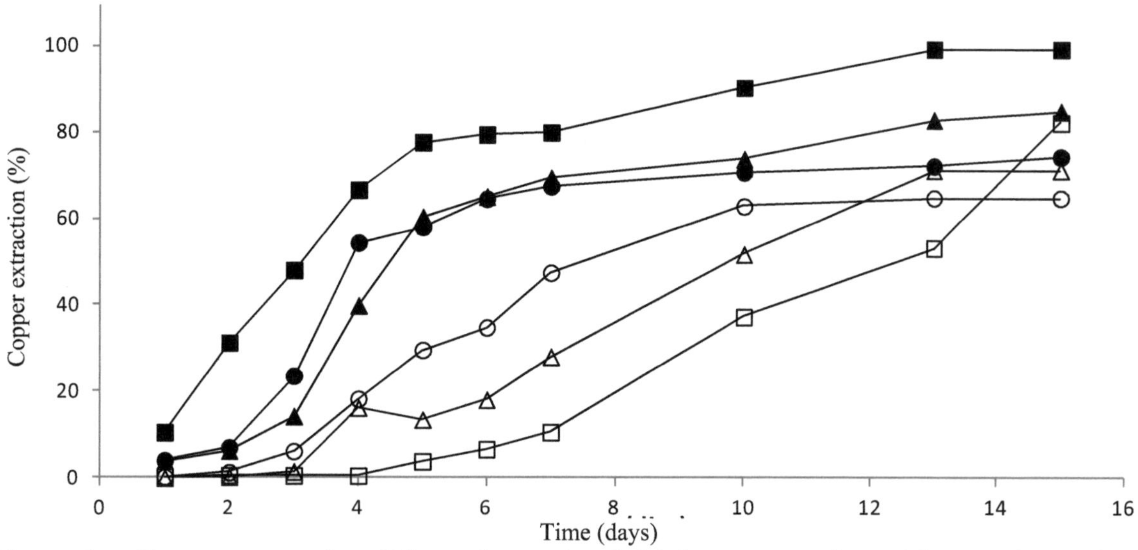

Figure 2 – Copper extraction (%) *vs* time using initial concentration of ferrous iron of (g L^{-1}): (■) 6.75; (▲) 13.57; (●) 16.97 and respective abiotic controls (□) 6.75; (Δ) 13.57; (○) 16.97 after 15 days.

Figure 2 shows that increase of initial concentration of Fe^{+2} didn´t increase copper extraction, as expected since bacterial activity is based on ferrous iron oxidation. According to Nemati et al.[16], concentration of Fe^{+2} higher than 5kg m^{-3} shows inhibitory effect on microbial growth. All studied conditions were up to 70%.

Choi et al.[15] studied copper bioleaching of printed circuit boards using different initial concentrations of Fe^{+2} (0-9g L^{-1}). Higher copper solubilization (5g L^{-1}) was obtained using 7g L^{-1} of Fe^{+2} initial concentrations. Similar results were obtained in this study (6,75g L^{-1}).

Same behavior was observed by Xiang et al.[17], copper bioleaching increased as the increased initial concentration of Fe^{+2} (0, 3, 6 and 9g L^{-1}) and as the decreased in the concentrations of 12 and 15g L^{-1}.

Copper bioleaching decreased with incubation time, being suggested that it occurs by Fe^{+3} precipitation forming a passivation layer on surface of printed circuit boards, which explains decreased of copper extraction in the studied initial concentrations of Fe^{+2} (12 and 15g L^{-1}).

Figure 3 and Figure 4 show, respectively, changes of pH and E_h *vs* EPH.

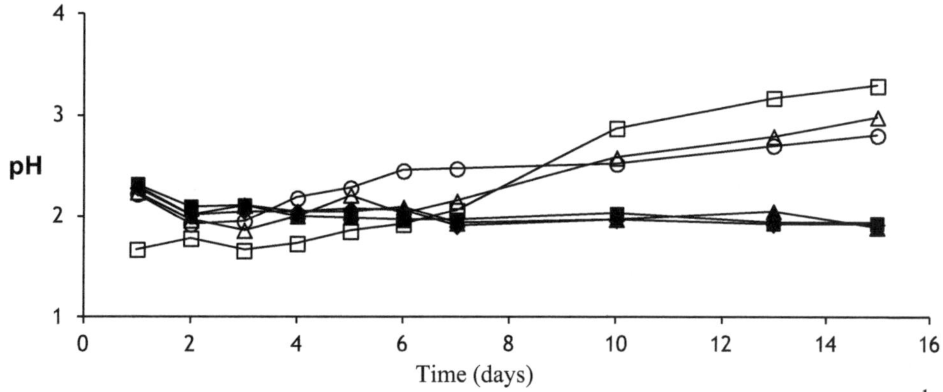

Figure 3 – pH Changes *vs* time using initial concentration of ferrous iron of (gL^{-1}): (■) 6.75; (▲) 13.57; (●) 16.97 and respective abiotic controls (□) 6.75; (Δ) 13.57; (○) 16.97 after 15 days.

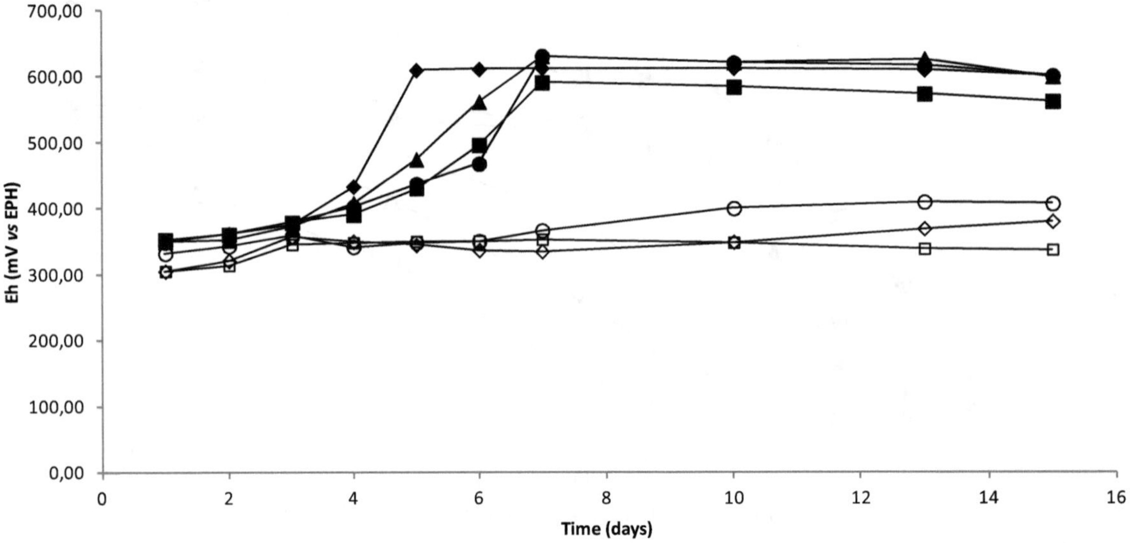

Figure 4 - E_h *vs* EPH changes *vs* time using initial concentration of ferrous iron of (g L^{-1}): (■) 6.75; (▲) 13.57; (●) 16.97 and respective abiotic controls (□) 6.75; (Δ) 13.57; (○) 16.97 after 15 days.

As can be seen in the Figure 3 and Figure 4, even with inhibition of bacterial activity in the initial concentrations of Fe^{+2} of 13.57 and 16.97 g L^{-1}, copper extractions were superior than 70% and it wasn´t observed significantly changes on pH and E_h in all studied conditions.

The E_h *vs* EPH of inoculated conditions keeps around 600mV *vs* EPH. Ferric iron (E^0_{red} +0,77V) solubilizes metals from printed circuit boards, including copper [17].

Figure 5 presents changes of Fe^{+2} concentration *vs* time.

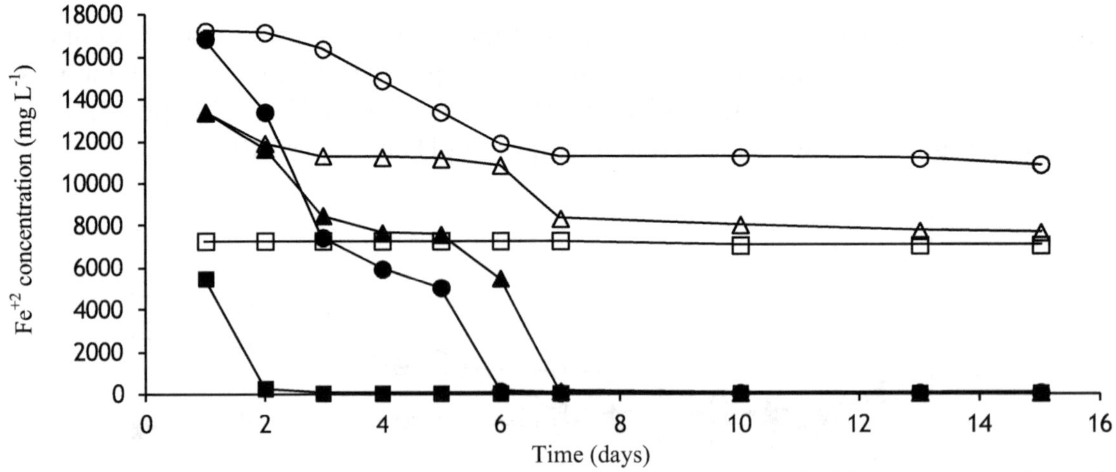

Figure 5 – Changes of ferrous iron concentration *vs* time using initial concentration of ferrous iron of (gL^{-1}): (■) 6.75; (▲) 13.57; (●) 16.97 and respective abiotic controls (□) 6.75; (Δ) 13.57; (○) 16.97 after 15 days.

In the Figure 5 is observed that the decrease of Fe^{+2} concentration in inoculated conditions (13.57 and 16.97 g L^{-1}) occurs slowly (until 7th day), which probably happened due to available

Fe^{+2} in the initial concentration of 6.75g L^{-1} to be enough to metals oxidation, based on obtained extraction.

In the abiotic conditions using Fe^{+2} initial concentrations of 13.57 and 16.97 g L^{-1} was observed that decay of Fe^{+2} concentration occurs slowly and consequently copper extraction is lower, unlike inoculated conditions. Results obtained by Meruane and Vargas[18] (144) in study about biological oxidation of ferrous iron by *A. ferrooxidans* ranging pH from 2.5 to 7.0 showed that the higher the pH of leaching medium (above 7), the greater the rate of chemical oxidation of ferrous iron in the abiotic conditions. In the pH range from 5.5 to 7.0 the rate remains almost constant and below pH 5.5, the oxidation decrease as a result of medium acidification.

Conclusion

This study demonstrated that ferrous iron initial concentration affects copper bioextraction and the experimental determination of optimal conditions increases metals mobilization.

Acknowledgements

The authors wish to thank the Fundação de Amparo à Pesquisa do Estado de São Paulo (FAPESP), São Paulo, Brazil for granting a PhD scholarship (Process n°08/53254-1) and for research financial (Process n°10/51009-0).

References

1. C. Gaylarde, "Bioextraction and Biodeterioration of Metals", *Cambridge University Press*, 388p, 1995.
2. K. Bosecker, "Bioleaching: Metal Solubilization by Microorganisms", *IEMS Microbiology Reviews*, 20, 3-4 (1997), 591-604.
3. Z. Guo, L. Zhang, Y. Cheng, X. Xiao, F. Pan, K. Jiang, "Effects of pH, pulp density and particle size on solubilization of metals from a Pb/Zn smelting slag using indigenous moderate thermophilic bacteria". *Hydrometallurgy*, 104 (2010), 25-31.
4. S. M. Mousavi, S. Yaghmaei, F. Salimi, A. Jafari, "Influence of process variables on biooxidation of ferrous sulfate by an indigenous *Acidithiobacillus ferrooxidans*. Part I: Flask Experiments". *Fuel*, 85 (2006), 2555-2560.
5. T. Yang, Z. Xu, J. Wen, L. Yang, "Factors influencing bioleaching copper from waste printed circuit boards by *Acidithiobacillus ferrooxidans", Hydrometallurgy*, 97 (2009), 29-32.
6. D. Bevilaqua, A. L. L. C. Leite, O. Garcia Jr., O. H. Tuovinen, "Oxidation of chalcopyrite by *Acidithiobacillus ferrooxidans* and *Acidithiobacillus thiooxidans* in shake flasks". *Process Biochemistry*, 38 (2002), 587-592.
7. H. M. Veit, A. M. Bernardes, J. Z. Ferreira, J. A. S. Tenório and C. F. Malfatti, "Recovery of Copper from Printed Circuit Boards Scraps by Mechanical Processing and Electrometallurgy", *Journal of Hazardous Materials*, B137 (2006), 1704-1709.
8. T. Rohwerder, T. Gehrke, K. Kinzler and W. Sand, "Bioleaching review part A - progress in bioleaching: fundamentals and mechanisms of bacterial metal sulfide oxidation", *Appl. Microbiol. Biotechnol.*, 63 (2003), 239-248.

9. L. Xia, L. Xinxing, J. Zeng, C. Yin, J. Gao, J. Liu, G. Qiu, "Mechanism of enhanced bioleaching efficiency of *Acidithiobacillus ferrooxidans* after adaptation with chalcopyrite", *Hydrometallurgy*, 92 (2008), 95-101.
10. P. K. Sharma, A. Das, K. H. Rao and K. S. E. Forssberg, "Surface characterization of *Acidithiobacillus ferrooxidans* cells grow under different conditions". *Hydrometallurgy*, 71 (2003), 285-292.
11. L. H. Yamane, V. T. Moraes, D. C. R. Espinosa and J. A. S. Tenório, "Recycling of WEEE: characterization of spent printed circuit boards from mobile phones and computers". *Waste Management*, 31, 12 (2011), 2553-2558.
12. Y. J. Park, D. J. Fray, "Recovery of High Purity Precious Metals from Printed Circuit Boards", *Journal of Hazardous Materials*, 164 (2009), 1152-1158.
13. O. Garcia Jr., "Isolation and purification of *Thiobacillus ferrooxidans* and *Thiobacillus thiooxidans* from some coal and uranium mines of Brazil", *Revista de Microbiologia*, 20 (1991), 1-6.
14. O. H. Tuovinen and D. P. Kelly, "Studies on the growth of *Thiobacillus ferrooxidans*. Use of membrane filters and ferrous iron agar to determine viable number and comparison CO_2-fixation and iron oxidation as measures of growth", *Archives of Microbiology*, 88 (1973), 285-298.
15. M-S. Choi, K-S. Cho, D-S. Kim and D-J. Kim, "Microbial Recovery of Copper from Printed Circuit Boards of Waste Computer by *Acidithiobacillus ferrooxidans*", *Journal of Environmental Science and Health. Part A – Toxic/Hazardous Substances & Environmental*, A39, 11 (2004), p. 1-10.
16. M. Nemati, S. T. L. Harrison, G. S. Hansford and C. Webb, "Biological oxidation of ferrous sulfate by *Thiobacillus ferrooxidans*: a review on the kinetic aspects", *Biochemical Engineering Journal*, 1 (1998), 171-190.
17. Y. XIANG, P. WU, N. ZHU, T. ZHANG, W. LIU, J. WU and P. LI, "Bioleaching of copper from waste printed circuit boards by bacterial consortium enriched from acid mine drainage", *Journal of Hazardous Materials*, 184 (2010) 812–818.
18. G. Meruane and T. Vargas, "Bacterial oxidation of ferrous iron by *Acidithiobacillus ferrooxidans* in the pH range 2.5–7.0", *Hydrometallurgy*, 71 (2003), 149–158.

REWAS 2013: Enabling Materials Resource Sustainability
Edited by: Anne Kvithyld, Christina Meskers, Randolph Kirchain, Gregory Krumdick, Brajendra Mishra, Markus Reuter, Cong Wang, Mark Schlesinger, Gabrielle Gaustad, Diana Lados, and Jeffrey Spangenberger
TMS (The Minerals, Metals & Materials Society), 2013

PGM RECYCLING FROM CATALYSTS IN A CLOSED HYDROMETALLURGICAL LOOP WITH AN OPTIONAL CERIUM RECOVERY

Stefan Steinlechner[1], Jürgen Antrekowitsch[1]

[1]Department of Nonferrous Metallurgy, University of Leoben,
Franz-Josef-Str. 18, A-8700 Leoben, Austria/Europe
(Corresponding author: Stefan.Steinlechner@unileoben.ac.at)

Keywords: PGM, cerium, hydrometallurgical recovery, car catalyst, petroleum catalyst

Abstract

Today PGM-recovery from spent catalysts is a standard procedure. Nevertheless still big amounts of these materials remain untreated. Furthermore the state of the art process is a very expensive solution and does not offer a recovery of all valuable materials.

The present paper introduces a hydrometallurgical recovery method for PGM from spent catalysts either from petroleum or automotive industry developed in cooperation with the company Railly&Hill.

It offers the opportunity for a locally small scale solution, with the advantage of short transport distances as well as low logistic costs. Moreover the energy consumption is lower than in other state of the art processes. Due to a closed loop of the leaching media, nearly no new residues are generated. Furthermore it offers the opportunity to recover rare earths from the wash coat, like cerium and with this forming an additional valuable product.

Introduction

The group of platinum metals is utilized in a wide range of application areas, like dental-, jewelry- and chemical industry but the most important area is based on its catalytic properties. Approximately 50 % [1] of platinum as well as palladium are utilized in automotive and industrial catalysts. In case of rhodium the usage in automotive catalysts is even higher; around 80 to 90 % [1] of the yearly produced metal is used as alloying element in the active metal layer. Especially in three-way-cats a mixture of platinum, palladium and rhodium is used to reduce harmful emissions, like carbon monoxide, hydrocarbons or NO_x.

One main reason for the high prices is the necessary amount of 300-900 kg of treated ore to process 1 g of PGM. Additionally the regional limitation of ore bodies leads to a strong dependence on mainly South Africa and Russia as PGM suppliers. Due to their limited natural resource deposits as well as their value, which is shown in table 1, the recycling of catalysts is of major interest.

Compared to the primary processing of ore the recycling of catalytic material, carrying approximately 1 to 15 g of PGM or even more, depending on the kind and size of car, seems to be mandatory from the economic point of view, the needed independence of supplier countries as well as ecological reasons. The average PGM content of car catalysts is 5-7 g per kg, while industry like petroleum refineries uses materials with average contents of above 10 g per kilogram material.

Table 1. Prices of platinum, palladium and rhodium [2]

	US$/oz	US$/g	€/g
Pt	1565	50.32	38.92
Pd	603	19.39	15.00
Rh	1150	36.97	28.60

Based on 50 different car catalysts of different brands the following figure 1 was obtained, showing the minimum and maximum as well as the average content of platinum, palladium and rhodium present in the active metal layer per kilogram of the monolith. The weight of the investigated monoliths was in a range of 230 to 2500 g, as mentioned above mainly depending on the size of the car.

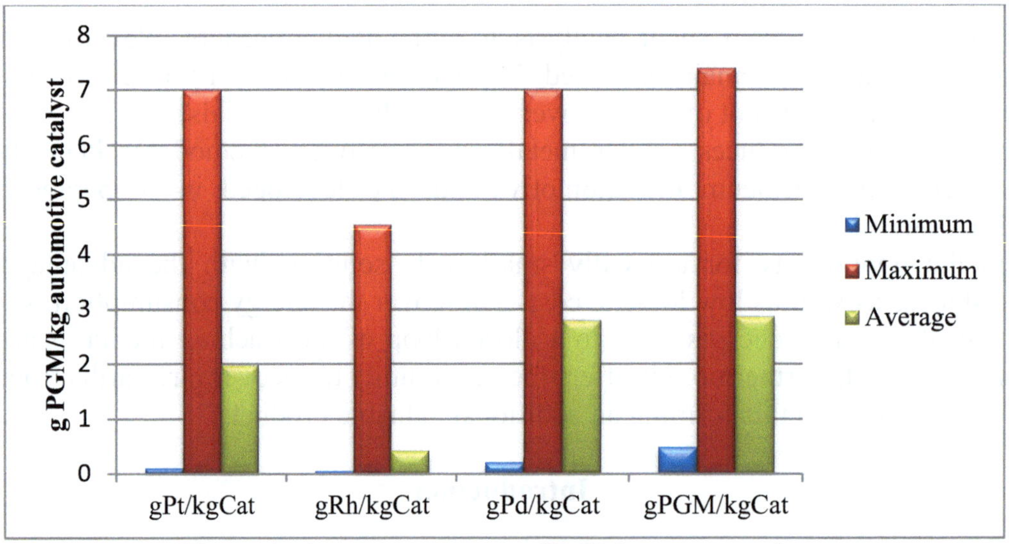

Figure 1. Average content of PGM on car catalysts

Figure 2 shows the average value of 1 kg of catalytic material based on the prices shown in table 1 as well as the average content calculated in figure 1 based on 50 different car catalysts. It can be seen on the total Euros per kg catalyst (see figure 2) as well as the amount of PGM on a catalyst (see figure 1) that the amount of platinum, rhodium and palladium are balanced; if there is less of one then there is more of another.

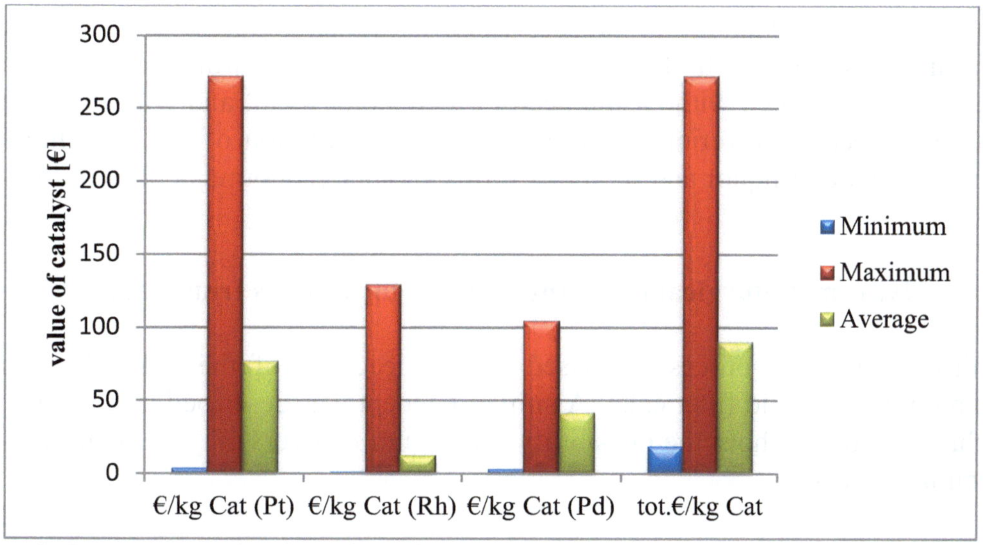

Figure 2. Monetary value based on the PGM prices shown in table 1 for car catalysts

Based on this high monetary value pyrometallurgical methods, which are consuming high amounts of energy, are also working economically. Even though hydrometallurgical processes seem to have different advantages, like lower energy consumption, lower emissions of greenhouse gases, selective leaching as well as the possibility to recover accompanying elements, the common method is the melting of the complete monolith in presence of fluxes and a collector metal, like copper, followed by the typical electrolytic purification obtaining the anode sludge, which is rich in the PGM's and can be further processed. The main reason therefore is that in a lot of hydrometallurgical methods huge amounts of waste water have to be treated.

The rarely applied hydrometallurgical recycling of car catalysts can be split in roughly two possible ways. The first is the leaching of the ceramic material, leaving the valuable metals as solids, with the disadvantage of low PGM-yields and no closed circuit leading to the already mentioned high amounts of waste water. The second method is the leaching of the valuable metal fraction. As a result of the very noble characteristic these metals have a high acid resistance and therefore need special conditions, like strong acids in combination with oxidation agents. Due to these conditions a dissolving of the monolith is going hand in hand with the PGM's and leads to the necessity of a regeneration of the liquor. In general the separation of the different platinum group metals can only be done by hydrometallurgical methods no matter if the recycling step is pyro- or hydrometallurgical.

Another interesting aspect is that below the active metal layer a so called wash coat is present on the monolithic material below the active metals. This coat carries elements like cerium but also indium and zirconium. The main aim is the increased surface area for the active catalytic material as well as the ability to buffer oxygen and therefore keeps the conditions for the chemical reactions more constant. Due to the fact that these increased contents of other valuable metals in the catalytic production to improve their properties started only some years ago, the returning car cats are nowadays typically low in their cerium content but in the next years the content of other valuable metals will increase, while at the same time the PGM content will decrease forming a metallurgical challenge for recovering as many metals as possible within the recycling processes. In common pyrometallurgical recycling steps for instance cerium could be found in the slags and with this will be lost.

Therefore the main advantage of the described hydrometallurgical process concept in this paper is that in the future it will also be possible to recover rare earths like cerium and that the PGM's are already separated without the need of a collector metal and with that no need for an energy intensive electrolytic refining is present. The consumption of acid is minimized by the closed loop of the mother liquor and with this forms another economic benefit of the investigated concept.

Hydrometallurgical recycling concept for PGM's from catalysts

The hydrometallurgical process is based on the leaching of the valuable metal fraction, leaving the monolith as solid filter cake. A similar concept was described in a former publication of Konetschnik [3] but without the possibility for cerium recovery. Figure 3 shows a flow sheet of the recycling process.

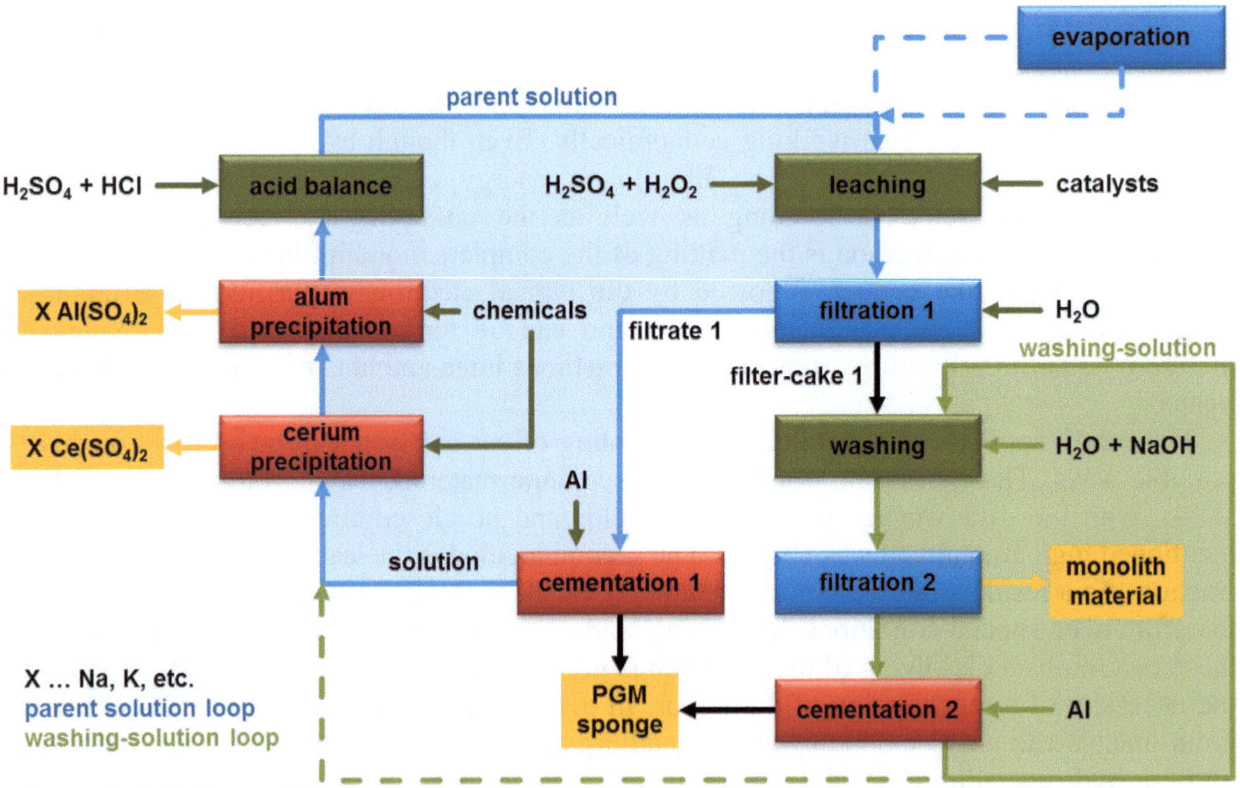

Figure 3. PGM recycling from catalysts in a closed hydrometallurgical loop with an optional recovery of cerium

Each step and their related chemical reactions are described in detail in the following paragraphs.

Leaching of Catalytic Material

Firstly, the catalyst is leached forming soluble chloro-complexes, like hexachloroplatinate (H_2PtCl_6). Therefore HCl as well as hydrogen peroxide are added and the temperature is increased to 70-90 °C. The best conditions were investigated by Sri Harjanto [4] describing also in detail the chemical reactions and necessary conditions for dissolving PGM's. Summarizing these special conditions are a low pH as well as a specific redox-potential. Equation 1 to 4 describe the simplified dissolution reaction in chloride media.

$$Pt_{(s)} + 8\ H^+ + 4\ O^- + 6\ Cl^- \rightarrow [PtCl_6]^{2-} + 4\ H_2O \quad (1)$$

$$Pd_{(s)} + 4\ H^+ + 2\ O^- + 4\ Cl^- \rightarrow [PdCl_4]^{2-} + 2\ H_2O \quad (2)$$

$$Rh_{(s)} + 6\ H^+ + 3\ O^- + 6\ Cl^- \rightarrow [RhCl_6]^{3-} + 3\ H_2O \quad (3)$$

$$Ce_{(s)} + 6\ H^+ + 3\ O^- + 4\ Cl^- \rightarrow [CeCl_4]^- + 3\ H_2O \quad (4)$$

As described by S.A. Cotton [5] the achieved pregnant liquor is red to brown as a result of dissolved $[PtCl_6]^{2-}$ as well as $[PdCl_4]^{2-}$.

Due to the strong acidic conditions a part of the monolith can also be dissolved, like that shown in equation 5. Typically the monolith is a mixture of aluminum, silicon and magnesium oxide, while firstly aluminum oxide is attacked.

$$Al_2O_{3(s)} + 6\ H^+ \rightarrow 2\ Al^{3+} + 3\ H_2O \quad (5)$$

As a result of the next step, the cementation of a PGM sponge with aluminum, it is no problem that a part of the monolith is also leached from the point of dissolved aluminum but leads to unwanted losses of H^+ ions, resulting in the necessity of an acid-balance step later in the recycling process.

The residue is mainly the monolith and can be easily separated by a filtration step. To prevent unwanted losses of valuable metals the cake is typically washed with a mixture of NaOH and hot water, followed by a separate cementation step.

Cementation of PGMs as Sponge

As mentioned above the pregnant liquor is further processed by dissolving aluminum and cementing the different platinum group metals. In general the cementation is a chemical process wherein a less noble element dissolves and replaces a nobler one in the liquor, which is squeezed out from the solution by forming a solid residue. Equation 6 to 8 show the specific reactions and that hydrogen is formed during this process step, which again is a further reason for the necessary acid-balance step later in the recycling process.

$$2\ Al_{(s)} + 2\ H^+ + [PtCl_6]^{2-} \rightarrow Pt_{(s)} + 2\ Al^{3+} + 6\ Cl^- + H_{2(g)} \quad (6)$$

$$4\ Al_{(s)} + 6\ H^+ + 3\ [PdCl_4]^{2-} \rightarrow 3\ Pd_{(s)} + 4\ Al^{3+} + 12\ Cl^- + 3\ H_{2(g)} \quad (7)$$

$$2\ Al_{(s)} + 3\ H^+ + [RhCl_6]^{3-} \rightarrow Rh_{(s)} + 2\ Al^{3+} + 6\ Cl^- + 1.5\ H_{2(g)} \quad (8)$$

As a product a PGM sponge is obtained, which could be sold to refineries or processed in the common way of PGM refining.

Precipitation of Dissolved Cerium

The aim to recirculate the solution as many times as possible makes it mandatory that the solution has to be cleaned from unwanted elements. These impurities are on the one hand the aluminum, which will be removed in the next step, as well as the cerium, which is removed by the described following equation 9.

$$[CeCl_4]^{2-} + 1.5H_2SO_4 + 0.5Na_2SO_4 \rightarrow NaCe(SO_4)_2 + 4Cl^- + 3H^+ \quad (9)$$

Both here called impurities are forming sellable products, which can be seen as a further advantage of the investigated process compared to the common pyrometallurgical cat recycling.

It can be seen from the equation that also a sulfate group SO_4^{2-} is needed to form a double sulfate with cerium. Therefore H_2SO_4 is consumed, which has to be added in the acid-balance step to supply the needed sulfate group next to a balancing of the pH-value. The reason for the use of sulfuric acid instead of hydrochloric acid is mainly the economics of the process because HCl would also adjust the pH but in general is more expensive than H_2SO_4. A further sulfate carrier is the chemical used for precipitation itself, which can be potassium sulfate or sodium sulfate.

The formed H^+ ions are partly replacing the lost H^+ ions from the cementation step by formed H_2 gas and reduce the required amount of added acid in the acid-balance step.

Precipitation of Dissolved Aluminum

As mentioned before the main impurity is dissolved aluminum from the monolith, which is separated in a precipitation step. Therefore alum, also called alaun, is formed by for instance potassium-chloride addition. The chemical mechanism therefore is described in equation 10.

$$Al^{3+} + 2H_2SO_4 + KCl \rightarrow KAl(SO_4)_2 + Cl^- + 4H^+ \quad (10)$$

The advantage of this step is the formation of a marketable product, next to the purification of the leach liquor. The formed alum represents also a double sulfate in the same way as the cerium precipitate does.

Acid Balancing of Parent Solution

The last necessary step before the parent liquor can be reused as leaching solution is the balancing of acid as well as hydrogen peroxide. The reasons for this step are partly chemical reactions of the recycling process and their formed products but also some losses of solution with filter cakes (monolith, PGM sponge, etc.), which have to be mentioned although there are washing steps but are generally very low.

But beside these small losses, a huge benefit of the closed circuit is that there are minimal metal losses to effluents additionally to minimized amounts of waste water due to the reuse of leaching solution.

For balancing the H^+ ions and the redox potential a mixture of sulfuric acid and hydrochloric acid is used next to hydrogen peroxide. There are mainly economic reasons because HCl is generally more expensive than sulfuric acid. The H_2SO_4 has as a second function, the balancing of SO_4^{2-} ions.

With this step the solution is now ready for reuse as leaching liquor for ground catalysts again.

Generated Products

The aim of the described process is to form products instead of residues. As a result there is more or less no waste stream, instead there are four marketable products formed. The main product is the cemented PGM sponge, which can be sold easily to refineries or can be processed by a common hydrometallurgical method to separated PGM salts or metals.

Secondly the remaining solids from the leaching step can be utilized in refractory production because the main compounds are SiO_2, Al_2O_3 as well as MgO, which are well known for their high melting points.

The alum has three main properties which influence its utilization. It can act as bleaching agent, the sulfate ions are antibacterial and it inhibits microbes. Therefore typical utilization areas can be found in cosmetics (deodorant, etc.), tanning, paper industry and medical applications.

The fourth product is the cerium double sulfate, which can be sold as it is or converted to an oxide by a temperature treatment.

Optional Pretreatment of Contaminated Cats

As illustrated in figure 4 it may be necessary to pretreat with organic material contaminated catalysts. This minimizes the chemicals consumption in the main hydrometallurgical recover process as well as offers the opportunity to accept also highly contaminated catalysts from petroleum industry. These catalysts from petroleum industry are higher in their specific valuable metal content and with this could have a positive impact on the economics of the process.

The aim of the pyrometallurgical step is the pyrolysis of organic compounds on the cat. These organic materials would lead to formation of carbon as well as sulfur dioxide in the leaching step and with this increase the chemicals consumption. Due to the consumption of oxygen mainly the hydrogen peroxide addition would have to be increased.

A crucial point is the treatment temperature and atmosphere because in case of unsuitable conditions the monolith material includes the valuable metal content and with this decreases the PGM yield in the recovery step. This has to be avoided by advisedly chosen parameters.

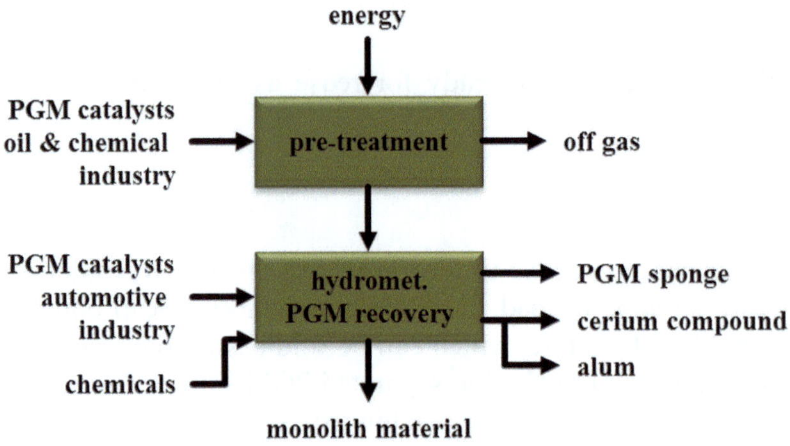

Figure 4. Pre-treatment of petroleum and contaminated automotive catalysts

Due to the focus on the hydrometallurgical closed recovery loop this pre-treatment step is not described in detail but has to be seen as mandatory not only for petroleum but also for contaminated automotive catalysts, otherwise problems in the leaching step will be the result.

Opportunity for Mini Mill Solution

The introduced and investigated process shows, based on estimated CAPEX and OPEX, promising results also in small scale, which offers a big opportunity in the area of mini mill solutions. For example regional limited scrap dealers often do not sell catalytic material and a huge part of valuable metals are lost, especially in the eastern regions of Europe. In such regions where the logistic for collection is not as distinct as in Western Europe innovative recycling concepts are needed, which do not need huge amounts of catalytic material to work economically.

Conclusion

Based on the increasing strictness of legislation for emission of cars as well as industry, catalysts are getting more and more complex in their material composition. This introduces further valuable metals in the recycling chain of catalysts, like cerium. Due to the non-noble character of cerium, which belongs to the rare earths with limited availability, the recovery in common pyrometallurgical process is not possible.

Therefore this work introduced a recycling process for catalysts, which meets the demands of future duties, like applicability in outlying regions with bad logistics, possibility for recovery of additional valuable metals, high yields of PGM recovery as well as low amounts of newly generated residues or waste streams, like effluents. The combination of a hydrometallurgical process with an optional pyrometallurgical pre-treatment step offers not only the possibility of treating catalyst varying in their valuable metal content but also the utilization of contaminated raw material from for instance the petroleum industry.

References

[1] C. Hagelüken, „Recycling von Autokatalysatoren“, *Heft 115 der Schriftenreihe der GDMB Gesellschaft für Bergbau, Metallurgie, Rohstoff- und Umwelttechnik des 43. Metallurgischen Seminars*, (2008), Dortmund, Deutschland, 69 – 86.

[2] Retrieved from http://www.kitco.com, October 31st 2012

[3] Konetschnik S., D. Offenthaler, J. Antrekowitsch and J. Sitter: Recovery of PGMs from spent catalyst material with a new HCl-H_2SO_4 leaching process. *Proceedings of EMC 2007*, (2007) 679-690.

[4] Sri Hrjanto, et.al.: Leaching of Pt, Pd and Rh from automotive catalyst residue in various chloride based solutions. *Materials Transactions, Vol. 47, No. 1*, (2006),129-135.

[5] S.A. Cotton: *Chemistry of precious metals* (Blackie Academic & Professional)

REWAS 2013: Enabling Materials Resource Sustainabili
Edited by: Anne Kvithyld, Christina Meskers, Randolph Kirchain, Gregory Krumdick, Brajendra Mishr
Markus Reuter, Cong Wang, Mark Schlesinger, Gabrielle Gaustad, Diana Lados, and Jeffrey Spangenberg
TMS (The Minerals, Metals & Materials Society), 20

A NOVEL PROCESS FOR RECOVERING VALUABLE MATERIALS FROM SPENT LITHIUM-ION BATTERIES

Gjergj Dodbiba[1], Yuta Yamaji[1], Kenji Murata[2],
Katsunori Okaya[1], Atsushi Shibayama[3], Toyohisa Fujita[1]

[1]Department of Systems Innovation, The University of Tokyo, Japan;
[2]Nippon Koki Co. Ltd., Japan;
[3]Faculty of Engineering and Resource Science, Akita University, Japan

Keywords: Lithium-ion battery, Underwater explosion, Separation, Leaching

Abstract

The demand for lithium-ion batteries has been increasing due to the increasing demand for laptop computers, cellular phones, automobiles, etc. The positive electrode of the lithium-ion secondary battery is mainly made of lithium oxides well as cobalt, nickel, manganese, etc. Thus, an effective recycling method not only would collect cobalt and lithium, but also would enable the separation of other materials from the spent batteries. In this work, a novel processing flow sheet is put forward and its efficiency is evaluated. The aim was to obtain pure fractions of various constituents.

Introduction

Lithium-ion battery production is constantly increasing, since the demand for batteries to be used in laptop computers, cellular phones and automobiles is rising. Therefore, a closed-loop recycling system for these batteries has to be developed. There are reports that indicate the production of lithium-ion batteries in Japan increased by about five times from 2000 to 2007 [1]. Generally speaking, lithium-ion batteries are fed into a rotary kiln and roasted, followed by crushing, acid leaching and solvent extraction, in order to recover cobalt, copper and iron. However, the recovered amount is still relatively small.

The composition of lithium-ion battery used in this experiment is shown in Figure 1. It should be noted that the main cathode of the spent batteries used in this were made of $LiCoO_2$.

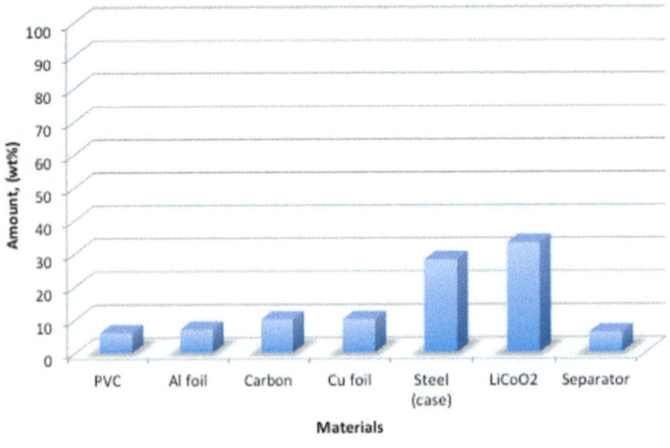

Figure 1. Typical composition of lithium-ion battery

Experimental

The flow sheet for recovering lithium ion battery is shown in Figure 2. At first, the lithium-ion batteries were disintegrated by means of a controlled explosion in water. Generally speaking, the main utilizations of industrial explosive are in large-scale public works such as dam construction, land development, mining and demolition of old buildings. The shock wave generated by explosion can cause extremely high pressure, and therefore can liberate, break or crush various kinds of materials regardless their hardness. In addition, this technique might be a feasible alternative (when compared with other conventional size-reduction techniques), not only because of its effectiveness in terms of the energy applied, but also because a possible explosion due to the presence of the flammable electrolyte in the batteries can be avoided as well as the emission HF gas can be contained. The following chemical reactions illustrate how HF gas can be contained when electrolytes of batteries (usually made of lithium-hexafluorophosphate $LiPF_6$) react with water.

$$LiPF_6 + H_2O \rightarrow LiF + POF_3 + 2HF \qquad (1)$$

If the water used for the explosion contains calcium ions, fluorine ion would then precipitate as CaF_2:

$$CaCl_2 + 2HF \rightarrow CaF_2 + 2HCl \qquad (2)$$

It should be noted that water used for controlled explosion could be re-used several times, before Li dissolved can be adsorbed by using MnO_2 as an adsorbent, a hypothesis that has still to be verified.

In experiment, both the sample and the explosive are placed in a steel container filled with water. After the explosion and the "liberated" parts felt above a 10 mesh net, which is than pulled up to collect the materials. As a result of the underwater explosion, the battery package was disintegrated and then introduced as feed into a cutter mill for further size reduction. Next the sample was classified into size fractions, namely: +1 mm and -1 mm.

The +1mm size fraction is separated by using a rare earth roll magnetic separator, of which the magnetic flux density was kept constant at 1.2 T. After recovering the magnetic fraction, which contains mainly iron parts, the non-magnetic fraction was further processed by using an eddy current separator. The non-conductive particles are mainly plastic film like PVC, etc. The conductive particles, on the other hand, are separated by air table [2] into lighter and heavier fractions. The aluminum plate and copper films are collected as low-density fraction, whereas the high-density fraction was mainly of relatively "big parts" that escaped the size reduction process, and therefore are returned to the cutter mill. The results of each separation process are given in term of grade, Eq. (3), and recovery, Eq. (4) of each product.

$$Grade_{"a"} = \frac{Mass\ \ of\ \ material\ \ "a"\ \ in\ \ the\ \ collected\ \ fraction}{Total\ \ mass\ \ of\ \ the\ \ collected\ \ fraction} \times 100\% \qquad (3)$$

$$Recovery_{"a"} = \frac{Mass \quad of \quad material \quad "a" \quad in \quad the \quad collected \quad fraction}{Total \quad mass \quad of \quad material \quad "a" \quad in \quad input \quad stream} \times 100\% \tag{4}$$

where "*a*" denotes one of the components of the mixture.

The -1mm size fraction, on the other hand, is fed into a floatation cell for recovering graphite powders as float product and cathode $LiCoO_2$ powders as sink product. The separated $LiCoO_2$ powders were then dissolved in sulfuric acid aqueous solution and neutralized by alkaline solution to precipitate cobalt as a hydroxide species. The cobalt hydroxide is then heated and Co_3O_4 is recovered. After the salts are removed from filtrate, Li ions were recovered by the adsorption, using blue green algae, MnO_2, or manganese nodule.

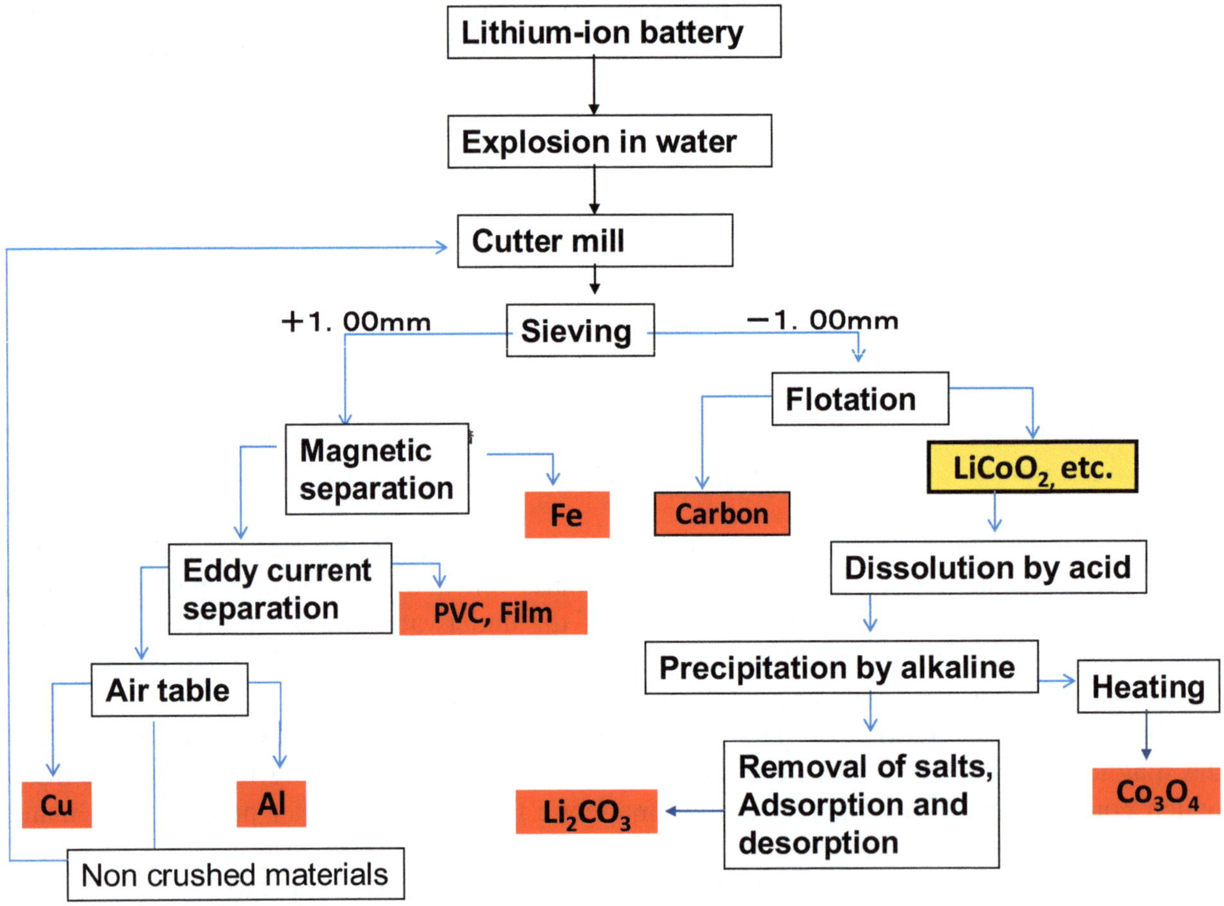

Figure 2. Flowsheet for processing the spent lithium-ion battery

Results and Discussions

During the underwater explosion, the shock waves in combination of the generated bubbles are very effective in liberating the materials of different densities. It should be noted that only a small amount of explosive (explosive dosage: 2 g/kg-batteries) was used for disintegrating the batteries. Due to the pressure generated by explosion (ca. 500 MPa) the batteries were disintegrated, making the handling of the materials an easy task. Next, the plastic case and inner parts are separated for further processing. Figure 3 shows the photograph of a Li-ion battery before and after explosion in water. Next, the separated parts are crushed by the cutter mill. Figure 4 shows the results of size reduction by cutter mill as a function of crushing time. It was found that only after 5 seconds about 40% of crushed particles had a size less than 1mm, and the amount of the undersize fraction was increased to 65% after 10 seconds.

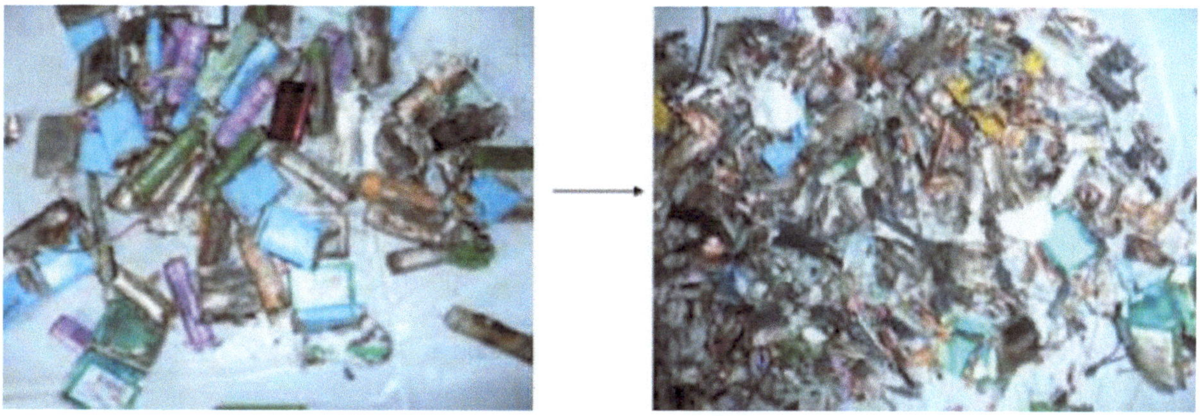

Figure 3. A lithium-ion battery before and after explosion in water.

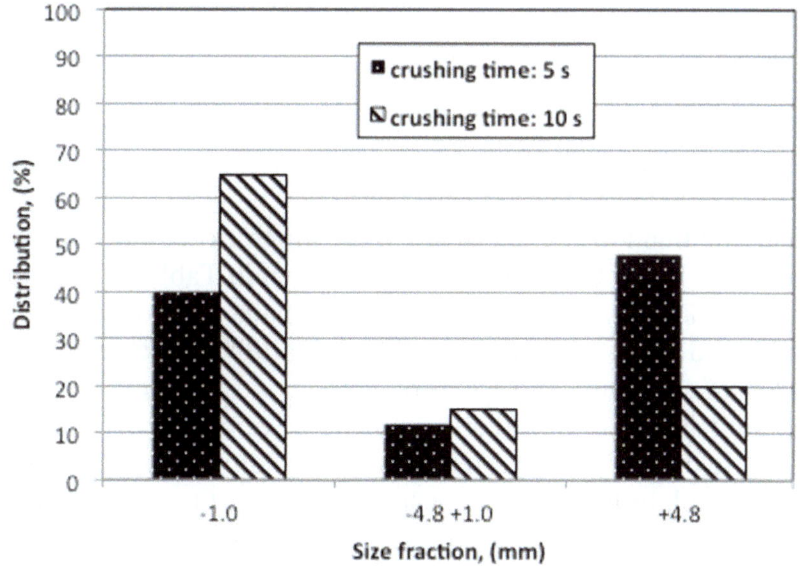

Figure 4. Distribution of the undersize fraction as function of crushing time.

Next, the fraction over 1 mm in size was subject to magnetic separation, of which the results are given in Table I. It was found that almost 100% of iron particles over 1mm size were effectively separated from non-magnetic particles. The non-magnetic fraction, on the other hand, was fed into an eddy current separator, which was able to recover about 70% of Al and 80% of

Cu from the size fraction +4.8 mm (Table II). In -4.8+1.0mm size fraction, the amount of conductive particles increased in the nonconductive fraction.

After the eddy current separation, the conductive particles are sorted by gravity separation using an air table. The results of air tabling are shown in Fig. 5. It was found that 90% of non Al or Cu particles are recovered as a low-density fraction. The high-density fraction contained some relatively big pieces, which in turn contaminated the Al and Cu fraction lowering their grade. Those big pieces were recovered by screening and re-introduced into the cutter mill.

Table I. Results of magnetic separation for +1mm size fraction of crushed lithium ion battery (Magnetic flux density on the belt surface: 1.2 T)

Magnetic separation		Grade, %		Recovery,%	
		Fe	Others	Fe	Others
	Feed	28.4	71.6	100	100
+4.8mm	Magnetics	88.3	11.7	96.0	10.2
	Non-magnetics	0	100	0	56.9
-4.8mm	Magnetics	18.1	81.9	3.8	4.8
+1.0mm	Non-magnetics	0.6	99.4	0.2	28.1

Table II. Result of eddy current separation for non-magnetics exiting magnetic separation (Drum rotation speed: 2500 rpm, Belt speed 0.78m/s)

Eddy current separation		Grade,%			Recovery, %		
		Al	Cu	Others	Al	Cu	Others
	Feed	7.3	10	82.7	100	100	100
+4.8mm	Conductive	26.3	47.4	26.3	68.7	78.2	13.3
	Nonconductive	2.7	2.7	94.6	4.5	4.2	46.8
-4.8mm	Conductive	27.3	18.1	54.6	10.8	4.8	6.3
+1.0mm	Nonconductive	10.9	13.1	76	15.1	12.8	33.6

On the other hand, the fraction smaller than 1 mm, which exited the cutter mill, was fed into a batch flotation machine [3]. The flotation results are shown in Table III. The graphite is collected as a float product at 95% recovery. The $LiCoO_2$, the main component of the positive battery electrode, was recovered as a sink product at about 90% recovery. The sink product is contaminated by 0.5% of aluminum and 1.4 % of copper.

In addition, the sink products of flotation powders were dissolved by using sulfuric acid aqueous solution. Next, the sodium hydroxide aqueous solution is added to precipitate cobalt hydroxide at pH 9. The solubility product constant [4] of $Co(OH)_2$ is 5.9×10^{-15} after the filtered cobalt hydroxide is washed and heated at 500°C for 2 hours. The structure of recovered cobalt oxide is Co_3O_4 as indicated by the X-ray diffraction. The chemical reactions are as follows:

$$2LiCoO_2 + 3H_2SO_4 + Na_2SO_3 \rightarrow Li_2SO_4 + 2CoSO_4 + Na_2SO_4 + 3H_2O \quad (5)$$

$$CoSO_4 + 2NaOH \rightarrow Co(OH)_2 + Na_2SO_4 \quad (6)$$

$$3Co(OH)_2 + 1/2O_2 \rightarrow Co_3O_4 + 3H_2O \quad (7)$$

After recovering the cobalt, the lithium ion is absorbed by using manganese oxide (MnO_2), manganese nodule, and, blue green algae's powder as adsorbent. The lithium ion has been adsorbed within 240 minutes as indicated in Table IV. We found that absorbing capacity of MnO_2 for Li ions is 1170 mg/kg after 240 mins.

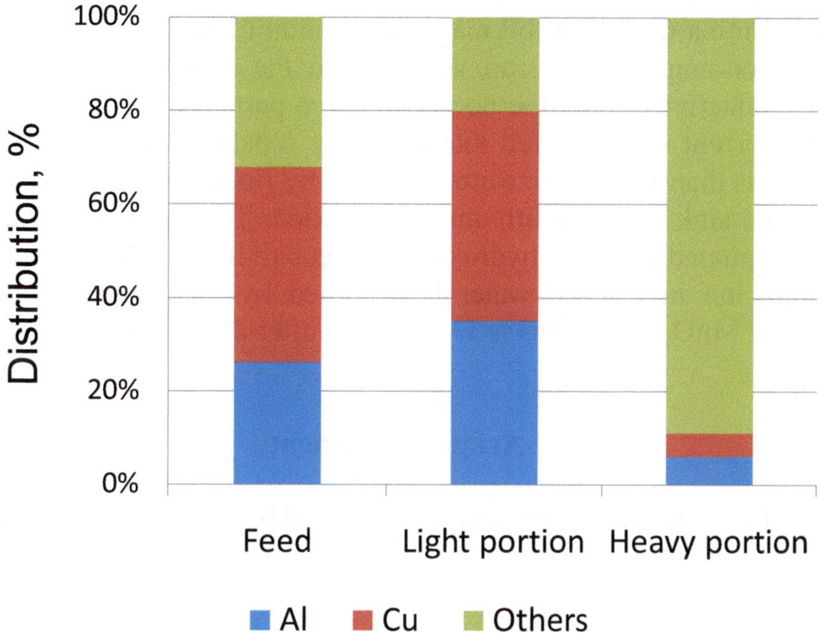

Figure 5. Results of air tabling
(Upward air velocity: 1.2m/s; Vibration frequency of the deck: 10 Hz)

Table III. Flotation result for -1mm size fraction of crushed lithium ion battery
(Pulp density: 10%, Methyl Iso-butyl Cabinol (MIBC): 0.14 kg/t, kerosene: 3kg/t)

Flotation	Grade, %			Recovery, %		
	Li	Co	Graphite	Li	Co	Graphite
Feed	(About 60% $LiCoO_2$)		about 30%	100	100	100
Float	0.7	3.3	90	15	12	95
Sink	3.8	24	5	85	88	5

Table IV. The lithium ion adsorption capacity (mg/kg)
as a function of the contact time and type of adsorbent

	Type of Adsorbent		
Contact time, (min)	Blue green algae	MnO_2	Manganese nodule
0	1170 mg/kg	1170 mg/kg	1170 mg/kg
10	1100 mg/kg	1090 mg/kg	1050 mg/kg
30	1060 mg/kg	1000 mg/kg	1000 mg/kg
100	1000 mg/kg	960 mg/kg	950 mg/kg
240	900 mg/kg	920 mg/kg	900 mg/kg

Conclusions

Lithium-ion battery recycling has been studied and a flow sheet for recycling has been put forward. The battery package was easily disintegrated by the explosion in the water, containing Ca^{2+} ion. Next, the disintegrated batteries are crushed by the cutter mill. The particles bigger than 1mm in size are fed into a rare earth roll magnetic separator and the iron has been separated as a magnetic fraction. Non-magnetic parts are separated by the eddy current. Aluminum and copper are recovered as conductive, while, the non-conductive parts are mainly plastics. Each fraction separated by eddy current was then fed into a gravity separator, i.e. an air table. On the other hand, the particles less than 1mm in size are separated by flotation. The float product was mainly graphite, whereas the sink one was lithium cobalt oxide. The lithium cobalt oxide is leached, neutralized and precipitated as cobalt hydroxide. The cobalt hydroxide is heated and recovered as Co_3O_4. The lithium ion in leached water is absorbed by manganese oxide. We found that absorbing capacity of MnO_2 for Li ions is 1170 mg/kg after 240 mins.

Acknowledgement

The authors gratefully acknowledge the support provided by JSPS Scientific Research (A), Grant No. 22246118.

References

1) JOGMEC, Material flow of mineral resources, Mineral resources information center, http://mric.jogmec.go.jp/mric_search/, (2008).

2) T. Furuyama et al., “Recovering PVC by triboelectric separation and air tabling”, Resources Processing, **53**(2006), 153-159.

3) Y. Kim et al., ”Recovery of $LiCoO_2$ from waste lithium ion batteries by means of flotation”, Shigen-to-Sozai, **118**(2002), 687-693.

4) D.R. Lide, “Handbook of chemistry and physics”, CRC press. 89th (2008), 8-123, 4-91.

REWAS 2013: Enabling Materials Resource Sustainability
Edited by: Anne Kvithyld, Christina Meskers, Randolph Kirchain, Gregory Krumdick, Brajendra Mishra, Markus Reuter, Cong Wang, Mark Schlesinger, Gabrielle Gaustad, Diana Lados, and Jeffrey Spangenberger
TMS (The Minerals, Metals & Materials Society), 2013

METAL RECOVERY FROM INDUSTRIAL SOLID WASTE – CONTRIBUTION TO RESOURCE SUSTAINABILITY

Yongxiang Yang

Department of Materials Science and Engineering, Delft University of Technology
Mekelweg 2, 2628 CD Delft, The Netherlands

Keywords: Metal recovery, industrial waste, resource sustainability

Abstract

Increased demand of metals has driven the accelerated mining and metallurgical production in recent years, causing fast depletion of primary metals resources. On the contrary, the mining and metallurgical industry generates large amount of solid residues and waste such as tailings, slags, flue dust and leach residues, with relative low valuable metal contents. On the other hand, end-of-life (EoL) consumer products form another significant resources. The current technology and processes for primary metals production are not readily applicable for direct metals extraction from these waste materials, and special adaptation and tailor-made processes are required. In the present paper, various solid waste resources are reviewed, and current technologies and R&D trends are discussed. The recent research at author's group is illustrated for providing potential solutions to future resource problems, including metal recovery from MSW incinerator bottom ashes, zinc recovery from industrial ashes and residues, and rare earth metals recovery from EoL permanent magnets.

Introduction

While striving for the solutions to CO_2 reduction, we are facing another increasing threat of resource depleting in minerals and metals. According to statistics, the economically exploitable metal minerals can mostly sustain between 20 and 100 years. A recent publication by Diederen indicated very severe global problem of metal minerals scarcity [1]. The depletion of the mineral resources is attributed by both the limited reserve in this planet and the rapidly increasing demand in materials consumption. In 2010, the whole world produced 1,414 mt of crude steel [2], 41,400 kt of primary aluminium and 16,200 kt of primary mined copper [3]. In total more than 1,500 mt of primary metals are produced in the world, among which almost 95% is steel, and the rest of the metals accounts for 5% (plus), according to the published figures for 2010 in USGS yearbook [3]. The production figure of crude steel has already included the contribution of steel scrap, while the nonferrous metals figure of more than 90 mt does not include the production from scrap. Figure 1 illustrates the evolution of world production (1960 – 2011) and consumption (1900 – 2012), according to Internal Copper Study Group (ICSG) [4]. It is clearly seen that in the past half century, the copper production has more than quadrupled, and the consumption figure shows even more clearly the fast increase since 1950s.

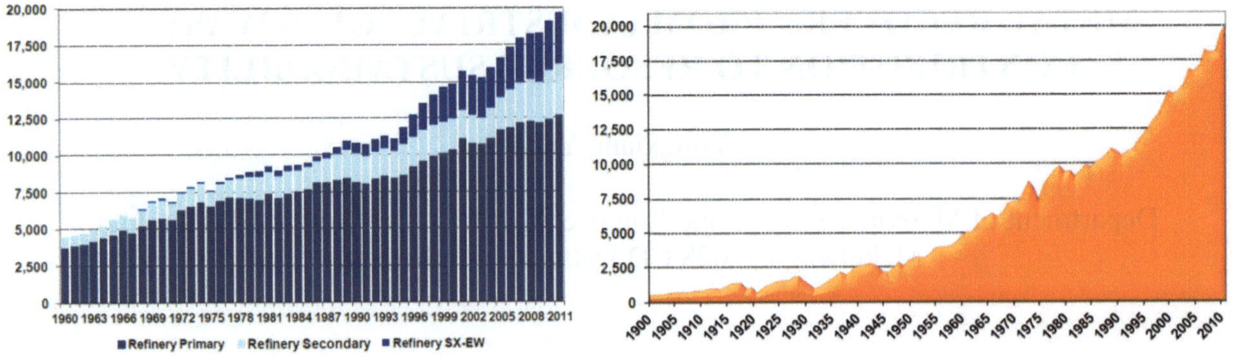

Figure 1. World refined copper production (left) and consumption (right) [4].

Looking into the total metal supply, recycling of metals has been playing a very important role in the total metal supply globally. According to various sources, over 35% steel and copper produced in the world comes from steel scrap, and about 30% of the aluminium and zinc supply originated from the scrap. This figure goes over 50% for lead. This means that for the total metal consumption, there is a significant contribution from the metal scrap recycled to the production.

Steel is one of the few magnetic metals, which is easy to separate from waste streams. About 80% of post-consumer steel is recycled, among which by sector steel recovery rates are estimated at 85% for construction, 85% for automotive, 90% for machinery, and 50% for electrical and domestic appliances [2]. According to World steel Association, over 20 billion tons of steel remains in use today in a variety of products which will be available for recycling [2]. According to International Aluminium Institute, now World Aluminium [5] in 2009 the estimated total aluminium products stored in use since 1888 is 662.9 million tons, which is available for recycling in the coming years depending on the life cycle of different types of aluminium products. According to Risopatron [6], recycling rate copper is very low, and an estimated old copper scrap accumulated in stock is about 270 million tons which is thought to be too expensive to recycle. However annually speaking, about 60% of available copper for recycling has been recovered in Europe and North America [7]. How much copper is still in use as products? Lifset et al. [7] estimated that as much as 85% of the copper mined and put into use throughout human history is in use today, because it has largely entered use in the past half century, and because many of the uses have lifetimes of several decades. According to International Zinc Association [8], 80% of recyclable zinc available in the world is recycled, and the 20% unrecycled zinc is mainly the flue dust from steelmaking processes.

It seems that the recycling rate, defined as the recycled metal compared to the total metal available for recycling, is relatively high for many metals. If we take another look at how much metals are getting lost during production and use phase, we need to do a lot more to conserve the metal resources. Table I illustrates the metal losses during production, manufacturing and use phases for zinc, copper, nickel, chromium and silver, according to Stocks and Flow analysis project STAF for a number of metals conducted at Yale University. The figures are taken for the years between 1994 and 2000 because of the complete statistics availability, and situation for recent years will maintain the same trend.

Taking zinc as example: In 1994, 7.21 million tons of refined zinc is produced. During mining over 1 million tons of zinc metal is lost as tailings, while during smelting 330 kt zinc is lost as slags and residues. That is to say, to produce 1 ton of zinc metal it will lose 184 kg zinc, and the total metal recovery rate during production (from mining to pure metal) is about 84%. If one looks further at the end use phase, more metals are getting lost, and 1.673 million tons of zinc is lost into landfill in the form of end-of-life product wastes. On one hand we are producing new metals, on the other hand we are throwing away metals through their end-of-life phase. Putting all kind of losses together, producing one ton of zinc, we are losing at the same time 420 kg the same metal. According to statistics (see Table I), this metal loss is significant for other metals also: 280 kg for copper, 380 kg for silver, 450 kg for nickel and 490 kg chromium. This calculation has taken into account the recycling of these metals, and the recycled metals are not counted as losses. This analysis indicate clearly that we will have a big potential to further recover the available metals, in particular from the low grade tailings and landfill mines/urban mines.

Table I. Global production of metals and the metal loss in the whole life cycle [9]

Metal	**Zn (1994)**	**Cu (1994)**	**Ag (1997)**	**Ni (2000)**	**Cr (2000)**
Mine metal (ore, kt)	7.800	9.490	20.200t	1.338	5.140
Refined metal (kt)	7.210	11.800	24.600t	1.120	3.900
Total metal loss (kt)	3.033	3.350	9.400t	509	1.910
Metal loss/Refined metal	0,42	0,28	0,38	0,45	0,49
Tailings (kt)	1.030 (34%)	1.400 (42%)	4.000t (42.5%)	167 (32.8)	740 (39%)
Slag (kt)	330 (11%)	150 (4%)	1.400t (15%)	74 (14.5%)	590 (31%)
Metal to landfill (kt)	1.673 (55%)	1.800 (54%)	4.000t (42.5%)	268 (52.7%)	580 (30%)
Reference	[10]	[11]	[12]	[13]	[13]

Taking zinc as example: In 1994, 7.21 million tons of refined zinc is produced. During mining over 1 million tons of zinc metal is lost as tailings, while during smelting 330 kt zinc is lost as slags and residues. That is to say, to produce 1 ton of zinc metal it will lose 184 kg zinc, and the total metal recovery rate during production (from mining to pure metal) is about 84%. If one looks further at the end use phase, more metals are getting lost, and 1.673 million tons of zinc is lost into landfill in the form of end-of-life product wastes. On one hand we are producing new metals, on the other hand we are throwing away metals through their end-of-life phase. Putting all kind of losses together, producing one ton of zinc, we are losing at the same time 420 kg the same metal. According to statistics of Table 1, this metal loss is significant for other metals also: 280 kg for copper, 380 kg for silver, 450 kg for nickel and 490 kg chromium. This calculation has taken into account the recycling of these metals, and the recycled metals are not counted as losses. This analysis indicate clearly that we will have a big potential to further recover the available metals, in particular from the low grade tailings and landfill mines/urban mines.

In contrast to the depleting natural resource of metal minerals, more and more waste materials are generated in our daily life and through industrial activities. However, the current technology and processes for primary metals production are not readily applicable for direct metals extraction from all these waste materials. In this paper, three examples are given for the recent research at TU Delft to illustrate the potential solutions to waste and resource problems: (1) metals recovery from municipal solid waste (MSW) incinerator bottom ashes, (2) zinc recovery from various industrial wastes, and (3) rare earth metals recovery from EoL magnet scrap from computer hard disk drives (HDDs). Technological development and challenges will be addressed.

Resource Recovery from MSW Incinerator Bottom Ashes

Incineration or waste-to-energy processing is proved to be an efficient technology to the solution of environmental problems for the disposal of Municipal Solid Waste (MSW). Waste combustion converts the solid waste into thermal and electrical power, at the same time it generates over 20% of solid residues in the form of bottom ash and fly ash. The bottom ash consists of both inorganic mixture minerals (SiO_2, CaO, Fe_2O_3 and Al_2O_3), and significant amount of metals. In the Netherlands and most of European countries, the metal content can be as high as 15-20 wt% of the bottom ash. As a European practice, about 60% of the metal content can be recovered through physical separation, and the rest of 40% metals are still trapped into the bottom ash due to limitations of the conventional technological. This is not only a loss of the valuable raw materials, but it also causes the problems for direct use as construction materials. The ideal solution would be the recovery of all valuable metals and generation of clean and high value-added construction materials.

When physical metal separation is approaching its limit, chemical and metallurgical separation becomes very important to extract the remaining valuable metals from the bottom ash. Although hydrometallurgical leaching of the bottom ash may recovery copper and other base metals such as ammonia leaching, the remaining ash contains still large amount of steel scrap particles and the ash itself is still prone of environmental pollutions. Sulfuric leaching would leach both base metals (Cu, Zn, Al etc.) and ferrous metals, and further separation of theses metals becomes a heavy burden. Leaching of metals is in general a slow process and a strong oxidant is normally required such as hydrogen peroxide, which will add significantly the processing cost.

As an alternative solution, a pyrometallurgical vitrification process has been investigated at the metals production and recycling group of TU Delft [14, 15]. Bottom ash vitrification is used for recovering the metal contents and for producing an environmentally clean slag product. The bottom ash from a typical Dutch MSW incinerator contains still about 8% metals after physical metal extraction through magnetic and eddy current separations. The bottom ash is a highly heterogeneous material containing both metals and a mixture of oxides and other compounds. The major elements are Si, Ca Fe, Al and Na. Vitrification proves to be an interesting option to separate metallic components and to generate a more stable glassy slag. At temperatures between 1360 and 1500°C, a homogeneous glassy slag product can be formed. On average 8.0% of metal was recovered after vitrification from the bottom ash already after the physical metal separation. Figure 2 illustrates the recovered metal alloy and the vitrified stable slag.

Figure 2. Bottom ash and the vitrified products: Cu-Fe alloy and stable silicate slag.

The recovered metal alloy consists of approximately 82% Fe, 12% Cu, and a few percent of other nonmetallic impurities (2.0% P, 1.4% C and 1.4% S), as well as some minor alloying elements. Although use of the alloy directly in a secondary copper smelter is possible, the oxidation of large amount of iron in the alloy scrap is not efficient solution and the iron value as steelmaking raw material is lost. Therefore, copper – iron separation is a better solution in order to produce two useful raw materials accepted by copper and steelmaking industry. Removal of copper at relatively high concentration (~12 wt%) in the iron-based alloy is possible by using FeS or the sulphide mixtures based on FeS. Carbon saturation is essential to the success of copper separation from iron by sulphide treatment. The use of fluxing agents (Na_2S or Al_2S_3) together with FeS will promote the removal efficiency of copper. A maximum of 93.9% of copper removal efficiency was achieved by using FeS (27.5 mol%)-Na_2S (72.5 mol%) mixture at a stoichiometric S ratio of 2 at 1500°C. Furthermore other cheaper source of metal sulphides such as pyrite (FeS_2) mineral or even industrial waste such as ZnS containing sludge were also tested. The results are illustrated in Figure 3, in comparison with pure FeS. It is clearly shown that both ZnS sludge and pyrite are more effective compared with pure FeS for copper removal with a removal efficiency of 92% for pyrite and 95% for zinc sludge at a S/Cu ratio of around 7 times of the stoichiometry. The separated ferrous metal can be sold as steel scrap, while the Cu_2S bearing matte phase can be used as secondary raw material in copper smelter.

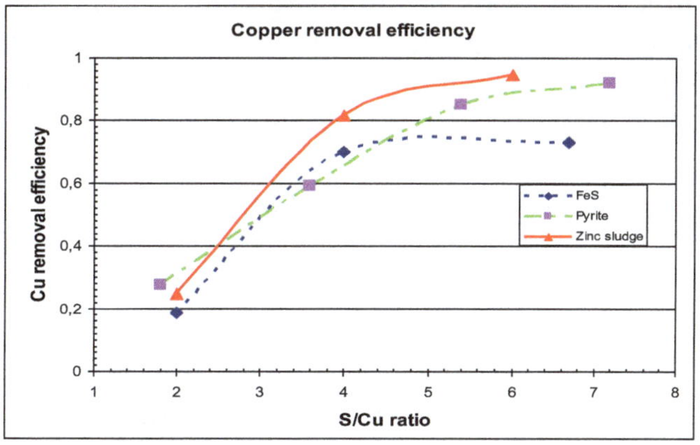

Figure 3. Copper removal efficiency using FeS, pyrite and zinc sludge from Fe-Cu alloy in graphite crucibles at 1500°C.

The vitrified bottom ash contains on average 52.1% SiO_2, 16.2% CaO, 12.2% Al_2O_3 and 7.7% Fe_2O_3, and the rest includes mainly Na_2O, MgO, TiO_2 and K_2O. The availability leaching test was conducted following the Dutch standard NEN 7341. The leaching values of the vitrified slag are significantly lower than its original bottom ash. Furthermore, equilibrium phase relations in the Na_2O-CaO-SiO_2 system at less than 50 wt% SiO_2 have been experimentally established at temperatures between 1200°C to 1400°C[16], which was missing in the thermodynamic database.

Zinc Recovery from Industrial Residues

Annually about 10 million tons of zinc metal is produced globally, among which 30% is produced from zinc bearing secondary resources. The zinc residues normally contain large amount of iron oxide which is often associated with zinc oxide in the form of zinc ferrite ($ZnO{\cdot}Fe_2O_3$) or even aluminate zinc spinel gahnite ($ZnO{\cdot}Al_2O_3$). Zinc ferrite, also called franklinite, is one of the main technological barriers for effective zinc recovery in particular for hydrometallurgical processing routes.

A lot of studies could be found in the literature for the treatment of zinc-bearing residues, in particular for EAF steelmaking flue dust. There are generally two types of processes: pyrometallurgical carbothermic reduction, and hydrometallurgical leaching based treatments. Waelz kiln process [17] is a common industrial process based on carbothermic reduction of zinc oxide and ferrite. and Ausmelt process applies Top Submerged Lancing (TSL) technology [18]. Some other developments were also reported such as INMETCO Process using Rotary hearth Furnace (RHF) [19], PISO Process using continuous channel furnace [19], to name a few. All pyrometallurgical processes produce upgraded zinc oxide as raw materials for zinc smelters. Some reported hydrometallurgical processes are EZINEX Process [20] based on NH_4Cl leaching, Modified ZINCEX Process (MZP) [21] based on H_2SO_4 leaching and solvent extraction. Both EZINEX and MZP processes produce metallic zinc through final electrowinning. However, all leaching agents such as acid, alkaline and ammonia or ammonium salt (such as H_2SO_4, NaOH, NH_4OH, NH_4Cl) are not possible to dissolve zinc ferrite ($ZnO{\cdot}Fe_2O_3$) and gahnite [22]. Concerning the difficulties of zinc ferrite leaching, studies have been reported on breaking down the ferrite into soluble zinc compound through roasting by use of caustic soda [23] and soda ash [24]. However, the reported roasting process is normally followed by acid leaching where removal of iron from acid leach solution is still a large burden. The studies at TU Delft focuses on the combination of roasting and NaOH leaching – zinc electrowinning process. The proposed process is selective for zinc – iron separation during leaching and is likely the best combination for zinc ferrite based residue materials for zinc and iron extraction.

Zinc Recovery from Residues Stored in Rotterdam Harbour

In Rotterdam harbour there is about 4700 tons of zinc containing materials, stored there for about 20 years without proper treatment solutions. These zinc-bearing materials have relatively low zinc content (17.5% and 24.8% in two types), and cannot be used directly in any zinc smelters (both hydrometallurgical and pyrometallurgical) as raw materials without zinc enrichment processing. Because of the presence of chlorine, lead and small amount of other heavy metals, it cannot be simply land-filled. Based on the chemical and structural analyses of the solid materials, as well as the preliminary experimental results from the laboratory tests, various alternative conceptual treatment flow sheets are proposed for in-depth experimental research and development at TU Delft. The synthesized technological routes are further studied through laboratory experiments in order to develop proper operating conditions and parameters which are needed for designing larger scale treatment facilities [25, 26]. Figure 4 shows the raw material (Zinc B), and the recovered metallic zinc through roasting - NaOH leaching - electrowinning route in the laboratory.

Zinc B

Electrolytic Zn

Figure 4. Zinc B from Rotterdam Harbour and produced zinc metal via leaching – electrowinning.

Figure 5 illustrates a general flow sheet, proposed and tested through lab-scale experiments. Based on the concept of soda ash roasting, caustic soda leaching and zinc electrowinning in alkaline solutions. Because of the high iron content in the feed (~20% and ~35%), caustic soda leaching has a great advantage of selective dissolution of zinc oxide, leaving the iron oxide in the leach residues.

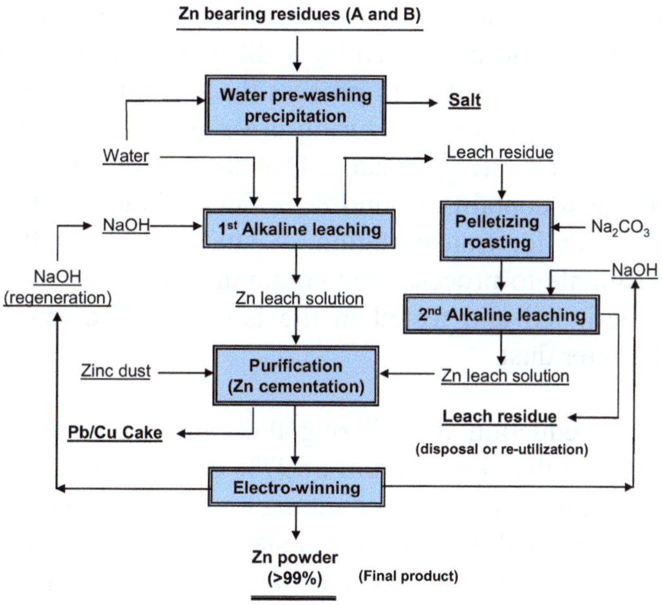

Figure 5. Conceptual flow sheet: combined pyrometallurgical and hydrometallurgical processing of zinc ferrite based solid residues.

According to XRD analysis it is evident that zinc ferrite is a major phase in both residues, in addition to free ZnO. Zinc ferrites are not soluble under normal alkaline and acidic leaching conditions. To achieve sufficient recovery of zinc, roasting of zinc ferrite to convert it into soluble zinc oxides has been investigated. Both caustic soda and soda ash have been tested for roasting, and in the end Na_2CO_3 at 950°C for 2 hours has been used in the roasting experiments. After two step leaching, about 75 – 80% zinc is reported to solution.

The NaOH leach residues contain mainly iron oxide and other minerals, and could be sold to ironmaking industry with minimum upgrading efforts (washing out caustic residues). Other by-products are Cu-Pb cake produced during solution purification, and could be sold to copper or lead smelters as raw materials. Some chloride salts are also generated during water pre-washing as a byproduct, which needs to be crystallized as either by-product or non-hazardous solid for safe disposal.

During electrowinning after impurity (Pb abd Cu) removal by cementation, a current density of 673 A/m^2 was applied with a cell voltage of 2.55 V on average. The specific energy consumption was estimated as 2.33 kWh/kg zinc. A current efficiency of 90% has been reached. At this stage, an electrolytic Zn of 95% purity (+4% Pb) has been obtained. After the overall processing, a total recovery of zinc is about 75%. Further process optimization is needed to increase the conversion rate from zinc ferrite to soluble free zinc oxide, to raise the leaching efficiency of zinc in caustic solution, to maximize the lead and copper removal based on cementation, and to find optimal electrowinning parameters for higher product quality with a lower energy consumption. Finally, adaptation of the developed process to the treatment of EAF steelmaking flue dust will make the technology more attractive to metallurgical industry.

Zinc Recovery from Brass Recycling Filter Dust

The residues generated at secondary brass smelters in the Netherlands are currently being transported for a long distance and processed by a different company abroad. However Dutch brass industry is looking for opportunities to process the secondary products on site, in order to reach a zero footprint on the environment. During brass recycling operation, certain amount of flue dust is generated from its melting furnace. Due to the low boiling point and high vapor pressure of zinc, the filter dust contains around 80% ZnO, 8% C, 4% CuO and 1.5% PbO. This dust is very fine and difficult for transportation. A study at TU Delft was performed on the processing techniques available to process zinc containing residues and from this study several treatment options were designed and tested in the lab [27]. The main focus was put on the recovery of zinc from the filter dust.

First the pyrometallurgical reduction by utilizing the carbon contained in the filter dust was tested at 1200°C. It was found that a maximum of 97% of the zinc could be recovered; a higher recovery at 1200°C was not possible as the remaining zinc was bound to gahnite ($ZnAl_2O_4$). The residue still contains 15% ZnO and did not form a slag. A combined reduction of both the filter dust and melting slag resulted in the recovery of 96.8% of the copper, 99.9% of the zinc (as ZnO) and 64.9% of the lead. The test was performed at 1400°C and it was found that at this temperature the gahnite could be reduced. Also by addition of lime and silica a suitable slag was formed which contained 2.18% CuO and 0.16% ZnO. Both the copper metal and zinc oxide produced had a purity of over 98%.

Hydrometallurgical treatment was also tested by using two different leaching agents, NaOH and H_2SO_4. It was found that the best leaching conditions for NaOH at a liquid solid ratio of 10 were 90°C using 320 g/L NaOH. After 15 minutes the maximum amount of zinc was already leached with 86% leaching recovery. Lead was leached for 68% and copper for 5%. Using 200 g/L H_2SO_4 at 65°C resulted in 96% zinc recovery and 90.5% copper recovery. Both residues contained mainly gahnite which is impossible to leach under atmospheric conditions. After an iron removal step for the acidic solution both leach liquors were purified with Zn dust cementation. A zinc product of 99.7% purity was produced through electrowinning for both solutions. The specific power consumption for the NaOH solution was significantly lower at 2.2 kWh/kg compared to 3.04 kWh/kg zinc produced from H_2SO_4 leach solution. Finally a conceptual flow sheet is designed for both a pyrometallurgical and a (H_2SO_4) hydrometallurgical processing in a brass smelting plant. Figure 6 shows the filter dust and the electrolytic Zn in H_2SO_4 system.

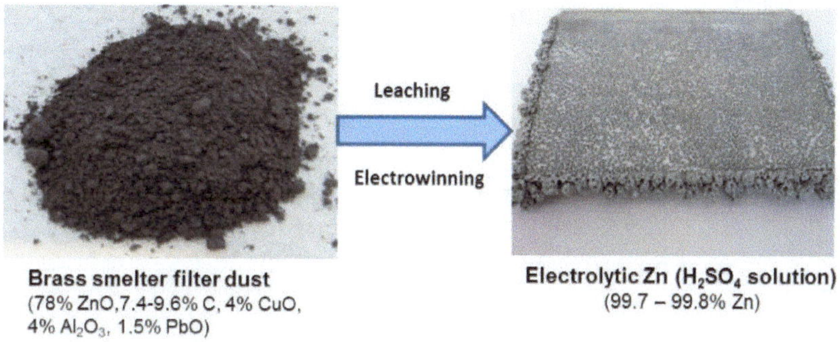

Figure 6. Illustration of the filter dust from brass recycling smelter and the electrolytically produced zinc through H_2SO_4 leaching and electrowinning.

Rare Earth Element (REE) Recovery from EOL Permanent Magnets

NdFeB based permanent magnets are widely used in computer hard disc drives (HDD), consumer electronics, wind turbines and electric-hybrid vehicles. REE-based magnet alone accounts for the largest REE consumption in terms of both volume and market values. The recent global supply shortage of REE has led to increasing incentives to the recovery of rare earth metals from the end-of-life (EoL) consumer electronics and industrial products. This study focuses on the REE recovery from NdFeB magnets in computer hard disc drives. For secure data destruction and difficulties of manual disassembling, shredding is the current industrial practice to process the EoL HDDs. The shredder residues, as a mixture of magnet and ferrous components, are strongly magnetic and extremely difficult for mechanical separations. The REE bearing magnet residues are normally ending up in steel plant as ferrous scrap. The valuable REEs (Nd, Pr and Dy) are lost into smelter slags. To recover the REEs, a combined mechanical and metallurgical processing route is developed at TU Delft, and is illustrated in Figure 7. Some recent experimental results are described below [28,29].

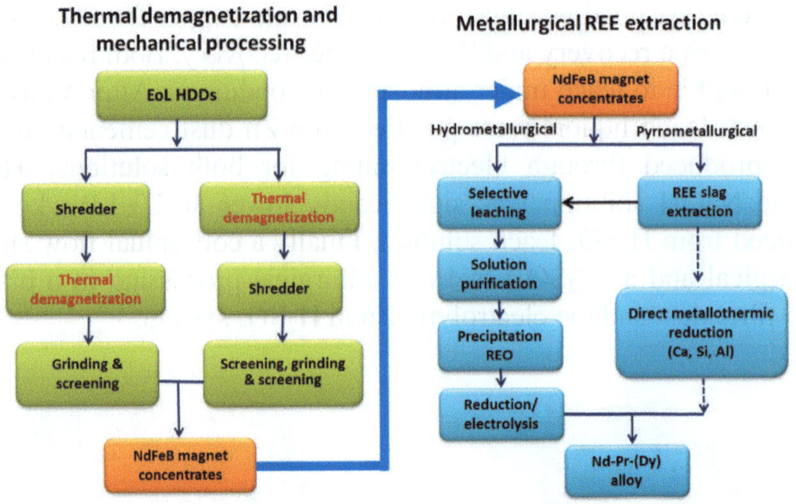

Figure 7. Combined physical and metallurgical processing of HDD shredder residues.

Thermal Demagnetization and Physical Separation

The strong magnetism of the shredder residue is a barrier to the physical separation of the magnet particles from the majority ferrous materials associated with the HDDs. Therefore, a thermal demagnetization and dissociation process has been developed to liberate the NdFeB particles from the resin-bonded bracket and the steel casing attracted by strong magnetic forces. After demagnetization at about 400°C in air atmosphere, a process is developed to separate the NdFeB particles from the steel fractions through grinding and screening, by making use of the brittleness of the sintered NdFeB magnets and the high ductility of the steel and other entrapped metals. After screening, the finer fractions are principally NdFeB magnet particles, and the REE concentration (Nd, Pr & Tb) is significantly upgraded from originally 2.3 wt% in the HDDs to an average of 16.6 wt % in the final fine fraction (< 1 mm) mixture which accounts for 63% of the mass of the shredder residue. The upgraded magnet scrap, the magnet concentrates, can be used as high quality secondary raw material for further metallurgical extraction.

Metallurgical Extraction of REEs

After demagnetization and physical upgrading, the recovery of rare-earths was subsequently tested with both direct hydrometallurgical and combined pyro- and hydrometallurgical processing routes. Almost 100% recovery of rare-earths (with less than 0.6 wt.% iron impurities) was accomplished by pyrometallurgical slag extraction with the two designed slag systems ($CaO-SiO_2-Al_2O_3$ and $CaO-CaF_2$) at 1500°C. At the same time all other metallic constituents are recovered in an iron-based alloy with a total 99% Fe recovery, and this ferrous alloy could be sold as steelmaking scrap with a total of 5.5 – 7.5% other elements. However, relatively low slag leaching recovery (around 60%) with either H_2SO_4 or HCl is obtained for REEs, which requires further optimization of the slag system, microstructure of REEs in the slag, and leaching conditions.

In the direct hydrometallurgical route, both rare earths and iron are first leached with 2 M H_2SO_4, showing high recovery (92-99% for REEs and 85% for Fe, respectively) within 8 hours, though selective leaching of REEs is preferred. Results at different temperatures (room temperature, 50°C) confirm that a lower temperature such as room temperature is preferred for REE leaching. REE-iron separation is later achieved by the addition of a precipitating agent NaOH, for selective precipitation of a rare-earths and sodium sulfate double-salt ($R_2Na_2(SO_4)_3 \cdot 2H_2O$, where R=Nd, Pr, Tb and Dy), provided that the pH is low enough (between 1-2) to avoid co-precipitation of iron. Using this method, optimal conditions allowed for high recovery of 98% rare-earth content and a high purity (98.4%) double-salt. The double-salt serves as a marketable intermediate product that can be converted to either RF_3 or R_2O_3, which are common raw materials in the primary production of REEs and the produced REEs can be used for magnets manufacturing again or other REE- based applications. Figure 8 illustrates the raw materials and intermediate and final products from this research.

Figure 8. From EoL HDDs to REE salt for metallurgical refining.

Summary

Broad opportunity has been explored for metals recovery from industrial wastes and EoL consumer and industrial products. Proper and efficient recovery of metals from these secondary resources will mitigate the problems of metals and minerals scarcity. Several research examples have been illustrated for addressing new challenges for efficient recovery of metal from large variety of waste streams and EoL products.

- The first example proves that much more metals could be further extracted from MSW incinerator bottoms ashes. Using metal sulfide in particular pyrite mineral or ZnS sludge, good separation of copper from iron could be achieved from the metal alloys generated with the ash vitrification. Two marketable raw materials, secondary copper concentrates and steel scrap, could be produced.
- The second example indicates that by using soda ash (Na_2CO_3) roasting, zinc ferrite minerals in the zinc bearing residues could be decomposed for more efficient hydrometallurgical processing. With alkaline (NaOH) leaching, more complete dissolution of zinc and good separation from iron oxide can be approached. After electrowinning in alkaline solutions metallic zinc can be produced with low energy consumption, and effective regeneration of NaOH as leaching agent.
- The last example shows that EoL HDD shredder residue can be a good source of REE (Nd, Pr, Dy, Tb), and hydrometallurgical (or combined pyro- and hydrometallurgical) processing will offer a good opportunity for REE recovery and separation from the major ferrous components.

These examples clearly indicated that the metal recovery from industrial solid wastes and EoL consumer products can make significant contribution to the materials sustainability.

Acknowledgments

The author thanks the contributions from all the current and past graduate (MSc. and PhD) students, and scientific staff of the group of Metals Production, Refining and Recycling at TU Delft for their contribution to the above illustrated research projects and activities. The financial support from governmental and industrial partners (KNAW and Van Gansewinkel Groep) is greatly acknowledged.

References

1. A.M. Diederen, "Metal minerals scarcity: A call for managed austerity and the elements of hope," *The Oil Drum: Europe* (http://www.theoildrum.com/node/5239, March 2009).
2. World steel association (http://www.worldsteel.org, accessed in 2011).
3. USGS, *Minerals Yearbook: Volume I.-- Metals and Minerals* (http://minerals.usgs.gov, accessed in 2012).
4. ICSG, The World Copper Factbook 2012. (http://www.icsg.org, accessed in 2012).
5. World Aluminium (http://www.world-aluminium.org, 2010).
6. C. R. Risopatron, "Global Copper Market Drivers 2010-2015" (Presentation at "Copper 2010, Hamburg, Germany, June 6-10, 2010).
7. R.J. Lifset, et al., "Where Has All the Copper Gone: The Stocks and Flows Project, Part 1," *JOM*, 54 (10) (2002), 21-26.
8. International Zinc Association (IZA) (http://www.zinc.org, accessed in 2012).
9. Y. Yang and C.E.M. Meskers, "Urban mining: contribution to the mitigation of resource scarcity,". *73e Jaarboek van de Mijbouwkundige Vereeniging* (Delft University of Technology, the Netherlands, 2011), 178-190.
10. R.B. Gordon, et al., "Where Is All the Zinc Going: The Stocks and Flows Project, Part 2," *JOM*, 56 (1) (2004), 24-29.
11. R.J. Lifset, et al., "Where Has All the Copper Gone: The Stocks and Flows Project, Part 1," *JOM*, 54 (10) (2002), 21-26.
12. J. Johnson, R.obert Gordon, and T. Graedel, "Silver Cycles: The Stocks and Flows Project, Part 3," *JOM*, 58 (2) (2006), 34-38.
13. B. K. Reck and R.B. Gordon, "Nickel and Chromium Cycles: Stocks and Flows Project Part IV," *JOM*, 60 (7) (2008), 55-59.
14. Y. Xiao et al., "Vitrification of Bottom Ash from a Municipal Solid Waste Incinerator," *Waste Management*, 28 (2008), 1020-1026.
15. Y. Yang, et al., "Metal Recovery and Refining from MSW Incineration Bottom Ash" *Proceedings of Global Symposium on Recycling, Waste Treatment and Clean Technology, REWAS 2008, eds. Dr. B. Mishra, Dr. C. Ludwig, and Dr. S. Das* (Warrendale, PA: TMS, 2008), 1285-1294.
16. Z. Zhang et al., "Phase Equilibria in the Na_2O–CaO–SiO_2 System," *Journal of the American Ceramic Society*, 94 (9) (2011), 3088-3093.
17. E. Saage and U. Hasche, "Optimization of the Waelz Process at the B.U.S. Zinkrecycling Freiberg GmbH," *World of Metallurgy – Erzmetall*, 57 (3) (2004), 138-142

18. J. Sofra and A.Heinz, “Effective Treatment of Zinc Bearing Dusts & Residues – Solution Ausmelt Technology,” *Proceedings of EMC 2003*, ed. Ulrich Waschki (Clausthal-Zellerfeld: GDMB Medienverlag, 2000), 491-505.
19. R.S. Rao, *Resource Recovery and Recycling From Metallurgical Wastes* (Amsterdam: Elsevier, 2006).
20. M.Olper, “Zinc Extraction from EAF Dust with EZINEX® Process,” *Third International Symposium on Recycling of Metals and Engineered Materials*, eds. P.B. Queneau and R.D. Peterson (Point Clear, Alabama: TMS, 1995), 563-577.
21. G. Diaz, D. Martin and C. Lombera, “Zinc Recycling through the Modified ZINCEX Process”. *Third International Symposium on Recycling of Metals and Engineered Materials*, eds. P.B. Queneau and R.D. Peterson (Point Clear, Alabama: TMS, 1995), 623-637.
22. C. Caravaca, A. Cobo and F. J. Alguacil, “Considerations about the recycling of EAF flue dusts as source for the recovery of valuable metals by hydrometallurgical processes”. *Resources, Conservation and Recycling*, 10 (1-2) (1994), 35-41.
23. D. K. Xia, and C. A. Pickles, “Caustic Roasting and Leaching of Electric Arc Furnace Dust,” *Canadian Metallurgical Quarterly*, 38(3) (1999), 175-186.
24. P.C. Holloway, T.H. Etsell, and A.L. Murland, “Roasting of La Oroya Zinc Ferrite with Na_2CO_3,” *Metallurgical and Materials Transactions B*, 38(5) (2007), 781-791.
25. Y. Yang, D. Kemperman and Y. Xiao, “Zinc Recovery from Zinc Ferrite - Bearing Industrial Residues,” *3rd International Conference on Engineering for Waste and Biomass Valorisation, eds. A. Nzihou and H. Liu* (Route de Teillet, France: Ecole des Mines d’Albi-Carmaux, 2010), 9 p.
26. D. Kemperman: Metallurgical processing of zinc bearing residues (MSc. Thesis, Delft University of Technology, 2010).
27. D. Brouwer: Smart Processing of Brass Smelter Residues (MSc. Thesis, Delft University of Technology, 2010).
28. Y. Yang, S. Abrahami and Y. Xiao, “Recovery of rare earth metals from NdFeB magnet residues of shredded hard disc drives” (Paper presented at ECI conference on Rare Earth Minerals/Metals - Sustainable Technology for the Future. San Diego, California, August 12-17, 2012).
29. S. Abrahami, “Rare-Earth Recovery from Post-consumer HDD scrap” (MSc. Thesis, Delft University of Technology, 2012).

REWAS 2013
Enabling Materials Resource Sustainability

Enabling Sustainability through Systems Modelling and Design

Session Chairs

Anne Kvithyld
Daniel Mueller

REWAS 2013: Enabling Materials Resource Sustainability
Edited by: Anne Kvithyld, Christina Meskers, Randolph Kirchain, Gregory Krumdick, Brajendra Mishra, Markus Reuter, Cong Wang, Mark Schlesinger, Gabrielle Gaustad, Diana Lados, and Jeffrey Spangenberger
TMS (The Minerals, Metals & Materials Society), 2013

ASSESSING THE CRITICALITY OF METALS

T. E. Graedel, E.M. Harper, and N.Nassar
Center for Industrial Ecology, Yale University

Extended Abstract

Today's technology employs virtually the entire periodic table. The stocks and flows of the major metals, essentially unknown a decade ago, are now reasonably well quantified. Those cycles can be used to generate on overview of societal metal use. A key issue is whether scarcity implies long-term shortages or unavailability. To address this issue, a detailed methodology for generating a reliable assessment of the criticality of metals has been completed, making extensive use of peer-reviewed datasets and analytical approaches from the fields of geology, international trade, political science, and international policy, among others. This criticality evaluation has three components – Supply Risk, Environmental Implications, and Vulnerability to Supply Restriction, each of which is itself the composite of several metrics, as shown below.

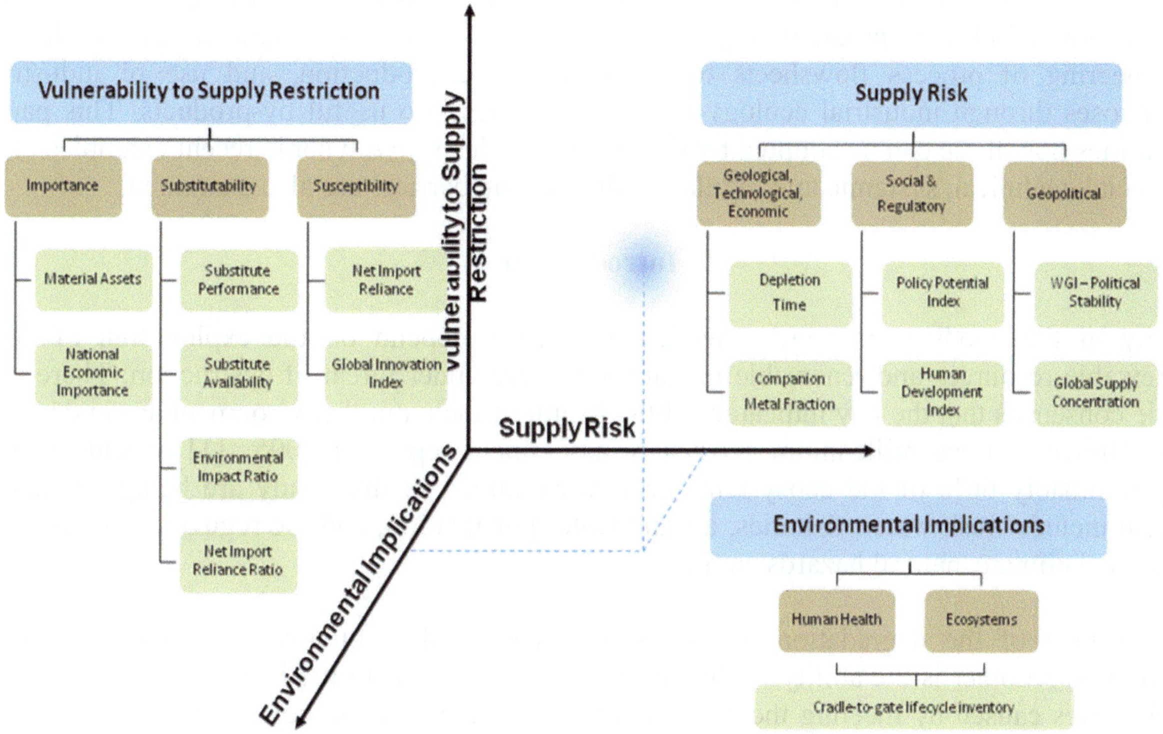

The results of applying this methodology to a number of widely used metals have generated loci in "criticality space" that reflect the degree to which metals in the ores of copper, nickel, rare earths, and several others are more or less critical than elements that could perform similar functions will be presented and discussed in some detail. The products of this research are expected to provide guidance for materials choice in product design and development, recycling potential, and the consideration of substitute materials in modern technology.

REWAS 2013: Enabling Materials Resource Sustainabili
Edited by: Anne Kvithyld, Christina Meskers, Randolph Kirchain, Gregory Krumdick, Brajendra Mishr
Markus Reuter, Cong Wang, Mark Schlesinger, Gabrielle Gaustad, Diana Lados, and Jeffrey Spangenberg
TMS (The Minerals, Metals & Materials Society), 201

Towards zero waste production in the minerals and metals sector

William J Rankin[1]

[1]CSIRO Process Science and Engineering, Box 312 Clayton South, VIC 3169, Australia

Keywords: primary metals, mineral wastes, cleaner production, industrial ecology

Abstract

The production of mineral and metal commodities results in large quantities of wastes (solid, liquid and gaseous) at each stage of value-adding – from mining to manufacturing. Waste production (both consumer and non-consumer) is a major contributor to environmental degradation. Approaches to waste management in the minerals industry are largely 'after the event'. These have moved progressively from foul-and-flee to dilute-and-disperse to end end-of-pipe treatments. There is now a need to move to approaches which aim to reduce or eliminate waste production at source. Modern waste management strategies include the application of cleaner production principles, the use of wastes as raw materials, the re-engineering of process flowsheets to minimise waste production, and use of industrial symbioses through industrial ecology to convert wastes into useful by-products. This paper examines how these can be adopted by the minerals industry, with some recent examples. The financial, technical, systemic and regulatory drivers and barriers are also examined.

Introduction

Many of the goods and services needed by society depend on the exploitation of non-renewable resources and renewable resources that are under threat of destruction. There is a wide consensus that the key indicators of the health of the natural environment are in decline. The United Nations millennium ecosystem assessment report of 2005 [1] concluded that approximately 60% of the ecosystem services examined in that study are being degraded. These include fresh water, fisheries, air and water purification, and the regulation of regional and local climate, natural hazards, and pests.

One aspect of the degradation of the environment is the potential depletion of natural resources. Another is the ability of the environment to cope with the impact of the emissions and wastes caused by meeting the ever-increasing material and energy needs of society. For example, in the United States more than 21 billion tonnes of resources of all kinds are consumed every year – about 80 tonnes per person per year, consisting of 76 tonnes of non-renewable resources and 4 tonnes of biomass [2]. Only 19 tonnes of these resources are used as direct inputs to processing; the rest is waste. Further quantities of wastes are produced during processing of the direct inputs and during the use and ultimate recycling or disposal of the products made from them. Other developed countries have similar patterns of consumption. The per capita consumption of materials in the European Union in the latter half of the 1990s, for example, was 49 tonnes per year [3].

This paper considers possible responses to the challenges posed by wastes from the mining and mineral processing industry.

Wastes and the waste hierarchy

A waste is commonly considered to be a thing or substance that has been discarded, or which will be discarded, and eventually sent to landfill or other disposal site. This, however, fails to recognise the possibility of it being useful again. Thus a better definition is: Something is a waste when it has no present use.

Waste reduction and elimination are not new concepts and many strategies for them have long been known. Our ancestors rarely wasted anything. Many products were remanufactured or reused and materials were recycled before being ultimately discarded. The present attitude towards wastes largely resulted from the increasing availability of cheap energy and materials following the industrial revolution. There has been a renewed focus on wastes only since the ever-increasing quantities of waste from energy and materials production and consumption began to have wide-scale environmental impacts, in the latter half of the 20th century.

The conventional waste hierarchy, reduce, reuse, recycle (the three Rs) lists waste management strategies in decreasing order of desirability. Thus, the most desirable strategy is to reduce the quantity of materials and wastes associated with a product, and to use fewer products. The next most desirable strategy is to reuse a product, and only as a last resort recycle the materials comprising the product. They focus on the reduction or minimisation of waste rather than on the elimination of wastes, and seem to have been developed with manufactured products, building and construction products, and domestic waste in mind. Their focus is on things that have had a useful life rather than on things or by-products that have not had a previous use, such as mining and processing wastes.

The focus on minimising waste started shifting towards eliminating wastes at source as a result of the European Union Council Directive 91/156/EEC of 1991 which established the hierarchy: waste prevention; recovery; safe disposal. Importantly, it addressed things that have had a useful life and by-products, such as mining and processing wastes, that have had no previous use. After several revisions, Directive 2008/98/EC of 2008 established the following hierarchy:

- prevention;
- preparation for reuse;
- recycling;
- other recovery (e.g. energy recovery);
- disposal.

Directive 2008/98/EC excludes wastes resulting from the prospecting, extraction, treatment and storage of mineral resources and the working of quarries, which had been included in earlier versions, since these are now covered specifically by Directive 2006/21/ EC which follows a similar hierarchy.

Strategies for minimising and eliminating wastes

Strategies for minimising and eliminating wastes in the production of mineral and metal commodities can usefully be grouped as follows:

- cleaner production;
- use of waste as raw materials;
- waste reduction through process re-engineering;
- industrial ecology.

These are arranged in order of increasing capacity to minimise or eliminate wastes, and hence form a hierarchy. This order also correlates with increasing degree of integration into the business of a company and the economy at large. Thus, cleaner production can be implemented at a single operation, whereas industrial ecology requires integration across companies and across industry sectors. These strategies are not mutually exclusive alternatives. There is considerable overlap between them and several strategies may be pursued in parallel.

Figure 1 illustrates the historical trend in approaches to addressing the environmental impact of wastes. Company behaviour has moved in recent decades from complying with regulations to corporate social responsibility and now needs to move to an industrial ecology approach. The drivers for this change have moved from being exclusively profit to include regulations and stakeholders. In parallel, the materials cycle focus has shifted from a focus on products only to by-products as well as products. It now needs to shift to the entire materials cycle and, ultimately, to the entire economy.

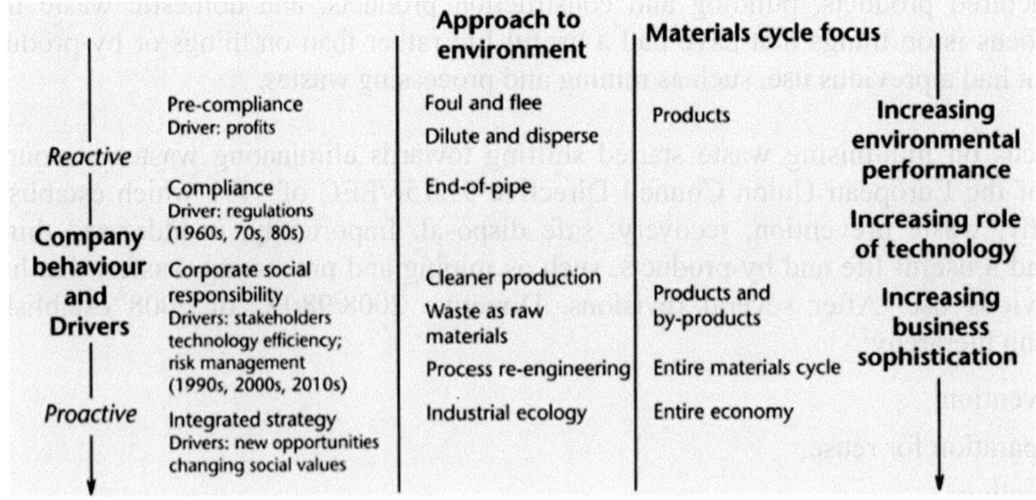

Figure 1. Historical trend in the approaches to the environmental impact of wastes (based on a figure by Giurco and Petrie [4]; with modifications).

Cleaner production

Cleaner production, also sometimes called pollution prevention, is the continuous application of an integrated preventive environmental strategy to processes, products, and services to increase overall efficiency, and reduce risks to humans and the environment [5]. Cleaner production can be applied to the processes used in any industry, to the products themselves and to various services provided in society. For production processes, cleaner production involves one or a combination of the following:

- conserving raw materials, water and energy;
- eliminating toxic and dangerous raw materials;
- reducing the quantity and toxicity of emissions and wastes at source during the production process.

Waste is considered as a product with negative economic value. Each action to reduce consumption of raw materials and energy, and prevent or reduce generation of waste, can increase productivity and bring financial benefits to an enterprise. The similarities between eco-efficiency and clearer production are numerous. Eco-efficiency includes cleaner production concepts and captures the idea of reducing waste through process change rather than end-of-pipe approaches. Like cleaner production, eco-efficiency goes beyond pollution reduction by emphasising value creation for the business and society at large.

The key difference between pollution control and cleaner production is that pollution control is an after-the-event, react-and-treat approach while cleaner production looks forward and attempts to anticipate and prevent. Cleaner production aims to minimise or avoid practices such as waste treatment (including stabilisation, encapsulation and detoxification), waste dilution to comply with regulations (e.g. releasing contaminated water into rivers or streams during high flow periods, blending arsenic-containing fumes with flotation tailings), and transferring hazardous or toxic substances from one medium to another (e.g. wet-scrubbing gases then disposing of the contaminants as waste water). Implementation of cleaner production requires a structured, holistic, common-sense approach using systems and people to both reduce environmental impact and improve the overall company performance [6].

Wastes as raw materials

Although processing of wastes could be considered an end-of-pipe environmental solution and therefore not particularly innovative, this is far from the truth. Many of the by-products from producing mineral and metal commodities, which are now considered wastes, contain much of value. Technologies for extracting that value are often technically sophisticated. There are several incentives for considering mining and processing waste as a raw material.

- A mining and processing waste is essentially 'free' (since it has already been mined) and is often in a form suitable for further processing (since it may have been crushed and ground).

- Use of a waste reduces the demand for new mined material since the product produced replaces product that would otherwise have to be mined.
- Use of a waste to make a saleable product reduces the volume of waste that must be stored or disposed of. This saves on storage and disposal costs and could lead to a reduction in the environmental impact. This is particularly advantageous for large-volume wastes such as beneficiation tailings.
- Producing a saleable product adds another source of income for a company.

Use of mining and processing wastes as raw materials is not a new concept and some wastes from mining and processing are already treated to produce saleable products. Sulfur dioxide in smelter gases is routinely used to make sulfuric acid. Slags from iron and steel production are used for making aggregate materials and as a raw material in cement manufacturing. However, mining and processing wastes are a largely untapped resource and there are many potential applications. An advance would be to optimise processes so that all output streams are useful products, rather than optimising processes around the principal product. In this way, all outputs would be optimised (in terms of composition and morphology) to maximise their effectiveness as inputs for further value-adding or, as a last resort, for disposal.

While utilisation rather than disposal of mining and processing wastes is preferable, it must be recognised that not all, or even most, mining-related wastes can be used productively. The quantities of mining wastes are so large that there are insufficient bulk applications, even in construction projects, to use significant quantities. This is particularly so for rock waste and overburden from mining. Furthermore, many mines are located in remote and/or sparsely populated areas and the transportation of low-value construction products (sand, aggregate) to populated areas for use in infrastructure projects is uneconomic. Hence, the focus is necessarily on wastes from mines and processing operations close to populated areas and/or on higher-value products which can be transported economically over long distances.

Some examples to illustrate the possibilities include utilisation of red mud from Bayer processing [7,8], spent pot lining from aluminium smelting [9], fly ash from power generation, slags from smelting operations and the use of wastes in geopolymer concrete [10]. There are many other examples in the literature.

<u>Process re-engineering</u>

The strategy of waste reduction through re-engineering aims to minimise the quantity of waste produced or to produce a by-product in a form that can be used more readily. This involves some process modification; often it may be necessary to completely redesign the flowsheet. There are three broad approaches:

- flowsheet simplification;
- use of novel equipment;
- use of novel processing conditions.

Flowsheet simplification involves the removal or combination of stages to reduce the overall number of stages required to produce a mineral or metal commodity. This reduces the amount of transport, handling and physical processing of material and can potentially reduce the amount of chemical processing. This saves energy (CO_2 emissions) and reduces the quantities of other wastes. The use of novel reactors or other equipment involves utilising unique characteristics of a reactor or other item of equipment to do something that was previously not possible, or was very difficult to do. The use of novel processing conditions involves utilising relatively standard reactors and flowsheet configurations with different reagents or processing conditions, such as temperature, pressure or concentration. While one of these approaches often predominates, a technological development usually combines aspects of two or all three. Some examples are listed in Table 1.

Table 1. Some examples of processes which reduce waste production and/or energy consumption.

Flowsheet simplification	Use of novel equipment	Use of novel processing conditions
Heap leaching	Ore sorting [11]	Electrolytic production of iron [12-14]
Finex; HIsmelt iron-making processes	Underground and in-pit pre-concentration [15]	The top gas recycling blast furnace (ULCOS program) [16]
TiRO™ process for Ti metal production [17]	Castrip™ process for thin strip casting of steel [18]	Use of biomass as fuel and reductant in metallurgical processes [19]
	Dry granulation of slag, with heat recovery [20,21]	
	Solar production of aluminium and other metals [22-26]	

Industrial ecology

The term industrial ecology refers to an industrial system that operates much like a natural ecosystem in which materials circulate continuously in a complex web of interactions. While ecosystems produce some wastes (substances that are not recycled), such as fossil fuels and limestone and phosphate deposits, they are largely self-contained and self-sustaining through the constant input of energy from the Sun. In a similar fashion, industrial ecology involves focusing less on the impacts of each industrial activity in isolation and more on the overall impact of all such activities. This means recognising that the industrial system consists of much more than separate stages of extraction, manufacture and disposal, and that the stages are linked across time, distance and economic sectors [27].

The concept of industrial ecology can be understood by considering the simple models of industrial systems in Figure 2. Figure 2(a) shows the familiar flow-through, or open loop, system. Industry takes in new materials and processes them using energy, and generates products and wastes. Both the products and the wastes are external to the boundary of the system. They are considered as externalities and their impacts are borne largely by society as a whole. New materials and energy come from outside the system and the impacts of their production are borne largely outside it. Some recycling may take place through recycling end-of-life products into the manufacturing system.

Figure 2(b) shows an industrial ecosystem. This is not quite a closed loop. New materials and energy still come from outside the system and some wastes still leave it, but products and process wastes remain within the system. Responsibility for products and for process wastes, and for the impacts of their use, is borne within the system. The unusable wastes which leave the system are of three main types:

- wastes generated during the extraction of new materials (e.g. overburden and waste rock from mining);
- wastes that escape from the recycling loop (since there are losses inherent in recovering and recycling materials);
- wastes lost through the use of products (e.g. by being discarded to landfill or incinerated).

Products that are currently in use or being held for recycling, and industrial wastes and other materials which will be reused at some time, constitute a reservoir (or stock) of materials available for use in the future.

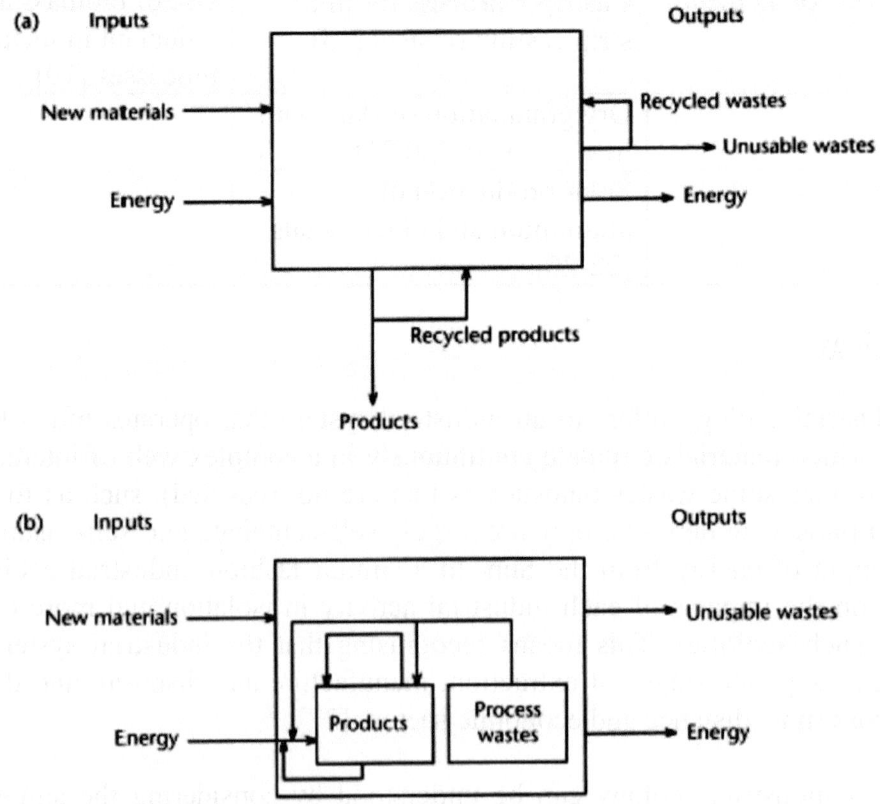

Figure 2. Open and closed material flow systems [27].

The cascading use of energy, which involves using the residual energy in liquids or steam emanating from one process to provide heating, cooling or pressure for another process, and the use of industrial by-products as feedstocks for processes other than the ones that created them, are major characteristics of industrial ecosystems.

There are many eco-industrial parks around the world and they are rapidly growing in number and complexity. Kalundborg (Denmark), Humberside (United Kingdom), Moerdijk and Rotterdam (the Netherlands) and Kwinana (Australia) are frequently cited examples [28]. Kalundborg, is the location of a highly evolved eco-industrial park [29] and is probably the best known example of the implementation of industrial ecology principles. The Kwinana Industrial Area (KIA) in south-western Australia is an eco-industrial park based largely on resource processing and is in a region which combines major resource processing operations with manufacturing, agriculture, aquaculture and recreational activities. The existing regional synergies are arguably more diverse and significant than those reported for other heavily industrialised areas [30]. Forty-seven regional synergies have been identified. Thirty-two of these are by-product synergies and 15 involve shared use of utilities. These initially developed in a largely unplanned way in response to perceived business opportunities and environmental and resource efficiency considerations. A more coordinated approach to identifying and developing linkages was adopted with the formation in 1991 of the Kwinana Industries Council.

Barriers and drivers to reducing and eliminating wastes

Very large financial investments are needed in the minerals sector for major changes in technologies. Established technologies have been refined over many years and operations usually give financial returns long after the capital costs have been depreciated. Introduction of new technologies introduces production risks which can, and often do, prove very costly. Minerals companies are reluctant to introduce new technologies unless it can be done in an incremental way with minimum risk to overall production. Furthermore, the relatively low cost of disposal of mining and mineral processing wastes in most mineral resource rich countries is a disincentive to do anything other than discard them.

Frequently, technical solutions are available or can be developed and implemented, but are too costly, risky or difficult to implement for systemic, organisational or regulatory reasons. There are no shortages of ideas and possible approaches even for radically new technologies, many of which would, if pursued, lead to more efficient or cleaner processes. But developing the right technology at the right time and in such a way that it can be introduced with minimum risk to production has proved a challenge for both technologists and business strategists in the minerals industry.

There is a large degree of entwinement between minerals companies and other industry sectors such as power generation, infrastructure (roads, rail, ports) and suppliers of reagents and other consumables. Technological changes in one area have implications that flow through the entire system. The co-production of multiple products (due to the complex nature of many mineral deposits), and the need to sell these to different markets with differing and changing demand cycles, adds another layer of complication. The range of solid, liquid and gaseous wastes produced and the technologies for managing them is another complication. These combine to make mineral and metal production companies technologically complex and this constrains the changes that can be made easily, cheaply and with little risk.

Companies often perceive themselves as in the business of making a particular commodity, such as steel or aluminium or copper. All other materials created in making their product are seen as wastes to be disposed of as cheaply as possible. Changing the culture of a company so that it perceives the resource in its entirety, not just part of it, as its greatest asset is a challenge which no minerals company has yet come near to tackling.

In most jurisdictions, regulations fail to promote closed loop systems and may actually discourage or prevent it. Of particular concern are regulations relating to the use of wastes or by-products as substitutes for virgin materials, and the assignment of liabilities. In many countries, defining a material as a waste or secondary raw material has consequences for what uses are permitted, what administrative procedures apply to its transport, export and processing, and what costs will be incurred. In some jurisdictions, a company that sells material classified as waste remains liable for any damages that may result from its use, even if it has been reused several times before the damage occurs.

Governments can help companies overcome market failure barriers by creating an environment that encourages adoption of cleaner production principles and that facilitates formation of industrial synergies, for example by:

- developing more appropriate regulations concerning wastes;
- entering into voluntary agreements with companies or industry associations on targets to achieve;
- applying market-based financial instruments such as tax concessions, taxes on emissions, emissions trading schemes and special purpose grants.

Historically, governments have responded to community expectations for better environmental outcomes through regulatory responses. The regulatory approach often prescribes conditions for resource access and use. Mining regulations often specify the maximum allowable level of pollution, minimum requirements for mine-site rehabilitation and the type of management processes that should be used to reduce environmental damage. However, in many situations the regulatory approach has failed to achieve the goals or has proved very expensive.

Market-based instruments (MBIs) for environmental management are increasingly being used for the management of natural resources and the environment. MBIs encourage behaviour through market signals rather than through explicit directives [31]. MBIs have been used successfully to control NOx and SOx emissions in the United States. They are appropriate where regulatory approaches have failed to stop ongoing degradation or where the cost is prohibitive. The focus in applying MBIs is on achieving outcomes through the self-interest of companies and individuals. MBIs have two potential financial advantages over more traditional instruments [32]. They allow different companies to make different adjustments in response to their unique business structures and opportunities; and they provide companies with an incentive to discover cheaper ways to achieve outcomes.

Concluding comments

The environmental challenges posed by non-renewable mineral resource extraction and use need to be addressed within the broader context of sustainability through an integrated strategy for managing the stocks of resources from which materials are obtained, the materials themselves, and the goods, products and infrastructure that contain materials [33].

Understanding within some industry sectors of this challenge, the need to transition to sustainability and the role of industry in this is growing, but much activity remains at the level of rhetoric. *Vision 2050* [34], developed by the World Business Council for Sustainable Development, represents a major step forward. It envisages by 2050 'a planet of around nine billion people, all living well – with enough food, clean water, sanitation, shelter, mobility, education and health to make for wellness – within the limits of what this small, fragile planet can supply and renew, every day'. The proposed pathway to achieve this vision involves fundamental changes in governance structures, economic frameworks, and business and human behaviour. It involves incorporating the cost of externalities (carbon, ecosystem services, water), halving carbon emissions worldwide (based on 2005 levels), and achieving a four- to 10-fold improvement in the use of resources and materials.

The minerals industry, through it's sustainability peak body the International Council on Mining and Metals has adopted many of the principles of sustainability and corporate social responsibility but is yet to fully incorporate sustainability thinking within its business models at all levels. The inevitable closing of the materials cycle and transition to sustainability will create new opportunities for companies prepared to adopt new business models. Minerals companies can help the transition to sustainability by working proactively with their stakeholders, particularly government and government agencies, non-governmental organisations and other business sectors, to implement zero waste strategies.

References

1. *Millennium ecosystem assessment, synthesis report: ecosystems and human well-being: general synthesis* (Washington DC: Island Press, 2005).
2. A.S. Adriaanse, Bringezu, A. Hammond, E. Rodenburg, D. Rogich and H. Schutz: *Resource Flows: The Material Basis of Industrial Economies* (Washington, DC: World Resources Institute, 1997).
3. S. Moll, S. Bringezu and H. Schutz, *Resource Use in European Countries* (Wuppertal, Germany: Wuppertal Institute, 2005).
4. D. Giurco and J.G. Petrie, "Strategies for reducing the carbon footprint of copper: new technologies, more recycling or demand management? *Minerals Engineering*, vol. 20 (2007), 842.

5. *Understanding cleaner production* (UNEP, 2010); www.unep.fr/scp/cp/understanding, 2010
6. *Cleaner production* (Canberra, Australia: Department of Environment and Heritage, Commonwealth of Australia, June 2000).
7. D. Cooling, "Improving the sustainability of residue management practices – Alcoa World Alumina Australia", In *Paste* (eds A Fourie and RJ Newell) (Perth, Australia: Australian Centre for Geomechanics, 2007), 3.
8. S. Jahanshahi, W.J. Bruckard and M.A. Somerville, *"Towards zero waste and sustainable resource processing"*, In *International Conference on Processing and Disposal of Mineral and Industry Waste (PDMIW'07)*, Falmouth, UK, 14-15 June 2007, 1.
9. K. Mansfield, G. Swayn and J. Harpley, "The spent pot lining treatment and fluoride recycling project". In *Green Processing 2002*, (Carlton, VIC, Australia: Australasian Institute of Mining and Metallurgy, 2002), 307.
10. J. Davidovits, *Geopolymer Chemistry and Applications*, 2nd ed. (Saint-Quentin, France: Geopolymer Institute, 2008).
11. N.G. Cutmore and J.E. Eberhardt, "The future of ore sorting in sustainable processing", In *Green Processing 2006* (Carlton, VIC, Australia: Australasian Institute of Mining and Metallurgy, 2006), 287.
12. D.R. Sadoway, "Electrochemical processing in molten salts: from green metals extraction to lunar colonization", *Sixth International Conference on Molten Slags, Fluxes and Salts*, Stockholm and Helsinki, 12-17 June, 2000.
13. A. Cox and D.J. Fray, "Electrolytic reduction of ferric oxide to yield iron and oxygen". In *Energy Technology Perspectives*, (eds G. Reddy, C.K. Belt and E.E. Vidal), (Warrendale, PA: Minerals, Metals and Materials Society, 2009), 77.
14. Y.Y. Xiao and D.J. Fray, "Molten salt electrolysis for sustainable metals extraction and materials processing – a review", *Molten salt electrolysis: Theory, Types and Applications*, (Eds. S. Kuai and J. Meng), (NY, New York: Nova Science Publishers, 2010), 255.
15. A.S. Bamber, B. Klein, R.C. and M.J. Scoble, "Integrated mining, processing and waste disposal systems for reduced energy and operating costs at Xstrata Nickel's Sudbury operations", *Mining Technology*, vol. 117 (3), (2008), 142
16. ULCOS, www.ulcos.org/en/about_ulcos/home/php, 2010.
17. C. Doblin and G.A. Wellwood, "TiRo^{TM} – the development of a new process to produce titanium", In *CHEMECA: Academia and Industry Strengthening the Profession*, New York, NY: Curran Associates, Red Hook, 2007), 280.
18. D.J. Sosinsky, P. Campbell, R. Mahapatra, W. Blejde and F. Fisher, "The Castrip® process – recent developments at Nucor Steel's commercial strip casting plant", *Metallurgist*, 52 (11-12), (2008), 691.
19. J.P. Birat and J. Borlee: "ULCOS, the European steel industry's effort to find breakthrough technologies to cut its CO_2 emissions significantly". In *Carbon Dioxide Reduction Metallurgy* (eds N.R. Neelameggham and R.G. Reddy) (Warrendale, PA: Minerals, Metals and Materials Society, 2008), 59.

20. D. Xie and S. Jahanshahi, "Waste heat recovery from molten slags" (Paper presented at the *4th International Congress on the Science and Technology of Steelmaking (ICS 2008)*, Gifu, Japan, October 2008, 674, Iron and Steel Institute of Japan, 2008.
21. D. Xie, S. Jahanshahi and T. Norgate, "Dry granulation to provide a sustainable option for slag treatment". In *Sustainable Mining* (Carlton, VIC, Australia: Australasian Institute of Mining and Metallurgy, 2010), 22.
22. P. Haueter, T. Seitz and A. Steinfeld, "A new high-flux solar furnace for high-temperature thermochemical research", Journal of Solar Energy Engineering – Transactions of the ASME, 121 (1) (1999), 77.
23. J.P. Murray, "Aluminum production using high-temperature solar process heat", *Solar Energy*, 66 (2) (1999), 133.
24. J.P. Murray, "Solar production of aluminum by direct reduction: preliminary results of two processes", *Journal of Solar Energy Engineering*, 123 (2001), 125.
25. M. Halmann, A. Frei and A. Steinfeld, "Carbothermal reduction of alumina: thermochemical equilibrium calculations and experimental investigation", *Energy*, 32 (12) (2007), 2420.
26. N.R. Neelameggham, "Solar pyrometallurgy – an historical review". *JOM*, February, 2008, 48.
27. R.A. Frosch, "Industrial ecology: adapting technology for a sustainable world *Environment*, 37 (10) (1995), 16.
28. B. Kurip, "Methodology for capturing environmental, social and economic implications of industrial symbiosis in heavy industrial areas" (PhD thesis, Curtin University of Technology, Perth, WA, December 2007); www.kic.org.au/files/biji
29. J.R. Ehrenfeld and N. Gertler, "Industrial ecology in practice: the evolution of interdependence at Kalundborg", *Journal of Industrial Ecology*, 1 (1) (1997), 67.
30. D. van Beers, "Capturing regional synergies in the Kwinana industrial area: 2007 status report" (Report, Centre for Sustainable Resource Processing, Perth, WA, 2007); www.kic.org.au/files/70724_csrp_capturing_regional_synergies_in_the_kia_2007_report_final
31. R.N. Stavins, "Experience with market-based environmental policy instruments" (Discussion Paper 01-58, Resources for the Future, Washington, DC, November 2001).
32. S. Whitten, M. van Bueren and D. Collins, "An overview of market-based instruments and environmental policy in Australia", In *Market-based Tools for Environmental Management: Proceedings of the 6th Annual AARES National Symposium (eds S. Whitten, M. Carter and G. Stoneham)*, 2004; www.ecosystemservicesproject.org/html/publications/docs/MBIs_overview.pdf
33. W.J. Rankin, *Minerals, Metals and Sustainability: Meeting Future Material Needs* (Collingwood, VIC, Australia: CSIRO Publishing, 2011), 367-381.
34. *Vision 2050: The new agenda for business* (Conches-Geneva, Switzerland: World Business Council for Sustainable Development, February 2010); <http://www.wbcsd.org/web/projects/ BZrole/Vision2050-FullReport_Final.pdf>.

REWAS 2013: Enabling Materials Resource Sustainabili
Edited by: Anne Kvithyld, Christina Meskers, Randolph Kirchain, Gregory Krumdick, Brajendra Mishr
Markus Reuter, Cong Wang, Mark Schlesinger, Gabrielle Gaustad, Diana Lados, and Jeffrey Spangenberg
TMS (The Minerals, Metals & Materials Society), 201

SCENARIOS FOR THE DEVELOPMENT AND IMPROVEMENT OF THE LIFE SUPPORT SYSTEMS OF THE ARCTIC ZONE OF RUSSIA

V.A. Tsukerman[1], S.V. Ivanov[1]

Kola Science Centre, Russian Academy of Sciences
Apatity, Murmansk region, Russia

Keywords: The Arctic, life support systems, sustainable development.

Livelihood systems are the key areas of the Arctic and depend on its natural and economic extremality. Development and improvement studies of livelihood systems of the Arctic Zone were carried out taking into account the expected changes in climate that lead to the strengthening of the role of marine factors, economic and social development. The paper discusses various scenarios of livelihood systems of the Arctic zone. Innovation scenario includes primarily the use of scientific and technical achievements and high-tech processes, development of natural resources of the Arctic shelf. Innovation scenario involves close cooperation of natural resources in the Arctic. The inertial scenario is based on conservative estimates of projected growth in key socio-economic indicators of the Arctic zone. To improve livelihood systems of the Russian Arctic there is needed an efficient interaction of all the subjects of economic and social development of the Arctic zone of Russia.

The Arctic can be considered as a research laboratory of the mankind, the territory of discovery, innovation search. The Arctic zone of Russia is a major training ground for basic and applied research in various fields of knowledge.

The Arctic as a whole is characterized by an extremely low density and high dispersion of population. However, the Arctic zone of Russia is featured by the highest urbanization: more than 80% of people live in cities and towns with population exceeding five thousand people.

Natural and economic extreme of arctic zone define an important role of the federal and regional public sector in the local economy.

Seaside character of the Arctic zone determines the originality of innovative economic systems formed here in the forecast period on the shelf and in the coastal zone. Possible climate changes, the shift of economic activities to the shelf zone of the Arctic seas could increase the role of marine factor in economic and social development of the Arctic zone [10].

The Arctic is characterized by extreme climatic conditions, availability of various and significant reserves of mineral and other natural resources, the concentration of economic facilities and social services on limited areas, remoteness and transport inaccessibility, and extreme vulnerability and slow recoverability of natural ecosystems.

The changes that are taking place in the Arctic are systemic, non-linear, rapid and irreversible. The most obvious examples are their unprecedented decline in seasonal minimum extent of sea ice in recent years and changes in the tundra ecosystem, where transitions from donor to acceptor in terms of greenhouse gas emissions are taking place. Nonlinearity is rapidly becoming a common feature of all the changes in the Arctic. These changes are not only irreversible due to the inability to return to baseline in the foreseeable future, but are systemic in nature, as it is about the processes that interact with each other, producing powerful feedbacks and l resulting in large-scale effects [11].

In our world, the increasing role is played by anthropogenic ecosystems in which large-scale biophysical processes are superimposed on equally large scale socio-economic transformations, leading to the emergence of complex dynamic systems with nonlinear, rapid and often irreversible changes [1]. In such circumstances, management priorities are uncertain, and there are many challenges that are very difficult to remove by the conventional approaches (eg, by signing formal multilateral agreements), prevailing today in solving the problems in the international arena. These conditions are nowhere more as evident as in the Arctic.

On the territory of the Russian North there is concentrated two-thirds of the resource potential of the Russian Federation. The region produces over 92% of natural gas, 75% of oil, 80% of gold, 90% of copper and nickel, significant amounts of other minerals. In the depths of the northern lands there lie 80% of the coal reserves in the country, which, according to the experts, will last for 800 years. Also the share of the North in the national production is 100% of diamonds, cobalt, platinum, molybdenum, apatite concentrate, as well as half of timber and fish products. Northern raw materials and fuel annually bring 25-30 billion dollars to the country's budget [12].

Effective economic coordination in the Arctic involves continuous negotiations between defense and economic activities: the use of military infrastructure for socio-economic development and resource development of the Arctic zone, accounting for the defense interests of economic and community activities, and other measures.

There are several possible development scenarios for the Arctic zone. *The inertial scenario* involves prolonging the currently acting trends in the key economic sectors in the Arctic, and is based on conservative estimates of projected growth of the key indicators of the Arctic zone.

The innovative scenario of the Arctic zone development involves close collaboration of polar countries in the joint development of large deposits of the Russian Arctic shelf and therefore much more rapid rates of its development than the inertial scenario. This scenario is based on optimistic estimates of developing the key industries and sectors of the Arctic economy including implementation of major investment projects in the energy sector and transport infrastructure.

The characteristic features of the innovation scenario are consistent realization of the existing competitive advantages based on the rational use of natural resources potential of the Arctic territories, a manifestation of a new quality of economic growth, based on the impact of new technologies in the various sectors of the economy and the rapid development of information and communication sector and sub-sector of the Arctic intellectual services.

Russia has the world's largest offshore area, a promising for oil and gas. The offshore oil and gas fields in difficult ice conditions is a high-tech process that makes the innovative development of sectors and activities involved the creation of reconnaissance, exploration, production, transportation and processing of oil and gas. Therefore, the study and development of resources of the Arctic continental shelf will catalyze the transition of the Arctic and the Russian economy to innovative development.

The Arctic transportation system includes the Northern Sea Route (NSR) as the historically single national transport link, complex of sea vehicles and river fleet, aviation, pipeline, rail and road transport and coastal infrastructure (harbors, navigation, hydrographic and meteorological support and communication).

The main focus of efforts to modernize the Arctic transport system in the next 10-15 years will be to the qualitatively change of the very nature of the polar navigation: turn (including constructive using of a global climate change) moving in the North Sea way to routine procedure with small risk.

The Northern Sea Route is closely related to the innovation process in the Arctic: all the innovations being introduced here will inevitably interact with other sectors of the economy of the Arctic and are involved in their modernization. This is a key specificity Arctic innovations aimed primarily at overcoming distances, overcoming space.

In the future, the Russian Arctic will become a place of pilot testing and implementation of new off-road transport. Rapid progress will occur in the implementation of wireless, satellite-based communication in remote sparsely populated Arctic settlements. Information communication technology will be introduced into the production processes in all large and medium-sized corporate structure of the Arctic zone, which will help improve the efficiency of the basic sector of the Arctic economy.

Significant changes will take place in another critical area of the Arctic life support – small-scale power plants. Wind farms on the coast, geothermal power plants new thermoelectric plants, steam and gas turbines will be set up [5].

The structural elements of the innovation infrastructure will ensure strengthening innovation in key resource complex of Arctic economy and in the field of life support [6].

The last two decades in the economy of the Arctic and northern areas of Russia there are have been major structural changes, which are primarily associated with a significant increase in the role of the service sector. Its share of total employment has increased everywhere, which can be seen not only as a symptom of the real sector collapse crisis of the economy, but also as the beginning of the process of long-term post-industrial transformation - the transition from the economy, the core of which were industrial production, oriented on large amounts of homogeneous production (in the North and in the Arctic - for extraction of large amounts of natural resources) to an economy in which the central role is played by knowledge-intensive service industry (production service) which determines the degree of innovation and the competitiveness of the manufacturing process.

In the future rates of mining structures Arctic production service will be superior to all other sectors of the Arctic economy, both in terms of the generated new jobs here, and on the added value created here.

Due to the nature of the economy across the Arctic, the Arctic production service inevitably has natural resource character. It clearly segregate two areas - subsector studying the environment properties, the climate, the Arctic landscape and subsector studying natural resources of Arctic land and water.

According to the concept of long-term socio-economic development of the Russian Federation up to 2020, a new model of the spatial development of Russia is based on the principles of polycentricism (multipolar) of regional growth centers, reduction of regional disparities (inter-regional and intra-regional differentiation in the level and quality of life), formation of new regional and interregional priority development zones, new urban centers and regional industrial clusters - sets of interrelated competitive high-tech industries on the territory, aimed at deep processing of raw materials and energy to ensure the development of the previously uninhabited areas.

Important for the Arctic zone development will be the effect of formation of urban agglomerations associated with the formation of transport complexes, logistics centers, data nodes, education and innovation infrastructure [9].

The complexity of natural processes in the Arctic areas of Russia requires modernization framework for the management of Arctic regions and municipalities in Russia at both the federal and the regional and municipal levels. Due to the very rapid climatic and socio-economic changes, the Arctic will control maneuvers among the extremely high risks [4].

To speed up the transition to the strategic goals and objectives of the Russian Arctic it is need an effective, conflict-free interaction of all economic and social development Arctic zone of Russia.

This involves the development of existing, forming new partnerships Arctic. Partnerships at different levels, different types and duration play a crucial role of the Arctic economy.

There is an ongoing intensive process of state registration of Arctic policy. As happened many times in the country's history, the activation of the state machine after realizing there was a real threat of the final loss country's leadership position of Russia in the Arctic.

Among the most significant challenges include environmental security Arctic territories. It must be recognized that almost all ecological problems, except the trans boundary transfer of pollutants into the Arctic from Southeast Asia, are the result of Soviet and modern Russian socio-economic policy.

Environmental issues are closely connected with the problematic state of the economy and social sphere of the Arctic territories. In this sense, the arctic regions are not much different from many of the Russian territories, located far away from the Russian Arctic Zone.

Until recently, it was decided to link environmental concerns of some areas in the Arctic exclusively with industrial development and complication of ecological conditions in different parts of the Arctic only in connection with the level of involvement of the industrial turnover. However as a vital link in the full range of problems accumulated in the Russian Arctic issues should recognize the organizational and legal regulation of the State of the Arctic Policy.

The Russian Arctic for a long period is considered as the region based on the use of nuclear power plants in order to obtain the power (NPP) and thermal (APEC) for the needs of cities and ports. The state's investment in the project of floating nuclear power plants with plans of their placement in the Arctic will require solutions to the issues of radiation safety of these facilities. The limited radio ecological information for the Arctic region is known, in spite of the presence in Russia pioneer research in radiation hygiene and radioecology dedicated for migration of artificial radionuclides (IRN) in the chain lichen-reindeer-man [7]. Appeal to this problem due to the high sensitivity of terrestrial and aquatic ecosystems to the impact of radiation factor [3].

The phenomenon of increased radiation dose to the herders mainly associated with environmental causes. Their analysis is important to evaluate the stability limits of natural systems of the North to the pollution by most dangerous radiological IRN.

As a result of intensive management in the Russian Arctic and adjacent areas the Arctic environment is under intense influence (including the expense of pollutants in the cross border transfer), resulting in the growing degradation of arctic ecosystems. Rise and development of severe weather, frost-geomorphological, ice and other adverse natural processes related to climate change contribute to the strengthening of such negative phenomena.

A particular problem is the potential contamination of the Russian Arctic, man-made radionuclides. The region has large facilities nuclear legacy associated with the activities of military and civilian nuclear fleet, and other radiation-hazardous objects.

At the same time despite the difficult climatic conditions of the Russian Arctic the socio-economic development of the Russian Federation in the medium and long term will be closely linked to the development of natural resources in the Arctic. In accordance with the "Principles of State Policy of the Russian Federation in the Arctic for the period up to 2020 and beyond" the use of the Russian Arctic as a strategic resource base is a fundamental national interest. In this case, the development of natural resources in the Russian Arctic should not lead to environmental degradation [8]. Furthermore, an activity must be accompanied by the elimination of accumulated environmental damage, rehabilitation of degraded ecosystems. In accordance with the "Principles of State Policy of the Russian Federation in the Arctic for the period up to 2020 and beyond" to the basic national interests of Russia in the Russian Arctic is referred "saving and providing protection of the Arctic ecosystems." One of the main purposes of a state policy of the Russian Federation in the Arctic is "the preservation and protection of the Arctic environment, the environmental consequences of economic activity in the face of increasing economic activity and global climate change." Finally, the "basic measures for the implementation of the state policy in the field of environmental security in the Arctic zone of the Russian Federation are: the establishment of special regimes of natural resources and environmental protection in the Arctic zone of the Russian Federation, including the monitoring of pollution, restoration of natural landscapes, disposal of toxic industrial wastes, provision of chemical safety, especially in areas where the population is compact [3].

Currently existing national environmental regulatory framework is about 40 federal laws, regulations, and about 1200 orders by the Federal Government of the ministries and departments. They apply to the whole territory of the Russian Federation, including the Russian Arctic. Therefore the lack of effectiveness of the system of environmental management and environmental safety in the Russian Arctic is fundamentally a consequence of the shortcomings of these documents [14]. However, for the Russian Arctic, a situation exacerbated by the fact that most of these documents does not account for the specifics of its climatic conditions. As a consequence, for economic activity in the Russian Arctic are set essentially the same environmental requirements as for other areas much less vulnerable to human impacts and much more easily compensate for harm suffered. Demonstration seriously the country's leadership to the problems existing in the Russian Arctic is the adoption of a number of strategic documents. Unfortunately, contained in these guidelines and requirements for the business activities are practically not implemented in specific legislative acts.

To correct these deficiencies and establish a fully-fledged legal and institutional framework that is capable to solve environmental problems of the Russian Arctic, without prejudice the economic interests of Russia, requires deliberate and consistent work both to improve the existing legislation and the development of new regulations [2].

Increasing human pressure on the natural environment of the Russian Arctic, including in the areas of passing the NSR, sets a task of creating an appropriate legal framework to ensure the safe and efficient sailing as the navigation and technical as well as from an environmental point of view. For this purpose, it seems appropriate to confirm the status of NSR as a national historical Russian transport communication in the Arctic. At the federal law level it is necessary to define the basic requirements for the admission of vessels to be used by NSR, especially sailing ships along the NSR, the ways and means to ensure the prevention of marine pollution from vessels in ice-covered areas. This should be taken into account requirements for the protection of the marine environment from pollution contained in the UN Convention on the Law of the Sea in 1982, according to which (Articles 211, 234, 235) a coastal State shall adopt laws and regulations to prevent and control pollution of the marine environment and the International Convention on Liability and Compensation in Connection with the Carriage of Hazardous and Noxious Substances in 1996, which required the presence of evidence of financial security of the civil liability of the owner of the vessel for damage caused by pollution of the marine environment [1].

Such measures to improve the environmental safety in the operation of NSR can be developed with the involvement of Canadian experience, regulating navigation and protection of the marine environment of the North-West Passage and taking the necessary legislation.

One of the main environmental problems in the Arctic is a need to eliminate the so-called "accumulated environmental damage" in particular in the areas of the former military sites.

In Russian legislation does not exist the legal mechanism of accumulated environmental damage delineating responsibilities in that area of the state (in the person of the Russian Federation, subjects of the Russian Federation) and business entities and promotes funding beyond the budget. To improve environmental safety of the affected areas, including the Russian Arctic, the gaps should be eliminated, which, through legislation should be resolved the whole complex of legal, financial, economic, organizational, technical, scientific and methodological issues.

The regulatory framework in the field of maritime safety in offshore operations can be divided into three groups: international regulations, laws and regulations of the Arctic region and Russian.

It should be noted that few of the international standards, dedicated to solving environmental problems, written specifically for the conditions of the Arctic, as such standards are usually prepared for global use, and existing maritime standards in respect of work in the Arctic are more technical and projecting character, indirectly protecting the environment.

While ensuring environmental safety work on the development of the Arctic shelf it should be guided by international and local regulations, given the experience of countries that are actively engaged in work on the shelf, particularly Norway. In this case, the legal regulation of security issues should be in the main areas: control of emissions into the atmosphere and into the marine environment, waste management, environmental control and monitoring. Currently, the International Maritime Organization (IMO) focuses on the protection of the atmosphere from emissions from ships and offshore structures for nearly all groups of harmful emissions [8].

North of Russia - an integral part of its socio-economic system, a vast, rich and unique world with a special civilization, harsh and beautiful nature. People-centered development of the Russian North, aimed at the reproduction of human and labor capacities to ensure high standards and improve the conditions of life - one of the global objectives of all Russians. Achieving this global goal essentially caused by solving the problem of improving the livelihoods of the northern regions of Russia [13].

The successful solution of the complex challenges of effective life support of the northern regions of Russia is possible in the implementation of evidence-based strategies for socio-economic development of the Russian North. Such a strategy of social and economic development should include and take into account the main factors and conditions that significantly affect the socio-economic development of the northern regions of Russia, based on the following conceptual assumptions: North of Russia is also in the long term should be retained as an integral part of an organic social and economic space Russia, whose role in modern geopolitics and the global economy will continue to increase. At the present stage of social and economic development in a globalized world it is need:

1) The systems approach to solving problems of social development, national economy and the economy of the northern regions;

2) The reorientation of priorities of socio-economic development of the northern regions of Russia with preference objectives of economic growth and objectives of social and labor development, including the improvement of livelihoods;

3) A significant increase in the role and effectiveness of state regulation of the economy, social and labor sphere and labor economics in the northern regions of Russia;

4) Strengthening the social orientation in the activities of business organizations, operating in the North;

5) The harmonization of systemic effects of market mechanisms and measures of governmental social and economic processes at the federal, regional and municipal levels;

6) Improving the livelihoods of the Russian North, provides for the modernization of its major functional systems, including provision of goods;

7) Further development of systems research and application development directed to the implementation of performance-based management methods in the realization of targeted programs for the development of the northern regions of Russia.

Improving livelihoods will create the conditions necessary for the development of human and labor potential, sustainable growth of the regional economy in the northern regions of Russia.

References

1. Avdiyskiy V.I., Kurmashov Sh.R. Forecasting and analysis of risks in activities of economic entities: scientific and practical bases/ Edited by Prof. M.A. Eskindarov: Financial Academy at the RF Government, 2003.

2. Averin A.N. Social policy of the state. M.: Academy of national economy at the Russian Federation Government. 2002

3. Averin A.N. Population migration. Textbook. M.: RAGS Publishing, 2006.

4. Averin A.N. New data on dynamics of indigenous nations in Russia// Sociological studies, 2005 No.2.

5. Averin A.N. Social problems of northern regions' development: Textbook M.: RAGS Publishing, 2006.

6. Aganbegyan A.G. Scientific-technical progress and acceleration of socio-economic development. M.: Economics, 1985.

7. Aganbegyan A.G. On the strategy of socio-economic development and orientation of the social reforms in Russia//Economic sciences in the contemporary Russia. No. 2, 2003.

8. Agafonov V.A. Analysis of strategies and elaboration of complex programs. M., 1990.

9. Agranat G.A. On development of Alaska by Russians/ Chronicle of the North V.5. M.: Mysl. 1971.

10. Agranat G.A., Koupriyanov A.G. Population and resources of the American North. M.: USSR Academy of Sciences Publishing House. 1963.

11. Addressed social aid: theory, practice, experiment / Edited by RAEN Academician N.M. Rimashevskaya M.: "ISEPN" Publishing House. 1999.

12. Akimov A.K. Social policy: regional aspects. Yakutsk. 1996.

13. Akimov A.K. Increasing labor motivation, employment as the basis for poverty decrease in the Sakha Republic (Yakutia). Yakutsk: Sakhapoligraphizdat. 2004. 552 p.

14. Alexeev Yu. P. Region. Management. Quality. M.: "Louch" Publishing", 2000.

REWAS 2013: Enabling Materials Resource Sustainability
Edited by: Anne Kvithyld, Christina Meskers, Randolph Kirchain, Gregory Krumdick, Brajendra Mishra, Markus Reuter, Cong Wang, Mark Schlesinger, Gabrielle Gaustad, Diana Lados, and Jeffrey Spangenberger
TMS (The Minerals, Metals & Materials Society), 2013

MODELING TO EVALUATE COORDINATION AND FLEXIBILITY IN ALUMINUM RECYCLING OPERATIONS

Tracey Brommer[1], Elsa Olivetti[1], Snorre Fjeldbo[2], and Randolph Kirchain[1]

[1] Massachusetts Institute of Technology,
[2] Hydro Aluminum

Key words: By product reuse, aluminum recycling, optimization and pooling

Reprocessing of aluminum production byproducts or dross for use in secondary production presents a particular challenge to the aluminum industry. While use of these non-traditional secondary materials is of interest due to their reduced energy and economic burden over virgin counterparts, these materials necessitate the use of particular furnaces, specialized handling and processing conditions. Therefore, to make use of them firms may pursue use of an intermediate recycling facility that can reprocess the secondary materials into a liquid product. After reprocessing downstream aluminum remelters could incorporate the liquid products into their aluminum alloy production schedules. Energy and environmental benefits result from delivering the products as liquid but coordination challenges result because of the energy cost to maintain the liquid. Further coordination challenges result from the need to establish long term recycling production plans in the presence of long term downstream aluminum remelter production uncertainty and inherent variation in the daily order schedule of the downstream aluminum remelters. In this context a fundamental question arises, considering the metallurgical complexities of dross reprocessing, what is the value of operating a coordinated set of by-product reprocessing plants and remelting cast houses?

A methodology is presented to calculate the optimal recycling center production parameters including the number and volume of intermediate products made in the recycling center, how these products should be allocated to downstream facilities given uncertainty in supply and demand. A long term production optimization model was used to evaluate the theoretical viability of the proposed two-stage scrap and aluminum dross reprocessing operation including the impact of reducing coordination on model performance. Then a dynamic simulation tool was used to evaluate the performance of the calculated recycling center production plan when resolved on a daily timeframe for varying levels of operational flexibility. The dynamic simulation revealed the optimal performance to correspond to the fixed recipe with flexible production daily optimization model formulation. Calculating recycled product specifications using the proposed simulation optimization method increased profitability when multiple downstream demand scenarios were included.

Reference

Brommer, T. "Recycling Production Designs: The Value of Coordination and Flexibility in Aluminum Recycling Operations." Ph.D. dissertation. Massachusetts Institute of Technology, Cambridge, MA. 2012.

REWAS 2013: Enabling Materials Resource Sustainabilit
Edited by: Anne Kvithyld, Christina Meskers, Randolph Kirchain, Gregory Krumdick, Brajendra Mishr
Markus Reuter, Cong Wang, Mark Schlesinger, Gabrielle Gaustad, Diana Lados, and Jeffrey Spangenberge
TMS (The Minerals, Metals & Materials Society), 201

IO-MFA AND THERMODYNAMIC APPROACH FOR METAL RECYCLING

Kenichi NAKAJIMA[1], Kazuyo MATSUBAE[2], Yasushi KONDO[3], Shinichiro NAKAMURA[3], Tetsuya NAGASAKA[2]

[1] Center for Material Cycles and Waste Management, National Institute for Environmental Studies, Tsukuba 305-8506, Japan

[2] Graduate School of Engineering, Tohoku University, Sendai 980-8579, Japan

[3] Faculty of Political Science and Economics, Waseda University, Tokyo 169-8050, Japan

Keywords: material flow analysis (MFA), thermodynamic analysis, metal, recycling

Recently, the issue of sustainable resource management has been increasingly recognized. In order to increase resource efficiency, Castro et al. (2004) pointed out an importance to understand the interconnections between the materials' processing routes and their thermodynamic constraints, and discussed losses due to contaminations during recycling. One of the dominant solutions to avoid such losses or contaminants is knowledge about the substance flows in material cycles. Material flow analysis (MFA) is a powerful tool to understand the resource consumption and material cycle in the national economy. Some advanced MFA studies discussed the complex web of metal flows and their linkages (Nakamura et al. 2007, 2008). Discussions on the limitations of impurity removal and the recoverability of elements in the recycling of EoL metal products, however, have been insufficient even in conventional MFA studies.

In their extensive works, Reuter and Verhoef (2004) suggested the importance of understanding metal linkages in natural resource processing by introducing the concept of the "metal wheel", and Castro et al. (2004) showed the compatibility of materials combinations for recycling. Van Schaik (2010) also pointed out the limitation of recycling based on thermodynamic consideration. Even these representative studies, however, do not provide us with the quantitative limitations of impurity removal and the recoverability of elements in the recycling of EoL metal products. To address this issue, we have been thermodynamically evaluated the distributional pattern of alloying elements among the metal, slag, and gas phases in major metal remelting processes (Nakajima et al. 2010, 2011).

This paper proposes a combined application of MFA with thermodynamic analysis for sustainable resource use, and conducts a detailed MFA of alloying elements for steel (nickel, chromium and molybdenum), typical impurities for steel (copper) and zinc for galvanized steel.

References

K.Nakajima, O.Takeda, T.Miki, K.Matsubae, and T.Nagasaka: "Thermodynamic Analysis for the Controllability of Elements in the Recycling Process of Metals", *Environmental Science & Technology*, Vol. 45, (2011), pp.4929-4936

K.Nakajima, H.Ohno, Y.Kondo, K.Matsubae, O.Takeda, T.Miki, S.Nakamura, and T.Nagasaka: "Simultaneous MFA of nickel, chromium and molybdenum used in alloy steel by means of input-output analysis", *Environmental Science & Technology*, (reviewing)

REWAS 2013: Enabling Materials Resource Sustainabilit
*Edited by: Anne Kvithyld, Christina Meskers, Randolph Kirchain, Gregory Krumdick, Brajendra Mishr
Markus Reuter, Cong Wang, Mark Schlesinger, Gabrielle Gaustad, Diana Lados, and Jeffrey Spangenberge*
TMS (The Minerals, Metals & Materials Society), 201

Development of Efficient Recycling System for Steel Alloying Elements in End of Life Vehicles

Hajime Ohno[1], Kazuyo Matsubae[2], Kenichi Nakajima[3], Shinichiro Nakamura[45],
Testsuya Nagasaka[2]

[1] Graduate School of Environmental Studies, Tohoku University
6-6-11-1004 Aoba, Aramaki, Aoba-ku, Sendai, Miyagi 980-8579, JAPAN
[2] Graduate School of Engineering, Tohoku University
6-6-11-1004 Aoba, Aramaki, Aoba-ku, Sendai, Miyagi 980-8579, JAPAN
[3] National Institute for Environmental Studies
16-2 Onogawa, Tsukuba, Ibaraki 305-8506, JAPAN
[4] Graduate School of Economics, Waseda University
Tokyo 169-8050, JAPAN
[5] EcoTopia Science Institute, Nagoya University, Nagoya, Japan

Keywords: MFA, WIO-MFA, Steel alloying elements, End of life vehicles (ELVs) recycling,

Abstract

Special steels, which have been imparted various properties by the addition of alloying elements, have become increasingly important as materials for human life in recent years. As the largest industrial consumer of special steel materials, the motor vehicle industry requires large volumes of various types of special steel for vehicle production. Scrap containing base metals and alloying elements is generated from end-of-life vehicles (ELVs). In current scrap treatment processes, alloying elements in steel materials are rarely considered, instead becoming impurities in steel recycling processes at electric arc furnaces (EAF) and losing their worth.

In this study, with the aim of avoiding loss of useful steel alloying elements, we discuss effective treatment of ELVs for efficient recycling and use of steel alloying elements using waste input–output material flow analysis (WIO-MFA).

Introduction

Recently, huge amount of steel materials are used with various combination of alloying elements (special steel) in our society. One of the major destinations of these alloying elements is the motor vehicle industry. Material properties required in the production of motor vehicles, such as high tensile strength, low weight, and mechanical workability, have been realized by the addition of various alloying elements to steel.

Meanwhile, demand for iron scrap has been increasing [1]. As a result, about 25% of total steel production now occurs in electric arc furnaces (EAFs). Special steel products such as motor vehicles, which are being manufactured in increasing quantities year on year, will likely be introduced into the EAF steelmaking process as scrap sometime in the future. Currently, iron scrap is the main source of iron in EAF steelmaking, and little attention is paid to the alloying elements in steel materials, except for the intrusion of copper or typical tramp elements and the separation of austenitic stainless steel. Thus, the elements in scrap may dissipate into slag or dust, or be introduced unintentionally into steel products that do not require them. Therefore, the supply and demand of scrap and the intrusion of alloying elements associated with scrap are of great concern [2, 3]. As scrap generated during the production of special steel products contains a relatively high proportion of manganese, nickel, and chromium, unintentionally high concentrations of alloying elements are found in the crude steel produced by an EAF or by processes (e.g., metal casting) that use iron scrap as the principal ferrous raw material [2]. Hence, the mass consumption of steel products with unidentifiable components can contribute to the dissipation and intrusion of precious alloying elements. Given these circumstances, investigation of the steel industry is essential for effective management of consumption and recycling of alloying elements.

Given this background, it is important to construct new recycling structures in which alloying elements in scrap are considered as important sources of alloying elements and not as impurities or waste. Furthermore, efforts should be made toward the recycling of sustainable resources and the effective use of untapped natural resources. For this purpose, it is necessary to identify the alloying elements present in a given steel product and scrap. In addition, estimation of the effect of efficient scrap treatment on conserving alloying elements in the present day is important. In this paper, focusing on the recycling of end-of-life vehicles (ELVs) as an example, we discuss the total amounts of recoverable alloying elements from ELVs and more efficient usage of ELV scrap in the steel recycling process in EAFs.

The steel material cycle has been the subject of many material flow analysis (MFA) studies based on the bottom-up approach [4-7]. While these MFA studies can provide valuable information about the existing amounts of targeted materials, it is difficult to trace the supply chain of all the materials used in products throughout the country using the bottom-up approach. Moreover, the representativeness and accuracy of the composition of highly fabricated products such as automobiles or machines can be highly problematic in this approach. Instead, we use the waste input–output MFA (WIO-MFA) model developed by Nakamura and Nakajima [8]. WIO-MFA is able to analyze inter-industrial and comprehensive material flow by combining MFA and input–output analysis (IOA). IOA is one of the most widely used tools for describing economy-wide activities and their environmental implications. As ferrous materials are some of the most widely used materials in our society, WIO-MFA is the most suitable technique to improve the understanding of the flow of steel materials including scrap. The recycling of motor vehicles is of great concern in the field of industrial ecology. Since the degree of fabrication of motor vehicles is very high, various kinds of alloys are contained in products and mixed as scrap by shredding and pressing during ELV treatment [9-11]. Once such scrap is melted in the metal recycling phase, it becomes difficult to avoid contamination and oxidation loss of metals due to their thermodynamic distribution characteristics [9, 12]. Amini et al. [10] noted that iron contamination in aluminum-based alloys can cause considerable exergy and dilution loss in current ELV treatments and metal recycling processes when such substances are recycled to form alloys of the same grade. To solve this problem, easily recyclable car parts designed with consideration of metal combinations have been discussed [9, 11]. However, many existing motor vehicles were not designed with this concept in mind, and scrap from these vehicles is yet to be treated. Furthermore, although the main focus of previous studies was contamination with base metals such as iron, aluminum, and copper, the actual situation is likely to be more complicated owing to the existence of different steel alloying elements. Thus, we focus on steel alloying elements and their treatment in ELV scrap in this study.

Methodology

Development of WIO-MFA Table for the Analysis of Flow of Steel Alloying Elements

The WIO-MFA table for the intended target is constructed and extended based on a national IO table for Japan. In previous studies, WIO-MFAs have been conducted based on the Japanese IO table for 2000 for base metals [13, 14] and polyvinyl chloride [15]. This method allows the compositions of materials and their flow routes in approximately 400 different products to be analyzed. Detailed methodology for WIO-MFA is provided in previous studies. In this paper, steel material and its resources, such as ferroalloys, are described in detail, and the material flow of steel and its alloying elements are determined based on the IO table for 2005 [16]. In addition, motor vehicle-related sectors are refined and effective recycling strategies discussed.

Disaggregation of steel material sectors

Steel-related sectors are disaggregated to analyze the flow of steel alloying elements as described in Table 1. In the product category, crude and hot-rolled steels are disaggregated by type to clarify the flow from materials to products. The ferroalloy sector, which is the primary source of alloying elements in the steelmaking process, is disaggregated into 9 sectors: ferromanganese, ferrosilicomanganese, ferrochromium, ferrosilicon, ferronickel, ferromolybdenum, other ferroalloys, metallic manganese, and molybdenum oxide briquette. Metallic chromium, nickel, and molybdenum are also separated from the sector of nonferrous metal.

Disaggregation of car parts sector

The car parts sector is disaggregated into 8 groups as described in Table 1. In the IO table for 2005, the automobile-related sector contains two final products (passenger cars and trucks, busses, and other cars) and 3 intermediate products (automotive body, internal combustion engine, and car parts). Of these, the car parts sector is one of the most highly aggregated sectors in the IO table. Because of this aggregation, material flow in the automobile-related sector is unclear. When various steel materials are collated in the aggregated car parts sector, the composition of the sector becomes an average of the various collated materials (Figure 1a), such that we cannot distinguish usages of steel materials. On the other hand, disaggregation of the car parts sector clarifies the flow of steel materials (Figure 1b). In addition, this disaggregation makes it possible to assess disassembly and collection of parts for ELV treatment.

Table I. Disaggregated sectors

Before disaggregation	After disaggregation		
Crude steel (BOF)	-Ordinary steel	-For section steel	-For other hot-rolled steel
Crude steel (EAF)		-For steel sheets and plates	-For steel pipes and tubes
		-For steel strip	-For cold-finished steel
		-For steel bar	-For others
	-Special steel	-For carbon tool steels	-For heat-resistant steel
		-For alloy tool steels	-For free-cutting steel
		-For high speed steels	-For piano wire rod
		-For other tool steel	-For high-tensile steel
		-For structural carbon steel	-For weathering steel
		-For structural alloy steel	-For low-temperature steel
		-For spring steel	-For high-tensile steel pipe
		-For bearing steel	-For stainless steel pipe
		-For ferric stainless steel	-For other special use steel
		-For austenitic stainless steel	-For others
Hot-rolled special steel	Carbon tool steels		Heat-resistant steel
	Alloy tool steels		Free-cutting steel
	High speed steels		Piano wire rod
	Other tool steel		High-tensile steel
	Structural carbon steel		Weathering steel
	Structural alloy steel		Low-temperature steel
	Spring steel		High-tensile steel pipe
	Bearing steel		Stainless steel pipe
	Ferric stainless steel		For other special use steel
	Austenitic stainless steel		
Ferroalloy	Ferromanganese		Ferromolybdenum
	Ferrosilico-manganese		Other ferroalloys
	Ferrochromium		Metal manganese
	Ferrosilicon		Molybdenum oxide briquette
	Ferronickel		
Non-ferrous metal	Metallic nickel		Metallic molybdenum
	Metallic chromium		
Car parts	Suspension		Interior
	Shaft		Transmission
	Steering		Breaks
	Electric equipment		Exhaust

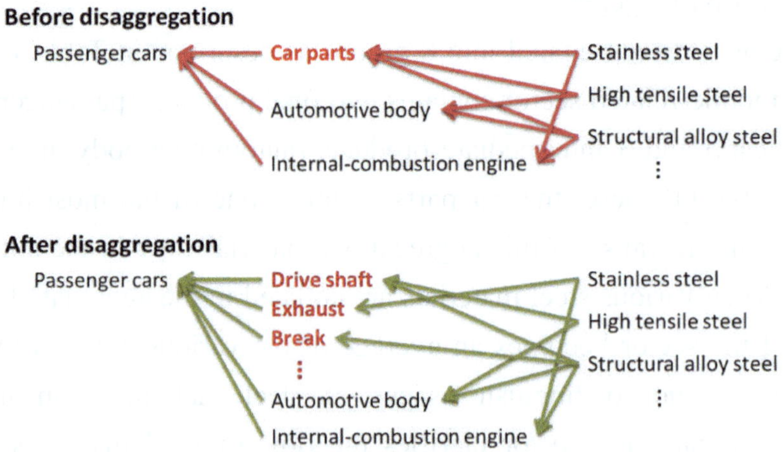

Figure 1. Schematic of the disaggregation of the car parts sector

Results

Final destination of steel alloying elements

Identifying the final destination of steel alloying elements is important for their effective life-cycle management. Using WIO-MFA, the concentrations and final destinations of manganese, chromium, nickel, and molybdenum are estimated. Results for manganese are also estimated; however, manganese is difficult to estimate accurately owing to its role as a deoxidizing agent in the steelmaking process; therefore, accuracy of results of manganese is lower than other results of alloying elements. Figure 2 illustrates the final destinations of manganese, chromium, nickel, and molybdenum in the top ten (49.3%) sectors. The largest accumulated sector for manganese, chromium, nickel, and molybdenum is the passenger car sector, followed by stainless steel materials, trucks, buses and other cars, and nonresidential construction. From the viewpoint of effective life-cycle management of alloying elements, it is important to consider domestic demand. Almost all alloying elements contained in steel materials are exported. Therefore, they cannot be managed and recycled in Japan. Conversely, alloying elements contained in around half of passenger cars and a large part those contained in trucks, buses, other cars, and nonresidential construction remain in Japan, and will be discarded there at the end of their lifetime. Overall, about 44% of final destinations were domestic in 2005. Therefore, there are considerable amounts of alloying elements that should be treated as precious "urban mines."

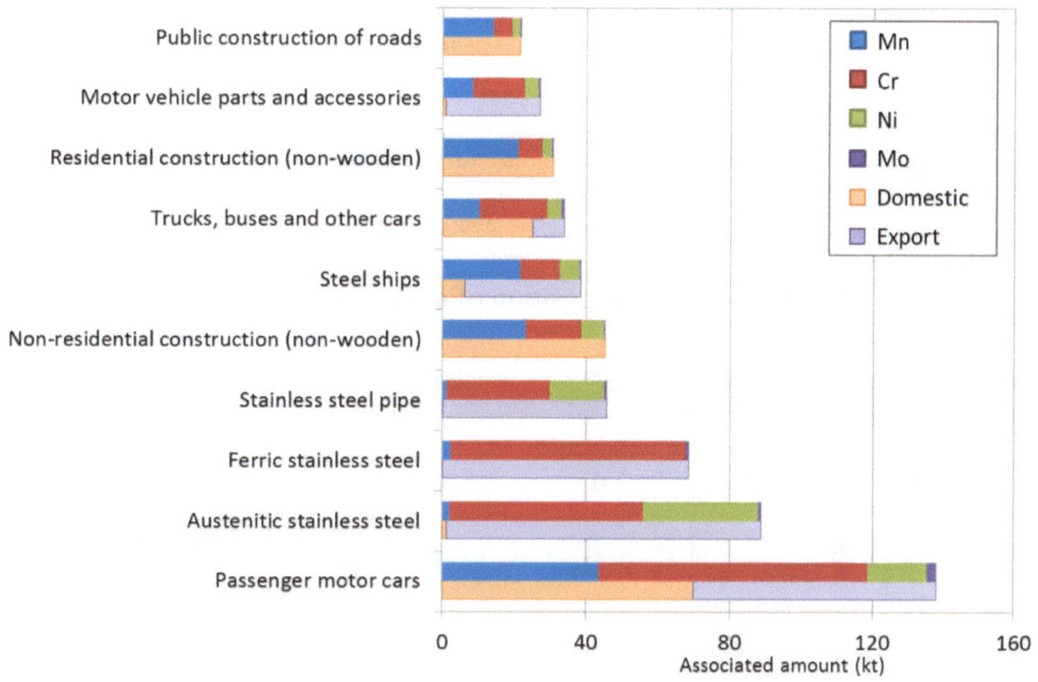

Figure 2. Final destination of steel alloying elements analyzed by WIO-MFA

Composition and alloying element contents of a passenger car

Motor vehicles display the greatest quantitative potential for recycling of steel alloying elements (Figure 2). We clarify the composition and content of alloying elements of a passenger car using disaggregated WIO-MFA and discuss efficient ELV scrap treatment and recycling.

Figure 3 illustrates the parts composition of a passenger car, which we obtained using WIO-MFA. The "direct" category includes all intermediate products (i.e., those not produced by automobile-related sectors) included in production of a passenger car. The weight of 1 passenger car unit is 1,049 kg, excluding tires and glass. The contents of alloying elements and copper in each part are illustrated in Figure 4. Amounts of chromium, nickel, and molybdenum in 1 passenger car unit are 6.34, 1.50, and 0.21 kg, respectively. Although the exhaust weighs only 29 kg, it contains large amounts of alloying elements owing to the use of stainless steel. In recycling steel materials from ELVs, copper contents are of great concern because the hot workability of copper-containing steel is lower than copper-free steel. Much of the copper content is present in the direct, engine, and electric equipment parts (Figure 4). Figure 4 also indicates that distribution of copper by parts is distinctly different to that of the other three steel alloying elements, although engines contain similar contents of all elements. The high concentrations of these elements in engines does not pose a problem, because engines are usually removed during the current ELV treatment process and treated separately [17, 18]. Therefore, copper and steel can be easily separated and steel alloying elements can be recycled in low-copper environments. Thus, our results suggest the effectiveness of detailed ELV disassembling for the recycling of steel alloying elements.

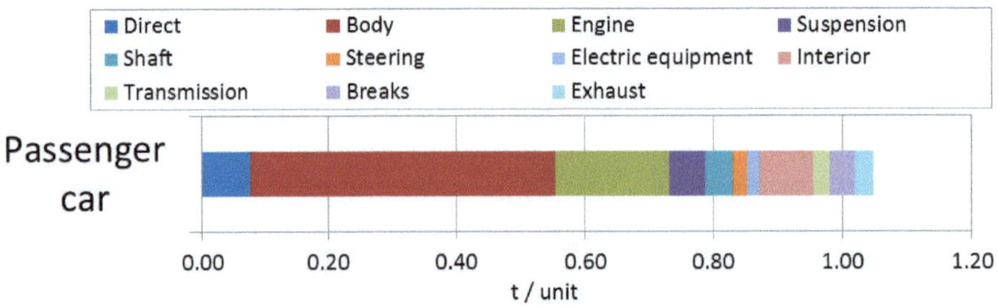

Figure 3. Parts composition of a passenger car

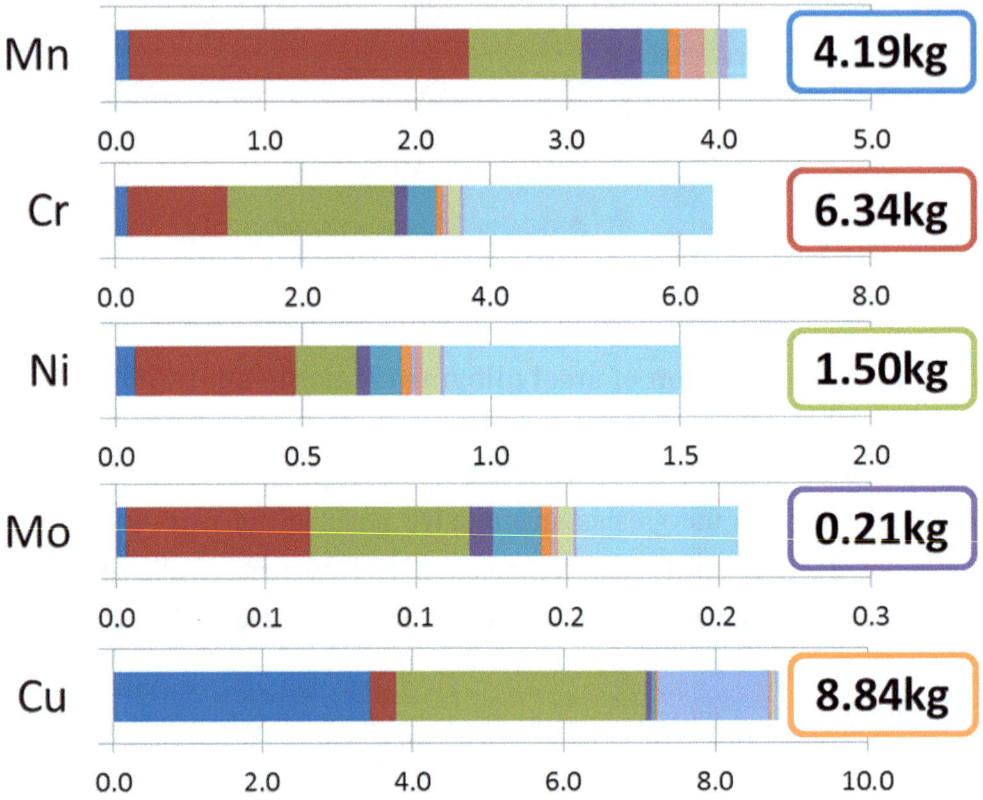

Figure. 4 Steel alloying elements and copper contents in each part of a passenger car

4 Discussion and conclusion

In many current ELV treatment processes, dismantling of parts is conducted to remove sources of copper contamination such as wire harnesses and electric cables [17]. However, after the removal of copper, dismantled parts are pressed and shredded together [18]. In this situation, the remaining alloying elements are mixed and diluted with iron matrix and are wasted. However, if dismantled parts were separated before shredding, they could be used to make steel products of similar composition.

For example, an exhaust contains large amounts of chromium and nickel despite its relatively low weight owing to the use of stainless steel. Concentrations of chromium and nickel in exhausts are 8.98% and 2.14%, respectively, a combination that is useful for stainless steelmaking.

We can use the results of this study to set scenarios for recycling steel alloying elements efficiently in the ELV treatment process. Focusing on the contents of alloying elements in each part will allow effective combinations of parts to be shredded and melted together. Additionally, we intend to propose a new ELV scrap recycling system for the efficient use of steel alloying elements.

References

1. The Japan Ferrous Raw Materials Association Pig iron, Iron and steel scrap supply and demand. http://www.tetsugen.gol.com/kiso/3jyukyu1nen.htm (Accessed 12 June. Month. 2010),

2. K. Nakajima, K. Yokoyama, T. Nagasaka, Substance flow analysis of manganese associated with iron and steel flow in japan. *Isij International*, *48*, (4),2008, pp 549-553.

3. A. Gleich, R. U. Ayres, S. Gössling-Reisemann, Eco-efficiency in industry and science. In *Sustainable metals management: Securing our future - steps towards a closed loop economy*, Springer: Dordrecht, 2006.

4. P. Michaelis, T. Jackson, Material and energy flow through the UK iron and steel sector - Part 2: 1994-2019. *Resources Conservation and Recycling*, *29*, (3),2000, pp 209-230.

5. P. Michaelis, T. Jackson, Material and energy flow through the UK iron and steel sector. Part 1: 1954-1994. *Resources Conservation and Recycling*, *29*, (1-2),2000, pp 131-156.

6. I. Daigo, Y. Igarashi, Y. Matsuno, Y. Adachi, Accounting for steel stock in Japan. *Isij International*, *47*, (7),2007, pp 1065-1069.

7. T. Wang, D. B. Muller, T. E. Graedel, Forging the anthropogenic iron cycle. *Environ Sci Technol*, *41*, (14),2007, pp 5120-9.

8. S. Nakamura, K. Nakajima, Waste input-output material flow analysis of metals in the Japanese economy. *Materials Transactions*, *46*, (12),2005, pp 2550-2553.

9. M. B. G. Castro, J. A. M. Remmerswaal, M. A. Reuter, U. J. M. Boin, A thermodynamic approach to the compatibility of materials combinations for recycling. *Resources Conservation and Recycling*, *43*, (1),2004, pp 1-19.

10. S. H. Amini, J. A. M. Remmerswaal, M. B. Castro, M. A. Reuter, Quantifying the quality loss and resource efficiency of recycling by means of exergy analysis. *Journal of Cleaner Production*, *15*, (10),2007, pp 907-913.

11. D. Froelich, N. HaoueS, Y. Leroy, H. Renard, Development of a new methodology to integrate ELV treatment limits into requirements for metal automotive part design. *Minerals Engineering*, *20*, (9),2007, pp 891-901.

12. K. Nakajima, O. Takeda, T. Miki, K. Matsubae, T. Nagasaka, Thermodynamic Analysis for the Controllability of Elements in the Recycling Process of Metals. *Environ Sci Technol*, *45*, (11),2011, pp 4929-4936.

13. K. Nakajima, S. Nakamura, Material flow analysis of metals based on the waste

input-output model (WIO-MFA model): Application to material cycle of iron and steel. *Journal of the Japan Institute of Metals*, *70*, (8),2006, pp 618-621.
14. S. Nakamura, K. Nakajima, Y. Kondo, T. Nagasaka, The waste input-output approach to materials flow analysis - Concepts and application to base metals. *Journal of Industrial Ecology*, *11*, (4),2007, pp 50-63.
15. S. Nakamura, K. Nakajima, Y. Yoshizawa, K. Matsubae-Yokoyama, T. Nagasaka, Analyzing Polyvinyl Chloride in Japan With the Waste Input-Output Material Flow Analysis Model. *Journal of Industrial Ecology*, *13*, (5),2009, pp 706-717.
16. Ministy of Industrial Affairs and Communications (Japan), 2005 input-output table for Japan.2009.
17. Japan Environmental Sanitation Center *Report for implementation of efficient and retional treatment for end of life vehicles*; Japan Environmental Sanitation Center: Japan, 2009.
18. Tohoku University *Material flow and element distribution anaylsis for efficient utilization of alloying elements associated with end of life vehicle scrap*; Tohoku University: Japan, 2012.

REWAS 2013: Enabling Materials Resource Sustainability
Edited by: Anne Kvithyld, Christina Meskers, Randolph Kirchain, Gregory Krumdick, Brajendra Mishra, Markus Reuter, Cong Wang, Mark Schlesinger, Gabrielle Gaustad, Diana Lados, and Jeffrey Spangenberger
TMS (The Minerals, Metals & Materials Society), 2013

PHOSPHORUS FLOW ANALYSIS FOR FOOD PRODUCTION AND CONSUMPTION

Kazuyo Matsubae[1], Kenichi Nakajima[2], Keisuke Nansai[2] and Tetsuya Nagasaka[1]

[1] Tohoku University

[2] National Institute for Environmental Studies

Key Words: Phosphorous, Food, Yield, Hybrid Input Output model, Substance flow analysis

Phosphorus is present only as a trace element on the Earth, but is one of the important strategic resources for agricultural food production and for the chemical industry. Natural phosphate ore is traded worldwide, mainly as a raw material for fertilizer. Approximately 147 × 10^3 kt of phosphate ore was mined in the world during 2005. Of this, 24.7% (36.3 × 10^3 kt) was produced in the USA, 20.7% (30.4 × 10^3 kt) in China, and 17.1% (25.2 × 10^3 kt) in Morocco, while there are essentially no deposits of phosphate ore in Japan or the EU (USGS,2012). It is of concern that, due to growing world demand for fertilizers, deposits of high-grade phosphate ore could be exhausted within the next 100 years (Vaccari 2009), and the average price of the ore in 2008 was approximately doubled that in 2007. Concerning the restricted supplies of phosphorus resource, it is important to consider the quantity and availability of phosphorus resources that currently remain untapped.

Various authors have analyzed P flow from the economical use and recycling perspective (Smil 2000), (Li, He et al. 2007) (Neset, Bader et al. 2008; Matsubae-Yokoyama, Kubo et al. 2010; Matsubae, Kajiyama et al. 2011). From these snapshots we might better be able to go beyond the "once-through mode of societal phosphorus metabolism" (Liu, Villalba et al. 2008). However, it is difficult to trace the supply chain of all the materials used in products throughout the country by using the bottom-up approach. The fact that phosphorus and other plant nutrients are one of the most widely used elements in our society calls for taking a bird's-eye view for a better understanding of the flow of phosphorus including agricultural products and meat products(Goodlass, Halberg et al. 2003). Input–output analysis (IOA) is one of the most widely used tools of industrial ecology for describing the economy-wide activities and their environmental implications.

With this in mind, we developed the Integrated Phosphorus Cycle Input Output (IPCIO) model to estimate the phosphorus requirement for economic activities and evaluate the recycling effects of reutilization of phosphorus resources which are currently untapped. The accounting framework of IPCIO has 4 natural resources and 25 phosphorus related commodities in physical term and 389 intermediate sectors of the Japanese economy in 2005 year. As empirical studies, phosphorus recovery and recycling scenarios are considered for future phosphorus resource management.

Acknowledgments

The present work is financially supported by Ministry of the Environment, Japan, and JST-RISTEX.

References

Goodlass, G., N. Halberg, et al. (2003). "Input output accounting systems in the European community - an appraisal of their usefulness in raising awareness of environmental problems." European Journal of Agronomy **20**(1-2): 17-24.

Li, B. X., Y. Z. He, et al. (2007). "Simultaneous determination of three organophosphorus pesticides residues in vegetables using continuous-flow chemiluminescence with artificial neural network calibration." Talanta **72**(1): 223-230.

Liu, Y., G. Villalba, et al. (2008). "Global phosphorus flows and environmental impacts from a consumption perspective." Journal of Industrial Ecology **12**(2): 229-247.

Matsubae-Yokoyama, K., H. Kubo, et al. (2010). "Recycling Effects of Residual Slag after Magnetic Separation for Phosphorus Recovery from Hot Metal Dephosphorization Slag." Isij International **50**(1): 65-70.

Matsubae, K., J. Kajiyama, et al. (2011). "Virtual phosphorus ore requirement of Japanese economy." Chemosphere **84**(6): 767-772.

Neset, T. S. S., H. P. Bader, et al. (2008). "The flow of phosphorus in food production and consumption - Linkoping, Sweden, 1870-2000." Science of the Total Environment **396**(2-3): 111-120.

Vaccari, D. A. (2009). "Phosphorus: A Looming Crisis." Scientific American **300**(6): 54-59.

AUTHOR INDEX
REWAS 2013: Enabling Materials Resource Sustainability

SUBJECT INDEX
REWAS 2013: Enabling Materials Resource Sustainability